Very truly yours,
Madison Cooper.

PRACTICAL

COLD STORAGE

THE THEORY, DESIGN AND CONSTRUCTION OF BUILDINGS AND APPARATUS FOR THE PRESERVATION OF PERISHABLE PRODUCTS, APPROVED METHODS OF APPLYING REFRIGERATION AND THE CARE AND HANDLING OF EGGS, FRUIT, DAIRY PRODUCTS, ETC.

BY

MADISON COOPER

REFRIGERATING ENGINEER AND ARCHITECT

AUTHOR OF "EGGS IN COLD STORAGE," "ICE COLD STORAGE," ETC.

SECOND EDITION

PUBLISHERS:

NICKERSON & COLLINS CO.

CHICAGO

1914

PRESS OF
ICE AND REFRIGERATION
CHICAGO

TO MY FATHER

As a tribute to his matchless enterprise and genius for practical and scientific research, and in acknowledgment of valuable assistance rendered, this work is affectionately inscribed.

THE AUTHOR

PREFACE TO FIRST EDITION.

The difficulties encountered in preparing a book on so broad a subject as practical cold storage have been so great as at times to discourage the author from continuing. He has endeavored to collect the greater part of his own writings and at the same time has compiled from all available sources. The present book has the many shortcomings usually found in every pioneer work, and there are many gaps in the chain of information given for the reason that detailed knowledge has in many cases been unobtainable. A large portion of the general matter which has appeared on the subject of cold storage is of little or no value as a part of a book on this subject, for the reason that it contains many repetitions and contradictions, and for the most part has been written by persons not familiar with refrigeration either from a practical or scientific standpoint. The matter and information which appeared prior to about 1895 is mostly valueless in the light of present information, as the earlier articles were generally incomplete and in part erroneous.

The immense amount of labor involved in digging through the great piles of chaff to find the few grains of wheat has been out of all proportion to the actual results obtained. Reliable scientific data and the records of tests have in many cases been difficult or impossible to obtain. Comparatively little along this line is in existence and some of it is jealously guarded by its possessors. Practical information on the handling, packing and storing of perishable products is obtainable only in a small way for the reason that comparatively few operators of cold storage houses have made any record of results and can put their experience in tangible form for the use of others. The author begs to acknowledge the assistance of his many friends among the engineers and cold storage men. It has been his aim to give due credit where any considerable amount of matter has been furnished by others, but some of the matter contained herein has been secured from sources the origin of which has been lost and not traceable, and if due credit is not given it is from no wrong intent on the part of the author.

This book is intended to cover the field of applied refrigeration with the exception of ice making, ice machines, and the technical and theoretical side of the mechanical production of refrigeration. These important matters are fully treated by several valuable and comprehensive works. The reader is referred to these books for the data, theory and information necessary to a full understanding of the principles of thermodynamics and refrigerating machine construction and operation.

There is much regarding the use of ice, both natural and artificial, as a practical refrigerant, even on a large scale, which has not heretofore been fully described. The possibilities of successful refrigeration by means of ice have not been carefully studied and given due consideration. If rightly applied, ice, either natural or manufactured, in combination with salt, will produce any results in the preservation of perishable products, which may be produced by any means of cooling; limited, of course, by the range of temperature which can be obtained. The importance and extent of this branch of the refrigerating industry has not been appreciated by those who have given their time to the study of refrigeration. The development of the mechanical systems of refrigeration came at a time when the use of ice as a refrigerant had not been reduced to a scientific basis, consequently our best talent was directed toward the perfecting and introducing of the ice machine. Nevertheless, there are many successful ice cold storage houses who are doing fully as perfect work as the best machine refrigerated houses. The value of products which are daily refrigerated by ice for preservation, exceeds by far those refrigerated by mechanical means. This statement is best appreciated when we consider that a large part of the output of the hundreds of ice factories is used in small refrigerators for the temporary safe keeping of fruit, vegetables, meats, dairy products, etc.; that the immense natural ice crop annually harvested is consumed in the same way; that an important portion of the eggs, butter, cheese, fruit, etc., are stored in warehouses cooled by ice, or ice and salt, and that perishable goods during transportation are kept cool by ice almost exclusively. From these facts it is evident that a description of the manner of securing and storing the natural ice crop and the best methods of utilizing ice, either natural or artificial, for cooling or freezing purposes, must be of considerable value to the users of refrigeration generally.

An important branch of cold storage design, and in fact all work in refrigeration, is the design and construction of walls which form insulation against heat, and built of such materials as may be had at a moderate cost. The chapter on insulation has aimed to give the results of the best information at present obtainable on this subject, both in the United States and in foreign countries.

The chapters on the practical operating of cold storage houses and the care and handling of goods for storage have been written largely from the author's practical experience, supplemented by information obtained from others. The business of cold storage has now assumed so vast a proportion and such a great variety of goods are now placed in refrigerated rooms for preservation, that no one person could possibly cover so vast a field. The author, therefore, acknowledges his indebtedness to many who have made a specialty and have had much experience with the various goods. General directions are given for the handling of a cold storage house without reference to any particular product, and if these are followed understandingly and care and judgment used, the cold storage manager may avoid many of the errors common to those new to the business. It must be remembered that a good house poorly handled cannot compete with an inferior house well handled. At least one-half is in the management and too much care cannot be exercised in looking after the details of a refrigerating installation, not only for the purpose of securing economy in operation of same, but also to insure the keeping of the stored goods in the best condition.

By far the major portion of what is printed in this book is from the original writings of the author; a portion of which has appeared in the columns of *Ice and Refrigeration; The Ice Trade Journal,* etc., as articles under the titles of "Eggs in Cold Storage," "Ice Cold Storage," etc. It was because of the success of the articles on "Eggs in Cold Storage," which were subsequently printed in pamphlet form, and the complimentary reception of same, which encouraged the author to undertake the present work. It is now submitted to the trade with a full appreciation of its imperfections and incompleteness. As far as possible these will be remedied in future editions. It is the earnest request of the author that those who find errors or omissions or can suggest in any way improvements, correspond with the author to the end that "Practical Cold Storage" may be made as complete and accurate as possible.

Any information which will further the interests of the business, will in turn benefit all who are engaged therein. For any one to believe that he is the possessor of secret information which is vital to his success over competitors, is in a great majority of cases the extreme of absurdity. Much of the matter appearing in this publication has at some time been considered as trade secrets. The false and narrow-minded position taken by some in connection with this matter is well illustrated by certain remarks made to the author in regard to the publication of this book. The following is a sample: "Now that you have this information accumulated, why not keep it for your own use instead of giving it away?" It is quite true that the author has expended in time, effort and money in connection with the preparation of the matter contained in this book, much more than he can be remunerated for in its sale. It is, however, here given for what it is worth and with the hope that it may be of substantial benefit to many readers.

It might not be out of place to call the reader's attention to the fact that, in practically all the original matter by the author contained in this book, reasons are given for statements made so far as practicable. This enables the new beginner or student to study intelligently the natural laws which govern the principles of refrigeration. Comment and criticism has been freely bestowed without fear or favor on the various ideas, systems and methods which do not meet the approval of the author. Matter which has been compiled or extracted from other sources has in some cases been changed or modified to suit the individual ideas of the author. Should the advocates of anything here criticised feel that they have not had a fair presentation the author will be glad to take the matter up and discuss the points involved.

While this work is in some respects imperfect and there is no doubt room for the addition of much information, reliable data, and the results of extended observations and tests, there has not heretofore been anything like as complete a presentation of the entire subject; and in consideration of this fact the reader is requested to be lenient in his criticism. If any errors or lack of details are noted, the author would be pleased to acknowledge same and will endeavor to explain the points at fault. No other object has been in mind in preparing this book than a furtherance of scientific knowledge on the subject of refrigeration as applied to the preservation of perishable products, and the great assistance rendered by those who have assisted is hereby acknowledged. The combination and comparison of information is beneficial, and if those who have further data or records of tests will only put them before others in their line of business, no loss will be sustained by the indvidual giving the information, while much general good will result.

PREFACE TO SECOND EDITION.

Since the appearance of the first edition of this book comparatively few improvements and changes have been made in practical applications of refrigeration, and development has been largely along the lines of perfecting and improving methods and systems already introduced. Many new applications of refrigeration have been found, and it has been demonstrated that the use of refrigeration for preventing destructive deterioration of perishable goods, so-called, as well as the controlling of chemical and other processes by supplying the low temperatures often needed, is even at the present time in its infancy. We may look to the future for a much wider application of refrigeration for many purposes, some, doubtless, at present not thought of. The chief use of refrigeration at the present time is in the preservation of perishable food products, but many other applications are being brought to light from year to year, among the most recent of which may be mentioned the following:

Curing tobacco, tempering watch springs, in the manufacture of rubber, drugs, syrup, soap, ink, paint, vinegar, isinglass, etc., in oil refineries, sugar refineries, chemical works, mercerizing works, photo material factories, in the manufacture of explosives, plows and other agricultural implements, optical instruments, electrical machinery, etc., in welding processes, for retarding growth of plants and vegetables, in laboratory work, hospital practice, shaft sinking and tunneling, for testing automobile parts, batteries, insulating material, paving material, etc.

The cooling of inhabited spaces, which means living and work rooms, during the heated term, is an application of refrigeration which has been much talked about, but with which comparatively little has been done. Most people prefer to "swelter" rather than provide a comfortable working temperature in warm weather. Even those who can well afford conveniences and comforts of any kind do not take much interest in this proposition, and very little progress has been made during past years in this direction. Some few theatres, hotel rooms, public halls, etc., have been cooled, but only experimentally as it were, and in a clumsy and cheap sort of way.

MADISON COOPER.

Calcium, N. Y., October 1st, 1914.

INTRODUCTION.

There is no authentic history of the use of refrigeration as applied to what is now popularly called "cold storage," and it is only within the past thirty or forty years that the practical usefulness of refrigerated storage has been appreciated by the world at large. In the year 1626 Lord Bacon is said to have taken a chill from the stuffing of a chicken with snow, in order to preserve it, which resulted in his death. It would seem that the death of so eminent a person from such a cause should have attracted attention to the possibilities of applied refrigeration, but either the poor success of the experiment, or the fatal result to its originator seems to have had a deterrent effect on further investigation along this line at that period.

It is doubtful whether any scientific demonstration or commercial enterprise of recent years has been of greater moment to the human race than the science of refrigeration and its practical application in the modern cold storage industry. When scientific inquiry had proven the efficacy of low temperatures in preventing decay and had demonstrated the possibility of obtaining and maintaining low temperatures at will, the cold storage business of today was but the natural evolution resulting from such demonstration. When it became apparent that profit was obtainable by placing perishable goods in cold storage during a period of glut or surplus and disposing of them at some subsequent period of comparative scarcity or increased demand, the building of cold storage houses and the perfection of machinery or apparatus for their economical operation became the inevitable result. The pioneers in the cold storage business were speculators of the extreme kind, but this cannot be said of those in the business today. Where in the early days the cold storage operator owned the goods he stored almost entirely, and his customers were uncertain, now the goods placed in cold storage are almost wholly owned by dealers, and are held for the supplying of their regular trade.

Refrigeration has four chief uses in the economy of nature and in commerce:

1.—To prevent premature decay of perishable products.
2.—To lengthen the period of consumption and thus greatly increase production.
3.—To enable the owner to market his products as needed.
4.—To make possible transportation in good condition from point of production to point of consumption, irrespective of distance.

First: Without refrigeration there woud be much actual waste from decomposition before it would be possible to place perishable food products at the disposal of the consumers. The immense fruit trade of the Pacific coast would never have been developed without the assistance of refrigeration, nor could the surplus meat products of the southern hemisphere have been brought half way around the globe to relieve the shortage in thickly settled England without its aid. Without the aid of refrigeration to create a constant market, the production of meats, of eggs, of fruits and other food products would be greatly curtailed.

Second: In many classes of produce the ordinary season of consumption was formerly limited to the immediate period of production, or but briefly beyond. Now nearly all fruits may be purchased at any season of the year and dairy and other products are for sale in good condition and at reasonable prices the year around.

Third: Instead of being obliged to sell perishable goods, when produced or purchased, at any price obtainable, the owner can now put away in cold storage a portion or all of his products to await a suitable time for selling. This not only results in a better average price to the producer, but places perishable food stuffs at the command of the consumer at a reasonable price at all times and greatly extends the period of profitable trading in such products.

Fourth: The certainty and perfection with which food products may be conveyed from the place of production to the large centers of population where they are to be consumed is one of the triumphs of refrigeration; yet the refrigerator car service is only in its infancy so far as perfection of results is concerned. It is safe to say that our immense Pacific coast fruit trade could not exist without it. The over sea carriage of products has also been developed along with the development of refrigeration as applied to this work.

Cold storage is a benefit to all mankind in that it allows of a greater variety of food during all seasons of the year. Health and longevity are promoted by the free consumption of fruits, and the placing of fresh fruits at the disposal of even the poorest of our citizens during every month in the year will certainly result in a wholesale benefit to mankind, so far-reaching in its effects as to be incalculable.

Physicians and scientists who have investigated the subject unite in praising the modern practice of refrigeration as applied to the preservation of food products and in arresting decay in all articles of value liable to injury by exposure to high or normal temperatures. A prominent English physcian* in an address before the Sanitary Institute at their Congress in Birmingham in 1898, after describing at length the various methods, namely: Drying, smoking, salting, sugar and vinegar, exclusion of air (canning), antiseptics, chemicals, etc., in use as food preservatives has this to say of refrigeration:

> This brings us then to the last of the modern methods of food preservation on the large as well as on the small scale, and as it is the last, so it is the best. The fishmonger avails himself of it in his ice well and on his stall. It is by its agency that all the perishable food on our great liners is preserved during even prolonged voyages, and it is used in the great food depots of many of our large towns. In this town tons of perishable foods are continually preserved by its action, and where such stores do not exist they ought to be provided. In this way all perishable articles can be kept until such times as they shall be required for sale and distribution.
>
> Formerly the methods of producing cold were complicated and dear, and had many drawbacks, but these have been overcome. * * * Cold acts not by killing the organisms that effect decomposition, but only by inhibiting their action; in which respect it differs from heat and certain chemical antiseptics, such as chlorine, for instance.
>
> Among the advantages of preservation by refrigeration may be mentioned:—
>
> 1—It has been proved the most effective as a preservative, surpassing in efficiency, salting, boric compounds, or any other practical method.

---

*Alfred Hill, M. D., F. R. S., Edin. F. I. C. Medical Officer of Health and Public Analyst to the City of Birmingham, Eng.

2—It adds nothing and subtracts nothing from the article preserved, not even the water, and in no material sense alters its quality.

3—It causes no change of appearance or taste, but leaves the meat or other substance substantially in its original condition, while it renders it neither less nutritious nor less digestible, which cannot be said of some other methods in common use.

My contention is that all additions to food whose influence on health is doubtful ought to be prohibited and their use supplemented by refrigeration.

Strong language like this coming from such an eminent authority not only vouches for the usefulness of refrigeration, but also for the perfection of its results, and to a thinking person offers an assurance that an industry established on so broad a basis must present an ever widening field of usefulness. New products are constantly being added to those which are placed in cold storage for safe keeping or preservation, and it seems not a wild prediction to say that at some time in the future the great majority of our food products and other perishable goods will be handled in and sold from refrigerated rooms.

Considering the importance the cold storage industry has already attained, its rapid growth and future outlook, the amount of accurate information available to those engaged in the business seems very meager. The difficulties to be overcome, the skill required, and the importance of a well designed structure are not usually explained by those interested in promoting new enterprises in this line, and consequently not appreciated by those making the investment. Financial disaster has overtaken many large companies who have erected costly refrigerating warehouses; those which have succeeded have in many cases been forced to install new systems, make expensive changes, and make a thorough study of the products handled. The experience of nearly all has been emphasized at times by heavy losses paid in claims made by customers for damage to goods while in storage, or the necessity of running a large house while doing a very small business. Those about to become interested in business may find food for thought in the above, and the history of a dozen houses, in different localities, will furnish valuable information for would-be investors.

The scarcity of knowledge on the subject in hand, while being partly the result of the partially developed state of the art until very recently, is also very largely owing to narrow-mindedness on the part of some of the older members of the craft who have largely obtained their skill by experience and study, some of them having expended large sums on experimental work. The same experiments have perhaps been made before, and are of necessity to be made again by others, simply because the first experimenter would not give other people the benefit of his experience. It seems that at the present stage in the development of refrigeration the improvements to be made during the next thirty years will be of very much less importance than those made during the last thirty years; trade secrets, so jealously guarded by some, must disappear, as they have in other branches of engineering. Storage men have been obliged to work out their own salvation in solving problems, sometimes, however, sending their most difficult points to be answered through the columns of the trade journals, and, perhaps, comparing ideas with those of their personal friends in the same line of business. It is to be observed that the most progressive and up-to-date manufacturing concerns give their contemporaries every opportunity to observe their methods, and are very willing and anxious to talk over matters pertaining to their work from an unselfish standpoint. So, too, the successful cold storage manager will be sure to make "visitors welcome."

In anything which appears in this book, it is not the author's intention to convey the idea that any mere theoretical knowledge which can be acquired by reading and study, or even by an exchange of ideas in conversation, can take the place of practical observation in actual house management; but there are applications of well known laws which are not generally understood by storage men and their progress is handicapped from lack of this theoretical knowledge. The two following illustrations, bearing on temperature and ventilation, are among the common errors made in practice, yet easily understood when studied and tested: Some storage houses formerly held their egg rooms at 33° F., fearing any nearer approach to the freezing point of water (32° F.), thinking the eggs would freeze. A simple experiment would settle this point, giving the exact freezing temperature, as well as the effect of any low temperature on the egg tissues. Eggs will not freeze at 28° F. Again, others have thought to ventilate by opening doors during warm weather. It never happens that storage rooms can be benefited by this treatment at any time during the summer months, and only occasionally during the spring and fall. The dew point of outside air is rarely below 45° F. during summer, and when cooled to the temperature of an egg room, moisture will be deposited on the goods in storage, causing a growth of mildew.

The question of the proper temperature at which to carry goods is of the first importance. Correct temperatures alone, however, will not produce successful results, any more than a good air circulation or correct ventilation would give good results with a wrong temperature. The common impression of cold storage is what the name implies—simply a building in which the rooms may be cooled to a low degree as compared with the outside air. Even those who manufacture and install refrigerating machinery and apparatus often show either gross carelessness or ignorance of the requirements of a house which will produce successful results. After a careful examination of some of the recently constructed houses supposed to be strictly modern and up-to-date, the author has the impression that the designers regarded temperature as the only requisite for perfect work. Some of the rooms in these new houses are simply insulated and fitted with brine or ammonia pipes, the proper location of same having received no attention whatever, being placed, in most cases, in convenient proximity to the pipe main, and in one or two instances, the top pipe of the cooling coils was fully two feet from the ceiling. The necessity for providing for air circulation seemed not worthy of consideration, to say nothing of the lack of anything like an efficient ventilating system for the furnishing of fresh air. These things are mentioned here for the purpose of cautioning against a superficial study of cold storage problems. It is advisable for everyone interested to understand the underlying laws which govern the results to be obtained. Read carefully the chapters on "Air Circulation," "Humidity" and "Ventilation."

Cold storage, if the right system is installed and properly handled, will produce some remarkable results in the preservation of perishable products. It must not be expected, however, that the quality and condition of the goods are improved by storage. Cold storage does not insure against a certain amount of natural deterioration. Goods for cold storage must be in prime condition and selected by an experienced person if it is expected to carry them to the limit of their possible life. A cold storage house successfully operated and managed will supply a uniform temperature at the proper degree throughout the storage season. It will regulate the humidity at the proper point and will supply fresh air properly treated to force out the accumulated gases. The storing of unsuitable, imperfect and inferior goods has led to much misunderstanding and some litigation between the man who stores the goods and the warehouse man. Both should, if possible, be familiar with the condition of the goods they are handling; the different stages of ripeness, quality and liability to deterioration. Cold storage cannot improve the physical condition of perishable goods and is in no way responsible for damage or decay which may arise from improper picking, grading, packing or handling before placing in the storage house. If these things are properly understood by all concerned much misunderstanding will be avoided. and greater satisfaction and profit will result to all concerned.

CHAPTER I.

HISTORICAL.

THE DEVELOPMENT OF COLD STORAGE.

Mother earth as a source of available refrigeration, is without doubt a pioneer. In the Temperate Zone at a depth of a few feet below the surface, a fairly uniform temperature is to be obtained at all seasons of about 50° to 60° F. In some places a much lower temperature is obtained. The same principle is true in any climate, the earth acting as an equalizer between extremes of temperature, if such exist. Caves in the rock, of natural formation, are in existence, in which ice remains the year around, and many caves are used for the keeping of perishable goods. The even temperature, dryness and purity of the atmosphere to be met with in some caves are quite remarkable, owing no doubt to the absorptive and purifying qualities of the rock and earth, as well as to the low temperature obtainable.

CELLARS.

Cellars are practically artificial caves and if well and properly built are equally good for the purpose of retarding decomposition in perishable goods. A journey through the Western states reveals many farmers who are the possessors of "root-cellars," (usually detached from any other structure,) and considered a first necessity of successful farming, the new settler building his cellar at the same time as his log house. A root-cellar is used partly as a protection against frost, but it also enables the owner to keep his vegetables in fair condition during the warm weather of the spring and summer months. The use of cellars for long keeping of dairy products is familiar to all. Many of us can recollect how our mothers put down butter in June and kept it until the next winter, and perhaps it will be claimed by some, that the butter was as good in January as when it was put down. It was not as good, far from it. If you think it was, try the experiment to-day and you will see how it will taste and how much it will sell for in January, in competition with the same butter stored in a modern freezer. The butter made years ago was no better either. No better butter was ever made than we are producing to-day. In short, cellars were considered good because they had no competition---they were the best before the advent of the improved means of cooling. Cellars are still of value for the temporary safe keeping of goods from day to day, or for the storage of goods requiring only a comparatively high temperature, but with a good ice refrigerator in the house, the chief duty of a cellar, nowadays, is to contain the furnace, and as a storage for coal and other non-perishable household necessities. To be sure cellars have their place as frost-proof storage in winter, but we are discussing the cooling problem here.

ICE.

The use of ice as a refrigerant during the summer months is a comparatively modern innovation, and not until the nineteenth century did the ice trade reach anything like systematic development. The possibility of securing a quantity of ice during cold weather and keeping it for use during the heated term seems not to have occurred to the people of revolutionary times. About 1805 the first large ice house for the storage of natural ice was built, and with a constantly increasing growth, the business rose to immense proportions in 1860 to 1870. The amount harvested is now much larger than at that time and constantly increasing, but the business is now divided between natural ice and that made by mechanical means.

The first attempt at utilizing ice for cold storage purposes was either by placing the goods to be preserved directly on the ice or by packing ice around the goods. These methods are in use at present as for instance in the shipping of poultry, fish and oysters, and the placing of fruit and vegetables on ice for preservation and to improve their palatability. The first form of ice refrigerator proper consisted merely of a box with ice in one end and the perishable goods in the other. This form of cooler is illustrated in the old style ice chests, which are now mostly superseded by the better form of house refrigerator with ice near the top and storage space below. On a larger scale small rooms were built within and surrounded by the ice in an ice house. These rooms were of poor design and did not do good work, largely the result of no circulation of air within the room. The principle of air circulation was recognized later, and by placing the ice over the space to be cooled, a long step in the right direction was taken. By this method the air was induced to circulate over the ice and down into the storage room. During warm weather a good circulation of air in contact with the ice purifies the air and produces a uniformly low temperature. Many houses on this system are still in existence, although rapidly being superseded by improved forms.

About the time when the overhead ice cold storage houses were being installed freely, mechanical refrigeration came into the field. Mechanical refrigeration in which the storage rooms are cooled by frozen surfaces, usually in the form of brine or ammonia pipes, was much superior to ice refrigeration, in that the temperature could be controlled more readily and held at any point desired and that a drier atmosphere was produced. Ice and mechanical refrigeration will be discussed fully in treating of construction and in discussing the value of different systems for different purposes. It may be remarked in passing that ice is at present and will probably always remain a most useful and correct medium of refrigeration, especially for the smaller rooms and, under some conditions, large ones as well. The invention and introduction of the Cooper brine system using ice and salt for cooling marked an important step in ice refrigeration. This system is described in the chapter on "Refrigeration from Ice."

MACHINERY.

The first method of mechanical refrigeration to come into general use, and one which is still largely in use on ocean going steam vessels, was by means of the compressed air machines. These operate by compressing atmospheric air to a high tension, cooling it, and expanding it. These machines are very uneconomical in that the compressed gas is not liquefied. Present practice in compression machines mostly employs either ammonia gas or carbon dioxide gas, both of which may be liquefied by pressures and temperatures readily obtainable. Other gases are in use also, but ammonia is the favorite as it liquefies more easily. The apparatus known as the absorption ammonia system is really a chemical rather than a mechanical process, but is usually classed along with the mechanical systems. In this system, the ammonia gas is driven off from aqua ammonia under pressure, by heating; the gas is liquefied by cooling, and the refrigerating effect obtained by expanding the liquid ammonia thus obtained through pipes surrounded by the medium to be cooled. This system is quite largely in use and preferred by many to the compression system, although the latter is used in a large majority of plants.

CHAPTER II

ORGANIZING AND STARTING A COLD STORE.

POSSIBILITIES OF THE BUSINESS.

As a means of preserving perishable food products, and in some cases other goods, from decay or deterioration, refrigeration has come into use with a rapidity that has surprised its most sanguine advocates. The author has been identified with the produce and refrigerating industries for nearly thirty years, and during the last half of this period has sometimes feared that the cold storage business was likely to be overdone. At present there seems no immediate prospect of such a condition, and it is probable that some years will elapse before there will be more cold storage space than goods to fill it. This seems the more probable when we consider the diversified products which are now stored in refrigerated rooms for preservation. Furs, as an illustration, are now placed in cold storage to prevent damage from moths, and to preserve the texture of the skins, and the best furriers report the results as greatly superior to the old method of treatment. Not only are the ravages of the moths prevented, but the furs come out of cold storage actually improved in appearance. Dried fruits are now perfectly kept during the warm months by placing in cold storage. Nuts are kept in the best possible condition by storing in cold rooms. Potatoes and cabbage are carried through the winter and turned out in a condition not thought possible years ago. These are only a few of the products comparatively new to cold storage. Each year finds something new in cold storage for safe keeping, and new uses are being found for refrigeration continually. There seems no limit to the possibilities of the business. It is certainly only a matter of time when the bulk of perishable products will be handled in and sold from cold storage, and kept under refrigeration from the time produced.

By far the greater number of the cold storage plants of small or medium capacity are built and operated by the producers of or dealers in perishable goods as an aid to their business. In fact, refrigeration is no longer considered only as a help, it is a necessity, and the perishable goods operator without suitable cold storage facilities is decidedly at a disadvantage as compared with his competitor who has. This chapter is not written for the man who has use for cold storage in the form of a private plant, but more for those inexperienced, or who might wish to become interested in a larger plant for general use.

The starting and building up of a commercial cold storage business requires all the business sagacity and ability usually necessary to success in any other line, and in addition there are some special qualifications which it may be worth while to consider. The formation of a company, the selection of a system of refrigeration, and the proper construction of the cold storage building are merely preliminary to the actual hard work and care necessary to success, and the cold storage business may develop into more of an undertaking than the average person has any idea of. Even after some investigation the business points are not always as plain as they should be. After the house is built business must be obtained, and satisfactory results given to customers or the venture will prove a failure. A cold storage should not be built, equipped and operated by a person with no knowledge of the perishable goods business, thinking that the business will come naturally. Cold storage is generally only an auxiliary of the perishable goods trade and must be considered as such.

There are many cold storage men now operating houses who complain of poor business, and think there is no demand for cold storage in their locality, when the simple truth is that they have not the proper facilities for the preservation of the goods they try to handle. They turn out musty eggs, strong butter and rotten apples, and consequently their customers only use the storage when they are compelled to. Cases may be cited where a properly-equipped house has been started in competition with the kind above described, and obtained a profitable business from the start. In progressive times like the present, when competition is sharp, it is poor business policy, if not positively suicidal, to go into business with anything except the best facilities. If you are going into the cold storage business, build a good house, and equip it with modern apparatus from designs by a practical and experienced man. A cheap house should not be considered.

An enterprising and self-reliant man is usually at the head of a new cold storage enterprise. It requires both these qualifications to establish a house where apparently little demand exists for such a concern, and generally this is about the situation where there is no cold storage house. There cannot, of course, be business done in the cold storage line where no cold storage house exists; but an intelligent canvass of the situation should indicate the probability or not of business following the erection of a house. If the situation shows fair prospects there can be no failure if the enterprise is handled with the same care and ability necessary for success in other lines of business. Cold storage houses have been constructed with small apparent demand for the space, but after being in business for a year or two to prove an ability to carry goods well, the house has done a good business. In not a single instance known to the author has a well-built, properly-equipped and carefully-operated cold storage house been a source of loss to its owners. In determining the advisability of erecting a house, it is well to have enough business assured, if possible, to pay operating expenses. If this much can be had the first season, the success of the business is no longer in doubt, and the house will generally pay nicely the second or third year. Should the owners be in the produce business, and buy and store enough goods to pay the operating expenses, they can demonstrate the success of the house the first year or two on their own account, and in future seasons obtain outside business very easily. Of course many houses are run for private use only, and the suggestions above do not apply to such cases. It is true that there have been a good many failures in the cold storage business, but they are invariably the result of a poor house or poor handling, with the resulting heavy claims for damage to goods in storage, or of overcapitalization and mismanagement.

PROBABLE CAPACITY REQUIRED.

It is difficult to determine what capacity plant to put up in a place of given size, but usually a smaller capacity than 50,000 cu. ft. should not be figured as a commercial enterprise, and a capacity of from 75,000 to 100,000 cu. ft. can be built and equipped so much cheaper in proportion that generally speaking a capacity of 50,000 cu. ft. is almost too small for economical construction and operation. It is quite often the case that the capacity of a cold storage plant is figured much too small, and it is seldom that the capacity is figured too large, and as a cold storage plant is not readily susceptible of increase of capacity it is advisable to build reasonably large to start with, allowing somewhat for natural growth of business. It is almost always the case that as soon as a cold storage house is built and demonstrates its ability to carry goods successfully, business develops which was not thought of before. The putting up of a cold storage practically creates to some extent a demand for refrigerated space and business for such a plant.

Very little reliable information can be obtained by those who contemplate the erection of a cold storage house, from people already in the business; especially if in the immediate vicinity of the proposed house. This is because those in the business already, regard the building of a new plant as more or less direct competition, and are quite liable to be biased in their views of the cold storage business in general, and of the proposed plant in particular. There is one thing which may be put down as unnecessary, that is the putting up of a small, cheap house as a trial, expecting, if it pays, to put up a larger and a better one. A small, cheap house, while not certain to be a failure, is more than likely to be so, and consequently the larger and better house is never built, and another is added to the ranks of those who think cold storage of no value, and a failure in a business way. Build well, if at all—it is not necessary to experiment, as this has been done repeatedly already, and, the results from a well-built cold storage house are to be depended upon. The population of a town or city does not always indicate its ability to support a cold storage warehouse. A large residential population has very little need for such an establishment, while a comparatively small wholesale center at once makes a demand for storage for perishable goods. A large town, located in a rich producing district, generally gives a good opening for the upbuilding of a business, particularly where the chief articles of production are eggs, butter, cheese or fruits.

COST OF PLANT.

The cost of a first class and fully equipped cold storage building is somewhat startling to many people who contemplate embarking in the business and who have their ideas of cost based on buildings of ordinary construction. The cost of a cold storage plant is two or three times as much as that of a non-refrigerated building of the same size. The shell of a cold storage plant is only a portion of the total cost, and seldom or never exceeds half the cost. In many cases it is only one-third the cost of the finished plant. This varies with the character of the structure, class of insulation, and type of refrigerating equipment. It may be stated as positive that there is no such thing as a cheap cold storage house which will at the same time do good work. Because of the cost of internal arrangements and equipment, a cold storage cannot be compared with any other kind of a building, and the reason why people are surprised at the cost is because they make comparisons with buildings of ordinary construction. Probably two out of three persons who investigate with the idea of building are deterred because of the expense running higher than anticipated. The reader, who has preconceived ideas on the cost of a properly-equipped plant, may safely prepare for a shock should he wish to obtain estimates.

The cost of a first class cold storage plant of 50,000 cubic feet capacity, allowing some space for receiving, shipping, storage of empty packages, office, etc., will be from $20,000 to $30,000. The sum of $20,000 will in some localities be sufficient to build such a plant of frame construction, properly insulated and equipped in a first class manner, and $25,000 to $30,000 will build a substantial brick building carefully insulated and equipped after modern methods. These estimates are approximate only, but are as near as estimates can be made without making accurate figures as applied to a particular locality. A plant double this capacity can probably be built at a cost of from $35,000 to $50,000. The cost is very much less in proportion as the capacity is increased. Of course the cost of such a plant depends on how much space is required outside of the cold storage rooms, into how many rooms the plant is divided, and the amount of freezer space needed. If any considerable part of the plant is required for low temperature or freezer storage it increases the cost considerably.

The above estimates are based, as stated, on first class construction and are estimated on average conditions. The Cooper brine system, using ice and salt for cooling, can usually be installed at a lower cost in small capacity than mechanical refrigeration if the brine circulating system is applied. In larger plants there is very little difference between the cost of equipping with the Cooper brine system and mechanical refrigeration. It may be stated very definitely in this connection that the best cold storage results require brine circulation in connection with the mechanical systems of refrigeration, and direct expansion piping should not be used for what are known as high temperature rooms at a temperature of 30 degrees F. and above. Direct expansion may be installed at lower cost, but it is not desirable, for reasons which are explained elsewhere.

It should be noted further that means of cooling are not the only requirement of a cold storage plant. The Cooper systems of air circulation and ventilation and the chloride of calcium process are applicable to any means of cooling, and these systems have proven their effectiveness wherever installed. This suggestion is offered for the reason that in estimating costs of cold storage plants, the cost of buildings, insulation of the rooms and the means of cooling are not the only costs. The Cooper systems referred to add comparatively little to the total cost of the plant, but yet they do add something. However, as they are applicable to any method of cooling, they should not be considered, when comparing the cost of the Cooper brine system, using ice and salt for cooling, with the mechanical systems of refrigeration.

An extract from the 1911 report of John A. Ruddick, Dairy and Cold Storage Commissioner of Canada, is interesting because he gives some figures on cost of cold storage houses, and further as showing what these houses are insulated with. From the table which follows it may be noted that mill shavings are used exclusively in three houses, and used in combination with other materials in three other houses; and that cork is used exclusively in five different houses; Lith in one; and hair felt in combination with shavings in two.

"Many inquiries are received at this office from those who desire to know the probable cost of a cold storage warehouse of given capacity. There are so many factors that have an influence in fixing the cost of a cold storage warehouse, such as the class of building, the character of the insulation, the proportion of high and low temperature space, the size of the warehouse, etc., that it is impossible to give any but an approximate answer to such questions.

"The cost, with some particulars of the construction, and the size in cubic feet, of some of the warehouses which have been subsidized under the Cold Storage Act, is given below to afford some information on this subject. All names and location of the warehouses are indicated by numbers only. The class of construction for the buildings, apart from the insulation, is represented by letters thus:

"A—Wooden building.
"B—Brick or concrete walls, 'mill construction floors.'
"C—Reinforced concrete.
"D—Brick, stone or concrete, ordinary floors.

"It will be observed that there is considerable difference in the cost of the warehouses in this list on a cubic foot basis. The difference is owing to the fact that some of them are wholly equipped for a very low temperature for fish freezing, while others have a large proportion of space intended for fruit or egg storage at non-freezing temperatures. The difference between the total space and the refrigerated space is represented by engine rooms, corridors, packing floors, etc.

J. A. RUDDICK,
Dairy and Cold Storage Commissioner."

Class of Building	Size of Building	Insulation	Refrigerated space	Cost Exclusive of Site
	Cubic feet		Cubic feet	Dollars
1B	974,622	Cork	700,224	167,000.00
2D	372,000	Hair felt and shavings	105,000	31,019.62
3A	142,218	Shavings	37,960	27,386.69
4C	744,488	Cork	346,538	160,500.00
5D	71,520	Cork	37,161	23,577.00
6B	202,262	Lith	64,000	65,000.00
7A	100,000	Shavings	50,000	18,682.00
8D	220,000	Shavings	33,600	20,000.00
9A	108,040	Cork and shavings	59,940	57,500.00
10B	356,400	Cork	225,000	158,043.00
11B	270,000	Cork	111,050	60,000.00
12D	200,000	Hair felt and shavings	131,510	49,000.00

COST OF COLD STORAGE HOUSES FOR APPLES.

Many figures have been given as representing the cost of cold storage houses for apples, per barrel of capacity.

These range from $2 per barrel for plants of 250,000 cubic feet, or with a capacity of 25,000 barrels as a minimum cost, up as high as perhaps $4 to $5 per barrel in a plant with a capacity of 500 to 1,000 barrels. It must be noted that there is nothing exact about these figures, as so much depends on cost of materials in different localities and type of building, but any estimates or figures are far better as a guide than none at all. The author believes it possible to build a 250,000 cubic foot cold storage plant for apples for $50,000, but this certainly would not be reinforced concrete nor would it be brick and heavy mill construction, nor any type of slow burning construction. It would mean an economically built building of frame and on a favorable building site, but there are many localities where a plant of the capacity stated could not possibly be built on a basis of $2 per barrel capacity. In other words, local conditions and everything else must be favorable in order to make it possible to build at this cost.

An important point influencing cost is the variety of apples stored. If summer and fall varieties are stored in large quantities requiring heavy cooling duty during the warm spells in late summer and fall, a much larger and more expensive refrigerating equipment is necessary, whereas if mostly winter varieties are stored, which go into the house during cool or cold weather when little refrigeration is needed to bring them down to carrying temperature, a comparatively light refrigerating equipment can be installed at proportionately lower cost.

Still another important factor is the arrangement of the building. If built on three or four floors, and few partitions are needed, (suppose, for instance, each floor is one big room), the cost of building is very much less. If the plant is built all on one or two floors requiring large superficial area, and divided into a large number of small rooms, the cost is greatly increased.

The amount of space needed for receiving, shipping, office, coopering, storage of empty packages, etc., in some cases is equal to the cold storage capacity, and this means increased cost per barrel of storage capacity. If it were possible to build a cold storage plant with nothing but cold storage and with no other space for other purposes, it may readily be seen that the cost would be much reduced.

It is probable that among the houses now actually in service there are few if any which could now be built at a cost of $2 per barrel of capacity. The larger ones would mostly cost from $2.50 to $3.50 per barrel, and some of the smaller ones from $3.50 to $5.00 per barrel.

CLASS OF GOODS PLACED IN COLD STORAGE.

The product which may be depended upon to furnish the largest portion of the business to a newly-established cold storage depends on the location. Some houses are built solely for cheese, others for eggs, and others only for apples; but generally speaking, eggs form the largest and best paying product which is handled in cold storage. Eggs are probably the most difficult of all products to successfully carry for a period of six or eight months. If they are stored in too dry an atmosphere they dry out or shrink, and in this condition decay more quickly. If the air is too moist the eggs will mold and become musty. There is more danger of having a room too moist than too dry, and the damage resulting from too moist a room is also much greater. The best temperature for eggs is 29° to 30° F., and they are carried at this temperature by the best houses. A forced circulation of air is beneficial, and the moisture in the air should be regulated to the proper degree. For testing the air moisture of a cold storage room an instrument called the sling psychrometer is used. The subject of humidity is rather complicated, and the reader is referred to chapter on "Humidity," and "Eggs in Cold Storage," for a more comprehensive treatment of this subject.

Butter is probably second in importance to eggs, and all cold storage houses have rooms fitted especially for this product. The correct temperature for carrying butter has not been definitely settled by a majority agreeing on some one temperature, and at present butter is held in cold storage at temperatures ranging from below zero to 25° F. The most common temperature now is between 10° and 15° F., and the author believes this to be low enough. Many practical men insist that zero is better, and some houses are carrying it at this temperature. Still others are holding temperatures for butter at from zero to 10° F. Butter storage room should be kept dry enough to prevent the formation of mold, and generally no attention is paid to the matter of humidity; the room being amply dry, nothing further is thought of it. If butter rooms are too dry, as they frequently are, it leads to a bad drying out of the packages, and of the surface of the butter as well, causing it to get "air-struck" or "strong" and shrink in weight. Butter, in order to keep well in cold storage, must be protected from contact with the air. Much has been said about freezing butter, but the butter fat practically has no freezing point, and it simply gets harder and harder the lower the temperature; so the idea that butter freezes at a temperature just under 32° F. is entirely erroneous. (See chapter on "Butter in Cold Storage" for more complete information.)

Cheese is not ordinarily considered so difficult a product as butter and eggs to refrigerate successfully, but this idea comes largely from the fact that cheese has only within the past fifteen years been well handled in cold storage, and the possibilities of refrigeration for this purpose have not been fully understood. Cheese will not spoil if stored in cellars or basements: nevertheless a properly-equipped cold storage room will quickly pay for itself in the improved results obtainable. Cheese should be carried at about the same degree of humidity as eggs, and at a temperature ranging from 38° down to 30° F. It is very common practice now to place cheese in cold storage when only eight or ten days old. At this age it is not properly cured, and should not be placed in a lower temperature than 38° F. The temperature may be gradually lowered after a month or two, and at an age of three or four months the temperature of the room should reach 30° F., but should not go any lower. If the temperature is carried much below 30° F. for any length of time it will injure the texture of the cheese, and even at 30° F. some claim that it makes the cheese "short" or brittle in texture. Cheese will freeze so as to be unfit for market at about 20° to 25° F. The reason why cheese should not be placed in too low a temperature while new, is that it may not ripen or "cure up" properly, and is liable to develop a bitter flavor. It must be remembered in considering this subject that cheese is of many different kinds and widely varying quality. What is said above refers to an average make of American cheddar cheese. (For further information on the cold curing of cheese see chapter entitled "Cheese in Cold Storage.")

Apples are stored in large quantities during the fall and winter months. The quality of the fruit should be prime, and not too fully matured. It is customary to place apples in egg rooms as fast as eggs can be removed in the fall, and no bad effect will result. Apples and eggs should not, of course, be placed in the same room together, but when a room is emptied of eggs it is customary to fill it with apples. After the apples go out and before again filling with eggs, the room should be thoroughly whitewashed. (See chapter on "Keeping Cold Stores Clean.") There are many different varieties of apples, and some of them require special treatment in cold storage, but the generally accepted temperature for apples for long-period storage is 30° or 31° F. Some apple men prefer higher temperatures, and get good results, but the lower temperatures are the favorite. Apples should not be quickly cooled when placed in cold storage. If a week or two is consumed in reducing them to the correct temperature so much the better. (See chapter entitled "Apples in Cold Storage.")

Oranges are very successfully cold-stored at temperatures of from 32° to 35° F. Lemons are very sensitive to cold, and may be seriously damaged if the temperature approaches near the freezing point. Fifty degrees is thought best for lemons. Lemons and oranges must be stored separately and isolated from products like eggs and butter. It is best not to handle these in the same building unless through a separate outside entrance, as much damage results to eggs and butter if flavored with the odor of citrus fruits. Some prominent cold storage houses have been very heavy losers from being obliged to pay for damage from this cause.

Dried fruit and nuts, flour, and other goods known as grocers' sundries, are now a large item for cold storage in some wholesale centers. This business comes largely from the wholesale grocers and commission men. These goods are stored at a temperature of 35° to 45° F. The storage of furs, woolens, etc., is an important and lucrative business in many cities, and where the volume of business is sufficient a room may be set aside for the purpose, and made to pay well. Any temperature below 40° F. is all that is necessary for this class of goods. Potatoes may be kept in cold storage at a temperature of 34° F., and carried until spring in prime condition. Potatoes freeze easily, and are entirely ruined when frozen, so the temperature must never touch the freezing point. Cabbage may be carried some time in a green condition, at a temperature of 31° F. Freezing will not damage cabbage materially if the frost is drawn out slowly. The freezing and storage of poultry is a remunerative business, and much poultry is handled through cold storage. The freezing may be accomplished at 12° to 15° F. with good results if stock is freshly killed and in small packages. For temporary holding without freezing a temperature of 30° F. is best. Poultry can only be held a few weeks at this temperature, a month to six weeks being the extreme limit. Beer and meat are handled by some houses. Beer should be held at 35° to 38° F., and meat at 30° to 38° F., depending on length of time it is to be carried.

RATES FOR COLD STORAGE.

The rates to be obtained for storing different products vary with the locality, competition, etc., but the following will serve as a guide. These rates are mostly higher than average rates on carload lots, but will serve as a guide to those not familiar with local rates. Each locality has its own rates to some extent:

	Per Month.	Season Rate.	Season Ends.
Eggs, per 30 doz. case	$ .12½	$ .50	January 1
Butter, per 100 lbs.	.20	.75	January 1
Cheese, per 100 lbs.	.15	.60	January 1
Apples, per barrel	.12	.50	May 1
Lemons, per box	.08	.35	July 1
Oranges, per box	.07	.30	July 1
Dried Fruit, per 100 lbs.	.07	.30	November 1
Nuts, per 100 lbs.	.08	.35	November 1
Furs, Coats, etc.	..	2.00	January 1
Potatoes, per 100 lbs.	.08	.30	April 1
Cabbage, per ton	1.50	4.00	April 1
Poultry freezing, per cwt.	.25	1.00	April 1

Beer, space rented at 15c. per cubic foot per year.
Meat, per 100 lbs., 15c per month.

EARNINGS OF COLD STORES.

To show the prospective earnings of a small house we will take one of 50,000 cu. ft. capacity operated on the Cooper system, and assume that we secure the first year half its capacity, or twenty cars of eggs. Twenty cars of eggs equal 8,000 cases. If we secure a season rate on all, at the carload-rate of 40 cents, this will give us a gross income of $3,200. Operating costs are difficult to obtain even with the simple ice and salt system owing to widely varying circumstances under which plants operate. An estimated cost of the ammonia or other mechanical systems is out of the question as the item of attendance alone is never uniform.

The operating expenses of a house of 50,000 cu. ft. of space conservatively figured will be in northern localities where natural ice may be secured cheaply and assuming that the plant is equipped with the Cooper brine system, using ice and salt for cooling, and operated the entire year, about as follows:

700 tons of ice at 50c per ton	$350.00
80 tons of salt at $5.50 per ton	440.00
6 tons of chloride of calcium, $15 per ton	90.00
Power for ice crushing and elevating, operating fans for air circulation and ventilation, and for driving freight elevator, average $25 per month	300.00
Labor, 6 hrs. per day for 200 days at 20c	240.00
	$1,420.00

The item of labor above does not include labor of handling goods in and out of the house, but is based only on the labor required for charging primary tanks of the Cooper brine system with ice and salt, looking after temperatures and running the house. In many places ice may be had for less than 50 cents per ton. In other places these costs average higher, but the above is conservative, and will apply to average cases.

From these figures it is seen that with our house half full of goods, the business would pay a fair profit above actual expenses. It may be well to note here that it costs practically as much to operate a cold storage house half filled with goods as it would if completely filled. The only difference is a small labor item of the handling, and the cost of cooling the extra quantity of goods in the first place to the temperature of the room, both very small items. The moral of this is that the cold storage manager should aim to have his house filled every year. If apples are to be had as the eggs go out in the fall, the income for the year is materially increased with little cost, as apples require only a small amount of refrigeration during the cool weather of fall and winter.

ADVICE TO THOSE NEW TO THE BUSINESS.

A few words of advice to prospective investors regarding the danger of experimenting in cold storage construction. It is dangerous from the fact that a failure means the damage of a very valuable product, and a consequent heavy money loss. The most absurdly foolish schemes have been tried by men with no practical or scientific information, and the result has been what any thorough-going cold storage man could foresee—either flat failure or no tangible results from the experiments tried. Sometimes it occurs that the would-be cold storage man thinks to save architect's and engineer's fees by planning his own building, or by taking some of the plans and ideas which appear from time to time in the agricultural or trade papers, and working them over to suit his case. It is the author's positive opinion that four times as much money is wasted in this way as there is saved. No two houses properly use the same construction and arrangement, and each case requires special study by the designer in order to do it justice, and he is a poor engineer indeed who cannot save twice his fees to his client. The above advice is given with an intimate knowledge of the subject, as the author has spent much money on experiments and tests of various kinds, and never expects to be properly reimbursed for the time and effort expended. All lines of industry are more and more specialized, and the planning and equipping of a cold storage house is just as much a special business as the buying and selling of produce.

As has already been pointed out, the results possible to attain by the use of ice are equally as good, within certain limits, as may be obtained by employing the ammonia or mechanical system. The ice and salt system has the advantage of being cheaper to install, cheaper to operate, and a better control of temperature is possible. These are all very good reasons why the ice and salt system should be adopted where ice is a sure crop, and can be put in the house at a moderate price. There is absolutely no question about the results obtained from storing goods in such a house, well-built and properly managed. The most perfect results possible in refrigeration may be obtained, and at a small cost as compared with the mechanical systems. Where manufactured ice is in use the small cold storage house, butcher, produce dealer, or any other business requiring refrigeration in comparatively small amounts, can in many cases obtain the best results at a lower cost by the use of ice and salt than by the installing of a small machine. Besides this they are absolutely safe against a breakdown.

The question is often asked, "How long will a cold storage house and its equipment of piping and iron work remain in good operating condition?" No positive answer can be made, as a great deal depends on the building and the apparatus, and the way it is handled and cared for. The average life of a cold storage building and the insulation should not be essentially different from that of an ordinary building of the same construction, and this means that it will last indefinitely. The equipment, with ordinary repairs, would do good service for from fifteen to twenty-five years, probably longer under favorable conditions. An ice storage room will remain in good condition for from fifteen to twenty-five years, and it is probable that it would be serviceable for the purpose for a much longer time.

CHAPTER III.

SYSTEMS OF REFRIGERATION.

INTRODUCTION.

In the first chapter under "Historical" a brief review was given of the introduction and development of various cooling methods for the preservation of perishable goods. It is not the intention in this chapter to give much more than the principles on which the various systems operate. The uses of caves and cellars for cold storage are so crude and unsatisfactory that they are hardly worth considering except from an historical standpoint. The various methods of cooling by means of ice may be found in the chapters on "Ice Boxes and Refrigerators," "Refrigeration for Retailers" and "Refrigeration from Ice," and the reader is referred to these chapters for more complete details. Mechanical refrigeration has been explained so fully in the "Compend of Mechanical Refigeration," "Machinery for Refrigeration," &c., that the following outline is only intended to serve in explaining the general principles of the various systems of mechanical refrigeration.

COLD AIR SYSTEM.

The cold air refrigerating machines operate on the principle that a compression of air generates heat, and its expansion afterwards absorbs heat or produces cold. The air is first compressed in an especially built pump or compressor, and the heat produced is removed by applying water for cooling. The refrigeration resulting from the expansion of air from the expansion cylinder is utilized for cooling purposes. Mechanical work and heat are convertible, and this law is utilized in the cold air refrigerating machine. It is necessary for the air when expanding to work against a piston in order to exhaust the stored heat of the compressed air. This leaves the expanding air in condition to do useful refrigerating work.

There are two systems of cold air machines; one being known as the open cold air system and the other the closed system. In the open system fresh air is taken in with each stroke of the compressor pump, and the air, after expanding, is discharged to the atmosphere. In the closed type of machine the same air is used over and over again.

The first cold air machine to be constructed in the United States was utilized for ice making, and was built by Dr. Gorrie in Florida about the year 1850. The heat of compression was removed by a spray of water introduced into the compression cylinder, and by expanding the cooled air a second spray of water was turned into ice.

Undoubtedly the largest number of cold air machines in use are of the Bell-Coleman make, which machine is of the Windhausen type; the Bell-Coleman machine being of improved design and construction. The Allen dense-air machine has been largely used in the United States, while in England the Linde Co. and the Haslams' have furnished a large number of installations.

Practically as well as in theory the cold air refrigerating machine requires more power than those machines which utilize a liquefiable gas in the cycle of operation. This means that large compression and expansion cylinders with the accompanying increased friction, as well as moisture in the air, &c., all operate to reduce the efficiency of the cold air machine. Improvements, however, have been made recently, and it is probable that a useful place will always be found for the cold air machine. On shipboard especially this system of refrigeration has been found very satisfactory for several reasons. It is much safer to operate, and the question of economy does not seriously enter the problem in connection with refrigeration or ice making on ocean-going vessels.

CARBON DIOXIDE SYSTEM.

This system is identical in principle with the ammonia compression system described further on. Carbon dioxide is a gas which is liquefiable at certain temperatures and pressures. A much higher pressure is required than with the ammonia system, and the system has not found favor for this reason, as the loss from clearance and friction is considerable and the system, speaking generally, is not as efficient as the ammonia system. There is, however, a great advantage in the carbon dioxide system from the fact that the gas is non-poisonous, and considerable quantities may be liberated in a closed space without danger to life or health. This qualification makes the system of distinct advantage in many places, especially in the confined areas of ships and for certain uses on land as well.

COMPRESSION AMMONIA SYSTEM.

Substances which are gases at ordinary temperatures may be changed to liquids at low temperatures and comparatively high pressures. Ammonia is such a substance and is in most common use for cold storage and ice making purposes. The gas is compressed by means of a power driven machine of suitable design and cooled by flowing water over coils of pipe containing the compressed gas. This reduces it to a liquid form. In changing a gas to a liquid much heat is given up which is extracted by the cooling water.

To obtain refrigeration the liquid is expanded to a gas again during which process heat is absorbed or cold generated. The expansion, controlled by means of a valve, is through a suitable system of pipe coils usually. The evaporation or expansion of the liquid into the form of a gas cools the pipes. and this in turn cools the surrounding medium, either air, brine or water, or whatever it may be. Ammonia piping placed in a cold storage room absorbs heat which means cooling, and ordinarily frost collects on the outside of the coils. Theoretically the same amount of heat is absorbed from the surrounding mediums during process of expansion as has been previously given up by the ammonia during process of compression and cooling. The ammonia is thus used in a continuous cycle, being returned periodically from a gas to a liquid state and vice versa.

While this system is very simple in general scheme, yet the necessary parts of the apparatus are quite complicated and various auxiliary machinery is necessary besides the compressor and condenser. It is, however, much more simple than the absorption ammonia system, which is about to be described, and for this reason, doubtless, it has come into more general use.

ABSORPTION AMMONIA SYSTEM.

The absorption ammonia system has an entirely different cycle of operation than the compression system. While the gas used is the same, yet it is used in a different way, and while the absorption system is generally classed as mechanical refrigeration, yet the process is more chemical or physical than it is mechanical. In the compression system the ammonia gas is known as anhydrous ammonia, and in the absorption system, the anhydrous ammonia is absorbed in water at the commencement of the cycle. The mixture of ammonia and water, known as aqua-ammonia is heated in a suitable still or generator, which evaporates or drives off the anhydrous ammonia, and the processes of condensation and expansion are just the same as they are in the ammonia compression system as already described. After expansion through a suitable system of coils and after doing the work of refrigerating or ice making, the ammonia gas (anhydrous ammonia) is absorbed back into the water again in a suitable tank or vessel known as the absorber.

There are four distinct stages in the process of refrigeration with the absorption system as follows: First, Generation of the gas or vapor; Second, Condensation of the gas; Third, Evaporation or expansion of the condensed gas from a liquid to a gas again; and Fourth, Absorption of the gas into water. As in the compression system there are some rather complicated parts and apparatus necessary to control all of these processes, and as the cycle is somewhat intricate rather closer attention is required and a better understanding of the underlying laws than is required in connection with the operation of the ammonia compression system.

It may be suggested in passing that the ammonia absorption system is especially adapted to situations where rather low temperatures are required and where condensing water of not above 60° F. is available. The efficiency and capacity of the absorption system does not fall off as rapidly when lower temperatures are required, as does the compression system.

CHAPTER IV.

GEOMETRY OF COLD STORAGE HOUSES.

BEST PROPORTIONS FOR COLD STORES.

An important factor in the cost of constructing and cost of refrigerating cold storage rooms, as independent rooms, or as a complete warehouse, is the relation of dimensions (length, breadth and height) to area of outside exposure. This point is often lost sight of in the design of refrigerated structures, and the desire to gain all the space possible on main floor sometimes leads to some very absurd arrangements from a theoretical, practical or business standpoint. The installing of first-class elevator facilities in a cold storage warehouse is very important and with a fairly high rate of speed and a commodious car, space on the floors above is practically as valuable as space on the ground floor. The idea that storage rooms should be low, say 7 feet to 9 feet, has often been carried to an unwarranted extreme. It is where rooms are to be used for temporary purposes only that it is desirable to have the rooms low to avoid unnecessary labor in handling the goods. Rooms for long period storage purposes as a general rule should be made from 10 feet to 12 feet in height; not only as a matter of economy of space and cost of construction, but the circulation of air in the room is much more perfect. This is especially true of direct piped rooms. The importance of this subject has been so often overlooked in the construction of cold stores that it has been thought advisable to direct attention to it here. The relation of the cubical contents of a building to its outside exposure or superficial area is readily appreciated by noting a few figures, as follows:

Take three rooms or buildings of equal storage capacity, with cubical contents of 1,000 cubic feet, and whose three dimensions vary. The cube with length, breadth and height each 10 feet (see Fig. 1) has an outside exposure of 600 sq. ft.

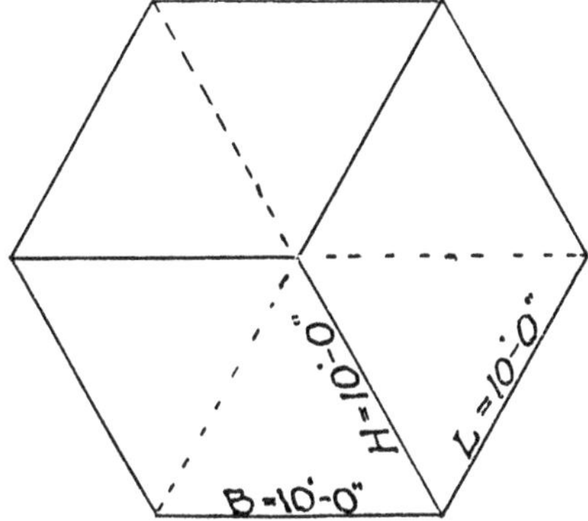

Fig. 1.—BxLxH equals 1,000 cubic feet.
10x10 equals 100; 100x6 equals 600 sq. ft.
Ratio of cubical contents to outside exposure 1,000 to 600.

Comparing with another rectangular space of equal capacity—whose breadth is 10 feet, height 7 feet 6 inches and length 13 feet 4 inches. (See Fig. 2.)

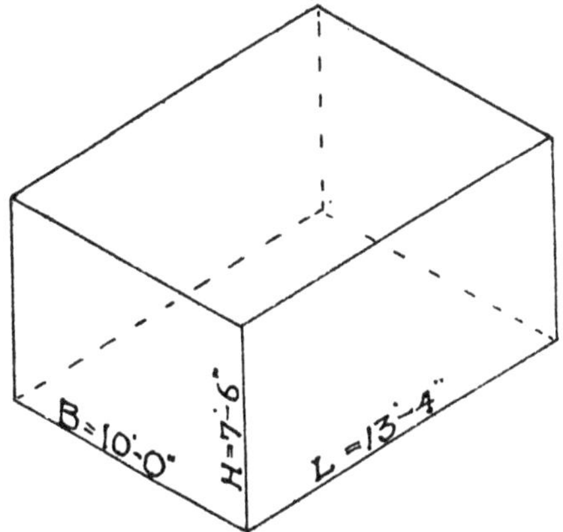

Fig. 2.—BxLxH equals 1,000 cu. ft.

10'x13'4"x2	equals 266 2-3	sq. ft.
10'x 7'4"x2	equals 150	sq. ft.
7'6"x13'4"x2	equals 200	sq. ft.
Total	..........616 2-3	sq. ft.

Ratio of cubical contents to outside exposure 1,000 to 616 2-3.

It will be noted that the change of dimension in this case is but slight from the cube, so the increase of outside exposure is only 2.77 per cent.

Taking another and more pronounced departure from the cube and still retaining the capacity of 1,000 cubic feet, where the breadth is 6 feet 8 inches, length 25 feet and height 6 feet (see Fig 3).

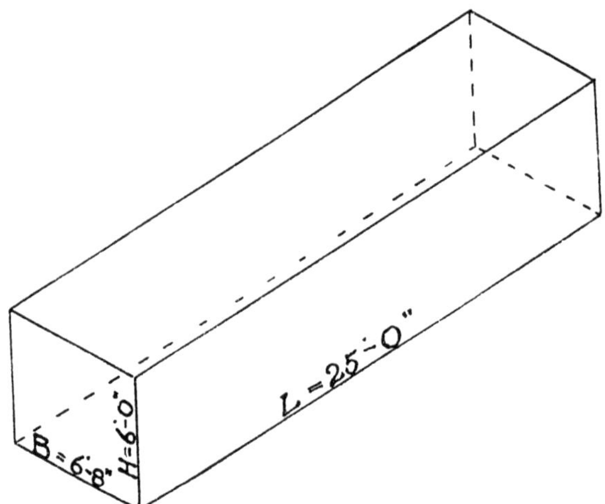

Fig. 3.—BxLxH equals 1,000 cu. ft.
6'0"x25'0"x2 equals 300 sq. ft.
6'8"x25'0"x2 equals 333 sq. ft.
6'8"x 6'0"x2 equals 80 sq. ft.

Total ....713 sq. ft.
Ratio of cubical contents to outside exposure, 1,000 to 713.

To sum up the comparison of the cube with the other two rectangular rooms or buildings would be as follows:

	Cubical contents in cubic feet.	Superficial area or outside exposure in square feet.	Percentage of increase over cube.
Fig. 1....................	1000	600	
Fig. 2....................	1000	616 2-3	2.77
Fig. 3....................	1000	713	18.83

The result is important in view of the fact that the loss of refrigeration from heat leakage through the walls is on the average probably three-fourths of the total amount necessary to supply and maintain temperature in cold storage rooms. The amount of heat leakage will be directly proportional to the exposed outside surface or superficial area of the room or house. The cost of insulation, which is usually figured by the square foot of wall surface, is also increased proportionately, and the cost of building is also greater. The cost of insulation and cost of cooling to make good the heat leakage will be 18.83 per cent greater if the room or building is built as in Fig. 3, than if built in the form of a cube, as in Fig. 1. Therefore, in the design of cold storage rooms or buildings, the nearer a cube may be approximated, the cheaper the first cost and cost of operation, other things being equal.

This must not be carried to an extreme which will make the conduct of the business laborious or expensive. Some classes of trade require much floor space and little height, while others may use a high room. For a business where many goods are handled in and out, daily, ground floor space is extremely valuable. In extreme cases it may be necessary, on account of expense and time consumed in handling, to arrange all storage rooms on the ground floor. To do this the advantages obtained must more than offset the increased cost of construction and operation. For a business where goods are mostly in for long-term storage a house of several floors is practically as convenient, costs less, is cheaper to operate and requires less ground space.

By the use of labor saving devices described elsewhere, such as the gravity carriers combined with elevating apparatus and spiral lowering parts properly arranged, there is no need of figuring to have all handling and storage space on one or even two floors. As an instance, may be cited the pre-cooling of oranges. The fruit is in storage for a matter of two to four days only, and the basement makes a most valuable location for the cold rooms. Main floor space for the grading and packing is necessary to secure air and light for the workers. The fruit is lowered into the cooling rooms by a spiral conveyor and raised by an endless chain elevator.

CHAPTER V.

INSULATION.

GENERAL CONSIDERATION.

The original matter comprising this chapter was prepared nearly eight years ago, and contained practically all of the scientific and practical information available on the subject up to that time. Since then comparatively little has been added to the general subject except by those who have been interested in making or selling special insulating materials, and by far the larger number of articles and papers which have been written on the subject have been by people who were thus interested rather than by engineers who should be able to gauge such matters from a disinterested standpoint. The educational influence, therefore, of the current literature on the subject of insulation tends to induce people to buy materials which are marketed in especially prepared forms ready for applying. It is the purpose in this chapter to present in a fair and unbiased manner the merits of different insulating materials and discuss ways and means of applying them regardless of outside influences or personal opinion. The author has had to do with the planning of many refrigerating installations during the past twenty-five years, and he believes that there has been no other influence in his recommendations than strictly merit, and it should be possible to say this of every legitimate engineer who expects to retain the respect and confidence of his clients.

Insulation as applied to the purposes of cold storage construction is the providing of a suitable wall to prevent the penetration of an unreasonable amount of heat, and perhaps also to prevent the penetration of cold during extreme winter weather. It is not possible to prevent the loss of refrigeration (the inflow of heat) or the coming in of cold (the outflow of heat) entirely, no matter how perfect an insulation is used; and the commercial aspect of the problem must be considered. It cannot possibly be profitable to provide an expensive insulation to save losses which will not pay interest on the increased investment. The selection of a suitable insulating material and its correct application is consequently of the utmost importance in the design of cold storage and refrigerating plants. It should be stated here that the illustrations contained in this chapter are not all by the author, nor does he in every case recommend the construction shown without qualification. The details shown are intended to set forth representative forms of insulation as generally used. The insulating values of different materials as shown by tests have to a great extent been accepted from the figures made by those who have tested the material in question, and the accuracy of these figures is not vouched for by the author. The apparatus used in making the tests is rather thoroughly described in detail, and sufficiently accurate information given so that it will be possible for the reader to make tests on his own account if desired.

The great variety of materials and combinations of materials which have been and are still used as cold storage insulation is accounted for largely by the fact that up to within comparatively recent years no proven standards of efficiency were available. The person having a given work in charge used his own ideas, and in most cases this resulted in poor insulation from an efficiency standpoint. There was altogether too much guesswork, individual ideas and popular prejudice. Some people would be satisfied with a very small quantity of the cheapest kind of material put up in any kind of a way, not appreciating the fact that insulation is the vital feature of a successful and economically operated plant. Others used plenty of material, but not knowing how to apply it properly much money and labor was wasted. As illustrating this idea the author has had occasion to remove as many as eight thicknesses of dressed and matched lumber composing the outer insulated walls of a cold storage building; each thickness of lumber separated from the others by a one inch furring strip and a layer of paper. Thus fully double the quantity of lumber necessary was used. By a little skillful design a better insulation could be put up for half the cost. As illustrating the other extreme a case is recalled where the only insulation consisted of a four inch air space formed between the outer boarding and the inner boarding on the studs of the building with an added two inch air space on the inside. The insulation thus consisted of two air spaces, one of four inches and another of two inches with three thicknesses of lumber and three thicknesses of paper. These two cases illustrate the great divergence of opinion as to what cold storage insulation should consist of. We know at this time that air spaces are fully understood to be out-of-date as insulation.

There has, however, been much intelligent pioneer work done in the designing and testing of insulation and in the selection of suitable materials. This work has at times been handicapped and difficult owing to inaccurate representations and claims made by manufacturers and salesmen of various special insulating materials, and the laboratory tests of the non-conducting properties of various substances have been distorted in some cases to suit one particular material. The tests, while perhaps correct, were often misleading to the customer so far as enabling him to form a correct impression of the real insulating value of the materials in question is concerned.

Another influence which has been at work has also delayed progress in developing of good cold storage insulation. The manufacturers of refrigerating machinery have usually devoted space in their catalogues to approved insulations, but they have seldom secured the services of skilled engineers in the design of same or suggested advanced or progressive ideas on the subject. The details of insulation which they recommended were as a rule entirely insufficient for economical operation, and they were satisfied to sell a larger refrigerating machine, as it was found easier to do this than to convince the customer that he should invest more money in better insulation. Besides, a bigger machine meant a bigger sale. It is a pleasure to say, however, that more judgment and fairness is now being used, and as knowledge spreads, we may look for better and more economically insulated cold stores in future.

THEORY OF HEAT.

Heat vitally concerns all cold storage and refrigerating problems, as it is the elimination of heat and the preventing of its entrance that makes refrigeration necessary. A brief outline of the theory of heat will be useful to a better understanding of the laws underlying the design and construction of insulation.

Heat is a form of activity or energy and does not possess substance. The sun is the one source of heat from which other sources are fed by direct radiation. The heat of the earth, chemical combination, electricity, friction, etc., are only manifestations, but are regarded as the lesser sources of heat. Heat tends towards equilibrium; thus a cold body is warmed by one of a higher temperature. If the temperature of the air is warmer on one side of a wall than on the other, heat flows through until temperatures are equalized. It is not possible to prevent this, but there is a vast difference in the heat retarding value of various materials which may be used to form an insulated wall.

The transmission of heat is effected in three different ways: first, by radiation; second, by convection; and third, by conduction. Radiation is the direct passage of heat through the air from one body to another without perceptibly heating the air, and is manifested to the senses by the heat which is felt when standing by an open fire. By radiation, heat is thrown off in every possible direction from every point of a hot body. In an inclosed air space with different temperatures as shown in Fig. 1, the radiant heat would pass from the high to the low temperature side directly across the space indicated by arrows. The scientific definition of radiant heat is that it is in the nature of a wave motion communicated through an exceedingly subtle ether, which is supposed to pervade all space, and that it is obedient to the laws of refraction, reflection, polarization, etc., the same as light.

Convection of heat is the transfer from one place to another by the bodily moving of the heated substance, such as when air, water or any other gas or fluid comes in contact with a heated surface; the particles touching the heated surface become warm and lighter, therefore ascending and giving place to the colder and heavier particles below. This action is illustrated by the heating of rooms with stoves; the air as warmed rises to the top of the room and its place is taken by the colder air from below. The principle of convection, or circulation as it is generally understood, is shown by Figs. 2 and 3, where the air in the inclosed space with one side warmer

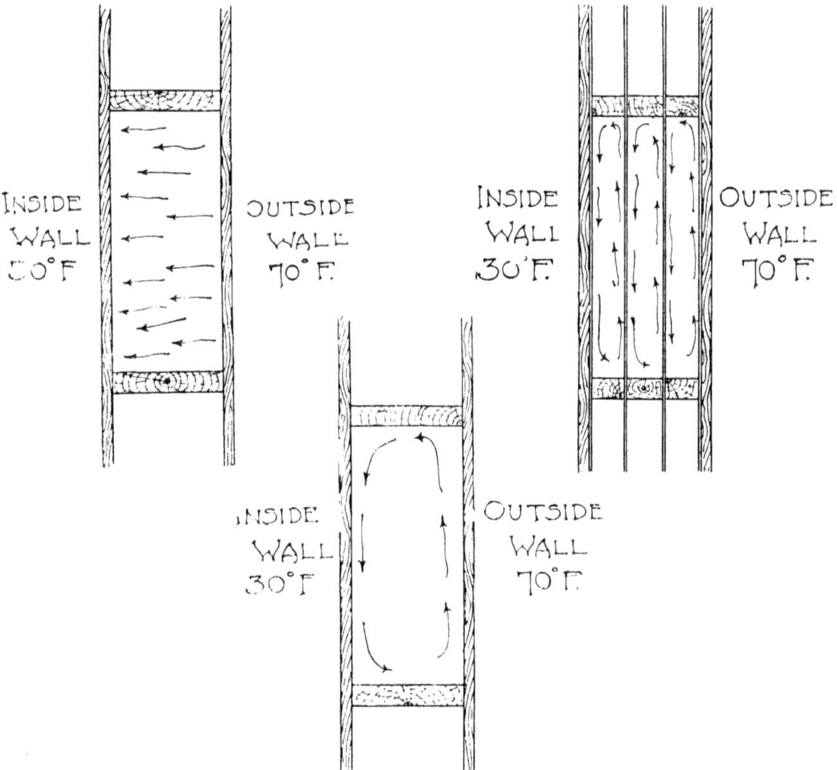

FIGS. 1, 2 AND 3—ILLUSTRATING WAYS OF HEAT TRANSMISSION.

than the other, being heated on that side, becomes lighter by expansion and rises; as it gets to the top of the confined space, it passes over to and down the cold side where it gives up heat; as it is cooled, it contracts and becomes heavier; it then sinks and returns to its original place. This circulation will continue indefinitely or until the temperatures on both sides of the space are equalized. Fig. 3 illustrates this principle when a wall is subdivided into a number of such spaces, and the circulation becomes more complicated and retarded, passing less heat for the same thickness of wall in a unit of time than a single space, as illustrated in Fig. 2.

Conduction is a term applied to heat flowing from a warmer to a colder part of a body, or if a solid substance is placed in contact with a body having a higher temperature, the particles of the substance nearest are warmed, and they in turn give up a portion of the heat received, to particles next to them and so on from particle to particle until the whole substance is heated; this is accomplished without any sensible motion. A more familiar example of conduction is putting one end of an iron poker in the fire; after a time, the other end will become heated and apparent to the sense of feeling.

As heat then is not a substance but a vibration of the molecules that compose a body, and as the rapidity of these vibrations is the cause of the difference of temperature, it is really improper to speak of heat and cold as such; but it is convenient to use these old familiar terms in describing the phenomena, just as it is said that the sun rises and sets, where it is in fact the earth that moves.

Theoretically, all bodies and substances transfer heat by radiation, convection and conduction at the same time, and this is called complicated transfers of heat. Scientists state that bodies at high temperatures will lose more heat by radiation than by convection and conduction, and that heat radiated by a coal fire is estimated to be about one-half of the total heat generated. At lower temperatures, such as is dealt with in refrigerating work, transmission of heat by radiation is very small, and practically, convection and conduction only need be considered in cold storage construction.

UNITS OF HEAT.

"Heat is measured quantitatively by the heat unit, which also varies in different places like other standards. The unit used in the United States and England is the British Thermal Unit (abbreviated B. T. U.), and represents the amount of heat required to raise the temperature of one pound of water 1° F. The French unit is the Calorie, and is the quantity of heat required to raise the temperature of one kilogram of water from 0° to 1° Celsius.

"Some writers define the B. T. unit as the heat required to raise the temperature of one pound of water from 32° to 33°. Others make this temperature from 60° to 61°, and still others define it as the amount of heat required to raise 1/180 pound of water from the freezing to the boiling point. The last two definitions give nearly the same result, and may be considered practically identical."*

The unit of heat transmission or insulating value is the number of B. T. U.'s that will pass through one square foot of a substance per hour, per degree difference in temperature between the two sides of the substance. Some engineers prefer (in refrigerating work) to use a time unit of one day (24 hours) instead of one hour in their values. This is perhaps more comprehensive, as refrigerating capacity is usually figured per day, and it also is an advantage in that the values are more likely to be expressed in whole numbers and less in decimals.

CONDUCTORS OF HEAT.

Many laboratory experiments conducted by noted physicists during the past century have given us tables of heat conducting properties of the metal, mineral, liquid and vegetable substances; these tables vary from one another, depending upon the methods used and the nature of the experiments. These experiments demonstrate that the metals are the best conductors of heat; that the vegetable and animal substances are the poorest conductors of heat, and that between these the minerals and liquids are all arranged in varying degrees of heat conductivity.

Laboratory tests of the heat conductivity of materials cannot be absolutely relied upon when these materials are to be used for cold storage insulation. These tests are usually

* Dr. J. E. Siebel, "Compend of Mechanical Refrigeration."

made under high temperature conditions and relatively low humidity, such as steam pipe covering. Such conditions do not obtain in cold storage work where the lower temperatures and relatively higher humidities are the conditions. Numerous articles and papers have been written for the trade periodicals and read before various associations on the subject of insulation. Some of these articles are very theoretical and are based altogether too much on laboratory test tables of heat conductors, which make them almost useless for practical application in cold storage construction.

The following table of the relative heat conductivity of a number of substances is taken from Sir William Thomp-

TABLE OF RELATIVE HEAT CONDUCTIVITY.

Article on "Heat" in Encyclopedia Britannica.

Substance	Conductivity
Copper	455.
Iron	80.
Sandstone	5.34
Stone	2.95
Traprock	2.075
Sand	1.31
Water	1.
Oak (across fiber)	.295
Walnut (along fiber)	.24
Fir (along fiber)	.235
Walnut (across fiber)	.145
Fir (across fiber)	.13
Hemp cloth (new)	.072
Wool (carded)	.061
Hemp cloth (old)	.0595
Writing paper (white)	.0595
Cotton wool	.0555
Eiderdown	.054
Gray paper (unsized)	.047
Air	.0295
Cork	.0145

Note: The figure for air has been fixed by J. Clark Maxwell's brilliant investigations. He gives its conductivity at 1/20,000 that of copper, as 1/3,360 that of iron, a determination reached by mathematical deductions from the kinetic theory of gases.

son's article on "Heat" in the Encyclopædia Britannica, reduced to a unit of conductivity of one for water; this includes authorities that he regarded as reliable on that subject. Part of this table was taken from experiments made by Péclet, whose table is also given in B. T. units.

In recent years many tests of composite insulations put together just as they would be erected in a cold storage house wall have been conducted and tables compiled therefrom by experimenters who have made the subject of insulation a study, and who have had much practical experience in its application in their capacity as designing architects and engineers. These tests show in many cases a wide variation in results, owing no doubt to the fact that the tests have been made under widely varying conditions and methods and also to the changeable factor of human error or personal equation in the observation of the tests. The work of these experimenters shows much painstaking care, and much good has resulted in raising the standard of the construction of scientific and practical cold storage insulation.

As to quantitative or rate of transmission, the following table from experiments made by M. Péclet* gives the amount

TABLE OF POOR HEAT CONDUCTORS.

By M. Péclet.

	Units of heat transmitted.
Gray marble, fine grained	28.
White marble, coarse grained	22.5
Limestone, fine grained	14.8
Limestone, coarse grained	10.5
Glass	6.
Brick	5.6
Terra cotta	4.8
Plaster of paris	3.6
Sand	2.2
Oak, across the grain	1.7
Fir, across the grain	0.75
Fir, along the grain	1.4
Walnut, across the grain	0.83
Walnut, along the grain	1.4
Guttapercha	1.37
India-rubber	1.36
Brick dust, sifted	1.33
Powdered coke	1.3
Iron filings	1.26
Cork	1.15
Powdered chalk	.86
Powdered wood charcoal	.63
Straw, chopped	56
Powdered coal, sifted	.54
Wood ashes	.5
Canvas, new	.41
Calico, new	.40
Writing paper	.34
Cotton, raw or woven	.32
Eiderdown	.31
Blotting paper	.26

*Péclet's "Traite de la Chaleur," IV Ed., Tome 1, Pgs. 542 to 555.

of heat in B. T. units transmitted per square foot per hour, through various substances one inch in thickness. He terms these poor conductors (to distinguish them from the metals). The results of these experiments are considered quite reliable, as they are used extensively by heating engineers of Europe in their calculations for the heating of buildings. The experiments were made by heating one side of the substances with hot water, and cooling the other side with cold water, the difference between the temperature of the two sides being 1° F.

In the latter part of the eighteenth century, Count Rumford, who did much work in the experimental study of heat, maintained that liquids had no conducting power at all, but gained heat by convection only. This was afterward found to be incorrect, as shown in the above table, and shows in fact that water stands next to the mineral substances in conductivity.

In an article written by Prof. John M. Ordway,* on "Non-conductors of Heat," which treats of insulation tests conducted on steam pipes, he subjoins the following table of non-conductivity of various substances. The figures in the last column are for covering, one inch thick, with a difference of 100° F. on each side of the covering. In most of the tests a stream of water at about 176° F. was kept running through the heater. In some cases the source of heat was steam at 310° F. as stated.

A careful study of these tables shows that still air is one of the poorest conductors of heat available for practical purposes. The distinction between confined air and still air, and the greater conducting qualities of the former has not been generally understood, and it is perhaps on this account that air space construction has been used so much for cold storage insulation. Note what Dr. Hampson, an English authority, has to say on air spaces. The conclusions reached by him have also been demonstrated by the author and other experimenters in this country, and the result is the present tendency to use materials which will subdivide the air into an infinite number

*_Ice and Refrigeration_, October, 1891, Page 216.

NON-CONDUCTORS OF HEAT.

Non-conductors one inch thick.	Net cubic in. of solid matter in 100	Heat units transmitted per sq. ft. per hour
Still air	....	43
Confined air	....	108
Confined air=310°	....	203
Wool=310°	4.3	36
Absorbent cotton	2.8	36
Raw cotton	2	44
Raw cotton	1	48
Live-geese feathers=310°	5	41
Live-geese feathers=310°	2	50
Cat-tail seeds and hairs	2.1	50
Scoured hair, not felted	9.6	52
Hair felt	8.5	56
Lampblack=310°	5.6	41
Cork, ground	....	45
Cork, solid	....	49
Cork charcoal=310°	5.3	50
White-pine charcoal=310°	11.9	58
Rice-chaff	14.6	78
Cypress (*Taxodium*) shavings	7	60
Cypress (*Taxodium*) sawdust	20.1	84
Cypress (*Taxodium*) board	31.3	83
Cypress (*Taxodium*) cross-section	31.8	145
Yellow poplar (*Liriodendron*) sawdust	16.2	75
Yellow poplar (*Liriodendron*) board	36.4	76
Yellow poplar (*Liriodendron*) cross-sec.	30.4	141
"Tunera" wood, board	79.4	156
Slag wool (Mineral wool)	5.7	50
Carbonate of magnesium	6	50
Calcined magnesia=310°	2.3	52
"Magnesia covering," light	8.5	58
"Magnesia covering," heavy	13.6	78
Fossil meal=310°	6	60
Zinc white=310°	8.8	72
Ground chalk=310°	25.3	80
Asbestos in still air	3	56
Asbestos in movable air	3.6	99
Asbestos in movable air=310°	8.1	210
Dry plaster of paris=310°	36.8	131
Plumbago in still air	30.6	134
Plumbago in movable air=310°	26.1	296
Coarse sand=310°	52.0	264
Water, still	....	335
Starch jelly, very firm, "	....	345
Gum-Arabic, mucilage, "	....	290
Solution sugar, 70 per cent, "	....	251
Glycerin, "	....	197
Castor oil, "	....	136
Cotton-seed oil, "	....	129
Lard oil, "	....	125
Aniline, "	....	122
Mineral sperm oil, "	....	115
Oil of Turpentine, "	....	95

of spaces. As insulators against heat Dr. Hampson, in a series of lectures at University College, Liverpool, England, sums the various substances up as follows:

Conduction and convection are best prevented by a totally empty space intervening between the external objects and the internal cold objects—in other words, by having a vacuum between two air-tight walls. Radiation can be to a great extent prevented by having a bright metallic surface between the inside and outside. The efficiency of this combination was shown in one of the silvered vacuum vessels designed by Professor Dewar, which contained liquid air which had been made half a week before. Where such an arrangement was impossible, the best thing to do was to fill the insulating space as far as possible with the substance that had the smallest capacity for conducting heat. Iron has about one-seventh the conducting power of copper, wood or other organic substances still less, ice has only about one two-hundredth, and air not more than a twenty-thousandth part of the conducting capacity of copper. Air, therefore, is the best insulating substance available; but its value depends upon its stillness, for if free to move in spaces of considerable size, it will be in constant circulation, convection currents carrying in heat from the warmer outside walls to the colder inside walls of the insulating spaces. These spaces should therefore be very shallow, so that the viscosity of the air, which is very small, will be able to prevent it from moving. It is their possession of a large proportion of air, prevented by septa or filaments from moving, that determines the excellence of the usual insulating materials, such as eider down, wool, feathers, hair, chaff, cork, slag-wool, asbestos, charcoal, wood, sawdust, etc.

A VACUUM THE POOREST CONDUCTOR.

Physicists seem to have proved that a vacuum (familiarly illustrated in the thermos, or vacuum bottle) is a poorer conductor of heat than air, and a reference is made to it by Dr. Hampson, as noted above. This was discussed by Dr. H. W. Wiley in an address before the American Warehousemen's Association convention at Washington, D. C., December, 1903, and as it is interesting in connection with the subject it is quoted in part as follows:*

There is one practical suggestion which these theories present, namely, that a vacuum is by far the best protection against radiation that has ever yet been discovered. Sawdust, shavings, cork, cloth, wood and many other substances have been extensively used to protect cold spaces against radiation, but none of these have anything like the obdurating properties of a vacuum. Liquid air and even liquid hydrogen contained in a vacuum receiver, that is a receiver surrounded by a vacuum, retain their liquid state for hours and even days. There is, of course, a loss by radiation and evaporation from the exposed surface, because pressure dare not be used in confining these bodies, but this loss is comparatively slow. The vacuum becomes an almost perfect protector against heat. If, therefore, the refrigerating rooms which you use could be surrounded with a vacuum

*Reported in *Ice and Refrigeration*, January, 1904, page 25.

space, it would most certainly reduce very largely the expense of maintaining the low temperature. There are, of course, practical objections to the use of a vacuum for this purpose of a very serious character. The two chief objections would be the difficulty of maintaining an air-tight space so that there would be no leakage into the vacuum and the enormous pressure upon the walls of the vacuous space produced by the atmosphere itself. It is easy to construct a steam boiler which will bear a pressure of from 400 to 600 pounds to the square inch, and it ought not to be difficult to construct a vacuous space around a refrigerating room which would resist a pressure of 15 pounds to the square inch. The expenditure and the energy required to evacuate this space and keep it practically free from air would, in my opinion, be profitably expended, providing the two conditions of imperviousness and pressure could be regulated. The idea is at least worthy of experimental trial and it is hoped that some of you will submit it to a practical test.

Mr. James Wills of New York once made a practical trial of a vacuum as insulation for brine piping with good results, but it has not been learned that the experiment has met with sufficient success to warrant its adoption on later work which was constructed under that gentleman's supervision.

Harold B. Wood in a paper on Ice Storage House Construction, suggests the construction of ice storage walls consisting of two six inch solid concrete walls with a six inch space between them, this space to be maintained as a vacuum. This suggestion was not made with a view of its immediate adoption, but as a food for thought as representing the possibilities at some time in the future. Vacuum insulation is possible but not at all practicable at the present time.

VARIATION OF HEAT TRANSMISSION.

M. Péclet proved experimentally that the rate of transmission of heat was directly proportional to the difference of temperature on each side of a substance, and was inversely proportional to the thickness. That is; if a substance one inch thick transmitted say, one B. T. U., the same substance two inches thick would transmit one-half B. T. U. under same conditions.

The results of later experiments, on poor conductors and on those used in combination (such as used in the construction of cold storage warehouses) show, however, that these conclusions are in doubt. John E. Starr, in an article* on results of tests, conducted by himself, states: "It is a well known fact that the amount of heat transferred per degree of difference increases somewhat with each degree of increase of difference of temperature." This same experimenter illustrates this principle graphically by a diagram of tests† showing ice meltage in ordinary domestic refrigerators at various differences of temperature between inside and outside. If the transmission had been directly proportional, the plotted curve on the diagram would have been a straight line.

This increase of transmission per degree of difference as the difference increases is also shown by a chart published in *Ice and Cold Storage* (British), March, 1901 (see Fig. 4, page 48), of results of tests with eight different constructions of the same thickness. It is a matter of regret that the methods of testing were not described in this case so that we could judge of their reliability. Referring to the chart it will be found that the line plotted for the rate of transmission of each material is a curve, having a range of temperature difference of 80° F. Calculating down to per-degree difference at each end of the chart and dividing the result by the range of difference (80° F.) shows co-efficients of increase, varying from 25 to 50 per cent.

The author, in conducting a series of tests in 1900 and 1901 (which will be described further on), obtained results that tended to prove the correctness of the observations cited above. This co-efficient of increase varies for different substances and combinations of materials, and to determine these co-efficients accurately would be a difficult task indeed. From the above facts it is obvious that the co-efficients of heat transmission obtained by tests of substances which were made under a temperature difference of only one degree, are too small for practical application, and they should be increased about 50% when used for designing cold storage insulation. This is of increasing importance when we consider that the tendency of modern cold storage practice is toward maintaining lower temperatures, often resulting in a difference of temperature of from 70° to 90° F. between inside and outside of walls. This

*"Non-conductors of Heat," in *Ice and Refrigeration*, July, 1891, page 37.
† "The Cost and Value of Low Temperatures," in *Ice and Refrigeration*, September, 1891.

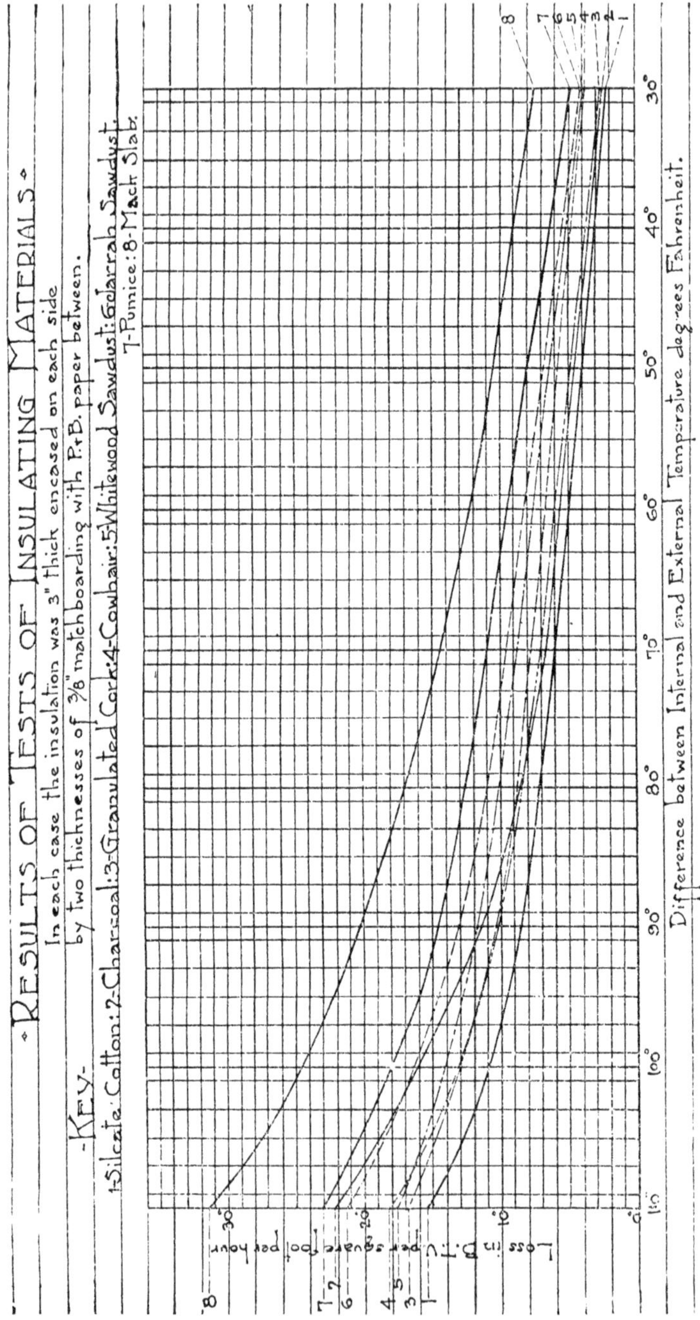

FIG 4.—CHART SHOWING RESULTS OF TESTS IN EIGHT DIFFERENT CONSTRUCTIONS.—ADOLPH BLOCK, GERMANY.

is comparatively a high range of temperature and the conditions to this extent are similar to heating work.

That transmission of heat through any substance is not inversely proportional to the thickness seems evident by an examination of the following table converted from the metric system by Chas. F. Hauss, Antwerp, Belgium. This writer states* that this table is used by Adolph Block of Hamburg, one of Germany's most reliable engineers:

TABLE OF CO-EFFICIENTS OF TRANSMISSION IN B. T. U. PER SQ. FT. OF SURFACE PER HOUR.

Cooling Surfaces	Thickness of Walls	Difference in Temperature—Fahrenheit														
		1°	5°	10°	15°	20°	25°	30°	35°	40°	45°	50°	55°	60°	65°	70°
	4¾"	0.48	2.40	4.80	7.20	9.60	12.00	14.40	16.80	19.20	21.50	24.00	26.50	28.80	31.20	33.60
	10"	0.34	1.70	3.40	5.10	6.40	8.50	10.20	11.90	13.60	15.30	17.00	18.70	20.40	22.00	23.80
Solid	15"	0.26	1.30	2.60	3.90	5.20	6.50	7.80	9.10	10.40	11.65	13.00	14.30	15.60	16.90	18.20
	20"	0.22	1.10	2.20	3.30	4.40	5.50	6.60	7.70	8.80	10.00	11.00	12.00	13.20	14.30	15.40
Brick	25"	0.18	0.90	1.80	2.70	3.60	4.50	5.40	6.30	7.20	8.10	9.00	9.90	10.80	11.70	12.60
	30"	0.16	0.80	1.60	2.40	3.20	4.00	4.80	5.60	6.40	7.20	8.00	8.80	9.60	10.40	11.20
Walls	35"	0.13	0.65	1.30	1.95	2.60	3.25	3.90	4.55	5.20	5.85	6.50	7.15	7.80	8.45	9.10
	40"	0.12	0.60	1.20	1.80	2.40	3.00	3.60	4.20	4.80	5.40	6.00	6.60	7.20	7.80	8.40
	45"	0.11	0.55	1.10	1.65	2.20	2.75	3.30	3.85	4.40	4.95	5.50	6.05	6.60	7.15	7.70
	12"	0.45	2.25	4.50	6.75	9.00	11.25	13.50	15.75	18.00	20.25	22.50	24.75	27.00	29.25	31.50
Solid	16"	0.39	1.95	3.90	5.85	7.80	9.75	11.70	13.65	15.60	17.95	19.50	21.45	23.40	25.25	29.30
Sandstone	20"	0.35	1.75	3.50	5.25	7.00	8.75	10.50	12.25	14.00	15.65	17.50	19.25	21.00	22.75	24.50
Walls	24"	0.32	1.60	3.20	4.80	6.40	8.00	9.60	11.20	12.80	14.40	16.00	17.60	19.20	20.80	22.40
	28"	0.29	1 45	2.90	4.35	5.80	7.25	8.70	10.15	11.60	13.05	14.50	15.95	17.40	18.85	20.30
For Lime-	32"	0.26	1.30	2.60	3.90	5.20	6.50	7.80	9.10	10.20	11.65	13.00	14.30	15.60	16.90	18.20
stone add	36"	0.24	1.20	2.40	3.60	4.80	6.00	7.20	8.40	9.60	10.40	12.00	13.20	14.40	15.60	16.80
10 per cent.	40"	0.22	1.10	2.20	3.30	4.40	5.50	6.60	7.70	8.80	10.00	11.00	12.00	13.20	14.30	15.40
	44"	0.20	1.00	2.00	3.00	4.00	5.00	6.00	7.00	8.00	9.00	10.00	11.00	12.00	13.00	14.00
	48"	0.19	0.95	1.90	2.85	3.80	4.75	5.70	6.65	7.60	8.55	9.50	10.45	11.40	12.35	13.30
Solid Plaster	1¾"-2¼"	0.60	3.00	6.00	9.00	12.00	15.00	18.00	21 00	24.00	27.00	30.00	33.00	36.00	39.00	42.00
Partitions	2½"-3¼"	0.48	2.40	4.80	7.20	9.60	12.00	14.40	16.80	19.20	21.50	24.00	24.00	28.80	31.20	33.60
Floors	Joists with double floors	0.07	0.35	0.70	1.05	1.40	1.75	2.10	2.45	2.80	3.15	3.50	3.85	4.20	4.55	4.90
	Stone floor on arches	1.20	1.00	2.00	3.00	4.00	5.00	6.00	7.00	8.00	9.00	10.00	11.00	12.00	13.00	14.00
	Planks laid on earth	0.16	0.80	1.60	2.40	3.20	4.00	4.80	5.60	6.40	7.20	8.00	8.80	9.60	10.40	11.20
	Planks laid on asphalt	0.20	1.00	2.00	3.00	4.00	5.00	6.00	7.00	8.00	9.00	10.00	11.00	12.00	13.00	14.00
	Arch with air space	0.09	0.45	0.90	1.35	1.80	2.25	2.70	3.15	3.60	4.05	4.50	4.95	5.40	5.85	6.30
	Stones laid on earth	0.08	0.40	0.80	1.20	1.60	2.00	2.40	2.80	3.20	3.60	4.00	4.40	4.80	5.20	5.60
Ceilings	Joist with single floors	0.10	0.50	1.00	1.50	2.00	2.50	3.00	3.50	4.00	4.50	5.00	5.50	6.00	6.50	7.00
	Arches with air spaces	0.14	0.70	1.40	2.10	2.80	3.50	4.20	4.90	5.60	6.30	7.00	7.70	8.40	9.10	9.80
Windows	Single	1.00	5.00	10.00	15.00	20.00	25.00	30.00	35.00	40.00	45.00	50.00	55.00	60.00	65.00	70.00
	Double	0.46	2.30	4 60	7.05	9.20	11.50	13.80	16.10	18.40	20.70	23.00	25.30	27.60	30.00	32.20
Skylights	Single	1.06	5.30	10.60	15.90	21.20	26.50	31.80	37.00	42.40	47.70	53.00	58.30	63.60	69.00	74.20
	Double	0.48	2.40	4.80	7.20	9.60	12.00	14:40	16.80	19.20	21.60	24.00	26.50	28.80	31.20	33.60
Doors		0.40	2.00	4.00	6.00	8.00	10.00	12.00	14.00	16.00	18.00	20.00	22.00	24.00	26.00	28.00
Dif. in Temperature		1°	5°	10°	15°	20°	25°	30°	35°	40°	45°	50°	55°	60°	65°	70°

*Paper read before American Society of Heating and Ventilating Engineers, New York, January, 1904.

This table is of limited value for cold storage work, but serves to show the great variation in results obtained by different experimenters. It will be noted that this table is based on M. Péclet's first proposition, viz: That the rate of transmission is proportional to the difference of temperature on each side of the substance.

The fact that the transmission of heat through any substance is not inversely proportional to the thickness is also shown by the following tables after Box,* where N is the value in B. T. U. transmitted per square foot for a difference of 1° F. between temperatures each side of wall in 24 hours.

½	brick	4½	inches	thick	N.	equals	5.5	B. T.	Units
1	"	9	"	"	"	"	4.5	"	"
1½	"	14	"	"	"	"	3.6	"	"
2	"	18	"	"	"	"	3.0	"	"
3	"	27	"	"	"	"	2.6	"	"
4	"	36	"	"	"	"	2.2	"	"

Stone	walls	6	inches	thick	N.	equals	6.2	B. T. U.
"	"	12	"	"	"	"	5.5	"
"	"	18	"	"	"	"	5.0	"
"	"	24	"	"	"	"	4.5	"
"	"	30	"	"	"	"	4.3	"
"	"	36	"	"	"	"	4.1	"

HEAT TRANSMISSION THROUGH WALLS.

The following formula for calculating the amount of heat that will pass through a wall of a certain area is by Dr. Siebel.*

If the number of square feet contained in a wall, ceiling, floor or window be f. the number of units of refrigeration, R. that must be supplied in 24 hours to offset the radiation of such wall, ceiling or floor may be found after the formula:

$$R = fn\ (t - t_1)\ \text{B. T. units.}$$

or, expressed in tons of refrigeration: $R = \dfrac{fn\ (t - t_1)}{284{,}000}$ tons.

In these formulæ t and t_1 are the temperatures on each side of the wall, and n the number of B. T. units of heat transmitted per square foot of such surface for a difference of 1° F. between temperatures on each side of wall in twenty-

*From "Compend of Mechanical Refrigeration," Page 181.

four hours. The factor n varies with the construction of the wall, ceiling or floor from 1 to 5. For single windows the factor n may be taken at 12 and for double windows at 7. (Box.) For different materials one foot thick we find the following values for n:

Pinewood	2.0	B. T. U.
Mineral Wool	1.6	"
Granulated Cork	1.3	"
Wood Ashes	1.0	"
Sawdust	1.1	"
Charcoal, powdered	1.3	"
Cotton	0.7	"
Soft Paper Felt	0.5	"

If a wall is constructed of different materials having different known values for n, viz, n_1, n_2, n_3, etc., and the respective thickness in feet d_1, d_2, d_3, the value, n, for such a compound wall may be found after the formula of Wolpert, viz:

$$n = \frac{I}{\frac{d_1}{n_1} + \frac{d_2}{n_2} + \frac{d_3}{n_3}}$$

The value of n may be obtained from any of the foregoing tables that are based on the transmission per hour by multiplying by 24, the number of hours in a day, and where the values given are for materials one inch in thickness, n and d should be in inches.

INSULATION OF COLD STORAGE WAREHOUSES.

The function of a cold store is to maintain temperatures suitable for the storage of perishable materials. This generally means that the cold storage rooms are held at a temperature below that of the surrounding air, but it may be also that the cold rooms are useful to keep out frost, such for instance as apple storage in winter. In the first case heat flows into the rooms, and refrigeration must be supplied to absorb it. In the second case it may be necessary to supply heat to prevent dangerous low temperature in the storage rooms. As ordinarily built the walls of a building do not offer sufficient resistance tc the passage of heat, and therefore in cold storage construction additional materials are used, and these are called "Insulation," as distinguished from the structural walls of the building.

A perfect insulation is impossible. No matter of what materials or how thick the walls are made, a certain amount of heat will pass through them, and this must be taken up by the refrigerating medium. If it were possible to stop all heat transmission through walls, doors, etc., no refrigeration would be necessary after the goods in storage had been cooled down to the required temperature. On the contrary, it is a well established fact that one-half to seven-eighths of the refrigeration applied to cold storage rooms is expended in removing the heat transmitted through the walls of the building, depending of course upon the amount of goods stored and the frequency with which they are handled in and out.

The great importance of proper and efficient insulation is evident when it is considered that all the heat passing through it must be taken up by the refrigerating apparatus, which, in the case of poor insulation, will need to be from 25% to 50% larger than if the insulation were first-class. This larger apparatus means a greater first investment than if a smaller apparatus could have been used, and this difference might better have been invested on the insulation. The additional operating expense of the larger apparatus would be continuous from year to year and would amount to many times as much as it would if first-class insulation had been constructed in the first place. The investment put into good insulation has to be made but once, while with poor insulation the loss of refrigeration through removing the greater heat leakage makes a continual heavy expense. Insulation should be considered in the light of a permanent investment, same as buildings and equipment, the returns of which should be based on the savings effected by the lower operating cost. It is a great deal cheaper to prevent heat from entering a building by providing efficient insulation than to remove it by means of refrigeration.

PRACTICAL FEATURES CONSIDERED.

It is agreed that air spaces are the basis of insulation. It is also agreed that the large air spaces such as are formed by the ordinary studding of a frame building or such as have been built up by means of several thicknesses of paper separated by furring strips, are inefficient and not worth their cost. It is also now agreed among the best posted engineers, and especially

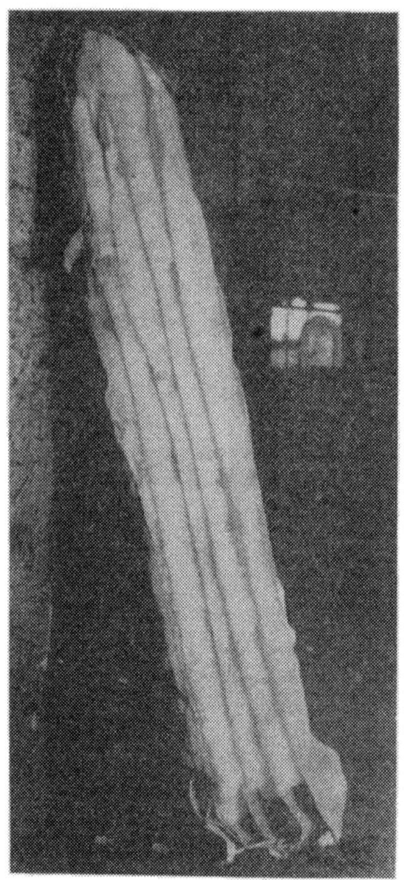

FIG. 5—SHOWING ICE FORMATION BETWEEN INSULATION NOT PROPERLY AIR-PROOFED.

those who have had a hand in tests and experiments on the various types, that insulation should consist of some light and porous material which holds within itself minute cells or small spaces of air. Cork is one of the best materials for insulating purposes on account of its fine texture and its waterproof qualities, and its property of non-capillarity.

The illustration (Fig. 5) shows clearly what happens when insulation is not properly air-proofed. The insulation in this case was formed by paper air spaces one inch apart, and was on the walls of a fish freezer held at 10° to 15° above zero. These paper air spaces, as shown, *are frozen full of ice.* The paper used was gray rosin sized, which is not air and moisture proof and which is not now much used for modern insulation.

It was early discovered in connection with the storage of natural ice, which was the real beginning of the cold storage and refrigerating industry, that sawdust was a very efficient protection against heat, and it is still being used for this purpose. Up to comparatively recent times practically all of the natural ice storage houses were either insulated in the walls, or the ice was covered with sawdust to protect it from the heat. Sawdust may be obtained almost anywhere, and as a cheap material which could be readily obtained it doubtless had a useful place in the early days of ice storage.

At the present time mill or planer shavings are being substituted to a great extent. Mill shavings have the advantage over sawdust in being obtainable in a fairly dry condition. Sawdust, with the exception of that obtained from box factories, etc., is generally from green lumber and most always damp or wet. It also moulds readily and thus has the elements of decay in it before it is ever placed in the building. On the contrary, mill shavings as now generally handled in bales, are nearly always obtainable in a thoroughly dry condition, being baled promptly from the planers and stored under cover. The shavings also come from the outside of the lumber and are thus from the dryest part of it, and in many cases the lumber planed is dry. Shavings in bales weighing about 80 to 100 lbs. are easily handled and can be shipped to some considerable distance at low freight rate, and have come into use as an insulating material not only in connection with natural ice, but in some of the very best cold storage plants.

There are, of course, many other materials used for the filling of spaces besides sawdust and mill shavings, and among them may be cited such materials as mineral wool, cottonseed hulls, chaff, leaves, cut straw, crushed coke, locomotive breeze, cinders, ashes, etc., and the permanency and efficiency of any

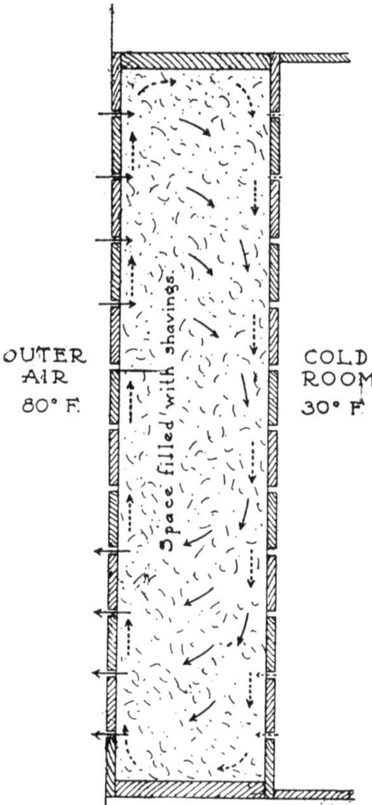

FIG 6—SHOWING ORDINARY SPACE BETWEEN STUDS OF A FRAME BUILDING.

of these materials which are used for filling spaces depends on a principle which is comparatively little understood and which it is here attempted to make plain.

Referring to Fig 6 which may represent the ordinary space between the studs of a frame building, note that the interior wall, or what would be the inside of a cold storage room, is exposed to a temperature of 30° F., while the outer wall which might represent the outside of the building is exposed to the atmosphere at a temperature of say 80° F. Heat is conducted from the outer or 80° wall to the inner or 30° wall, as we have already seen, by three means:

First—Conduction.

Second—Radiation.

Third—Convection.

First—Conduction is of small consequence in a wall of this kind except through the solid studs of the building, as there are so many different pieces or particles of the sawdust or shavings, etc., that the route of travel for the heat would be extremely tortuous and long and very little heat would pass from the outer wall to the cold wall by this means.

Second—As radiation is the direct travel of heat from one surface to another without interference from any solid being interposed, we may assume that the amount of heat transferred by radiation from the warm wall to the cold wall is practically negligible.

Third—This leaves us only convection, which in plain language means circulation, and we can reduce it to a still plainer term by calling it air circulation, and the illustration shows the path of circulation, conveying the heat from the warm wall to the cold wall.

As above stated, the circulation of air in a filled space such as we are considering is nowhere near as rapid as it would be were the space open without anything in it, but the circulation is there just as surely, and as applied to the principle we are considering is fully as destructive as though it were more rapid.

Assume further that the space we are considering is contained between a layer of boarding on the outer or warm wall and on the inner or cold wall, and that these layers are composed of rough lumber, which necessarily would mean that there would be cracks or openings between the boards through which the air might pass. It will readily be seen, then, that air would penetrate near the top of the wall, circulate to the inner or cold side of the wall, down the cold side, and eventually gravitate out at the bottom as shown by the solid line arrows. There would, of course, be also some circulation (convection) within the space between the studding independent of this in and out circulation referred to, and this is shown by the broken line arrows. In addition there will be some circulation *into* the cold room at the top and *out* of the cold room at the bottom as shown by the small broken line arrows.

Now, what is the result of this? Take a warm day in the summer with high humidity and with air flowing in through the filling material and coming in contact with the cold side of the wall. There is a condensation which will cause a wetting of the inner boards, the inner edge of the studs, and a wetting of that portion of the filling materials (sawdust or shavings, etc.) which lies nearest the inner boards. There is no possible guesswork about this. It has been demonstrated in a very large number of cases. This wetting, of course, is not present at all seasons of the year, but only during warm weather. In cold weather the filling material would naturally dry out, and this doubles the damage. Alternate wetting and drying will in a very few years cause a rotting and detoriation of the filling material and the frame and sheathing of the building. Of course, not many cases would be as extreme as this, and in most cases paper would be used, but some of the paper is of such a character that it allows the passage of air with its contained moisture, and the result while not perhaps as destructive, is nevertheless enough to be serious and damaging.

Numerous cases have come to the attention of the author where the ceiling boards on the upper floor of a cold storage plant have become wet and saturated with moisture, and some of these cases have been so aggravated that the nails have pulled out of the joists and allowed boards to fall from ceiling. Other cases may be cited where the same result has occurred because of the circulation of air through a brick wall and in between the joists. The necessity for a perfect air seal especially on the outer surface of an insulated wall is unquestioned, but this fact is not as well appreciated by the trade in general as it should be. Where these bad effects occur those who come in contact with the results are likely to condemn the material used, and it is this unfair condemnation of useful materials to which it is desired to call attention at this point.

The results of the use of filling materials as outlined, in the hands of inexperienced or incompetent people who do

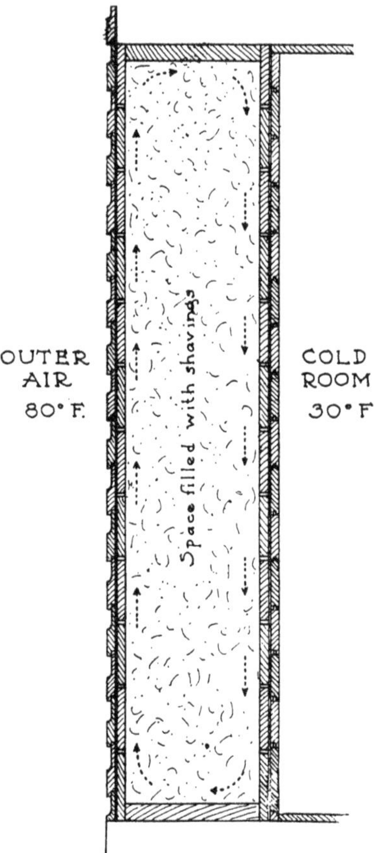

FIG. 7—SHOWING WALL INSULATED WITH SHAVINGS.

not understand the underlying laws governing the design of insulation and the practical features thereof, has to an extent, brought such a useful and valuable insulating material as mill shavings into undeserved disgrace. Other materials like mineral wool, sawdust and granulated cork have similarly been discredited, and the author has personally seen some materials removed from cold storage construction which might lead a person not well versed in the subject to condemn these materials for insulating purposes. It is really not the materials that are at fault, but the manner in which they are used.

We will refer to another sketch (Fig. 7) showing a wall insulated with what we have called a "filling material" such as granulated cork, sawdust, mill shavings, mineral wool, etc. Instead of the studding being boarded on the exterior and interior with rough boards and protected perhaps with a poor quality of porous paper, this time we use a layer of surfaced boards, then special insulating paper which is perfectly air-tight and water-proof, and then matched boards; thus having the paper tightly clamped between two smooth boards, making practically a perfectly sealed surface both on the outer and on the inner surface of the insulated wall. Please note details of construction. In such a wall there can be no circulation of air through the outer boards nor through the inner boards. The circulation or convection as shown by the arrows is confined entirely to the space within the air-tight surfaces resulting from the use of insulating paper on the outer and inner sides of the wall as stated. Thus there is no penetration of warm, moisture-laden air and no rotting, as there would be if the space between the studs were not made air-tight.

The author has in his experience seen some very extraordinary conditions and some very bad conditions too. These bad conditions embrace practically every known material which is used for insulation, and to make it more emphatic it should be repeated that it is not in most cases the fault of the material, but the method of application. Any insulating material to be durable and permanent must be sealed from air and moisture and this includes cork as well as other material. Cork, while being non-capillary to an extent, is such to an extent only, and if exposed to moisture conditions will rot out and deteriorate. Mineral wool will not rot, but will lose its insulating value to some extent if damp, and the supporting studs and boards will soon rot if dampness occur in the manner explained. Other materials deteriorate and decay if exposed to dampness. Air tight protection to the insulating substance is the key to permanent insulation regardless of the particular kind of material used.

Any material used as cold storage insulation must be protected from penetration of air. This statement may be laid down as the underlying principle of all insulation, and there is positively no exception. While it is claimed that corkboard is non-capillary and will not absorb moisture, this is only true to a certain extent, and corkboard or cork in any form will rot out just as certainly as will sawdust if exposed to the same conditions, and it will rot out just as quickly, too.

The illustration (Fig. 8) explains the action of moisture in connection with shavings when they are not properly protected from air and moisture contact. The sketch represents

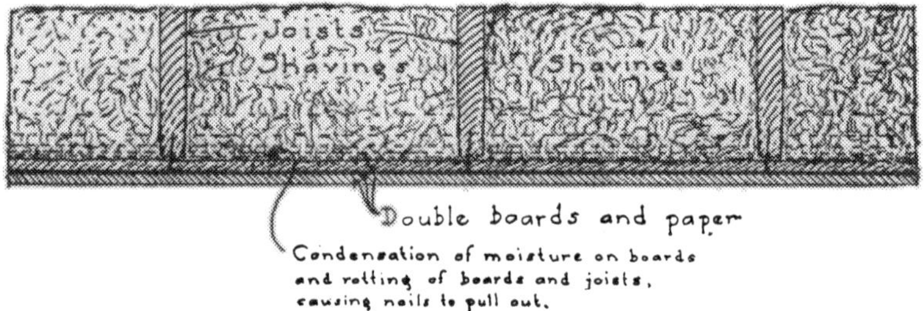

FIG. 8—SHOWING ACTION OF MOISTURE ON BOARDS AND JOISTS.

the ceiling joists of a cold storage room. The joists are boarded on the bottom edge with double boards with paper between, and the joists are filled in between with shavings as shown. The top edge of the joists are exposed and the shavings are not protected on top by paper or other covering material. The temperature of the room below being say at 30° F. and the temperature of the air above it at 80° F.; the air in circulating into the shavings would carry with it moisture which would be condensed on the boards and bottom edge of the joists during extremely warm weather; and then again during cold weather this moisture would evaporate or dry out. This being repeated for several years results in a rotting of the boards and the lower edge of the joists to an extent which will cause the nails to pull out and the boards to fall off, and this has actually occurred. The condensation could have been entirely prevented by covering the shavings with water-proof paper and boarding, the same as applied to the lower edge of the joists. Leakage of moisture through the ceiling of cold storage rooms is quite common in some of the older plants, and this has been attributed to leakage of the roof in many cases, as the condensation may be so great as to actually cause a dripping.

An attempt has been made to explain why cold storage insulation may become deteriorated after a time and why it is necessary to understand the natural laws governing, to be able to properly design insulation. There is no material in use which is exempt from deterioration unless properly protected, and even though the material be to an extent non-capillary, such as cork, yet the deterioration will be just as certain as though the material were greatly absorptive like hair felt or mill shavings. All insulation should be protected from air circulation and air contact not only on the face of the wall toward the refrigerated room, but just as carefully protected on the exterior face of the wall toward the outside air; in fact the protection of the exterior face of the wall is more important than on the inner face of the wall.

Another element in cold storage practice that demands the construction of first-class insulation is that goods should be carried at a uniform temperature throughout every part of the room. With poor insulation this is not possible, no matter how large the cooling apparatus may be, as the parts of rooms nearest the outside walls will be higher in temperature than those nearest the cooling surfaces. This condition often results in a part of the goods being carried at a higher temperature than they should be, on account of danger of freezing those that are nearest to cooling surfaces.

The value of the insulating materials depends upon their efficiency in preventing the transmission of heat from the outside to the inside of the building. A study of the heat conducting properties of the various materials and substances as shown by tables here given leads to the conclusion that with but few exceptions we must turn to the vegetable and animal substances for this efficiency. In selecting materials of this class for practical insulation, we are limited by many requirements besides non-conductivity of heat. These are enumerated below in the order of their importance, viz:

1. It should be odorless, so as not to taint the perishable goods stored.
2. It should have the minimum capacity for moisture, and in case it should become damp, it should not rot or ferment.
3. It should be vermin proof, and give no inducement for rats or mice to nest within it.
4. It should not be liable to inherent disintegration or spontaneous combustion.
5. It should be of light weight, not so much on account of lightness itself, because buildings are usually built sufficiently heavy where they are to be used for warehouse purposes, but because the lighter materials are usually better non-conductors of heat.
6. If used as a filler, it should be elastic so that when it is once packed firmly, it will not settle further and leave open spaces which will be almost impossible to find and costly to repair.
7. It should be reasonably cheap and economical of labor so as not to be prohibitive for general use.
8. It should lend itself to practical application in general work.

In addition to the above requirements, water or moisture proof qualities are desirable and in certain classes of structures fireproof qualities are needed. The best natural non-conductors of heat are neither fireproof nor waterproof, and therefore these qualities must be secured by proper design and application, as has already been pointed out. A consideration of the various materials commonly used and illustrations of the application of same will therefore be in order.

MATERIALS.

From the tables already given, it will be noted that still or perfectly motionless air is one of the best insulators against heat. But to keep it motionless it is necessary to confine it in very small spaces to prevent circulation and convection of heat. This is best accomplished by properly constructing spaces and filling them with some sort of material in bulk. The value of these fillers depends upon the number of minute spaces into which they divide the air. Their value follows closely upon their specific gravity; that is, the lighter the material, the better insulation it is, owing to the microscopically confined air in the cells or structure of the material itself. Again, the value of these fillers depends upon the density to which they are packed; it has been found that if they are packed too loosely they will permit air circulation, and if packed too closely, the conduction of heat will increase. With nearly all the materials at present in use, the best results seem to be obtained when packed to a density of from eight to ten pounds per cubic foot. Starr gives a specific gravity of about .160 as being the lowest density to which a material should be packed. This corresponds to about ten pounds per cubic foot, which is, in the experience of the author, heavier than such materials as straw, wood shavings or cork shavings can be packed in actual practice. In using fillers in walls, attention should be given as to whether or not the materials of which they are composed are in their natural state good or poor conductors of heat. Mineral wool, for instance, is made from furnace slag or rock which are considered comparatively good conductors. If materials of this nature are packed very tightly, their value as insulators is greatly lessened. Materials which in a raw state are poor conductors, such as straw, sawdust, wood shavings or cork may be packed very tightly without decreasing their insulating value. In fact, the insulating value of such materials is generally increased by close packing.

STRAW, CHAFF, ETC.

Such materials as chopped straw and hay, dried grass and leaves, chaff and hulls of the various grains have all been used as fillers, as described above, and under certain conditions they are fairly efficient as non-conductors of heat. They are frequently abundant and cheap, but as the proper protection of same from access of air and moisture is not well understood, they are seldom used at the present time. In country locations and on the farm they are often used to considerable advantage as a packing material for temporary ice houses, fruit houses, etc., their availability, far from manufacturing centers, making them naturally fit for such purposes. As the scientific design of cold storage insulation becomes better known these common materials will no doubt come into more general use as their efficiency and low cost entitle them.

SAWDUST.

Sawdust as an insulating material practically belongs with those noted above, but it is used to such a large extent for various purposes connected with refrigeration that it deserves separate mention. There seems to be no preference for the sawdust of any particular wood, as they are all about the same in insulating value. This value is very high when the sawdust is dry and clean, but if damp, it will rot, ferment and heat, and in this state will disintegrate and settle down, leaving spaces at top for leakage of heat. The most undesirable feature developed by the use of sawdust when damp, is the liability of a moldy or musty condition of the rooms, and this may affect the goods in storage. Nearly all sawdust available is from green lumber, and this is very undesirable for insulating purposes. As already pointed out, if sawdust is thoroughly dry and kept so by proper air and moisture-proofing methods, it will make good and efficient insulation for many years. The use of green or damp sawdust in contact with lumber or woodwork should not be permitted under any circumstances.

There has long prevailed an idea that sawdust or similar materials would harbor rats and mice, and that this made such materials undesirable as insulation. As a matter of actual fact none of the common insulating materials, not even mineral wool, is exempt from this criticism; but there is little likelihood of trouble from this cause if the building is well built and the houses kept in good repair. In all the author's long experience, rats or mice have never been a serious bother. If they get in they usually quickly succumb to the cold and cleanliness of premises.

The most useful application of sawdust is for packing ice in houses storing natural ice, where it is open to the action of the air at all times and renewed each year, or as it rots out, and for this purpose green or damp sawdust is nearly as useful as if dry. However, with the growing tendency to store ice under refrigeration there will be less demand for sawdust for this purpose.

SHAVINGS.

Shavings or chips from the planing mill have largely superseded sawdust as a common material for insulation, as they are free from many of the objections that have proved most undesirable in the use of sawdust. Shavings are specified by the author in the composite insulations designed by him, and he believes that when they are properly used and protected from air and moisture there can be no objection made to them. But they should not be used in large bulk (nor should any filler for that matter), but rather in combination with several other materials, as illustrated further on. Shavings will not rot, ferment or settle down under similar conditions as rapidly as will sawdust, because the fibrous structure

FIG. 9—BALE OF SHAVINGS.

of the wood has not been destroyed. Shavings are elastic and clean to handle, and if properly packed (about 8 to 9 pounds to the cubic foot) will remain in position for an indefinite period. They should be delivered to the building reasonably dry and clean, but if there is some mixture of dry sawdust, this is not objectionable.

Many firms, particularly in the eastern and part of the middle western states, make a practice of putting up shavings in bales. This is a great advantage, both for shipping and handling, as it permits of their use at points distant from their manufacture. They are put up in compressed bales weighing 80 to 120 pounds, ten and fifteen cubic feet per bale.

(after shaking out, and when repacked in the insulated wall) and as supplied have the appearance shown in Fig. 9. The demand for shavings for fuel and other purposes makes them extremely hard to obtain in some localities during the fall and winter, and this difficulty will no doubt increase with time, as the settled portions of our country are being rapidly denuded of forests. The shavings of the soft woods are preferable, as they are less brittle and lighter than those from the hard woods. It is also preferable to use shavings from some odorless wood, such as spruce, hemlock, whitewood, etc.

If shavings come to hand which are damp or have been wet, they may be dried in a short time by spreading out under cover in a warm dry room. If they have begun to mold or ferment they should not be used.

MINERAL WOOL.

A material which is much used is commonly known as mineral wool, granite rock wool, rock cotton or rock cork in this country, and as silicate cotton in England. Mineral wool is usually made from the slag of blast furnaces, with limestone added; and the rock wool or rock cotton, from granite and limestone. The principles involved in manufacture are the same in either case and the process is comparatively simple. The rock is first crushed, then mixed with coke and fed into furnaces, where it is fused at a high temperature, about 3,000° F. The molten slag or lava is then run out at the bottom of the furnace through a high pressure steam blast which blows it into fleece or wool, much resembling sheep's wool, except that the fibers are brittle. These fibers are very fine, and interlace each other in every direction, forming innumerable minute air spaces. In common slag wool about 92% of the mass consists of air spaces and in the best rock wool the proportion is about 96% when it is very lightly packed. It will be seen that for this reason it is a very good insulator, regardless of the fact that it is made from a material having a comparatively high conductivity. Used as an insulator, it should be free from "shot" and all other solid pieces, as much as possible. It has the qualities of being fairly vermin and fire proof and is not liable to decay, but if it is packed too tightly in the walls, its brittleness will cause it to break up, which decreases its insulating value. It should not be packed closer than about twelve pounds to the cubic foot. Mineral wool will absorb moisture quite freely, if not properly moisture-proofed, and it is stated by some authorities that if it becomes wet and then freezes, the water that has penetrated the air cells between the fibers, will expand and break the structure of the material into a granulated mass, which will settle or pack down, and in this state it is a poor insulator. One of the chief objections to mineral wool as a filler is its difficulty in handling, as the fibers will prick the skin and in a very short time will cause the hands to become sore, but the most important objection is the minute particles of wool floating through the air as it is handled, making it bad for the eyes

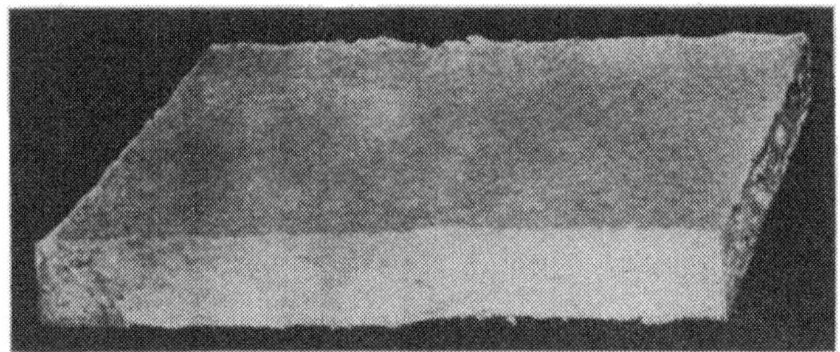

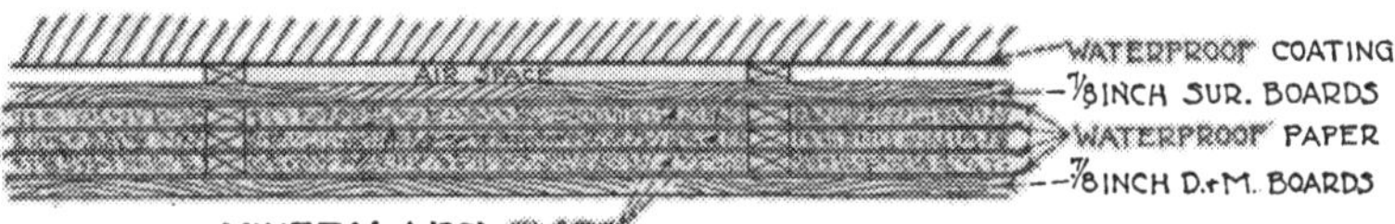

FIG. 10—MINERAL WOOL SLAB AND ONE METHOD OF APPLICATION.

and injurious to breathe. It is for this reason that workmen dislike to handle it, and this dislike indirectly causes the work to be slighted and poor insulation may result. Owing to its nature, mineral wool or any of its manufactured products are very desirable as a retardant to rats and mice, and it is valuable to use in protecting other materials from their ravages. Two inches of this material on the exterior of an insulated wall makes it reasonably mouse and rat proof.

MANUFACTURED FORMS OF MINERAL WOOL.

There has been, in the past few years, a tendency to manufacture insulating material that would be portable, easily handled and put in place, not liable to settle, etc. This has been accomplished by making the material into compressed slabs or sheets to a density and stiffness sufficient to be easily handled, sawed and fitted same as if it were lumber. Slabs made of mineral wool are thus manufactured by several different firms in this country, and have the appearance shown in Fig. 10. These slabs are usually made in standard sizes of 18x48 inches and 36x48 inches and from one to three inches in thickness, the manufacturers being willing to cut these slabs to any size smaller than this, if specified. These slabs are a great improvement over mineral wool in bulk form, as they can be adapted to modern construction where it is the purpose to stratify or laminate the materials to form a composite insulation, as such is now considered the most efficient in retarding heat transmission. Mineral wool in this form may also be properly protected from moisture and finished with cement for inside lining of rooms if desired.

Many methods of applying this "felt" or mineral wool slab have been devised by the manufacturers, but these are more or less impracticable on account of the assumption that these slabs are sufficiently strong to hold nails and support the construction; and the fact that these boards are not air or moisture proof is overlooked and therefore the construction must make good these necessary requirements. Fig. 10 shows a method used by the author in applying this material. It will be noticed that the slabs are not necessary to the solidity of the construction, but they are placed between battens or furring and lightly tacked in place; waterproof paper is placed on each side and between each slab, thus preventing any leakage of air or moisture through the wall. Fig. 11 (see following page) shows a method, recommended by the manufacturer, of applying mineral wool slabs to brick or stone walls in the construction of fireproof insulation. The wall is first coated with waterproof cement put on hot, or Portland cement, into which the slabs or sheets of insulating material are set. Two or more courses of two or three inch slabs may be used, with the cement between. After setting the slabs, another coating of waterproof cement is applied and the surface plastered with Portland cement troweled down to a smooth surface.

Another application of mineral wool to cold storage insulation is to pack it into rectangular galvanized iron cans of suitable size and thickness with soldered joints, and build these cans into the wall of the building. In one such application,

FIG. 11—METHOD FOR APPLYING MINERAL WOOL SLABS.

consisting of a twelve inch wall of salt-glazed terra cotta blocks, a four inch and an eight inch, with five inch mineral wool filled cans inside, and finished with a three inch terra cotta wall, all the members were set in Portland cement and the wall plastered with the same material. This construction would certainly be permanent, but the insulating efficiency could not be high. Two separate layers of the galvanized cans separated by an air space would increase the efficiency greatly.

CHARCOAL.

Charcoal is described as a more or less impure form of carbon obtained from various vegetable and animal materials by their partial combustion out of contact with air. That most in general use is obtained from wood and is a hard and brittle black substance which in a granulated or flaked form is used to a large extent in England and in Europe for insulation. It is used as a filler and applied in the same manner as sawdust, mineral wool or shavings. Charcoal has not been used to any considerable extent in this country for insulation, except for the ordinary family refrigerator. Its use is not to be commended; and on account of its black, dusty nature, it is very dirty, to say the least. The abundance of many other materials at hand, equally efficient, does not warrant giving it even a trial.

CORK.

Granulated cork is considered one of the most efficient and high-grade fillers for insulating purposes that we have available, and it is odorless, clean, elastic, durable and does not absorb moisture readily, but like all other fillers, is subject to attack by rats and mice, unless properly protected. Cork is the bark of a particular tree growing on the coasts of Northern Africa and Southern Europe. Spain furnishes by far the greater portion of that imported into this country. This bark is deprived of its non-elastic and impure parts, after which it is cut up into proper sizes for commercial use. The granulated cork is the waste product in the manufacture of stoppers, handles, etc. When filling spaces with cork, it should be rammed in tightly, so as to reduce the size of the air spaces between the particles, and to prevent future settling. Granulated cork mixed with hot pitch or asphalt has been used and is considered by the author to be a good insulator around brine mains where they pass through masonry walls or are laid under ground. With this material, molds or forms are placed around pipes and the mixture poured in hot, thus completely surrounding the pipes and making a permanent covering.

Cork has also been made up into compressed sheets, bricks, etc., of various sizes. The appearance of the sheets is shown in Fig. 12. They are usually made 12x36 inches in size and vary from one inch to three inches in thickness. These sheets are made by compressing the granulated material or shavings of cork in iron molds and baking in a temperature of about 500° F. This is done without the addition of any cement or binding material, but the process liquefies the natural gum of the cork sufficiently to bind the granules into a solid mass. In some processes a cementing material is used, making what are termed impregnated cork sheets. These boards are more or less porous, and therefore to apply them practically the author has used them in constructions similar to mineral wool slabs as shown in Fig. 11, set between furring strips and with waterproof paper between each layer of sheet cork.

FIG. 12—SHEET CORK INSULATION WITH CEMENT PLASTER FINISH.

The manufacturers evidently recognized the difficulty of applying the sheets (otherwise than shown in Fig. 11) without nailing through them. This was impracticable because the sheets lack sufficient strength to hold nails and nails are also objectionable on account of tearing the paper and cork. Consequently the two inch and three inch thicknesses of sheet cork can now be obtained with inserted nailing strips of wood, as shown in Fig. 13. This is unquestionably a good improvement, as it gives a more solid construction for nailing, and does away with furring strips to some extent. Referring again to Fig. 13, the author would consider it impracticable

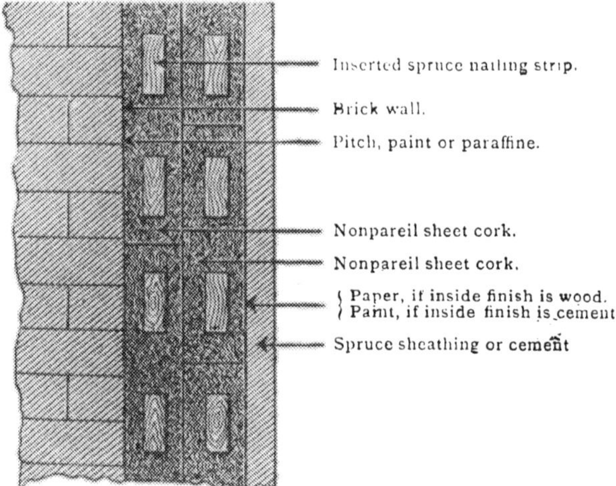

FIG. 13—SHEET CORK WITH INSERTED NAILING STRIPS OF WOOD.

and difficult to fasten the first sheet to the brick wall as shown. A better method would be to set horizontal nailing strips of wood in the brick wall every sixth or seventh course, or nail horizontal furring strips to the inside of the wall, set 18-inch centers and set the sheets vertically with joints lapped over the furring strips, as shown in Fig. 14.

Another method of erecting cork sheets, which possesses several advantages, is to cement them solidly to brick or tile walls in a bed of Portland cement. A single course of the cork sheets either two or three inches thick is used, or a double course with cement between, as shown in Fig. 12, according to the severity of the conditions. The interior finish may be either matched boards, which are nailed to the inserted wood strips in the cork sheets above referred to, as shown in Fig. 13, or a fireproof cement finish of either Portland cement or White Marble (Magnesian) cement may be applied directly to the exposed surface of the cork sheets, as shown in Fig. 12. This method gives an efficient insulation which is both waterproof and fireproof and is being used at present more largely than any other with very satisfactory results.

As above stated compressed cork is made in shape and size resembling brick, which, for partitions and inside walls,

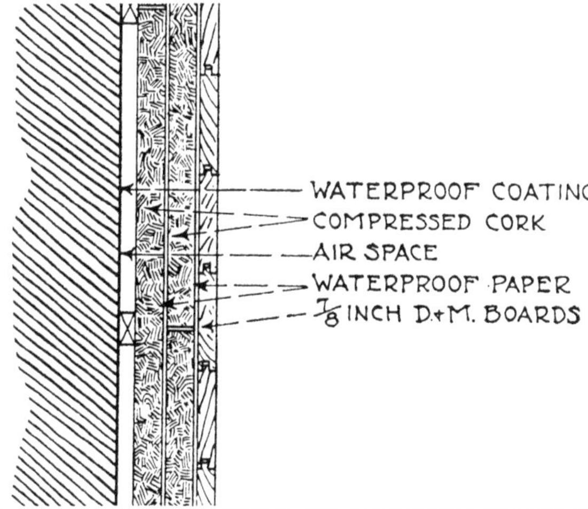

FIG. 14—COOPER'S METHOD OF APPLYING SHEET CORK.

are laid up in the same manner as brick with liquid or asphalt cement as a binder for the joints. Cork bricks are also made that are impregnated with hot asphalt or pitch so as to surround each particle, the purpose being to produce an article that should be water-proof. This treatment would without doubt decrease the insulating value of the cork bricks and its purpose is therefore questionable.

Cork sheets or blocks set in and plastered with Portland cement, are being successfully used in constructing partitions without other supporting means or the regular structural members. The life of such partitions is necessarily the life of the cork, and therefore extreme care must be used in water-proofing. Caution is urged against depending too much on the waterproof and non-capillary qualities of cork. It must be properly protected from air and moisture by a permanent tight covering of some kind. The cement plaster cannot be depended upon for this purpose unless coated with neat cement put on with a brush, and great care used in the work.

HAIR FELT.

Hair felt material has very appropriately been termed "Nature's Insulation," as there is no question but that nature

FIG. 15—HAIR FELT.

created hair for the chief purpose of protecting animal life from the changes of temperature. It is one of the most indestructible materials with which we have to deal, and when properly applied it is one of the best insulators available. Cattle hair as it comes from the tanners is thoroughly washed and air dried, put through pickers and blowers until all dirt, etc., is removed and the hair thoroughly deodorized. It is then put through felting machines where it is formed into sheets of one-quarter of an inch to two inches in thickness, put up in rolls twenty-four inches to seventy-two inches wide and fifty feet long. This felt has the appearance shown in Fig. 15. In specifying this material the author requires it to be furnished in narrow widths (preferably 24 inches) and applied between furring strips and paper set vertically as indicated in Figure 16. The sheets should run from floor to ceiling continuously and may be held in place by nails driven into side of furring strips at an angle and then bent in as shown in Fig. 16. No nails should be driven directly through the hair felt and papers, as that destroys the air and waterproof qualities to that extent and thereby decreases the value of the insulation. In applying the sheets of hair felt to the ceiling, it has a tendency to sag. This can be avoided by nailing temporary cross cleats to the furring strips every five or six feet, as the sheets of felt are put into place, and these can be removed as the inside wood finish is put on. The use of twine as shown

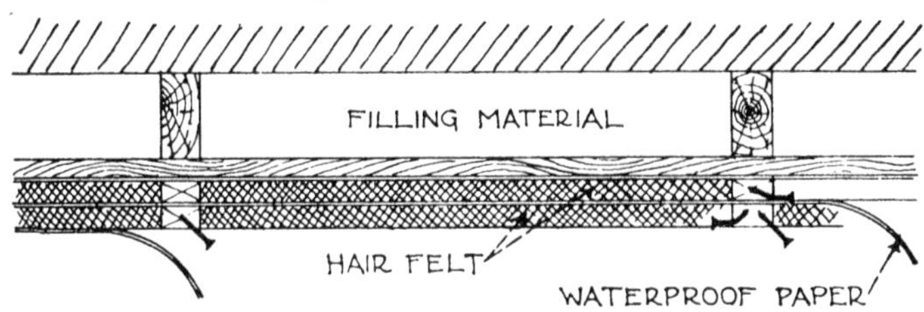

FIG. 16—METHOD OF APPLYING HAIR FELT.

in the sketch has proved practicable in many cases. A little practice and patience is needed. If the hair felt is ordered the proper width for use, there will be very little cutting to be done except to cut off the lengths as needed. The best method of cutting hair felt is with a long bladed sharp knife or chisel guided along a straight edge held down firmly; some workman with accurate aim can do a fair job with a sharp hand axe.

Besides being used as it comes from the manufacturer, hair felt is put up in many ways, by applying paper, etc., to the surface. In some situations this would be very desirable.

QUILT INSULATORS.

Those insulating materials known as "quilts" are in the nature of a felt held in place between two papers and stitched together, and are usually made in one-quarter and one-half inch thicknesses, thirty-six inches wide, put up in rolls of from 100 to 500 square feet. These quilts were originally designed and manufactured for deafening purposes, viz.: to absorb and dissipate the sound penetrating through floors and partitions in dwellings, etc., where with proper construction they serve both as deafeners and insulators.

There are various filling materials used for making up these quilts, such as hair felt, mineral wool, flax fibre and eel-grass, all of which are very durable, each possessing qualities that recommend them for use. The nature of the hair felt and mineral wool has already been touched upon. The so-called flax fibre, recently introduced, is made from flax straw, that has been crushed, picked and deodorized, and the sap or gum

FIG. 17—CABOT'S INSULATING QUILT.

removed, leaving a light fibrous material that if properly protected makes a good insulator. Eel-grass is used in "Cabot's Quilt" exclusively and has been on the market for a number of years as a deafening material. The quilt has the appearance shown in Fig. 17. This eel-grass, or sea weed, as it is often called, is a long, grass-like material of great durability.* It has great resistance to fire, and owing to the large percentage of iodine (common to all sea-plants) which it contains, it is repellant to rats and mice.

For the application of these quilts to cold storage and refrigerator car construction, they have been made in thicknesses up to one-half inch, and waterproof papers have been placed on one or both sides of the quilt instead of the common build-

*"A sample of eel-grass, 250 years old and in a perfect state of preservation, may be seen at Mr. Cabot's office."—F. E. Kidder in "Building Construction."

ing papers. Some makers have coated one side of the quilt with a waterproof asphalt coating, this to be turned toward the damp side of the wall. These improvements have enhanced the value of these quilts for insulating purposes and they compare very favorably with other materials for practical use.

The common method of applying is to place a layer of quilt between two sheathings of flooring and nail through it,

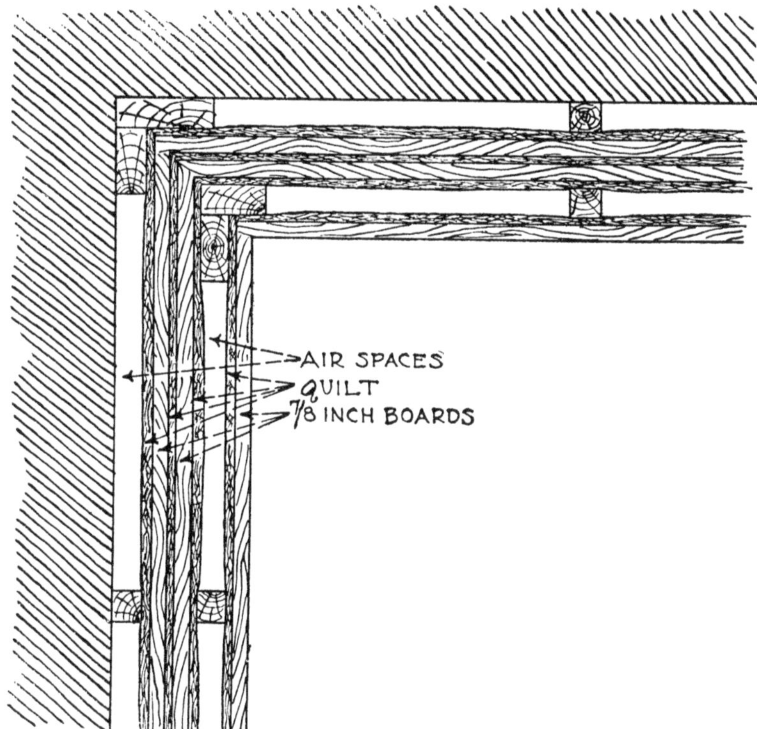

FIG. 18—METHOD OF APPLYING CABOT QUILT.

and then apply more sheathings and quilt as shown in Fig. 18. While fair results can be obtained by this construction, it is somewhat impracticable on account of the elastic nature of the quilt, and is also wasteful of lumber. A better method of applying these quilts and saving lumber and increasing the insulating value would be as recommended by the author and shown in Fig. 19, on following page. In case it is desired to omit the shavings, the wall may be furred with ⅞-inch strips and the quilt then applied, as shown, to the number of thicknesses desired.

INSULATING PAPERS.

As we have already seen, those materials having any considerable insulating value, are extremely porous, and therefore

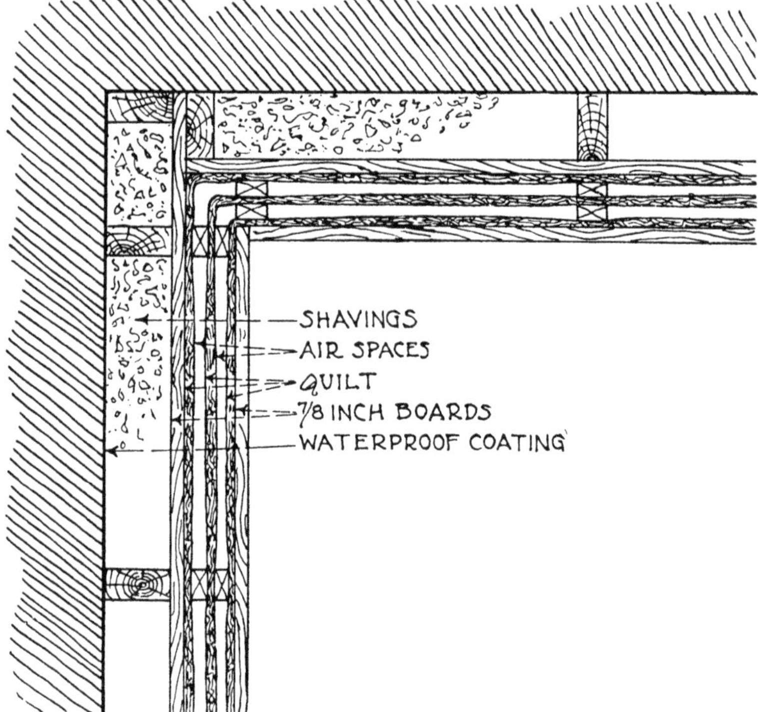

FIG. 19—ANOTHER METHOD OF APPLYING CABOT QUILT.

air under even a light pressure will flow through them quite easily. To prevent this flow of air and the penetration of moisture is absolutely necessary, otherwise the insulation will become damp, and in time, regardless of materials used, almost worthless on account of deterioration and decay. It is the general practice to use air-tight and waterproof papers on each side of the insulating materials for the purpose of preventing this flow of air and moisture through the walls.

There has been a widespread impression that papers possess a high insulating value, and consequently many expensive and complicated insulations have been constructed, using paper as the chief material. It is now generally recognized by refrigerating engineers that the chief value of paper in an insulated wall is in its resistance to the passage of moisture and air through the walls. Its use also tends to make an insulated wall more composite without increasing its thickness, as it changes the density of the insulation and thereby retards the transmission of heat. Besides the requirements of being air and water proof, papers must be odorless, have strength and durability, and should not be brittle and liable to crack in low temperatures, as this makes them difficult to handle and results in leaky insulation.

There are a great many insulating papers on the market, some of which are reliable and durable, but all so-called rosin-sized, oiled and tar coated or tar saturated papers should be avoided on account of their odor, and the rosin-sized papers also avoided on account of the positive certainty of disintegration, because they carry their own destructive elements. It is also advisable to avoid all so-called "coated" papers, that are coated on both sides and have a light-colored center, as they will disintegrate sooner or later, if unfavorable conditions arise. Papers should be selected that have been saturated and thoroughly impregnated with pure asphalt or similar material or have a center layer of asphalt, as thus they are practically indestructible when used for cold storage insulation; these qualities, combined with the requirements above stated, make them superior to all others for insulating purposes. These high grade papers are more expensive in first cost, but their durability makes them cheaper in the end. As the cost of using good papers is usually less than 5 per cent of the total cost of the insulation, and waterproof and air tight paper is the vital feature of good insulation, it is poor economy to select an inferior grade. Insulating papers are usually manufactured thirty-six inches wide and come in rolls of 500 or 1,000 square feet.

The papers should be applied with the greatest care in lapping around corners, etc., all joints should be lapped at least two inches and under severe conditions these joints should be cemented. It should be kept in mind that the proper application of the papers is one of the most important parts of the insulating work, because, as already noted, the insulation must be air and water proof to remain efficient for any length of time. If the workmanship is poor, the advantages of using first-class papers are neutralized.

WOOD FOR INSULATION.

Wood has, of course, played as important a part in constructing insulation, as it has in general building operations, because of the ease with which it can be procured and worked, together with its strength, lightness and durability. In addition to its general use for framing and floor construction in brick warehouses (except in thoroughly fireproof structures), it is used for forming the air spaces or filled spaces and inside finish of the insulated rooms. Wood has been regarded as a good insulator and this belief has, in many cases, tended to its excessive use in constructing insulations, such as, for instance, the use of from six to ten thicknesses of boards in one wall. Considering the greater insulating values of filling materials over solid wood, it is the general opinion of most refrigerating engineers that many thicknesses of boards built up with air spaces in such a manner, is not only extremely expensive, but it is not efficient as insulation in proportion to the cost and space occupied. One of the chief requirements in the use of wood, already stated as essential to other insulating materials, is that it should be odorless. This applies particularly to the inside sheathing and finish of cold storage rooms used for sensitive goods such as eggs. This requirement restricts the kind of woods available to a very few, of which spruce, hemlock, basswood and whitewood are the most desirable. Spruce is to be preferred on account of being easier to work and not so liable to have loose knots and shakes as hemlock, but it cannot be easily obtained at reasonable prices in large quantities except in the eastern states and the far west. Hemlock is abundant in all the northern, middle and eastern states and Canada, where it is used extensively in all building operations, but it is cross-grained, rough and splintery, and liable to split when nails are driven into it. It is claimed that owing to its splintery nature, hemlock is practically mice and rat proof. White pine may sometimes be used, when the other kinds are not available, but it should be as free as possible from pitch and thoroughly seasoned. For a warehouse in Butte, Mont., designed by the author, it was necessary to use tamarack for the inside finish as the only native wood available that did not have a stronge odor, and its use in this case was very satisfactory. Where it is necessary to use a wood that may have a slight odor, it should be given one or two coats of properly prepared whitewash or other deodorizer as soon as the walls are finished. Directions for making and applying whitewash may be found elsewhere in this book.

That lumber should be thoroughly dry to get the best results in efficiency and durability of the insulation it is almost unnecessary to state. If the lumber is even partially green, it carries a considerable amount of moisture with it into the insulation, causing more or less injury, and the use of underseasoned lumber should therefore be properly guarded against. In erecting insulation during cold weather it is very necessary to keep fires going so as to have all materials as dry as possible. This is often neglected to the great detriment of the work.

All boards for sheathing should preferably be dressed and matched, as it gives more air-tight and stronger work, and particularly for inside finish, as it has a much better appearance. Rough boarding or surfaced boards may often be used for the interior of the insulated walls where the joints are properly protected, and rough boarding has often been used for inside finish solely for the purpose of giving a rough surface for whitewashing, as it is claimed that whitewash will peel or flake off of a wall of dressed lumber. This is, however, not considered a valid reason by the author, as whitewash properly prepared will not peel off. (See chapter on "Keeping Cold Stores Clean.")

NAILS.

The use of any particular kind of nail may, on first thought, seem to be of little importance, but when we consider that the efficiency of the completed work depends upon every detail of construction and that nails are good conductors of heat, there is no question that the kind of nails and the manner of their use is of some importance. The author usually specifies that "cement coated wire box nails" be used. These nails have a smaller diameter than the ordinary wire nails, but the cement coating gives them greater holding power. This fact permits the use of a smaller size nail for the same class of work, for instance: Where 8d and 10d common nails are used for sheathing, 6d and 8d cement coated may be used for the same work. It is therefore evident that using cement-coated nails not only reduces the heat transmission on account of their smaller diameter, but also on account of being able to use nails one-half inch shorter, as indicated above. The cement coating also protects the nails to some extent from rusting.

COMPOSITE INSULATION.

Strictly speaking, all constructions are composite (except solid wood or masonry construction), as they are necessarily made up of materials having different densities and different values as insulators. An English authority divided insulated walls into two classes, which he calls the "forced" and "optional" insulation. The former is of course the masonry or frame wall of the building proper, and the latter the "lining" or material added as insulation. Where the building is a frame structure, the whole wall may be termed optional insulation, because the space between the studding of walls may be insulated with any filling material desired.

As already stated, even if the insulating value of one inch or one foot in thickness of this or that material be known, it gives no practical basis on which to design insulations, except as a guide as to what materials may be practically used. To calculate the value of composite insulations by the use of formulae is extremely inaccurate, as account should be taken of the papers used which have more or less value and the division of the material into spaces by the paper acts to largely increase the insulating value. It is for the purpose of determining the value of composite insulations that various testing apparatus have been designed by different experimenters. It may not be out of place to describe these briefly, so as to be able to judge the reliability of the results obtained in the different cases.

INSULATION TESTERS.

The most common and inexpensive apparatus used is a box constructed of the material that is to be tested and provided inside with a water-tight tin box having a drain pipe with a trap connected at the bottom, similar to that shown in Fig. 20. Top of box is provided with a removable tight cover. This complete box is then placed in a room where a constant temperature is maintained and a known quantity of ice is

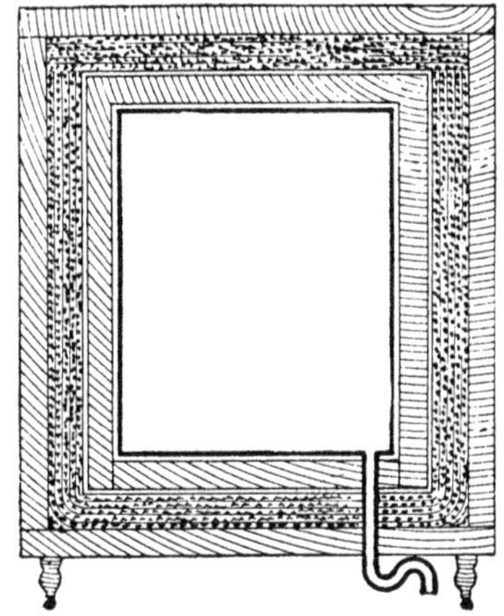

FIG. 20—INSULATION TESTING APPARATUS.

placed inside the box. At stated intervals the ice meltage can be determined by weighing the water coming from the ice through the drain pipe in bottom. To get comparative results boxes of various materials, but of the same size and thickness, can be fitted up, and all tested under the same conditions. The quantities of ice melted in each case can be compared and the relative efficiency of each material judged.

To determine the rate of heat transmission in B. T. Units for this kind of tester, it will be necessary to consider time, difference of temperature, square feet of surface in box and quantity of ice melted. This can be illustrated by an imaginary test case as follows: Take a tester two foot cube inside measurements with walls four inches thick, the inside and outside temperatures being respectively 32° F. and 70° F. (assuming the inside temperature to be the same as the ice) and the meltage of ice per day (24 hours) is 50 pounds, we then have:

Inside surface	24 sq. ft.
Outside surface	42.6 sq. ft.
	2)66.6
Mean surface between inside and outside of box..........	33.3 sq. ft.
Difference of temperature between inside and outside of box..	38° F.
Ice melted per day (24 hours)...............................	50 lbs.
142x50 equals B. T. U. transmitted per day	=7100
B. T. U. transmitted per sq. ft. per day.....	$\frac{7100}{33.3} = 213.2$
B. T. U. transmitted per sq. ft. per hour per deg. difference..........................	$\frac{213.2}{38x24} = 0.233$

Testers similar to the one above described with changes in details, spaces provided for thermometers, etc., have been designed with the purpose of getting more accurate results, if possible. Riege and Parker* designed and used a tester which they described as follows:

In testing the value of insulating walls or partitions there should be some means of determining accurately the rate of flow of the heat through the wall or partition. This can be done with accuracy in several different ways, though in the following test the apparatus used consisted essentially of a wooden box, a tin box, thermometers, and a pail to catch the drip from the ice, the agent used in cooling. The boxes were respectively forty-four inches and twenty-five inches cube on the inside, the tin box containing the ice to be melted. The wooden box was made of ⅞-inch white pine, tongue-and-groove boards, nine inches wide, and really was a double box, the boards of the inner one running at right angles to those of the outer in order to make the box as air-tight as possible. The lid of the box, which was twelve inches deep on the inside, had a 12-inch band around the edge where the lid rested on the box, and to make the joinings of the two as air-tight as possible, the edge was lined with felt. Both the lid and the box were then lined with what is known as builder's paper or "sheathing," which secured a still better protection against outside changes of temperature. The tin box rested on a couple of wooden horses, twelve inches high, so that when placed in position, there was a space of twelve inches all around the tin to the sheathing. From the bottom of the tin box led a tube which passed into an inch pipe, this latter pipe extending through the wooden box, and the lower end being immersed in a can of water, prevented any outside air from entering either box. All fittings were air-tight. Thermometer tubes were let in from the

**Ice and Refrigeration*, January, 1895.

four sides to within six inches of the tin box, as well as a long tube from the wooden box cover through the tin box cover to within a couple of inches of the ice.

Before starting a test the ice was allowed to melt until the drip from the can showed a regular flow, thereby allowing the true weight of ice melted during the test to be determined. During a test temperatures were noted on all four sides of the wooden box, also within six inches of the tin box, and also inside the latter, the readings being taken half hourly and the weight of ice melted hourly. When the thermometer tubes were not in use, they were closed with corks. Comparison tests were made of each substance, each test lasting from six to twenty-four hours. Tests were made of air, shavings and cork, first at ordinary temperature and later with a steam coil an inch above the wooden box to represent the effect, when steam was circulated, of the sun on the roof of a storage house.

A later improvement on the above described tester, which is used by some experimenters to-day, where simplicity and cost must be considered, is to construct it similar to a domestic refrigerator and about the same size, with an ice bunker and air ducts installed so as to get a uniform circulation and temperature inside. Small windows with three or more thicknesses of glass are placed in sides of testers to read the thermometers which are hung inside. The heat transmission is calculated in the same manner as described for small tester above.

The comparative results that can be obtained with the above described testers are quite accurate when the different materials tested are of the same thickness in each case. But in comparing different thicknesses of material, there is a chance for great error unless the inside dimensions of tester are changed so as to give the same proportionate mean surface. This fact is readily seen if the tester shown in Fig. 20 is taken and the walls made eight inches thick instead of four inches. The mean surface of the tester in that case would be 45.3 square feet, as against 33.3 square feet in the first case. This difference in mean surface would favor the thicker insulation when calculated in B. T. U.

John E. Starr, in an article on "Non-conductors of Heat,"* describes the testing apparatus he designed and used at that time, which was quite complicated as compared with those above described, but was no doubt intended for greater ranges of temperature than could be obtained with them. He describes his tester as follows:

The writer, in investigating the value of insulating construction, has used a rather simple but effective apparatus for accurately meas-

**Ice and Refrigeration*, July, 1891.

uring the flow of heat. He has a box carefully constructed and thoroughly insulated on the top, bottom and two sides. The two ends remaining (exposing an area of something over four square feet) were used as the test ends, and the various styles of construction to be tested were built in these ends. In this way two tests could be made at the same time. Directly against these two ends were placed water reservoirs of thin galvanized iron of the same superficial area as the test ends, that is to say, something over four feet square. These reservoirs were about one inch wide and each held from twenty-five to twenty-seven pounds of water. Outside of these reservoirs was another very thick insulation against the outer air, all except a small opening in the middle of the top for thermometer readings. A steam coil was placed inside of the box, and connected at its inlet with a steam boiler and at its outlet with a steam trap. By regulating the steam pressure the interior of the box can be kept at any desired temperature; and the construction is such that any heat that finds its way into the water must come through the insulation to be tested, and that all the heat that comes through the insulation must find its way into the water, as the water exactly covers the insulation. The tests, therefore, can be made quantitative, as well as qualitative, by observing the rise of temperature of the water, and taking into account its weight. Readings are taken at regular periods of the temperature inside the box, and of the water in each cap, or reservoir, and of the surrounding atmosphere. The water caps, however, are so thoroughly insulated from the surrounding atmosphere that unless the temperature of the water in the caps rises to a very high degree, and unless the test is of very long duration, only a small amount of heat escapes from the water and passes into the air. The value of the insulation surrounding the caps being known, however, a correction can be made, if necessary, for such escape of heat from the water.

What is probably the most valuable and scientifically accurate testing apparatus in use was constructed by the Nonpareil Cork Manufacturing Company at their factory at Bridgeport, Conn. A description of this apparatus was published in *Ice and Refrigeration*, June, 1899, and is reproduced here as follows:

Their apparatus consists of an insulated room, 12x10x8 feet, the temperature of which can be held at any point desired from zero Fahrenheit up, by means of a W. M. Wood compression machine operating with direct expansion. A uniform temperature throughout the room is secured by forced circulation, an electric fan being used to drive the air up over the expansion coils which are inclosed at one end of the room. The air passes out and down through a false ceiling having graduated perforations arranged to allow a uniform amount of cold air to fall in all parts of the room. This is the method employed in various refrigerating plants with entire success, by Mr. John E. Starr. In the center of the room is an insulated box, 3x3x6 feet inside measurement, having one side removable. It contains an electric heating coil and a small electric fan arranged to give a circulation of air and insure uniform temperature in all parts of the box. Standard thermometers, both mercurial and recording air pressure, reading 1/10° F., are placed so as to give the temperature of the refrigerated room and the heated box, the readings being taken outside the room. This obviates any necessity of the operator entering the refrigerated chamber while the test is being made. The

Weston standard ammeter and voltmeter measure the electricity supplied in the fans and heating coil, and a suitable rheostat regulates the amount of the current.

The method of determining the heat conductivity of any insulating construction is as follows:

The temperature of the room and box are respectively lowered and raised until they conform to the conditions under which the proposed insulation will be used. Then the amount of heat or electricity supplied to the box is gradually diminished by means of the rheostat, until the point is reached where the temperature in the box remains constant. It is evident that at this point the radiation must exactly equal the amount of heat supplied, or there would be a rise or fall in temperature, as the case might be. After the supply and radiation have been maintained constant for two hours, readings are taken every five minutes for three hours more. If they do not vary more than 1/10 of 1° F., they are considered practically exact. The average is taken as the permanent radiation of the box under the given conditions. The box contains 100 square feet of surface, measured at the center of insulation, consequently 1/100 of the total radiation is the rate per square foot. This rate being obtained, the removable side of the box is replaced with a side constructed of the insulation whose value is desired. The test is then repeated, and the total heat loss from the changed box will be greater or less, as the case may be. The removable side contains twenty square feet surface, therefore eighty square feet of the box remain unchanged, and the radiation through this must be the same as before. This amount is at once determined from the previous tests, and the difference between the total heat loss from the box in its changed condition, and this amount must give the radiation through the twenty square feet, comprising the new side which has been put in. One-twentieth of this amount is of necessity the rate per square foot, and this divided by the difference in temperature between the room and box, will give the exact radiation per square foot of surface per degree of difference in temperature.

A testing apparatus which has been used by the author is based on the same principles as that above described, but somewhat smaller in size and of simpler construction. The tester was built with one side removable and arranged so that any kind of insulating material could be tested in same. This tester was placed in a cold storage room where a temperature of 15° F. could be obtained and which was equipped with an air circulating system. The inside of the tester was heated by eight incandescent electric lamps, each controlled by a button switch from the outside. The temperatures inside of the tester were observed by means of a specially made long-stem thermometer projecting through top of tester and read from the outside. The thermometer was graduated to read to 1/5 of a degree. The apparatus is shown in Fig. 21.

The incandescent lamps used were 110 and 52 volts of 16 c.p. on a 52-volt circuit. These would give approximately

50 B. T. units and 200 B. T. units respectively per hour, but they were accurately calibrated by a water calorimeter test. This consisted of an insulated tank, capable of holding twenty-five or more pounds of water, which was allowed to stand in a con-

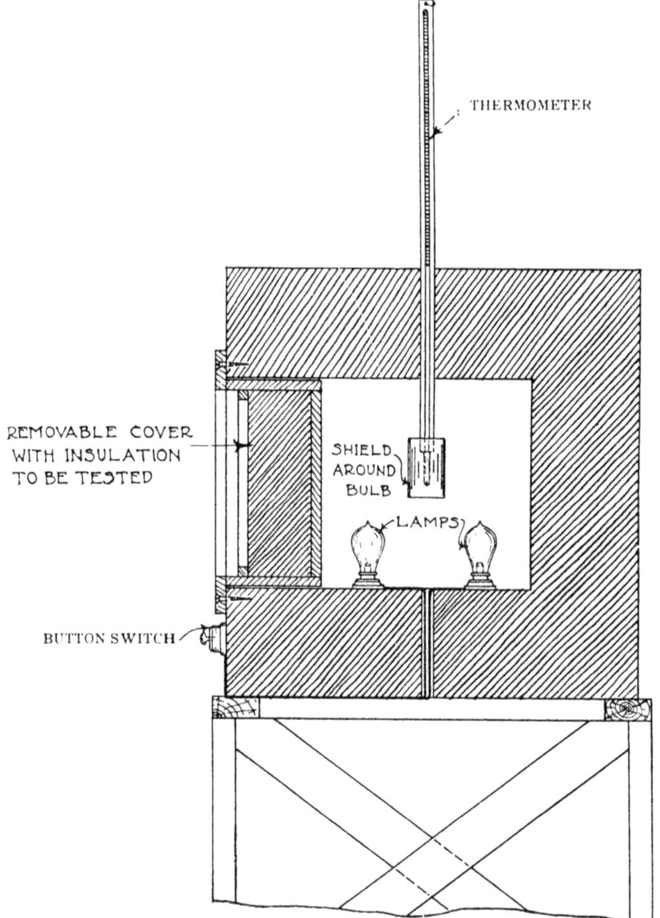

FIG. 21.—COOPER'S INSULATION TESTING APPARATUS.

stant temperature until the temperatures of the water and tank were equal. Under such conditions each lamp was put in a water-proof socket and tested separately by immersing in the water and noting time taken to raise twenty-five pounds of water 1° F. In this way the B. T. units of each lamp per hour were obtained, and with each lamp controlled separately, the temperatures in tester were controlled at will up to 150° F. Owing to the fact that the life of incandescent lamps is limited, they will, after a certain time, decrease in power. This made it necessary to retest them periodically. The author believes that the results obtained with the above apparatus are as accurate as those that have been obtained by other experimenters.

INSULATING VALUES OF COMPOSITE STRUCTURES.

The results obtained by the different experimenters have been illustrated and published in various trade papers, pamphlets and catalogues. The author has assembled and illustrated these results in the following figures as accurately as information permits. Figs. 22 and 23 give the result of tests made by John E. Starr (some of which were made for the Nonpareil Cork Mfg. Co.), at different times and for different purposes. Most of these were presented by him in a paper read before the eleventh annual convention of the American Warehousemen's Association in October, 1901.*

Figs. 24 and 25 give the results of other tests made by the Nonpareil Cork Manufacturing Company with their elaborate testing apparatus, described above, comparing their material, for the most part, with the wood board and air space construction.

Fig. 26 is a reproduction from an article in *Ice and Refrigeration,* September, 1896, on "Cold Storage Buildings." The drawings are there credited to the Fred W. Wolf Company, who give the heat transmission as having been determined from practical experience, but do not describe how or by whom these tests were made.

Fig. 27 gives the result of tests made by the author with the testing apparatus above described as designed by him. These show the tests on a variety of materials and were not made in the interests of any of them in particular, but were made chiefly to determine the value of air space construction as compared with filled spaces and sheet material.

*Reported in *Ice and Refrigeration*, November, 1901.

The above described tests are all based upon the amount of heat transmitted per square foot, per degree of difference between inside and outside temperatures, per day (24 hours).

Figs. 28 and 29 are reproductions from drawings made by George H. Stoddard and accompanying his paper on "Insulation," which was read before the eleventh annual convention of the American Warehousemen's Association.* The nature and form of testing apparatus which was used to obtain the results shown in these drawings were not given by him in his paper. In explanation of these drawings we quote in part from his paper as follows:

It may be of interest to consider how the transmission of heat takes place through an outside wall, such as is often used for a cold storage warehouse. Fig. 28 shows a section of such a wall. Starting from the outside, it is made up as follows: 24 inches of brick, 2-inch air space, two ⅞-inch matched spruce sheathing with paper between, then twelve inches of spruce mill shavings, then two ⅞-inch spruce sheathing, with paper between. We will assume a temperature of 92° F. for the air and objects outside, and a temperature of 32° F. for the air and objects inside of the warehouse. Heat is transmitted through this compound wall as follows: To the outer surface of the brick by radiation and contact of air, through the brick by conduction, across the air spaces by radiation and contact of air, through the inner wall of sheathing and shavings by conduction, and from the inner surface of the wall by radiation and contact of air.

Knowing that the rate of transmission must be the same to and through and from the wall, it is of interest to note the temperatures of the different faces of the wall, and see how they vary from the outside temperature of 92°, to the inside temperature of 32°. There is only the difference between 92° and 90.7° between the outer air and the outer surface of the wall, then the temperature drops to 81.8° at the inner surface of the brick, to 80.1° at the other side of the air space, to 76° at the inner surface of the outer double sheathing, to 37.4° after passing the shavings, to 33.3° at the inner surface of the inner sheathing, and then to the temperature of 32° in the room.

In Fig. 29 are shown curves representing the rate of transmission through walls similar to that which we have just considered, with the brick wall varying from eight to twenty-eight inches in thickness, combined with inner walls having the shavings from two to twelve inches thick. The vertical distances in all of the similar diagrams represent the B. T. U. transmitted per square foot per hour per 1° F. difference in temperature between the inside and outside of the wall, and also the equivalent pounds of ice melted per square foot per twenty-four hours for a difference of a little over 59° F.

Fig. 30 shows a similar curve for a partition of typical construction (see Fig. 31), with the thickness of shavings varying from two to twenty-four inches.

In Fig. 31 are shown partitions made up of sheathing and with one, two, three and four air spaces, and also one made up of sheathing and paper with nine air spaces, and the rate of transmission of heat

*Published in *Ice and Refrigeration*, November, 1901

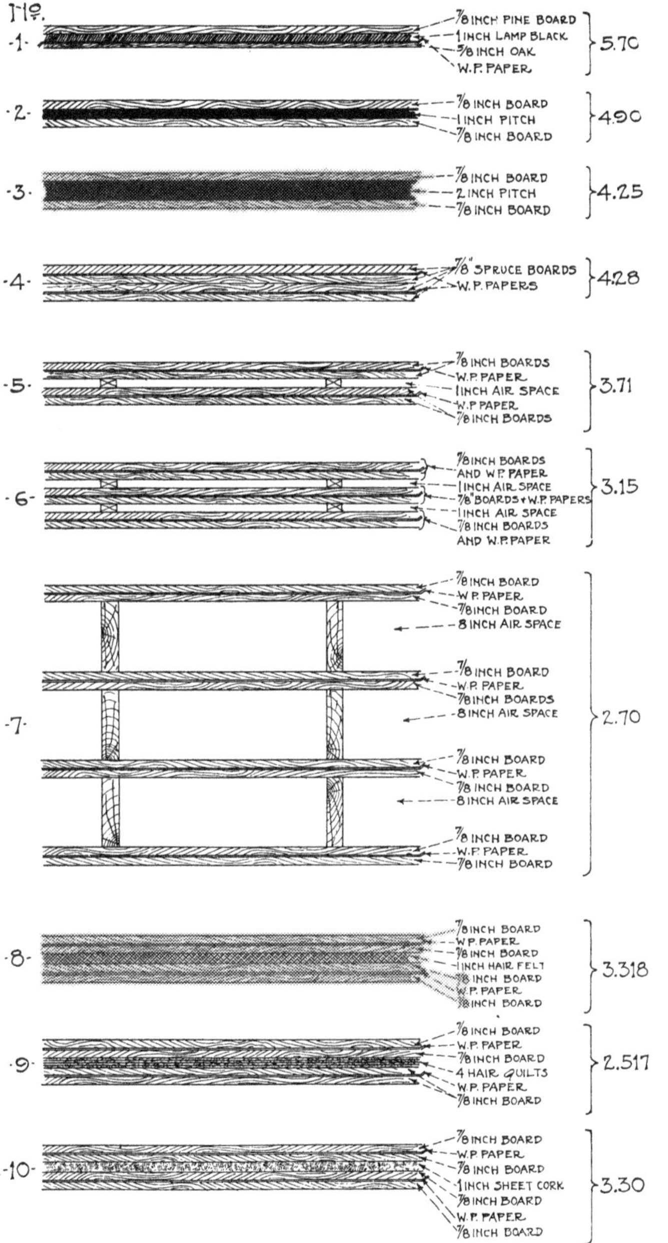

FIG. 22.—B. T. U. TRANSMITTED PER SQ. FT. PER DAY, PER DEGREE OF DIFFERENCE OF TEMPERATURE.—STARR'S TEST.

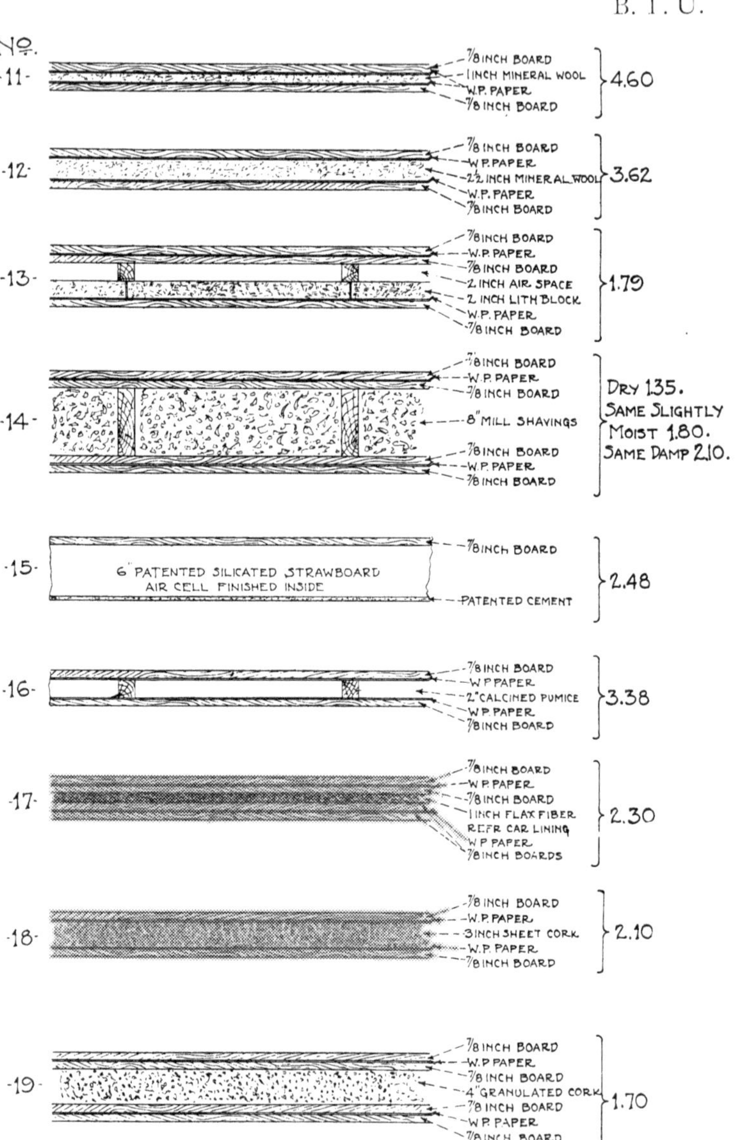

FIG. 23.—B. T. U. TRANSMITTED PER SQ. FT. PER DAY, PER DEGREE OF DIFFERENCE OF TEMPERATURE.—STARR'S TEST.

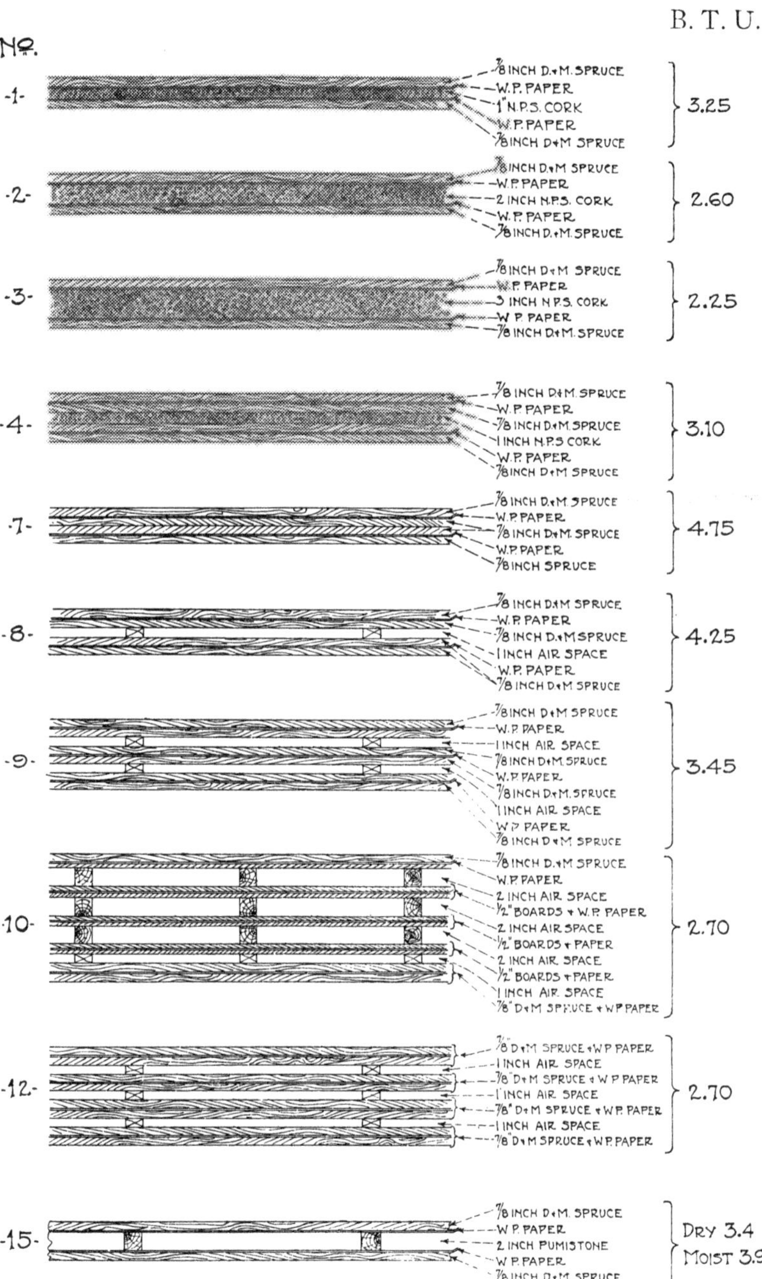

FIG. 24.—B. T. U. TRANSMITTED PER SQ. FT. PER DAY, PER DEGREE OF DIFFERENCE OF TEMPERATURE.—NONPAREIL CORK MFG. CO.

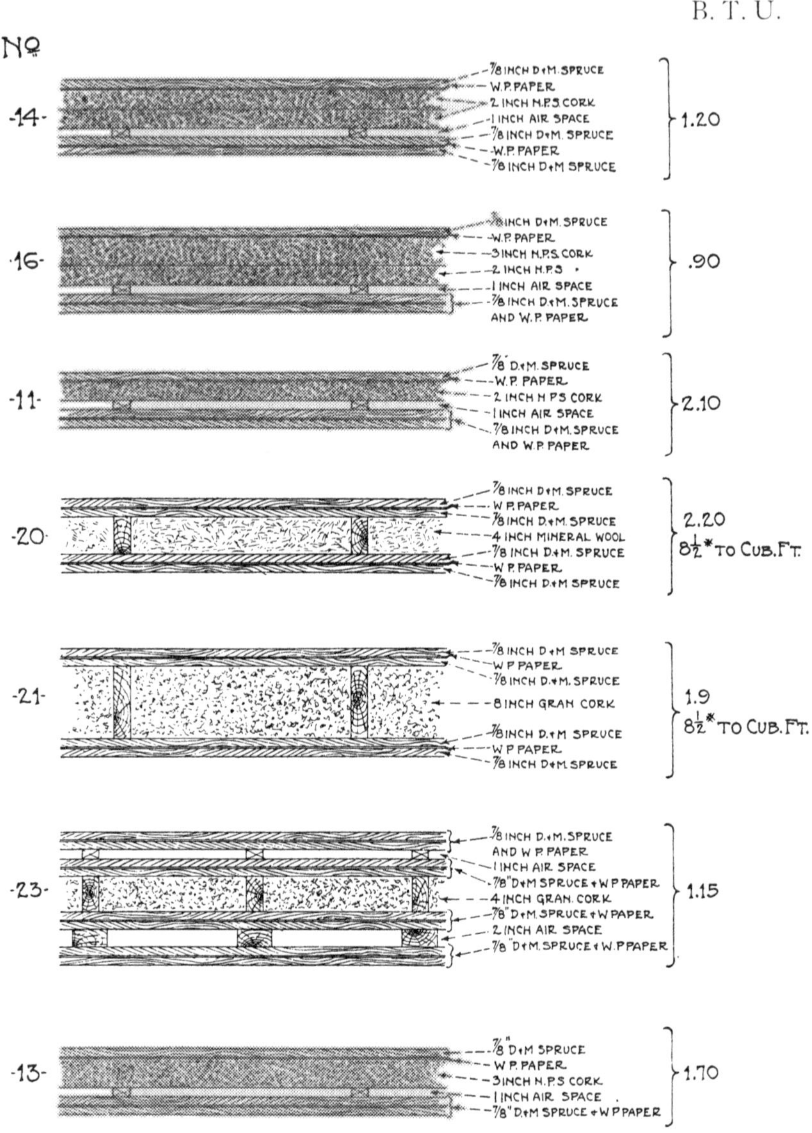

FIG. 25.—B. T. U. TRANSMITTED PER SQ. FT. PER DAY, PER DEGREE OF DIFFERENCE OF TEMPERATURE.—NONPAREIL CORK MFG. CO.

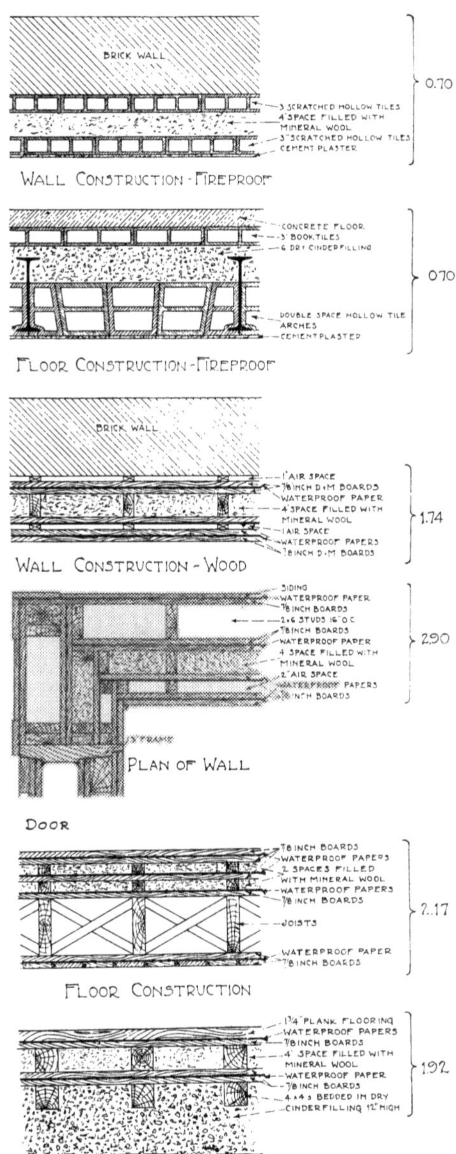

FIG. 26.—B. T. U. TRANSMITTED PER SQ. FT. PER DAY, PER DEGREE OF DIFFERENCE OF TEMPERATURE.—FRED W. WOLF CO. TEST.

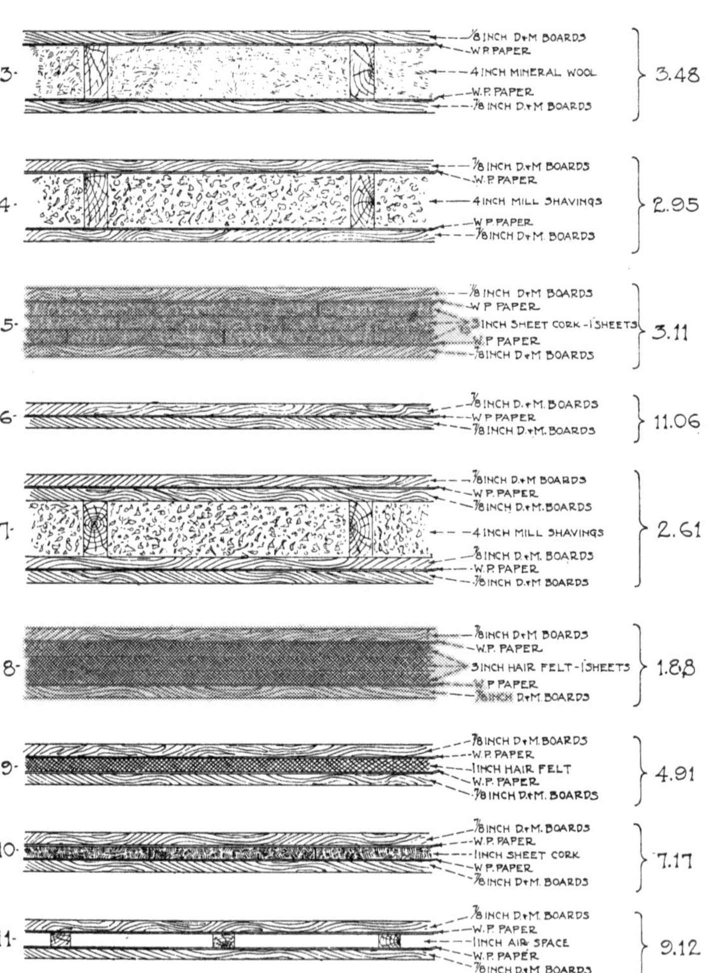

FIG. 27.—B. T. U. TRANSMITTED PER SQ. FT. PER DAY, PER DEGREE OF DIFFERENCE OF TEMPERATURE.—COOPER'S TEST.

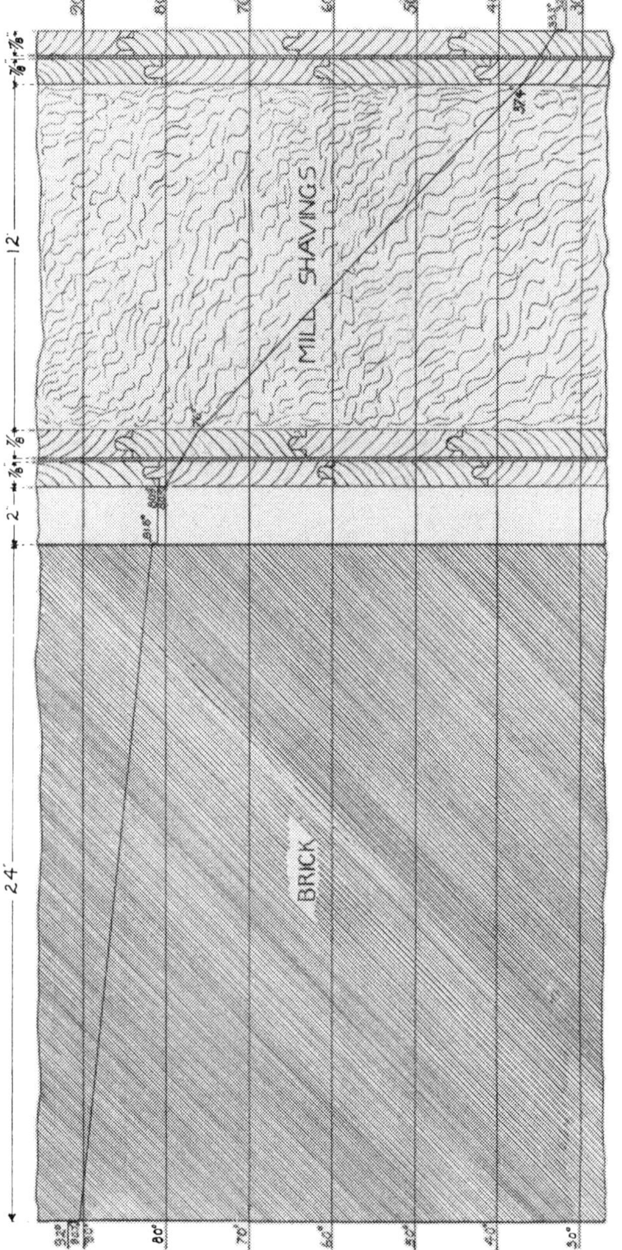

FIG. 28.—STODDARD'S DIAGRAM ILLUSTRATING TRANSMISSION OF HEAT THROUGH AN OUTSIDE WALL.

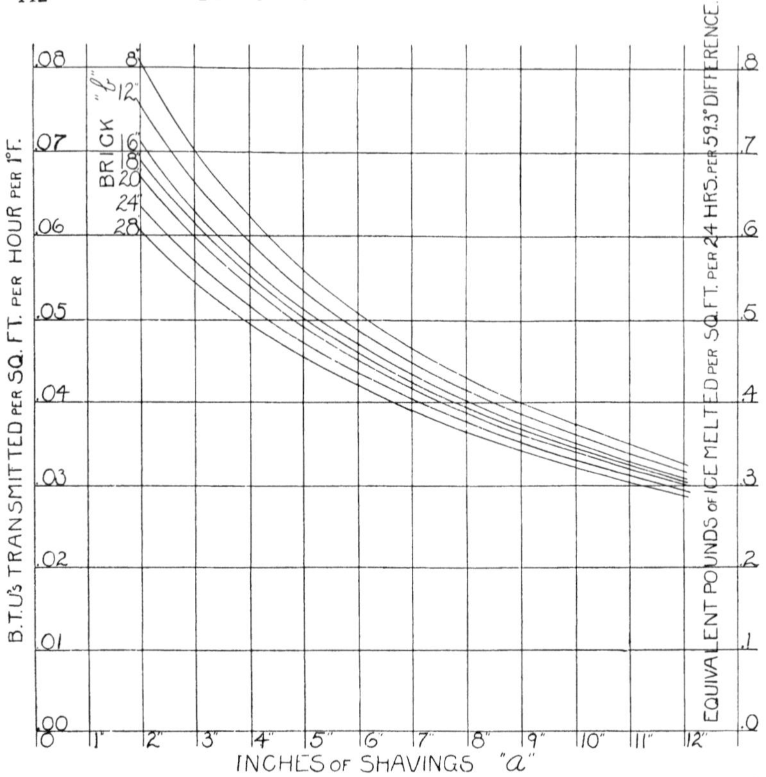

FIG. 29.—STODDARD'S DIAGRAM SHOWING RATE OF HEAT TRANSMISSION THROUGH A WALL.

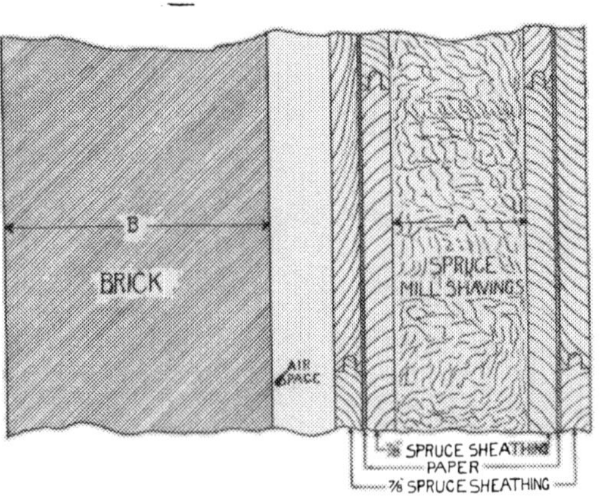

FIG. 29a.—STODDARD'S DIAGRAM SHOWING CONSTRUCTION OF ABOVE WALL.

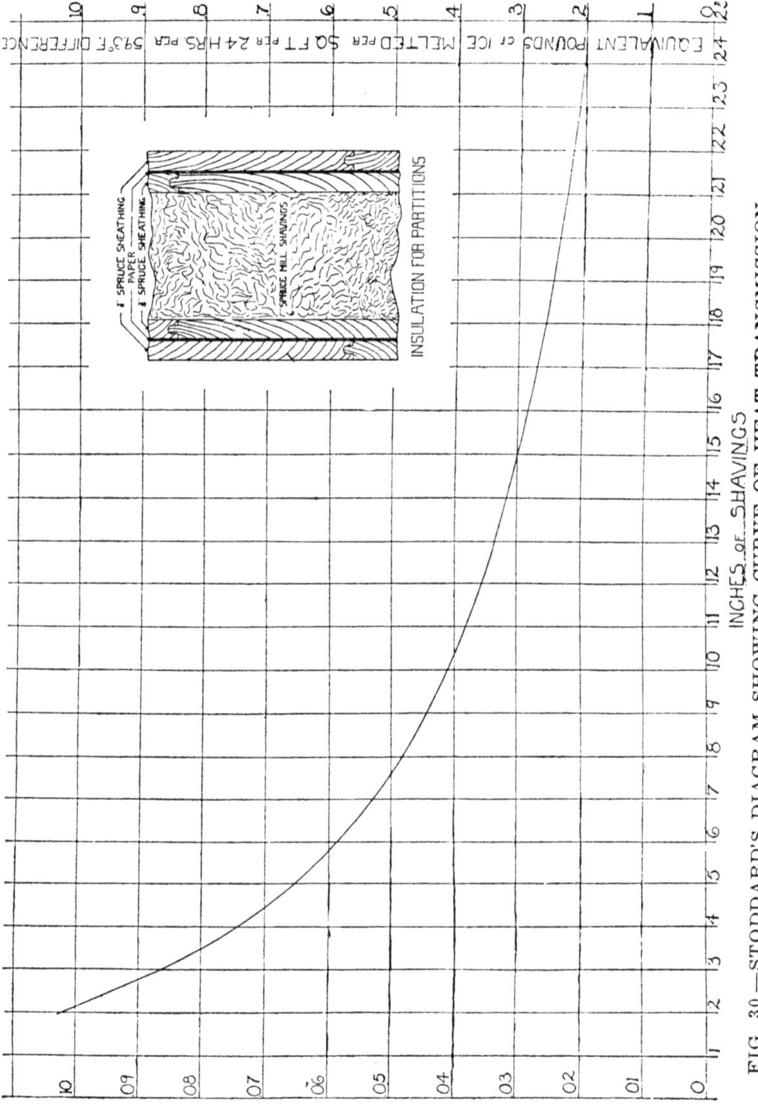

FIG. 30.—STODDARD'S DIAGRAM SHOWING CURVE OF HEAT TRANSMISSION.

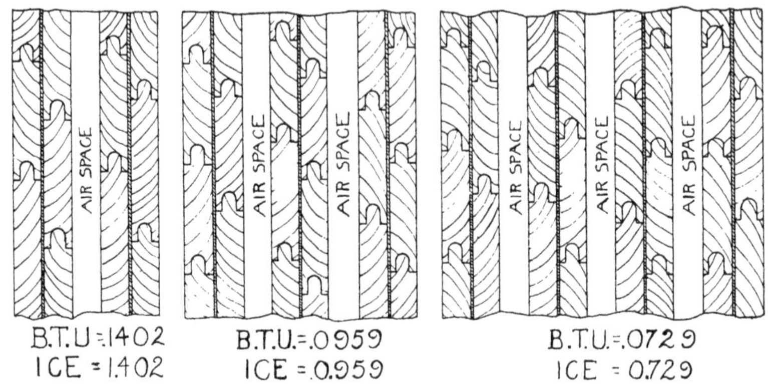

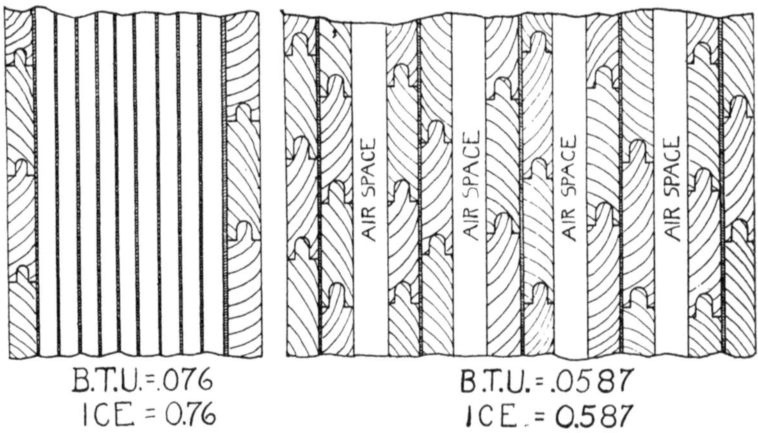

FIG. 31.—PARTITION OF SHEATHING AND AIR SPACES SHOWING RATE OF HEAT TRANSMISSION.—STODDARD.

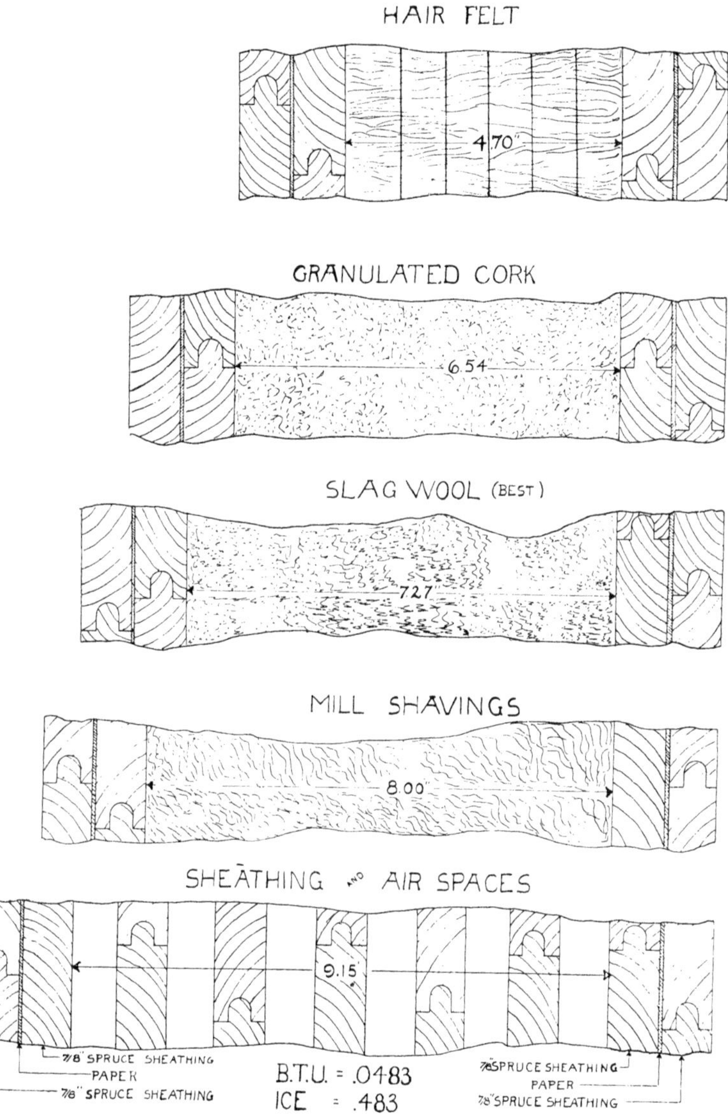

FIG. 32.—RELATIVE THICKNESS OF PARTITIONS PACKED WITH VARIOUS INSULATING MATERIALS.

through such insulation is given in B. T. U. and ice melted. In Fig. 32 is given the actual thickness of different partitions packed with various insulating materials of such a thickness that all of the complete partitions shall be of the same insulating value. There is also shown one made up of sheathing with air spaces. From this is seen how much more space is taken up by one form of insulation than by another.

An examination of Figs. 31 and 32 will show the comparatively small value of air spaces for the purpose of insulation, and it may be stated, that, for this purpose, a wide air space has no greater value than a narrow one, and that any space over one-half inch in width, if it can be kept dry, will be of greater value if filled with an insulating material as good as mill shavings, than if left as an air space.

AIR SPACES.

It is evident from the results shown with the various constructions, that those built up out of wood boarding and air spaces, or air spaces formed with battens and paper make the poorest showing when the space occupied and cost in labor is taken into consideration. The author considered the one-half inch air spaces formed by battens and paper as shown in his tests to be efficient until practical experience and the tests conducted by him proved otherwise. The workmanship in building such spaces is usually poor, as unusual care, not appreciated by the average workman, must be taken so as not to puncture them when under construction. Air space construction is difficult to erect so as to be air and moisture proof.

Another extensive use of air spaces has been between the brick wall and the insulation, as shown in Figs. 29 and 33. The alleged purpose of its use in this position has been, first, proof against moisture entering the insulation; second, for the insulating value it may have. With the growing disbelief in the use of air space construction, this second purpose can be considered of little value. The prevention of moisture entering the insulation from brick walls by the use of air spaces is only partially true, as may be readily understood. The moisture will enter the air space and will eventually affect the insulation more or less. There are sometimes local conditions that would warrant the use of an air space between the brick wall and insulation, but in the opinion of the author, such design should be avoided wherever possible by waterproofing the brick wall and placing the insulation against the waterproofing. This method saves both space and material for the same

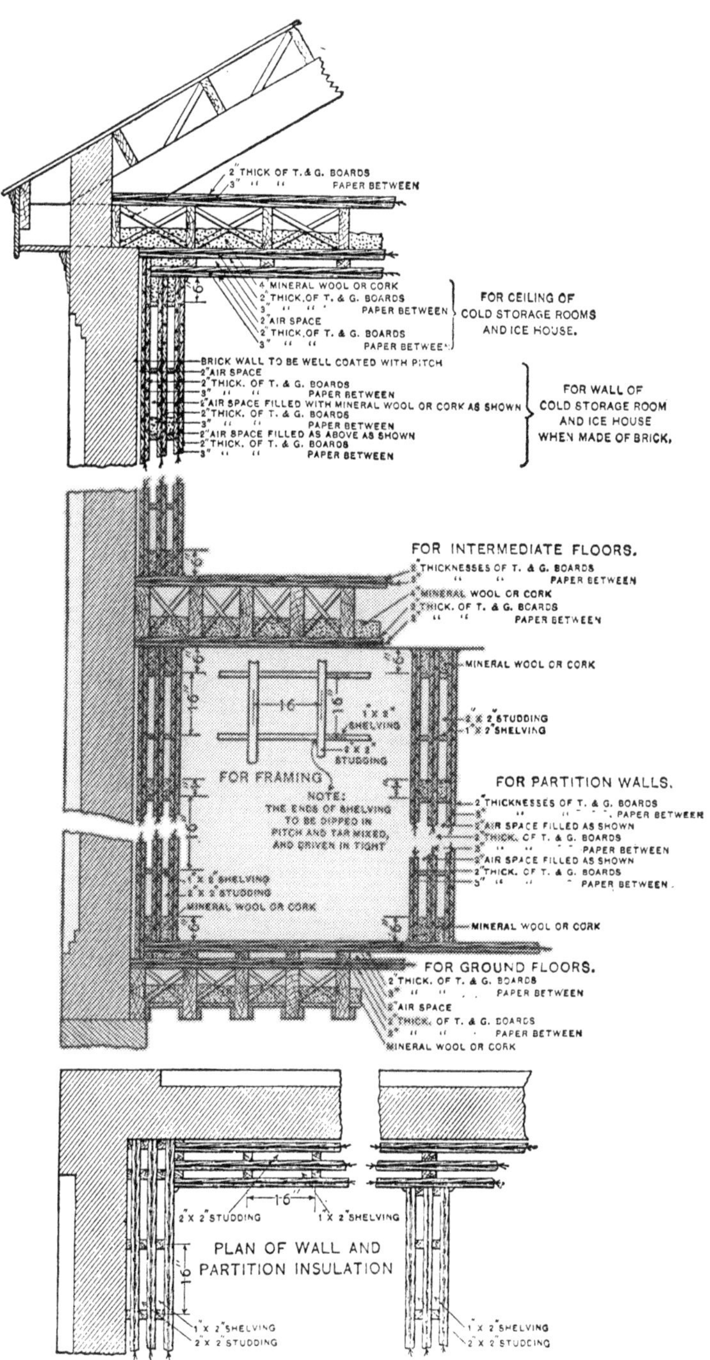

FIG. 33.—AIR SPACE INSULATION.—FRICK CO.

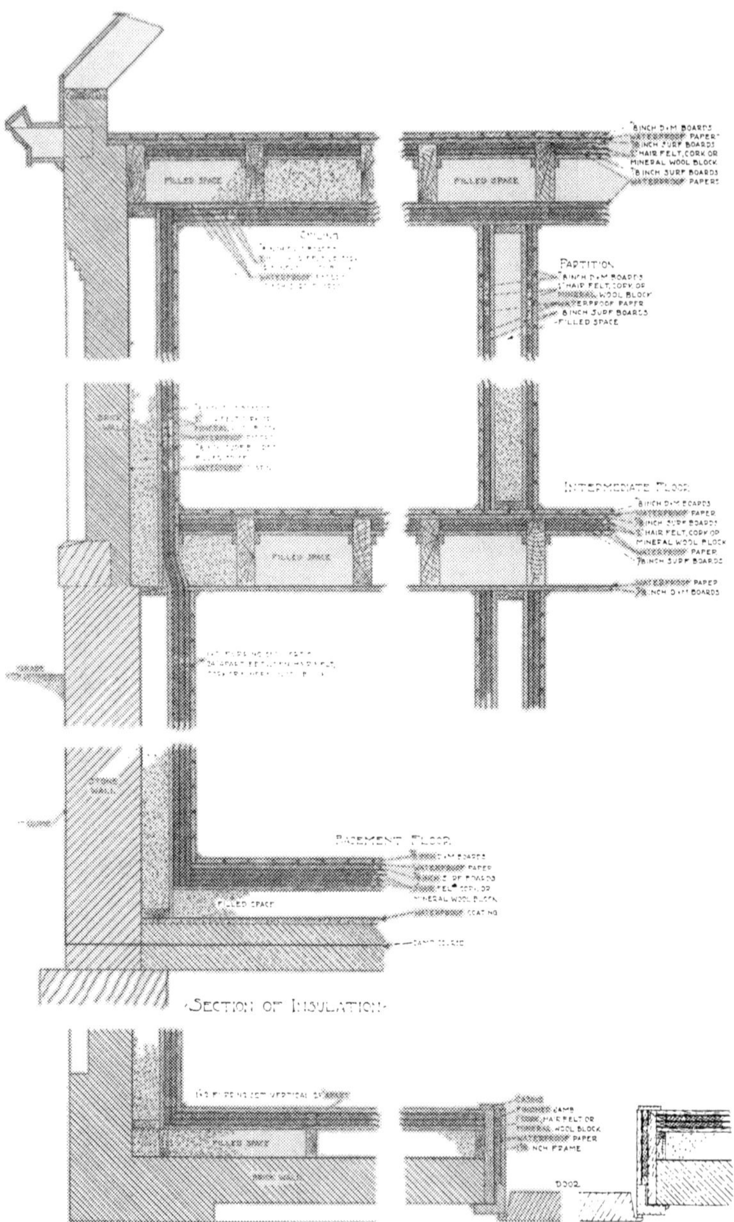

FIG. 34.—COOPER'S DESIGN FOR INSULATED CONSTRUCTION.

insulating efficiency obtained. The question of the amount of space occupied by the insulation is of importance, as it represents a certain money value, both in first cost and as storage space, and it should be the designer's aim, within practical limits, to use the best insulation and that requiring the least space.

TYPES OF INSULATION.

Fig. 34 illustrates a construction used to a great extent by the author in his cold storage work. The inside of the masonry walls are waterproofed and the filled space of from six to ten inches is placed against it, then the sheathing, sheet material and papers are placed inside, next to the storage rooms. With a ten-inch filled space and four-inch sheathing and sheet material as shown, a total of fourteen inches, we have an insulation for a storage temperature of 30° F., and the basement with a total of fifteen inches, for a temperature of from 20° to 25° F. For sharp freezing purposes there should be an eight inch and a six inch filled space and an additional thickness or two of sheet material.

Figure 35 illustrates the construction and insulation of a frame building suitable for a temperature of 30° F. The space between the studs should be sub-divided, as shown in Fig. 35a. This lessens the liability of settling of the filler and the penetration of moisture. The use of wide filled spaces such as shown in Fig. 28 is not considered good practice, as heat passes through a construction of uniform density more rapidly than through one made up of successive layers of different densities, therefore the thickness of the wall in the former case will be greater than that in the latter to obtain the same insulating value. This is evident by referring to Figs. 28 and 34, the former with twelve-inch brick wall, air space and twelve-inch shavings, as shown, a total of thirty inches in thickness, transmitting ¾ B. T. U. per square foot per degree difference between inside and outside temperature. The latter, with twelve-inch brick wall, eight-inch shavings and five inches of sheathing, sheet material and paper, as shown, totals twenty-five inches in thickness, will transmit same

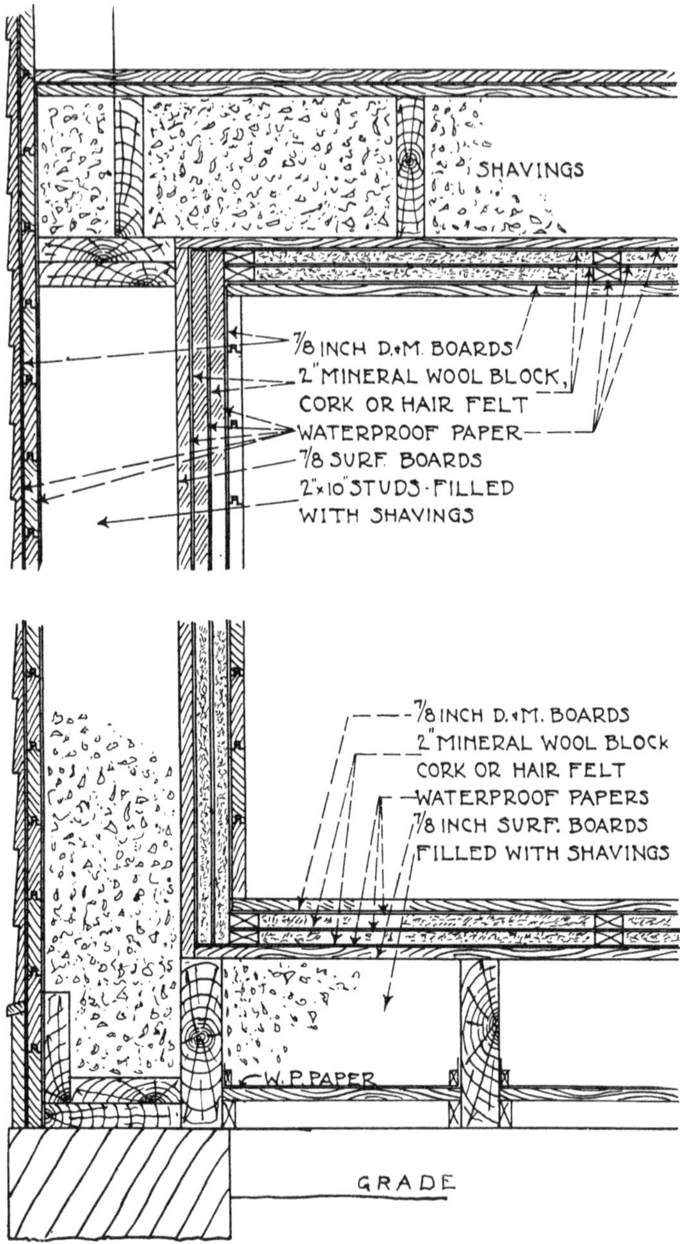

FIG. 35.—SECTION SHOWING INSULATION FOR A FRAME BUILDING.

amount of heat. Here is a saving of five inches in storage space which will earn the difference in cost between the two constructions in a short time. Dividing the shavings space into two separate spaces is advisable to increase insulating efficiency.

INTERNAL CONDENSATION.

Another feature in the construction shown in Fig. 34 of equal importance to the space saved, is durability. This is accomplished by placing the indestructible materials, such as water-proof paper, and mineral wool block, sheet cork or hair felt in successive layers on the side next to the storage room where the conditions are most severe. These severe conditions are caused by a tendency of the enmeshed air in all insulating materials to condense the moisture held in suspension when subjected to low temperatures. This moisture will impair the durability of some materials, such as sawdust, shavings, etc. That such condensation does take place within the insulation, near the low temperature side, was demonstrated to the author in the tests made by him. Between each test the removable cover was unscrewed from the tester, which was located in the cold storage room, and taken into a room having an ordinary temperature, where the material in the cover was changed for the next test. The enmeshed air in the material put into the tester, was of an ordinary temperature and held a certain proportion of moisture in suspension, and when the material was reduced to the low temperature in the cold storage room, this moisture would condense on the cold side of the layer confined by the waterproof paper. In some cases where the room temperatures were very low the condensed moisture would freeze on the cold side. These conditions obtained, depending on the dryness of the material when it was put into the tester, but would show more or less moisture in almost every case. The moisture condensed would be greatest in the layer nearest the cold side of the partition and would gradually diminish in each layer toward the high temperature side, where it would be perfectly dry. This was not moisture that had been carried into the insulation by the leakage of air, but was the condensation of the moisture held in suspension by the air enmeshed in the material. Air space construction showed the greatest condensation, wide filled spaces came next and the high grade of sheet materials divided by waterproof paper showed the least. From the above it is evident that high-grade materials should be used next to the storage rooms, as they will not deteriorate as rapidly with the presence of moisture. This construction also protects the loose filler in the filled space, which is removed further from the inside wall, therefore making its duty less severe as regards the action of moisture. This construction is shown in Fig. 34 as used by

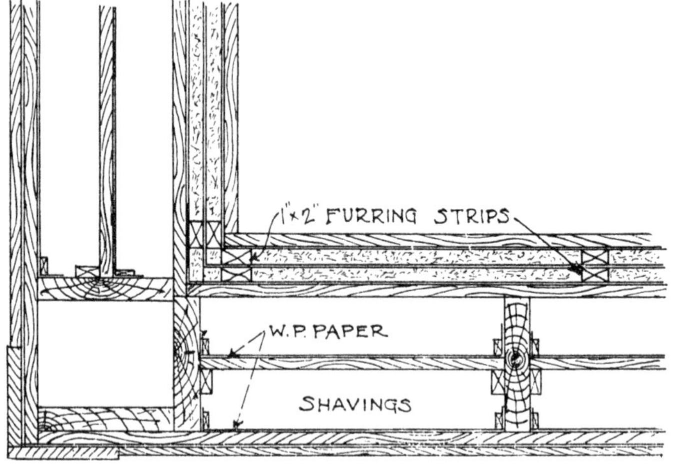

FIG. 35a.—PLAN OF SUBDIVISION OF SPACES AS SHOWN IN FIG. 31.

the author. It is also evident from the above described action of moisture in the insulation, that the importance of using dry materials cannot be urged too strongly.

PRACTICAL POINTS.

Where the storage space occupies two or more stories, it has frequently been the practice to insulate the intermediate floors out to the masonry walls and then erect the wall insulation independently for each story, as shown in Fig. 33. This is a questionable practice, as it increases the number of joints in the construction and consequently the chance of air leakage into the rooms. Cases illustrating the damage from air leakage between joints and above ceiling or beneath floors of storage rooms come to the author's attention frequently. The last one was so pronounced a case that the entire ceiling of an upper room became so rotten in a few years' time that it practically fell, making the entire renewal necessary. The damage was caused by leakage of air, and consequent deposit of moisture on the cold ceiling. Where the building is a fireproof structure with steel beams and masonry floors, it cannot of course be insulated otherwise than by treating each floor independently, or in a similar way to that shown in Fig. 42.

A better method than that shown in Fig. 33, would be to make the wall insulation continuous from floor of lower story to ceiling of upper story, as shown in Fig. 34. Where the ends of joists bear into the wall, the insulation should be scribed and closely fitted around each joist. This is easily done where the construction is properly designed to allow a wide spacing of the joists. Insulation constructed in this way decreases the chances of leakage to the minimum.

Referring again to Fig. 33, it will be noticed that the spaces between joists are but partially filled with the filling material, leaving an empty space in each case. This would have been of greater insulating value if packed full to the top. The spaces between the lower floor joists should first have been lined with waterproof paper before the filler was put in, as shown in Fig. 35, to prevent the penetration of air and moisture.

TANK INSULATION.

The insulation detail for a steel ice freezing tank, as shown in Fig. 36, is of the general form used by most designers. Attention is here again called to the use of high-grade insulators at what will be the coldest point of the insulation for reasons already discussed. In the light of our present information, the use of a 2-inch or 4-inch air space next to the steel tank would be considered a waste of space. This would much better be filled with granulated cork and hot pitch or asphalt, as shown in detail, as this would prevent the condensation of moisture and protect the steel tank from corrosion. The space under bottom of tank should first be filled level with top of floor cleats and then the tank set in place, after which the space around sides can be poured full from the top. A frequently neglected detail is covering the top edge of tank insulation with galvanized iron or other waterproof material, as shown in Fig. 36. This detail should not be overlooked or slighted, as the unavoidable spilling of water and dripping of brine in connection with ice making will eventually damage the insulation.

FIREPROOF INSULATION.

The tendency of modern building construction, especially in our large cities, is toward the solution of the problem of fireproofing, so as to decrease the danger and risk of fire. It is

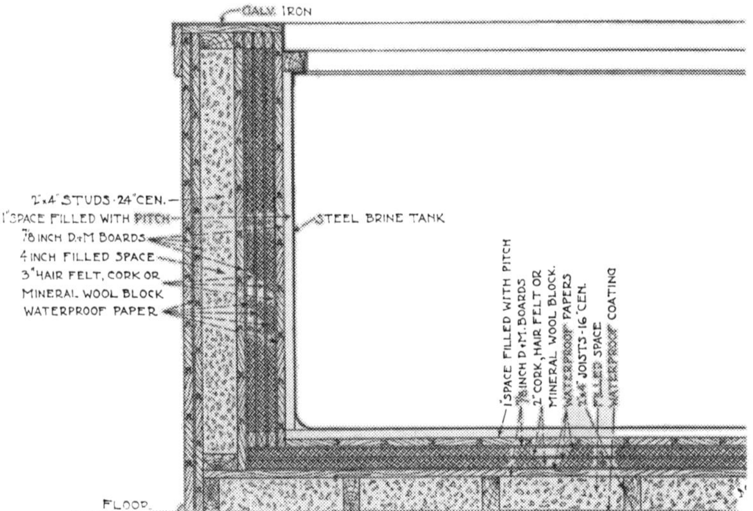

FIG. 36.—INSULATION OF BRINE TANKS.

natural that this same problem should confront the cold storage man, but the difficulties of providing insulation that would be fireproof, and remain at the same time insulation in fact, is a problem not easily solved.

In the article on "Insulation for Cold Storage," read before the eleventh annual convention of the American Warehousemen's Association, Starr stated regarding fireproof insulation, as follows:

I cannot refrain from alluding to the subject of fireproof insulation, more in the hope of drawing out information than adding to it.

Steel frame work has in the past been a considerable bugbear to cold storage men, but the time is already at hand when the problem of entirely fireproof construction, both as to building and insulation, must be solved. As to the insulation end of the problem, the difficulty is not so much with the question of obtaining a fireproof filling material, but to find a substitute for wood, to hold the material in position. Cementing insulating material in the shape of blocks has been experimented with, but if the cementation is enough to give sufficient ruggedness to the block the insulation qualities are, as a rule, impaired, and the cost as well as the space occupied by the insulation is increased.

If a semi-fireproof insulation that might be classed as fire-retardant construction is permissible, the problem is somewhat simpler, as several materials are at hand that can be classed as retardants, such as compressed cork, hair felt, silicated paper, air spaces, etc., but if the whole structure, studs, filling material and wearing face, are to be fireproof, we are practically restricted to mineral wool, mica and calcined pumice for filling material; and for retaining and wearing wall, to brick or some of the various cement fireproof boards on the market. The use of the latter would seem somewhat experimental, and the question of fireproof studs for supporting them, so far as I know, has not been answered.

As nearly all building materials available that can be called fireproof are poor insulators against heat, it is evident that the walls must be extremely thick to have the same insulating value or the machinery must take up the extra heat transmitted through them. The former course, with the present method of constructing fireproofing, is almost prohibitive on account of first cost and the space occupied; the second course necessitates a continuous heavy operating expense. The advisability of fireproof insulation, especially in small houses, is questionable, as the items of interest on investment, space lost, and increased operating expenses will oftentimes equal the difference in insurance rates obtained between a fireproof and a well designed "mill construction" warehouse.

It is the observation of the author that the greatest number of fires occuring in cold storage warehouses originate outside of the storage rooms, except in occasional instances where it is caused by defective electric wiring. Therefore, if the cold storage portion were surrounded and divided by fire walls, openings properly protected and the electic wiring installed according to approved methods, it would seem that the fire risk was cut down to a minimum, as the nature of the machinery and the goods usually stored are not of an inflammable character. This point must, however, be appreciated by the fire underwriters to make it of any value to the storage man in the way of lower rates. That they will appreciate it in the near future there is no doubt, and then the requirements of fireproof insulation will seem unnecessary, except perhaps in the large warehouses in the large cities.

Cold storage warehouses have been erected on the lines indicated above with "mill construction" and wood insulation and have obtained a very low insurance rate, considering the general attitude of the fire underwriters toward the cold storage warehouse business.

The necessity of fireproof insulation is felt in storage vaults for furs and fabrics, and justly so, as these articles are usually of great value and oftentimes could not be replaced if lost or damaged. This class of storage will permit a greater operating expense to offset the poorer insulating value of the fireproof insulation. Figs. 37 and 38 are details of the wall and ceiling insulation in the storage vaults of the Lincoln Safe Deposit Company, New York. The plaster blocks were fastened to the brick walls, every alternate block in every alternate row with iron anchors. The ceiling blocks were supported by tee irons, which in turn were suspended from the brick arches, as shown. These plaster blocks usually consist of plaster-of-paris and some binding material, such as manila fibre or common straw, and in the event of a severe fire they would probably fail and allow the filling material to fall out. The failure of plaster blocks was fully demonstrated by the Baltimore fire, where, in every case noted, partitions erected of them were completely destroyed. The company above named, in making later extensions to their plant, used cork blocks applied directly to the brick wall and plastered inside, as shown in Fig. 13.

Constructions such as shown in the two upper details of Fig. 26 are strictly fireproof and may be used where the temperature difference between the inside and outside would not be more than 25° to 30° F.

Figs. 39, 40 and 41 are reproduced from illustrations designed by Alfred Siebert.* These constructions were intended for brewery refrigeration where the temperatures required are

*In "American Handy-Book of the Brewing, Malting and Auxiliary Trades."

comparatively high, as their heat transmission would be prohibitive for cold storage work. The difference of detail between Figs. 35 and 36 is in the method of bonding the tiles to the main wall; in the former case, some of the tiles are laid head-

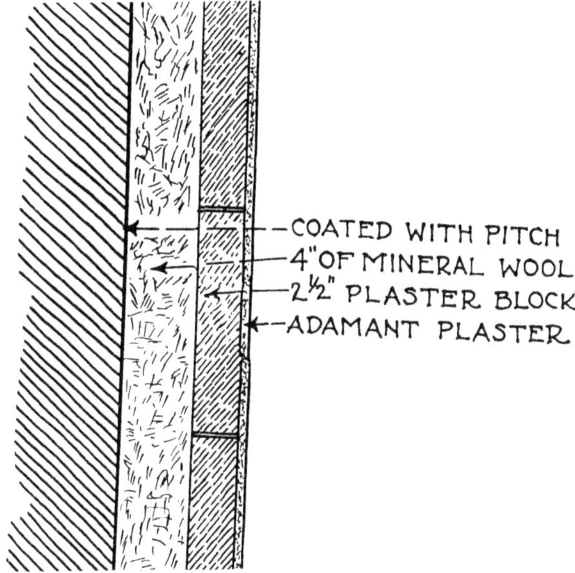

FIG. 37.—DETAIL OF WALL INSULATION.

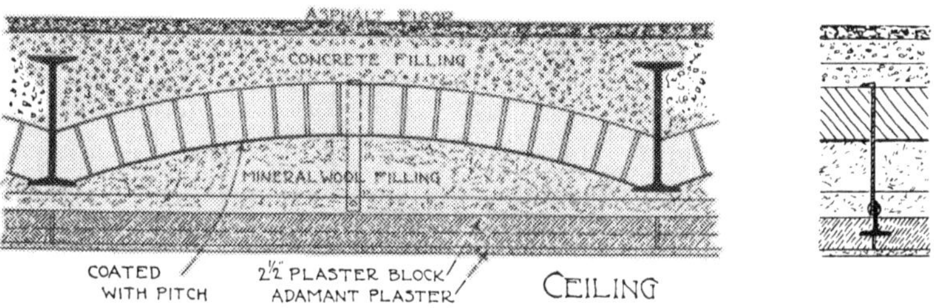

FIG. 38.—DETAIL OF CEILING INSULATION.

ers with one end secured in the brick wall and in the latter case iron anchors are used.

If a fireproof building is to be insulated where a slow burning or fire retardant material can be used, a construction as shown in Fig. 42 is the most practical and with the proper sheet or block material would be almost thoroughly fireproof.

Such insulation can be finished inside with cement or plaster, as shown.

The inside finish on the walls of the rooms is of some importance as a fire retardant. With the use of hard oils, varnishes, shellacs, etc., the spread of a fire would be rapid, as these materials are very inflammable, but with the use of preparations such as cold water paint or whitewash, the spread of fire would be retarded, as these mixtures are not inflammable, and would give some protection to the woodwork on that account.

BRINE PIPE INSULATION.

The importance of thoroughly protecting the pipes that carry the cooling agent to the various parts of the storage building is as great as insulating the rooms. On account of the low temperature of these pipe surfaces, they condense much moisture, and if the covering is poor and not well protected on the outside from air leakage, a dripping and soggy condition is sure to follow each time the cooling agent is shut off. If this condition is once obtained the value of the covering is permanently impaired.

There are some pipe coverings on the market, especially suitable for brine piping, made of cork or mineral wool in block form in the same manner as already described for wall insulation, and are made sectional to fit any size pipe or fitting, having the appearance shown in Fig. 43. Some of these sectional coverings are provided with canvas cemented to the sections with ample lap at the joints, and these laps are cemented together as the sections are put in place. Directions for putting on are usually sent with the material by the manufacturers. Hair felt is also a good material to use if properly applied, as indicated in Fig. 44, and can be handled very well, if cut in lengths of five or six feet, wrapped around the pipe, and thoroughly wired with galvanized or copper wire. If a second layer is to be put on, waterproof paper should be put between the two layers and wired on, and the second layer then applied in the same manner as above. The outside layer should have waterproof paper wired on and then covered with strip canvas, binding it on spirally with a good lap at the

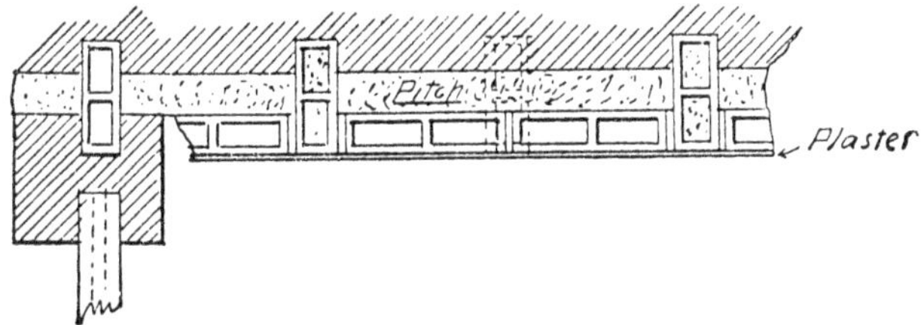

FIG. 39.—SIEBERT'S BREWERY INSULATED CONSTRUCTION.

joints. The canvas must be bound on tightly. The covering should then have at least two coats of a good elastic waterproof paint.

It is of primary importance that the pipes should be dry

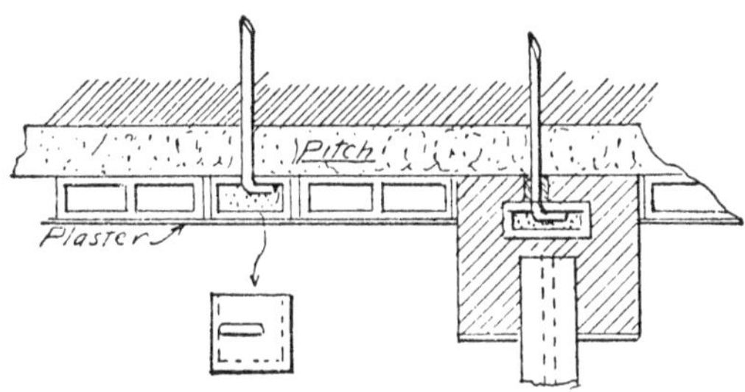

FIG. 40.—SIEBERT'S BREWERY INSULATED CONSTRUCTION.

and should be given a coat of paint before covering is put on. The author recommends that the layers of covering should be thin, not more than one inch in thickness, and that at least two thicknesses be used, having waterproof paper between each

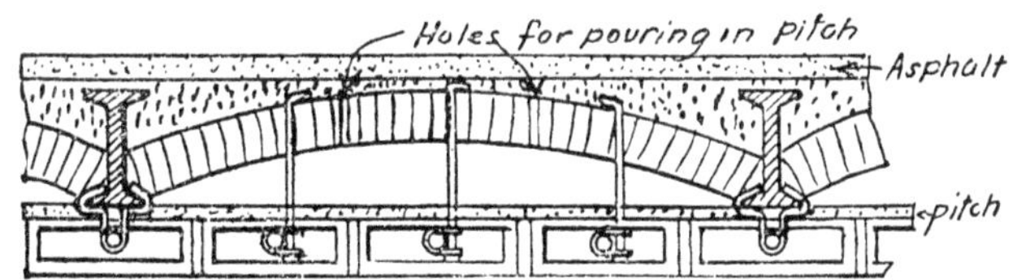

FIG. 41.—SIEBERT'S BREWERY INSULATED CONSTRUCTION.

layer with cemented joints, so as to insure the air-tightness of the covering.

For brine mains laid under ground, through brick walls or up through partitions, a covering of granulated cork mixed

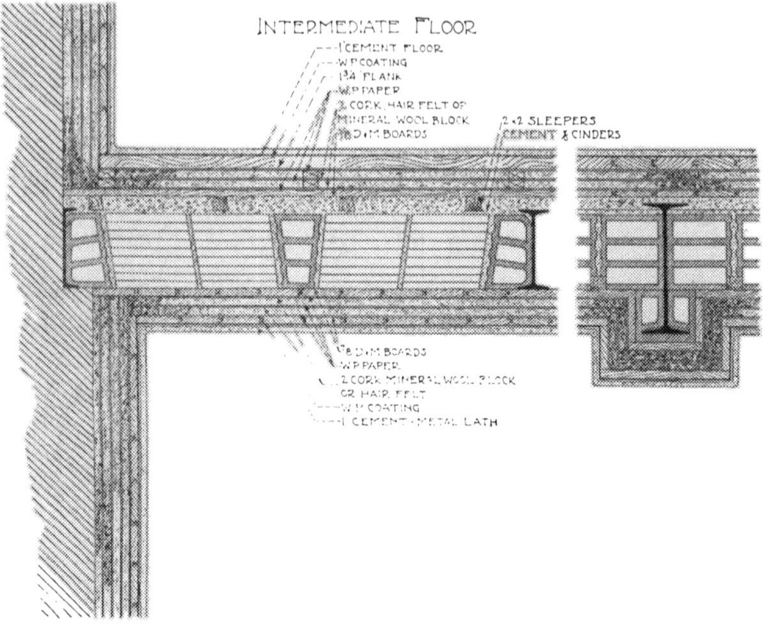

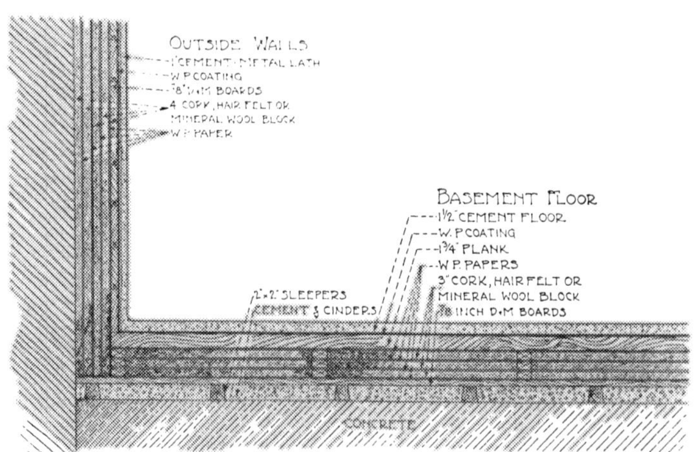

FIG. 42.—INSULATION OF FIRE PROOF STRUCTURES.

with hot pitch or asphalt is best, as described under cork materials. This method was used by the Quincy Market Cold Storage Company of Boston, Mass., in running a street pipe line from one of their buildings to the other. The pipes were laid in creosoted plank boxes of proper size to permit sufficient space around them, and the mixture of cork and pitch was then poured in.

Fig. 45 illustrates a form of tunnel for underground brine pipes that has been used by the author. In this case, as shown, the tunnel was constructed of brick, waterproofed both inside and outside and the top constructed so as to be removable in

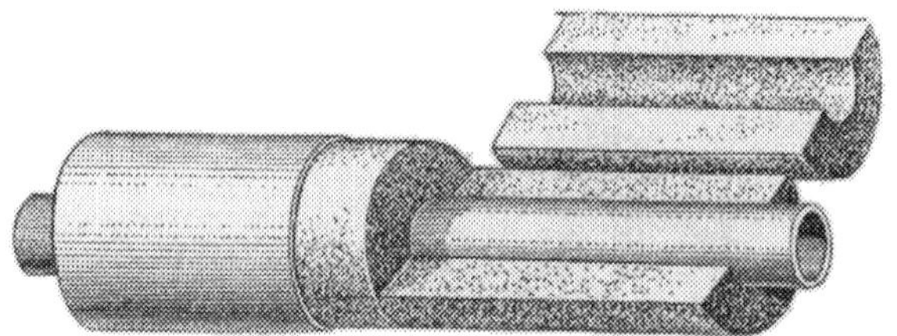

FIG. 43.—CORK BRINE PIPE INSULATION.

case of necessity. The brine mains inside were covered in the usual way, leaving an unfilled space around them in the tunnel.

WATER AND DAMP PROOFING.

The results of the penetration of moisture into the insulation has already been discussed under the various subheads; and the functions of waterproof paper in the interior of the insulation to stop this moisture, should by this time be pretty well understood. But the penetration of moisture through the masonry walls to the insulation must be prevented by special treatment.

The tendency of masonry to absorb moisture is fully recognized and provided for in the building trades. It frequently happens in heavy and driving rain storms, of some duration, that the water will be driven through a 9-inch and even through a 13-inch brick wall. This is counteracted in general building operations, if it is desired to plaster on the inside of the wall, by constructing a 2-inch air space in the masonry wall. This space will prevent the passage of moisture sufficiently so as not to damage the plaster. A second method is to line the inside of a solid masonry wall with hollow brick or porous terra-cotta blocks. The third and most common method is to form an air space on the inside of the wall by vertical furring, and the lath and plaster is then put on. All of these methods have been used in cold storage warehouse construction, especially the last, as has been shown by the illustrations given.

Basement walls are usually coated on the outside with cement and pitch or asphalt to prevent the moisture in the

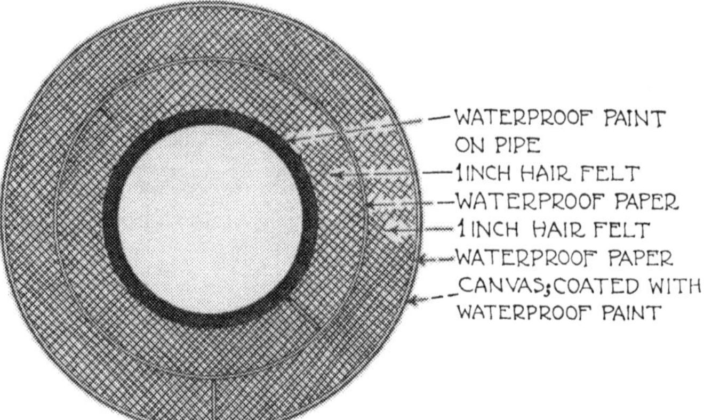

FIG. 44—HAIR FELT BRINE PIPE COVERING.

soil from penetrating to the inside. If the soil is very wet and there is danger of the water level reaching above basement floor at some periods of the year, as is often the case in some localities, there should be a dampproof course extended under basement floor, through the masonry walls and up on the outside of them to grade. This work belongs to building construction rather than to our present subject and it is therefore unnecessary to treat of it in detail. The position of this damp course is indicated in Fig. 34.

The common method of protecting the insulation from the moisture in the masonry walls is to coat the walls on the inside with various preparations, such as paraffin, pitch, asphalt, etc. These are usually put on hot in a liquid state. No preparation having a strong penetrating odor, such as coal tar, should be used, as it is liable to taint the goods in storage. Pitch, if properly put on, makes a fair coating, but on account of its quick hardening and brittleness, it is very difficult to apply in cold or even cool weather, and when cooling it will contract and fine cracks will appear running in every direc-

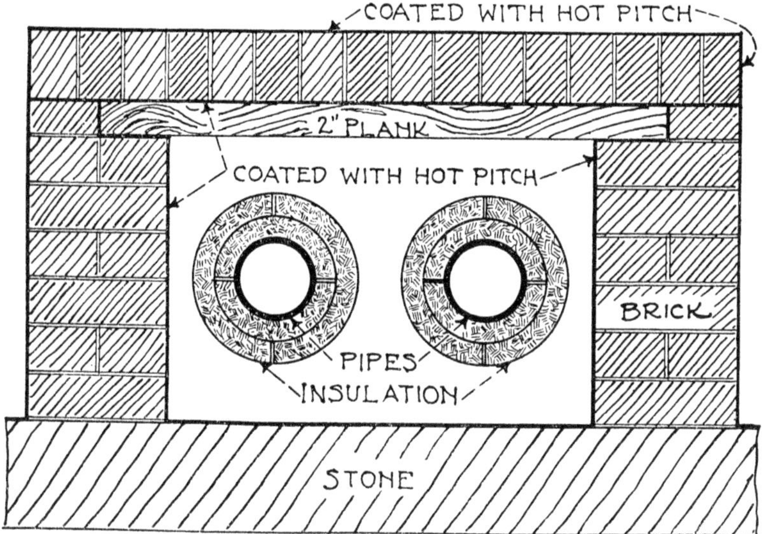

FIG. 45.—TUNNEL INSULATED CONSTRUCTION.

tion. To avoid this, the roofing men will mix coal tar with it to give elasticity, but it is then, of course, unfit for the inside walls of cold storage rooms on account of the odor, as stated.

The best material for coating inside walls is pure asphalt, and it is specified almost exclusively by the author for this purpose. This material is odorless after it is applied, the odor given off when subjected to heat is not penetrating and quickly disappears. Unlike coal tar or pitch, which are products of distillation from gas works, pure asphalt is a natural mineral bitumen, and although it is similar in appearance to pitch, it is not so dense or brittle and it has sufficient elasticity so that it will not crack when cooling. Besides the commercial paving asphalts which are very impure, there are also refined asphalts on the market which are claimed to be over 90 per cent pure. These are the product of distillation from the oil wells of Texas and California, and because they contain a higher percentage of bitumen are more elastic than the paving asphalts. Asphalt is difficult to apply to cold storage walls on account of quick hardening, but not so much so as pitch. The chief difficulty, especially in small cities, is to obtain a pure asphalt and also to get workmen who have had experience in applying it. The local roofing men have little or no need of pure asphalt, as the common material for flat roofs in this country is pitch and coal tar, and consequently they do not carry asphalt in stock. In fact, many of them are under the impression that asphalt, pitch and coal tar are the same thing and will sometimes attempt to use the latter materials when asphalt is specified.

The commercial paving asphalt comes as a solid cake in barrels weighing from 500 to 550 pounds and containing, when melted to a liquid, about fifty gallons. The refined asphalts come also in 250-pound barrels, containing twenty-five gallons. Asphalt is melted in large kettles, such as used by roofers, without the addition of any oils or coal tar. Care should be taken not to boil the asphalt, as its natural oils are thereby evaporated, and when cooled down it will become more brittle. The hot asphalt should be applied to the surfaces with string mops to get the best results, the process is slow and tedious on account of the heavy consistency and its quick cooling. The surface should afterward be examined and all holes and crevices pointed up with asphalt. If the walls are dry and the weather warm, a gallon of asphalt will cover about thirty square feet of ordinary brick surface; in cold weather a gallon will cover about twenty square feet, or 6,000 or 4,000 square feet per ton, respectively. Where the walls are very rough or constructed of rubble masonry the asphalt coating will not cover much more than 3,000 square feet per ton. The surfaces that are to be coated must be free from frost or ice, and should be thoroughly dry to obtain the best results.

While a good coating of asphalt on inside of wall will prevent moisture from reaching the insulation, it does not waterproof the brick wall itself. Brickwork full of moisture is a much poorer insulator than when dry, and as we should get the greatest insulating value possible out of the construction, it is evident that the outside of the walls should also be waterproofed. There are a great many preparations on the market that are being used for waterproofing external walls with more or less success, but as they will all oxidize and disintegrate in time, the coating has to be renewed at intervals to prevent the absorption of moisture. The coating may receive proper attention when applied for the first time, just after the building is erected, but it is very likely that necessary future coatings will be neglected or forgotten; on this account it is not safe to rely upon the outside coating only, the inside walls should also be waterproofed as indicated above.

Boiled linseed oil is often used on external walls with very good results. If three coats are first given, one coat applied every three to five years thereafter will be sufficient. The oil does not change the color of ordinary brickwork to any extent, but tends to give it a darker and richer appearance.

White or red lead, ground in boiled linseed oil, is more durable than the oil alone, but it entirely changes the appearance of the building and in most cases would not be permissible on that account. New work should not be painted until the walls have been finished two or three months, and at least three coats should be given the first time. The above two preparations are probably as good, if not better, than any of the patented preparations on the market.

Cabot's Brick Preservative, made in Boston, Mass., has been used in general building operations as a waterproofing quite extensively, and, it is claimed, with good success. This preparation is made both colorless and with a red color so as to be adaptable to any color of brick, and it is applied with a brush in the same way as oil, no heat being necessary.

Mr. Stoddard, in his paper on "Insulation," previously referred to, describes in detail tests on the waterproofing of brick, using various preparations and materials. These tests are about as complete as anything that has been attempted in this line, and being pertinent to the subject, are given in full, as follows:

During the summer of 1899 a large variety of paints, oils, varnishes, cements and so-called waterproof coatings were tested for a cold storage company in the hope of finding some coating that would make waterproof and airproof the brick walls of its warehouses. The tests were made with quarter bricks with good, fair surfaces, free from large holes, and, as nearly as possible, like those used in the exterior walls. Quarter bricks were used instead of whole bricks, so that sensitive balances could be used for the different weighings. All weighings were made to within one-thousandth of a gram. The results of the more satisfactory tests are tabulated below, and besides these, many other tests were made, but they were either unsatisfactory or the materials tested of no value for the desired use. The quarter bricks to be tested were immersed in water of a temperature of about 70°, the brick being placed on its side, with one inch of water over it. Weighings were made as follows:

Of the brick before coating.
Of the brick after coating.
Of the brick after immersion 24 hours.
Of the brick after immersion 48 hours.
Of the brick after immersion 72 hours.
Of the brick after immersion 96 hours.
Of the brick after immersion 120 hours.

At the end of each twenty-four-hour period the quarter bricks were taken from the water, the outer surfaces carefully dried by cloth and blotting paper, and then the bricks were immediately weighed before any evaporation could take place from the pores of the brick. This was repeated in most of the tests until the bricks had been immersed for a period of 120 hours. After this continued immersion the bricks were taken from the water and their surfaces examined in order to see what change, if any, had taken place in the coating. In some cases the coating had softened, in some shriveled, and in one case the coating, naphtha and a paraffine-like substance, which before immersion was evidently well into the pores of the brick, had gradually worked out into the water.

The nature of the substances tested varied greatly. Some were in the nature of paints and varnishes, and were retained mostly upon the surfaces of the bricks. To this class belonged the materials used in tests marked A, B, D, G, L, O, P and Q. Other substances were more in the nature of a paste or coating applied upon the surface of the bricks. In this class may be included the substances used in tests marked C, I, K, N, R, S, T and U. Another class of substances was supposed to soak into the bricks, and by filling the pores exclude moisture. To this class belonged the substances used in tests E, F and J. Other coatings consisted of two substances, which, when combined, were supposed to form an insoluble compound or compounds which would fill up the pores of the brick. The tests of this class are marked H, M and V.

Some substances which were submitted for test could be applied to the bricks only by soaking, and so were not available. Some bricks offered for test were soaked full of the so-called waterproofing, and of course would not absorb water or anything else while in that condition, as the pores of the brick were already filled. Many resins, gums and oils were tested, but were of no practical use.

Pitch, asphaltum, etc., were objectionable, because of their odor and color. The results of the tests giving the most favorable results are as shown in following tables:

In regard to the result of the tests it is worthy of remark that some of the substances that have been considered as among the best waterproof materials proved to be either of little value or very inferior to some of the other substances.

TESTS OF WATERPROOFING BRICK.

1	2	3	4	5	6	7	8	9	10	11	12
	WEIGHT—GRAMS				INCREASE IN WEIGHT BY ABSORPTION OF WATER					COMPARED TO BARE BRICK	
Sample.	Bare Brick.	Coated Brick.	Coating.	Per Cent Increase.	24 Hours.	48 Hours.	72 Hours.	96 Hours.	120 Hours.	Per Cent Inc. by Coating and Water.	Per Cent Inc. by Water.
A	630.32	639.10	8.78	1.39	0.30	....	1.10	....	1.50	1.63	0.24
B	556.71	571.11	14.40	2.59	1.39	....	2.16	2.49	2.89	3 11	0.52
C	578.43	581.92	3.49	0.60	1.15	2.18	3.25	3.49	5.13	1.14	0.89
D	527.80	537.70	9.90	1.88	1.00	....	2.88	4.00	5.10	2.84	0.97
E	616.10	637.60	21.50	3.49	2.10	5.55	7.15	....	9.99	5.11	1.62
F	633.80	706.87	73.07	11.53	4.75	12.13	12.83	....	14.13	13 75	2.23
G	584.40	588.92	4.52	0.77	4.88	7.48	9.68	11.43	14.38	3.23	2.46
H	499.52	551.00	51.48	10.31	7.30	9.70	11.30	13.30	15.32	13.36	3.07
I	504.12	523.40	19.28	3.82	3.73	6.33	....	12.12	15.63	6.93	3.10
J	666.94	670.07	3.13	0.47	20.33	21.13	21.63	23.13	....	3.94	3.47
K	607.29	610.90	3.61	0.59	7.00	8.60	9.30	....	21.83	4.19	3.59
L	519.68	527.34	7.69	1.48	3.76	5.78	....	12.68	21.72	5.66	4.18
M	652.50	692.99	40 49	6.21	24.78	....	27.16	....	28.24	10.53	4.33
N	510.20	529.10	18.90	3.70	23.10	....	23.80	....	23.72	8.35	4.65
O	570.87	586.20	15.33	2.69	26.98	28.00	28.00	....	28.70	7.71	5.03
P	496.20	503.00	6.80	1.37	24.85	....	28.75	28.71	28.72	7.13	5.79
Q	502.87	515.12	12.25	2.44	29.08	30.70	31.28	....	32.03	8.85	6.37
R	....	543.60	....	....	3.72	5.00	6.15	7.15	....	....	*1.32
S	....	602.20	....	....	3.10	5.55	7.35	9.20	....	....	*1.53
T	....	606.31	....	....	2.35	4.69	8.07	10.21	....	....	*1.68
U	....	581.16	....	....	6.46	9.69	12.69	15.64	....	....	*2.69
V	....	621.85	....	....	21.15	29.60	31.02	....	32.15	....	*5.17
W	....	....	....	....	....	....	....	....	....	....	....
X	....	....	....	....	....	....	....	....	....	....	....
Y	....	....	....	....	....	....	....	....	....	....	....
Bare	Brick	489.04	....	....	21.26	....	39.69	39.69	42.43	....	*8.68

* Compared to coated brick. 1 gram equals 15.43 grains; 28.35 grams equals 1 ounce avoirdupois.

KEY TO TESTS OF WATERPROOFING BRICK.

KEY TO TESTS OF WATERPROOFING BRICK.

A.—Bay State air and waterproofing.................3 coats.
B.—Red mineral paint, ground in oil................2 coats.
C.—Spar varnish with plaster of paris...............2 coats.
D.—Spar varnish ....2 coats.
E.—New York sample, No. 2........................Soaked.
F.—New York sample, No. 1........................Soaked.
G.—Shellac ....1 coat.
H.—Portland cement, 1 coat; soap and alum, 3 coats..4 coats.
I.—White enamel paint.............................3 coats.
J.—Paraffine in naphtha...........................3 coats.
K.—Hot paraffine ....3 coats.
L.—Water paint ....3 coats.
M.—Portland cement mixed with Ca Cl_2, 1 coat.
Water glass, 3 coats..............................4 coats.
N.—Portland cement ....2 coats.
O.—Black varnish, No. 2...........................3 coats.
P.—Spar varnish ....1 coat.
Q.—Black varnish, No. 1...........................3 coats.
R.—Waterproofing, No. 1.
S.—Waterproofing, No. 4. Similar to "R."
T.—Waterproofing, No. 3. Similar to "R."
U.—Waterproofing, No. 2. Similar to "R."

V.—Bi-chromate potash and glue—exposed to sunlight.

The Sylvester process, H, soap and alum, proved to be of little value, even when applied to a surface made as smooth as possible with Portland cement. This process was also tried without the cement, but was even less effective. Hot paraffine has often been used to waterproof walls; but, under the conditions of these tests, it proved to be very far from waterproof. Portland cement is another substance which did not prove to be as good as its reputation.

Of all the materials tested, those used in tests A, B, C and D rendered brick, to which they were applied, more nearly waterproof. Spar varnish, used in tests C and D, was very good under test; but it is a very expensive material, and will withstand exposure to the weather only for a rather limited time.

The material used in B was a common mineral paint ground in oil. It was very good under test; but the best authorities on paint predicted for it a very short life in actual use, as it would disintegrate after a short time by the oxidation of the oil.

The substance used in test A not only proved to be the best waterproofing substances of any tested, but it seems to have all the qualities necessary for the coating of the outside of brick walls. It is moderate in price, and is easily and quickly applied, being put on with a brush the same as a varnish or paint. When applied to a brick wall, it forms a glossy, hard, transparent coating, and, instead of defacing the wall, it greatly improves its appearance, making the common brick look like enamel or glazed brick. The substance is a specially prepared and highly oxidized oil that has been and is used in the best varnishes. As it is thoroughly oxidized in its preparation, exposure to air should affect it but little, and it should not need to be renewed for many years. The brick walls of a number of large warehouses were coated with this substance one and two years ago, and the coating is apparently as good as when first applied. One gallon will cover from eighty to 100 square feet of surface with three coats, the first coat taking considerable oil, but each successive coat taking less. A brick wall should be as dry and warm as possible when the coating is applied. It should not be applied to a damp wall just laid, or when the outside temperature is below 40° F. This oxidized oil is known commercially as "Bay State Air and Waterproofing."

If the coatings of this substance continue to wear as well in the future as they have in the past two years, the substance will prove of the greatest value for airproofing and waterproofing the brick walls of cold storage warehouses. Any efficient waterproofing that can be applied to the outside surface of a cold storage warehouse is of the greatest importance, as there is where the entrance of moisture would best be stopped; but this outside coating should not be depended upon alone to prevent the entrance of moisture into the warehouse, and there should always be inner layers of some air-tight material, like an air-tight paper, with the joints cemented.

If we make use of a durable insulating material of good efficiency, apply it carefully and of a proper thickness, and make it air tight and moisture proof, we have done all that is practical to well insulate a cold storage warehouse.

A better method than using preparations will, in the opinion of the author, be used in the future for waterproofing external walls. This is to face them with glazed brick or salt-glazed terra-cotta blocks, laid with thin joints of rich cement mortar. The glazing is absolutely waterproof and would last for an indefinite time, but the present cost of glazed brick would make their use almost prohibitive, as they cost from $80.00 to $100.00 per thousand. Glazed terra-cotta on tile in the form of hollow building blocks can now be obtained, and are used as a facing for outside walls in the same manner as pressed brick. In this position these blocks, if properly laid, will practically prevent the absorption of moisture, and would cost about the same, laid in the wall, as selected common brick.

COST.

There are very little reliable data available on the cost of constructing insulation. This is owing mostly to the fact that this kind of construction is comparatively new in the building trades, and is usually done by the cold storage men with day labor. As a rule no separate accounts of costs are kept, as it is not apparent to the owners what future service such information would yield—they do not expect to build any more cold storage houses. There is also the variable factors of labor and material which may affect each locality differently, often to the extent of 50 per cent difference in cost. This is of course true of all building operations, but especially so of constructing insulation, as the work is new and unfamiliar to workmen generally. All these conditions make it difficult to determine the cost of any particular insulation, without knowing exactly the conditions of each individual case.

The advantages of sufficient and properly constructed insulation will usually appeal to the prospective cold storage man until the question of cost is brought up. It is a mistaken idea in general that when the building proper is finished, the greater part of the investment necessary for a complete cold storage house is expended. The construction of a cold storage house may be divided into three general operations; first, construction of the building proper; second, insulation; third, machinery or cooling apparatus. The additional cost of the insulation may generally be taken as one-half to two-thirds the cost of the building proper.

Generally speaking, the cost of insulation, erected in place, for temperatures of 30° F. down to 0° F., will be from about 25 cents up to 50 cents per square foot, in proportion to the above temperatures. The Nonpareil Cork Manufacturing Company gives the cost of the construction, shown as style No. 20 in Fig. 21, as about 22 cents per square foot; that shown as style No. 13 in Fig. 21 as about 38 cents per square foot, and that shown as style No. 16 in Fig. 21, as about 48 cents per square foot. A construction shown in Fig. 24, with air space next to the brick wall, four ⅞-inch boards and twelve inches of shavings, will cost from 20 to 25 cents per square foot. A construction shown in Fig. 30, with eight inches of shavings and two inches of hair felt, sheet cork or mineral wool blocks, may be constructed for 25 to 30 cents per square foot. This construction is suitable for temperatures of from 30° to 35° F. This construction shown in the lower part of Fig. 30, which is suitable for a temperature of 20° to 25° F., may be constructed for 28 to 32 cents per square foot. A construction of the same character suitable for temperatures of from 5° to 10° F. may be built for about 40 cents per square foot. Referring to the five constructions shown in Fig 28, giving the same insulating value for various thicknesses of different materials, and comparing the hair felt with the air space and wood board construction, there is a total thickness of eight inches with the hair felt partition, and a total thickness of thirteen inches with the board and air spaces; giving a difference of five inches in thickness with the same insulating value. The hair felt construction would cost from 35 to 40 cents per square foot, and the board and air space construction would cost 30 to 35 cents.

The waterproofing of the brick walls has been included in the estimates given above. The cost of waterproofing with hot asphalt, when that product can be obtained at $40.00 per ton, will be about 2½ cents per square foot. Waterproof and odorless papers cost from $2.50 to $5.00 per roll (1,000 square feet), depending on the thickness and quality.

The insulating material in the form of blocks or sheets, such as mineral wool block, sheet cork and hair felt, varies in cost from four to six cents per square foot per one inch thick. This does not include freight, which would increase the cost, depending on the locality. Mineral wool is sold by the pound or ton and can be obtained at from $25.00 to $30.00 per ton.

The cost of planer mill shavings is variable, depending upon the proximity to the mills, season of the year, etc. In some cases known to the author they have been obtained for the mere trouble of hauling them away, but in most cases they are sold, either by the load or by the bale. The cost per bale of 80 or 100 pounds varies from 15 to 25 cents.

SUPERINTENDENCE.

On account of the special character of cold storage insulation, the work should be carefully and frequently inspected to see that the materials are of the quality specified and that the work is executed according to details. The construction of insulation requires more care in the way of tight joints and first-class workmanship throughout, than is usually obtained in ordinary buildings. The labor required is mostly such as belongs to carpenters, and as they are accustomed to do work along certain lines common in ordinary building operations, it is sometimes difficult to train them into the high class work necessary for cold storage insulation. It must be constantly kept in mind that the insulation must be air and water proof. The materials and the combination in which they are used, no matter how excellent they may be, are much decreased in insulating value if these points are neglected. The materials, as they arrive at the work, should be inspected to determine if they are dry, and they should be kept under cover until used, to prevent them from becoming wet or damp. Planing mill shavings are sometimes damp when they arrive at the work and the bales should be loosened up and spread out in the building to allow them to air dry. The materials should be delivered sufficiently in advance to admit of proper inspection and of being replaced with new material, if found unsatisfactory.

The superintendent should see that all filling materials, such as granulated cork, mineral wool, shavings, etc., are properly packed into the spaces to about the proper density. (See Materials.) The prevention of the future settling of the filler is mainly a question of personal care in seeing that it is properly packed, and all corners and tops of filled spaces, which are difficult to pack, will need particular attention. The waterproof papers, as already stated, are used to prevent the passage of air and moisture and their application, therefore, is of prime importance. All joints should be lapped two or more inches, and each course of papers should be lapped around corners and angles of rooms. In case the paper should be torn by the workmen, it should be replaced or another sheet should be placed over it. All sheathing and matched boards should be free from large or loose knots, should be fitted up close in all corners and angles, and nailed at bearings only. No nails should be driven through boards and paper, and project into the filled spaces or into the sheet material. As a proper finish for the inside corners and angles of rooms and around door jambs, the author recommends and uses ¾-inch or ⅞-inch quarter-round mouldings as giving air-tight, neat-appearing and serviceable finish.

CHAPTER VI.

DOORS AND WINDOWS.

It was for many years the custom to build the doors required for cold storage rooms directly "on the job" where used, applying such hardware as was available. Some of the doors resulting from this practice were very poor and seldom if ever could a passable job be obtained. At present no one thinks of building cold storage doors on the job as these may be obtained from manufacturers who make a specialty of this class of work. The home-made door if it fits when new will seldom remain tight for any length of time. Ordinarily they are made with a long bevel fitted to a corresponding bevel of the frame, and the least swelling or settling will result in difficult operation and air leakage. Sometimes packing of canvas or other material is applied to the bevel, but it is almost impossible to make a tight fit in this manner. Nearly all doors made on the job sooner or later stick in the frame and refuse to open without many persuasive kicks and surges—we all know how it is.

The special cold storage and freezer doors made by firms who make a specialty of this line shut tightly with small pressure, forming a practically perfect air seal, and open readily when the handle is grasped. The time saved in opening and shutting these doors will soon pay for the additional cost over the home-made article. The prices for these doors are reasonable considering the excellent workmanship, insulation and material used. Any insulation may be had by specifying it when ordering. The author recommends these special cold storage doors to those who desire first class work. If it is necessary, owing to remoteness from transportation, or other reason, to build the doors "on the job" the chief aim should be to build them tight at one or two points all around, and not on a long bevel. Sometimes a gasket or packing may be used between door and frame to make a better seal.

It is very generally known by those who are familiar with cold storage work that windows are a bad proposition from an insulation, mechanical or a practical standpoint. The increased fire exposure is of some consequence, too, and with the low cost of electric light, windows should not be thought of for cold storage work. Barring the small amount of heat given off, the incandescent electric lamp is an ideal device for lighting cold storage rooms, as the air is not vitiated as when using gas, kerosene or candles. Nevertheless, there are many situations where windows are very desirable in connection with refrigerating work. Artificial light is often difficult to locate properly and in connection with ice storage rooms, where more or less moisture is nearly always present and artificial light therefore troublesome, natural light through windows is greatly to be preferred.

In considering methods of construction for windows in connection with cold storage work, the setting of the sash in the frames is one of the most important and at the same time difficult parts of the problem, and we might as well forget all about trying to make windows removable in any way if used in connection with cold storage work. It is practically out of the question for the reason that they must be fitted tightly and set so as to be air-tight. The frames in which the sash set must be of heavier and more substantial material than ordinary. Where ordinarily a 2-inch frame is used, 3-inch is better for cold storage work, and it should be preferably of hard wood, or nothing less substantial than hard pine. Frames must be planed on both sides so as to make it possible to make an air-tight joint by fitting insulating paper tightly. The frame must positively be set in the wall before the insulation is put in place, and it must be wide enough to allow the insulating material, whatever it may be, to be fitted tightly around it when the insulation is being placed. If wide planks are not available, the frames should be made of plank grooved to receive a spline or tongue, making a comparatively tight joint and substantial work. Wherever joints are made in the frames they should be thoroughly coated with thick lead and oil.

The sash to be used should be what is known as double-glazed. Sash of ordinary thickness may be used but they are made so as to receive glass on both sides, leaving an air space of about one-half to three-quarters of an inch between the two glass. The glass should, of course, be set in lead and oil and carefully puttied in so as to form an air-tight joint. In fitting the sash into frames care should be taken that they fit reasonably tight and they should be set in fresh lead and oil paint, and stops driven up tightly so as to form an air-tight job.

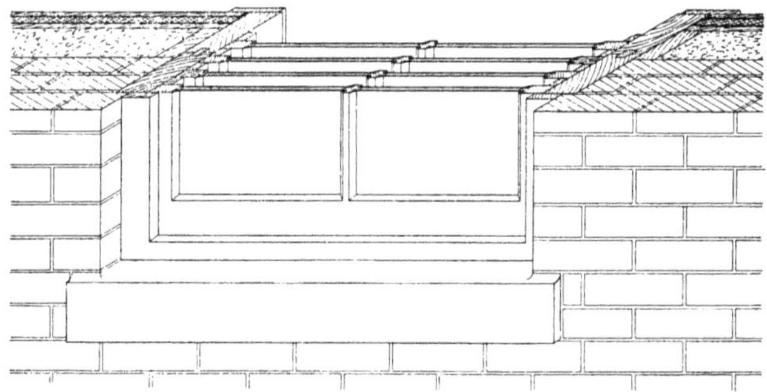

FIG. 1—AN APPROVED METHOD OF CONSTRUCTION AND SETTING OF COLD STORAGE WINDOW.

If careful and experienced workmen are employed in the construction and setting of the frames and in the setting of the sash in the frames as above directed, and if the insulation is carefully fitted around the frame, a job which is not quite nearly impervious to air, but which is absolutely air tight may be obtained, and one of the most important objections to windows as used in cold storage rooms is thereby eliminated. Another most important objection is the loss of refrigeration on account of heat transmission through the glass, and through the air spaces formed thereby. This objection can be partially eliminated by using a sufficient number of sash. No less than four double-glazed sash should be used for temperatures of about 30° F., and five or six in the better class of work or for lower temperatures. Use as large panes of glass as is practicable, to avoid the shadow and obstruction of light caused by a multiplicity of muntins. It is not practicable to put in sufficient sash to make the insulation equal to that of the balance of a well insulated wall, but if the above directions are carefully followed and a sufficient number of sash used, the loss of refrigeration through the comparatively small area of the windows ordinarily used may be reduced to a small amount. The isometric drawing herewith shows a detail of construction and relation of parts to each other.

What is said above in reference to the setting of windows in cold storage rooms applies, of course, equally well to the setting of windows in cold storage doors, which is often a very desirable thing to do. If the door opens, however, from a refrigerated corridor or air-lock into the cold storage room proper, a large number of sash or thicknesses of glass are not necessary, and two sash or four glass will usually be ample.

If windows are wanted which may be opened they should be ordered from the makers of cold storage doors described above. A window for cold storage purposes, which will open, is practically a door with glass in it and is made in essentially the same way.

CHAPTER VII.

AIR CIRCULATION.

IMPORTANCE OF PROPER AIR CIRCULATION.

A circulation of air is necessary to produce the best possible conditions in a cold storage room, and this necessity is now realized by the most progressive people engaged in the business. Considerable controversy has taken place between those who advocate the cooling of rooms by piping placed directly in the room, and those who have adopted some form of fan or forced circulation in which the pipes are placed in a coil room or entirely outside the storage room, and the air distributed through the room by means of air ducts. The people who have been the longest in the business do not like to believe that any improvement can be made on placing the pipes in the room, and insist that they can turn out as good stock as their more progressive competitors who use some form of forced circulation. To substantiate this argument, they refer to So-and-so who tried fans and had to put pipes in the rooms to hold his temperature, and claim that the results from the forced circulation system are no better than from the old methods of gravity air circulation. This argument is not sound, and it is proposed in this chapter to show clearly why a circulation of air is necessary, and also why a positive circulation, by means of fans, with a proper system of air distribution, is better than direct piped rooms, or any circulating system which depends on a difference of temperature in the air in different parts of the room for its operation.

Notwithstanding the attention which this subject has attracted, and the resulting discussion, there is yet much which is but imperfectly understood, such as the confusing of the terms, "air circulation" and "ventilation." The two are as distinct as can be, and it should be borne in mind to begin with that ventilation is what the name implies—the introducing of fresh air from an outside source for the purpose of purifying the room. Circulation refers only to the movement of air within the room, and in no case should the term, "ventilation," be applied to this subject in connection with refrigeration. Ventilation is mentioned only in explaining the difference between the two, and is not under consideration here, but is taken up in a separate chapter. Our present subject for discussion is air circulation in refrigerated rooms—the same air over and over—and has no connection with the supplying of outside air. To the end that the misunderstood features of the subject may be cleared up somewhat, the history and underlying principles of refrigeration and air cooling will be taken up, to show as clearly as possible the gradual development of the industry leading to the systems and methods of cooling now in use. The advantages of a forced circulation of air in cold storage rooms will be so plainly demonstrated that any thinking man must acknowledge them.

HISTORICAL.

The most primitive form of cold storage consists in employing the comparatively low temperature to be obtained in cellars or caves for the keeping of products subject to rapid decomposition. In this way they are protected from the extreme heat of summer, and to this extent preserved by a natural source of refrigeration. In this crude form of cold storage, air circulation was unknown, and if any existed it was by accident. Articles placed in a cellar or cave are cooled by radiation or conduction from the earth altogether, and not by a circulation of air. After caves and cellars, natural ice was employed for cooling purposes, and came quickly into general use, for the reason that lower temperatures and a dryer air were to be obtained. For cooling purposes, ice was first stored in underground pits dug in the earth, with the idea that the melting of the ice would be retarded. Goods for preservation were placed on or within the mass of ice. This was an improvement over the use of cellars in the matter of temperature only. Even after the ice house was placed above ground and provided with insulated walls, the favorite method was to build a room within the ice house, and surrounded on three sides by the ice, for the storage of goods to be preserved. Circulation of consequence did not exist, and goods placed therein quickly deteriorated, caused by a growth of mold and a musty condition of the air, induced by a very moist atmosphere.

A bit of personal experience will serve to illustrate some of the early phases of ice cold storage. About the year 1875

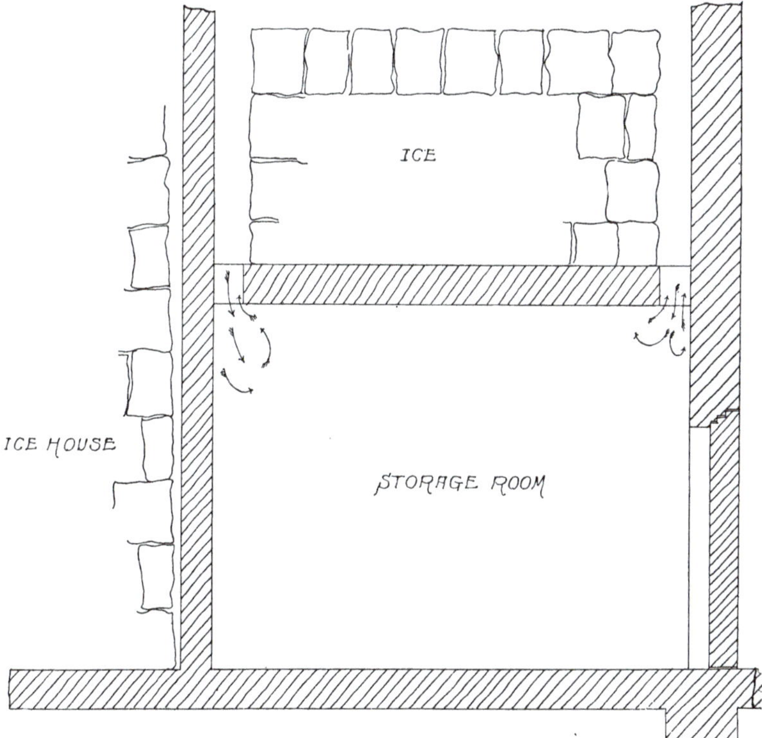

FIG. 1.—DIAGRAM IMPROPERLY CONSTRUCTED ICE COLD STORE.

the author's father constructed a large ice house adjoining a cheese factory and creamery. In one corner of the ice house, and opening into the creamery, was built a fair sized room for the storage of butter. The ice was placed on top of this room and also against two of its sides. Openings were provided at the top for the cold air from the ice house to come into the room, but no circulation of consequence took place, because the laws governing air circulation were not given proper attention. A large part of the cooling in the room was by direct conduction through its walls. The room carried fairly cold, at about 37° F. A large block of fine creamery butter was stored in the room for about three months. When removed, the tubs were very moldy, and the butter as well; the butter, even during the short time stored, being decidedly injured in flavor. This room was very damp, the ceiling and walls showing very wet, and moldy to some extent. In the light of present experience, this method of storing butter seems absurd, and it is mentioned simply to illustrate how a lack of circulation and some

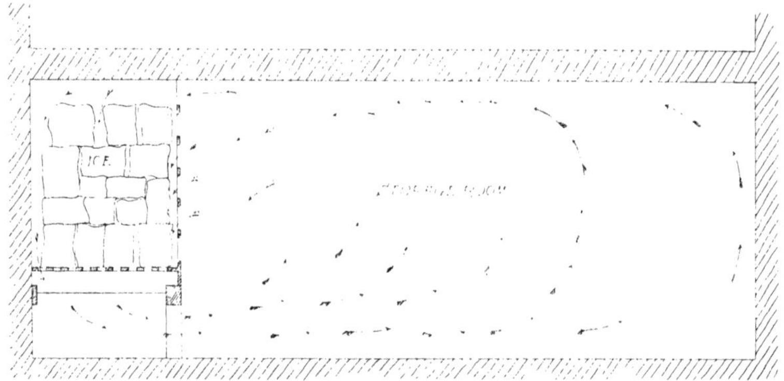

FIG. 2.—FULL ICE RACK WITH GOOD AIR CIRCULATION.

means of absorbing the moisture will cause bad symptoms in a cold storage room in a comparatively short time. Fig. 1 illustrates the construction of this room, in the corner of the ice house. It will be noted that no flues were provided to conduct the warm air to the top of the ice house, and the cold air toward the bottom of the storage room. Openings from the ice chamber only were provided, and this will not promote a circulation of air except under accidental conditions.

Shortly after the above related experience, a large room in the basement of the stone store building was fitted up for the storage of cheese. This was built on the side icing plan, the ice being placed in a rack or crib along one side of the room, which was about twenty-five feet wide. The room was insulated by studding and sheathing against the walls, and filling behind with sawdust. It was surprising to see the ice disappear, and the temperature could not be held below an average of 45° F. This room was superior in one respect, however, to the butter storage room just described. It had a fairly strong circulation of air as long as the ice rack was kept full, and cheese came out in fair condition, though moldy, after a three or four months' carry. A serious drawback to the successful working of the room was that when the ice was partly melted in the ice rack, the top of the room would become much warmer than near the floor. This was especially noticeable

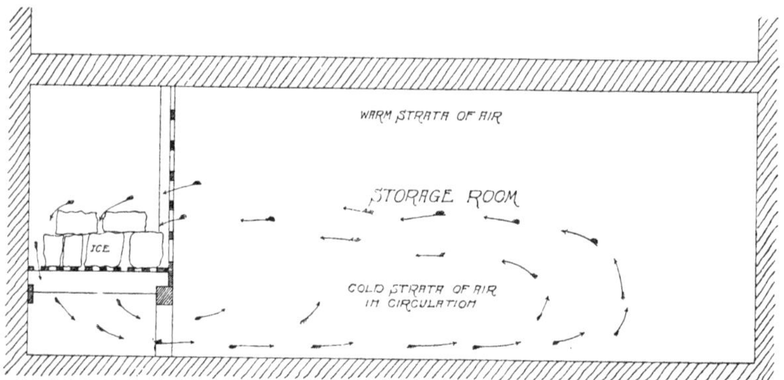

FIG. 3.—SHOWING SLUGGISH AIR CIRCULATION.

during warm weather. When the ice rack was full this condition was greatly improved, but when the ice was much reduced, the air at the top of the room became warm and dead. Fig. 2 illustrates a full ice rack and a comparatively perfect circulation of air to the top of the room. Fig. 3 shows a sluggish circulation, with a dead stratum of warm air at the top of the room, resulting from the small quantity, and location of ice in the rack.

As a natural improvement on the side icing plan mentioned above a structure two stories high was constructed, with ice at the top and storage space below. The ordinary domestic refrigerators are mostly built on about this plan, and this idea has been developed to the fullest possible extent. Many patents have been granted to inventors for improvements in details of construction and the promoting and control of circulation in cold storage rooms with overhead ice. Fig. 4 shows why overhead ice produces a good circulation, if properly designed, with up and down flues. Prominent among the old

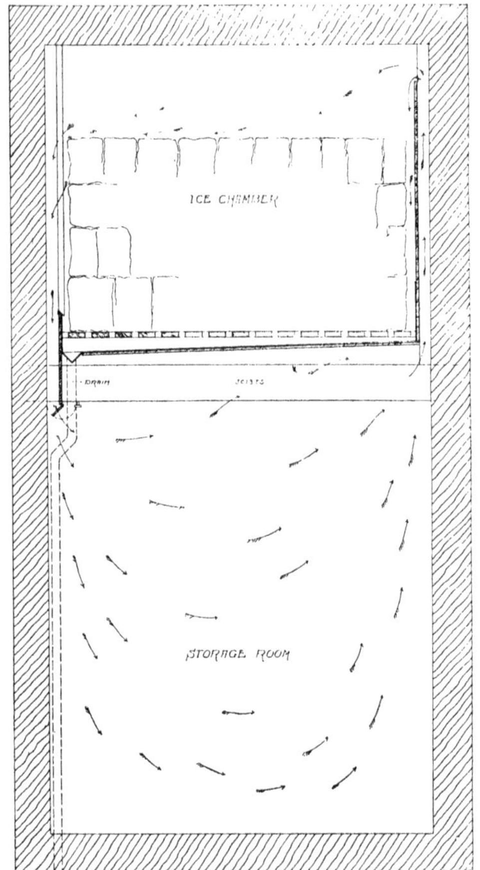

FIG. 4.—OVERHEAD ICE WITH GOOD AIR CIRCULATION.

overhead ice systems are the Jackson, Stevens, McCray, Dexter, Nyce and Fisher. These systems, as compared with any method of end or side icing, are markedly superior, and many of these old houses are still in service. Any system using natural ice only as a cooling agent is now considered obsolete, when compared with the present day methods of air cooling by means of chilled pipe surfaces in the form of brine or ammonia piping, but in the early days of cold storage these old systems were very satisfactory. Circulation of air may be mentioned as the keynote of whatever success was attained by the overhead ice systems. So much for the value of a circulation of air in any room cooled by ice. It has been proved in practice that a circulation of air is necessary in such a room. It is equally true of a room cooled by metal surfaces through which a refrigerant at a low temperature circulates.

CIRCULATION PURIFIES THE AIR.

A penetrating and fairly strong circulation of air is absolutely necessary in cold storage rooms because it is a part of the process which purifies the air. Nearly all goods which are ordinarily placed in cold storage for the purpose of retarding decomposition give off moisture. Along with the moisture given off are impurities in the form of finely divided decomposed matter from the surface of the goods. Gases resulting from surface decomposition, and the ripening of the goods in some cases, are also present. Besides the moisture given off by the goods, other moisture is continually finding its way into cold storage rooms by the opening of the doors, leakage through the insulation, and from the lungs of persons present in the rooms, all of which contains a greater or less percentage of impurities. These last sources are small in comparison with the amount of moisture and impurities given off by the stored goods, but, nevertheless, are quite large in some cases, and worth considering. To prove beyond a question that goods give off large quantities of moisture and impurities, it may be well to consider what would be the result should the moisture and impurities be allowed to accumulate in the storage room. Let us assume an absolutely tight room, cooled from an outside source without exposed pipe surfaces or other means of taking up the moisture and impurities which are contained in the air of the room, say a room within another room, the outside room being cooled, and taking up all heat from the inside room. An experiment conducted by the author, described in chapter on

"Eggs in Cold Storage," under the heading of "Packages," illustrates fully the necessity of taking up moisture as given off by the stored goods. These experiments demonstrate conclusively what would result if goods were placed in a refrigerated room which did not contain means for absorbing the moisture and impurities that are given off by the stored goods. It is imperative that the moisture be continually removed from a cold storage room containing moisture-giving goods.

The relation between moisture and impurities in cold storage rooms is very close, as these elements are united to a large extent. It is a well known fact that water has a great affinity for impurities of various kinds. The same is true of water in the form of vapor or moisture in the air of cold storage rooms, which has a great attraction for the gases and impurities which are given off by the stored goods. In fact, it is probable that the greater part of the impurities never part company with the moisture when they are both exhaled by the goods. It is, then, easy to understand that a room which has means of absorbing moisture also has means of purifying the air, and that the air is purified to a large extent in proportion to the thoroughness with which it is circulated and brought in contact with the means for absorbing moisture. It must not, however, be understood that the air of a cold storage room is absolutely purified by having the moisture removed. There are gases which have little or no affinity for moisture which cannot be disposed of in this way. Fresh air must be supplied to maintain perfect conditions in cold storage rooms where goods are stored for long periods. (See chapter on "Ventilation.") If a cold storage were perfectly purified by the removal of moisture there would be no odors of consequence present in such a room. How many cold storage rooms has the reader ever seen that were free from noticeable odors?

Probably the worst form of impurity which is met with in cold storage rooms is the germs which produce a growth of fungus, or mold. These germs are no doubt present in the atmospheric air everywhere. Their presence is manifested only under certain favorable conditions of moisture and temperature. Conditions of excessive moisture in the presence of decaying animal or vegetable matter, together with a moderate degree of heat, are favorable for a very rapid growth of fungus. It is a well known fact that in the dry mountain districts of California or Colorado freshly killed meat may be hung in the open air without decomposition. The air contains so little moisture that the germs will not propagate. Fresh meat exposed in the same way in the moist, tropical climate of Florida or Cuba would be quickly decomposed so as to be unfit for food. Germs of mold and decay flourish in a warm, moist atmosphere, but quickly succumb where it is dry and cool. As the moisture is absorbed and removed from the air of a cold storage room, with it are largely removed the germs and other impurities. Low temperature pipe surfaces freeze the moisture from the air, and in this way a large portion of the impurities is disposed of. It may already have occurred to the reader to ask what all this has to do with air circulation in cold storage rooms. We have discovered that a room may be cooled from an outside source and still be an unfit place for goods when no means of taking up the moisture are present. Even should the pipes be placed directly in the room, the results would be bad unless there is a circulation of air. A circulation of air is absolutely essential to a perfect cold storage room, because the air must be continually moving in contact with the pipe surfaces or other means of absorbing moisture. The question of what means are the best for removing the moisture from a storage room is not under discussion. Our problem is to ascertain the best means for circulating the air in contact with the means for absorbing the moisture.

METHODS OF PIPING THAT HINDER CIRCULATION.

When mechanical refrigeration first came into the field, the arrangement of cooling surfaces and a provision for air circulation was neglected about as it was by the pioneers in natural ice refrigeration. The cooling pipes were placed almost anywhere, regardless of the laws of gravity which control air circulation. At first the ceiling of the room was a favorite place for locating the coils of pipe for cooling the room. The ceiling was utilized because thus the pipes were out of the way in piling up goods, and also on the theory that "cold would naturally drop." Cold, or, more accurately speaking, cold air, will naturally drop, but placing the pipes on the ceiling of a room will not assist the circulation; it will, in fact, produce practically no circulation at all if the whole ceiling of the room is covered with pipes uniformly. Ceiling pipes have generally been abandoned for the more rational method of placing the pipes on the side walls of the room. Fig. 5 shows ceiling piping, and should make plain why no circulation is created when the pipes cover nearly the whole top of the room.

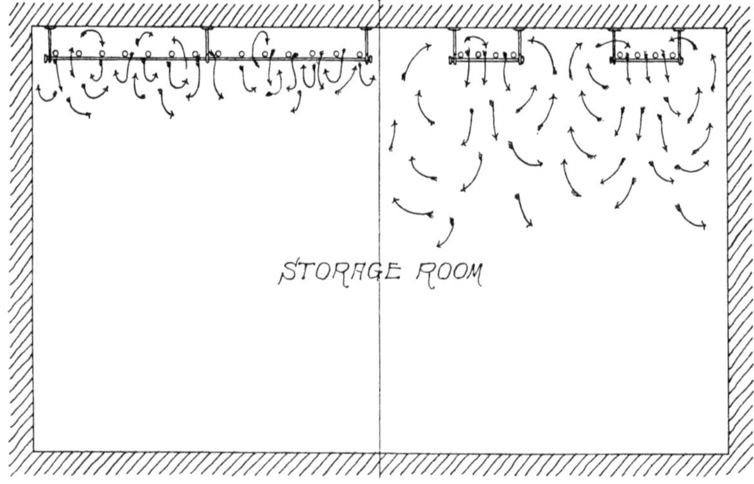

FIG. 5.—SHOWING CEILING PIPE WITH IMPERFECT AIR CIRCULATION.

The left half of the diagram shows the pipes covering the entire ceiling, the right half in two sections. Note the arrows showing the resulting circulation in each case. As is well known, cold air is heavier than warm air and, if free to move, the cold air will seek a lower level than the warm air. This movement of the cold air downward and the warm air upward is what is known as gravity air circulation. A slight difference in the temperature will cause a circulation of air if the warm and cold air are separated from each other and not allowed to mix, which would cause counter-currents and retard the circulation. In a cold storage room, the air in contact with the cooling coils, as it is cooled, flows downward toward the floor by reason of its greater specific gravity. The comparatively warm air above is drawn down to the pipes, where it is in turn cooled, and the flow is continuous. If the entire ceiling is covered with pipes, what results? The air in contact with

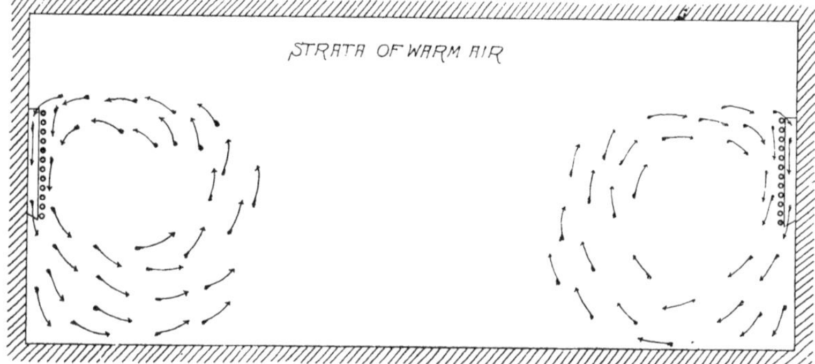

FIG. 6.—SHOWING SIDE WALL PIPING.

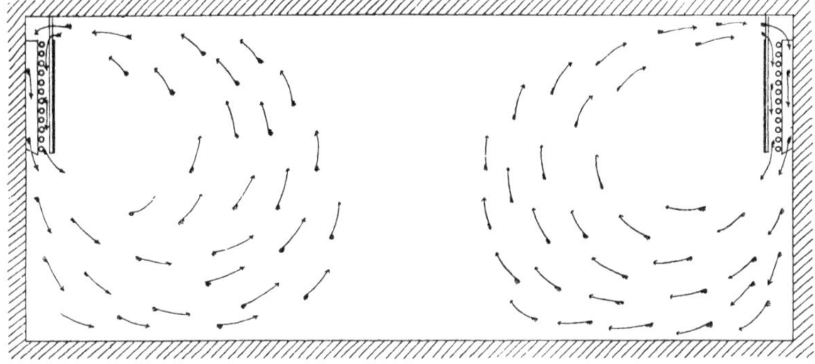

FIG. 7.—AIR CIRCULATION WITH DIRECT PIPING.

the pipes cannot fall because it cannot be replaced by warm air from above. The result is that practically no circulation of air takes place in such a room. A slight local circulation in the vicinity of the pipes is all that results, except under unusual or accidental conditions. The goods are cooled for the most part by direct conduction and radiation; the top tier of goods would be cooled directly from the pipes and each tier under successively from its neighbor above in the same manner. Goods are cooled by radiation by the passage of heat from the goods directly to some colder object without the heat being conveyed by the movement of the air, as it should be, and as it is where a good circulation is present in the room. In a room in which the goods are cooled by radiation mostly, the moisture instead of being deposited entirely on the cooling pipes, as it should be, is also likely to be deposited on the walls or ceiling of the room, or on the goods themselves. The result of such a condition may be serious. This cooling by radiation, as compared with cooling by a circulation of air, may seem like a very finely spun theory to some, but let the skeptic watch his house for a demonstration. Is there any practical cold storage man now in the business who has not noticed an accumulation of frost or moisture on goods if they were piled too near to the exposed cooling pipes? What causes this result? Radiation—nothing else.

METHODS OF ASSISTING GRAVITY CIRCULATION.

The bad effects of radiation cannot be altogether overcome by placing the pipes on the sides of the room, but it is counteracted to some extent by the resulting circulation of air. Fig. 6 shows side wall piping and the resulting circulation, which is confined largely to a small space near the coils. The arrows show approximately the path of circulation. If the room is wide, no circulation at all will take place near the center. In some cases pipes have been carelessly placed two or three feet down from the ceiling, as shown in the illustration. This results in the air of the room becoming stratified—a warm layer of air in the top of the room resting on a cold layer beneath. Figs. 2 and 3 illustrate this clearly. This may be operative to such an extent as to cause a difference in temperature between floor and ceiling as great as 10° F. A case has come to the author's notice with exactly these conditions. Another bad arrangement of side wall piping was that of a room more than fifty feet square piped completely around on the side walls from floor to ceiling, with the exception of the doors. No circulation could penetrate to the center of such a room, and conditions were very poor, in consequence.

The placing of a screen in front of the side wall piping, hung well up toward the ceiling of the room, as illustrated in Fig. 7, marks the first scientific step toward a betterment of air circulation in a room with direct piping. It prevents the action of radiation, and assists the volume, velocity and area of circulation, but does not well take care of the center of the room, although the increased velocity forces the air to cover a greater area and flow to a greater distance from the coils. The screen or apron should be of wood or any moderately good non-conductor. By separating the warm from the cold currents of air, the velocity is increased on the same principle that a fire burning in a flue creates a greater draft than when burning in the open air. Radiation is prevented in the same

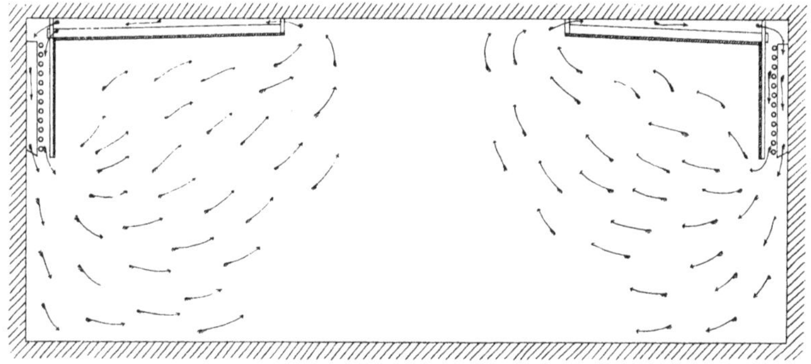

FIG. 8.—SAME AS FIG. 7 WITH FALSE CEILING.

way that a fire screen protects one from a too hot fire in a grate, only the radiation, as already explained, is in a reverse direction.

In Fig. 8 the same arrangement of apron is shown as in Fig. 7, but added thereto is the false ceiling extending out toward the center of the room. This addition to the perpendicular apron causes the air, after circulating over the coils, to spread out more toward the center of the room and cover the cross-sectional area much more uniformly. While it decreases the velocity proportionately, it is considered a superior arrangement to the perpendicular apron alone, placed in front of the coil. The false ceiling should have a slant of about one foot in twenty, and the opening on the outer edge near center of room need not be over four or five inches in depth in most cases. Without the false ceiling some space must be left for a circulation of air at the top of the room; with it, the goods may be piled close up to the false ceiling, so no space of consequence is wasted in using it.

The arrangement shown in Fig. 9 was first originated by Mr. C. M. Gay, and was described in the August, 1897, issue of *Ice and Refrigeration.* Barring the space occupied, it is by far the best arrangement of room piping now in use. The following is quoted from Mr. Gay's description: "Upper pipes of box coils should be about ten inches below ceiling of the room, to prevent sweating. (Sweating in such a case is caused by

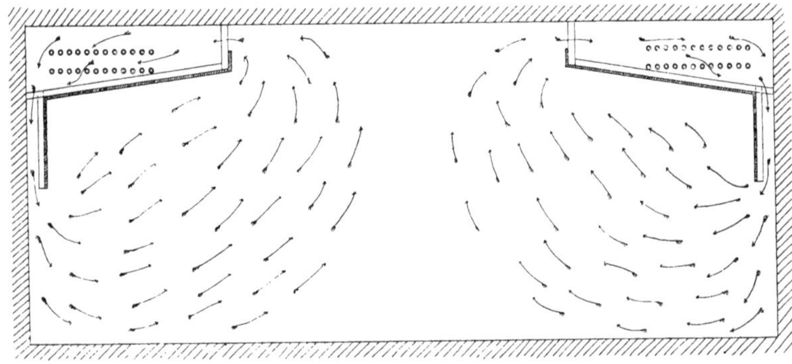

FIG. 9.—MR. GAY'S ARRANGEMENT OF ROOM PIPING.

radiation, as already explained.) When brine or ammonia is turned into these pipes the cold air around the pipes seeks an outlet downward, and passes between the false partition and the side wall of the room, thus displacing or pushing along the air in center of room, the cold air naturally seeking the lowest point and the warm air the highest point, each by reason of its relative gravity. Thus, as the cold air falls from the cooling surfaces it is replaced by the warm air from highest point in center of room. This secures a natural circulation and a dry room, there being no counter-currents nor tendency to precipitate moisture on walls or ceiling." Mr. Gay's remarks regarding his system apply with still greater force to the St. Clair system, and to a greater or lesser extent to any system which provides for a removal of the cooling pipes from the room.

The St. Clair system, illustrated in Fig. 10, is sometimes called the pipe loft system, because the cooling coils are placed above the storage room in a pipe loft or coil room. This is a favorite arrangement where an overhead ice cold storage house is remodeled and equipped with the mechanical system. In this case the pipes are placed in a portion of the old ice room, and perhaps the old air ducts used for air circulation. If the

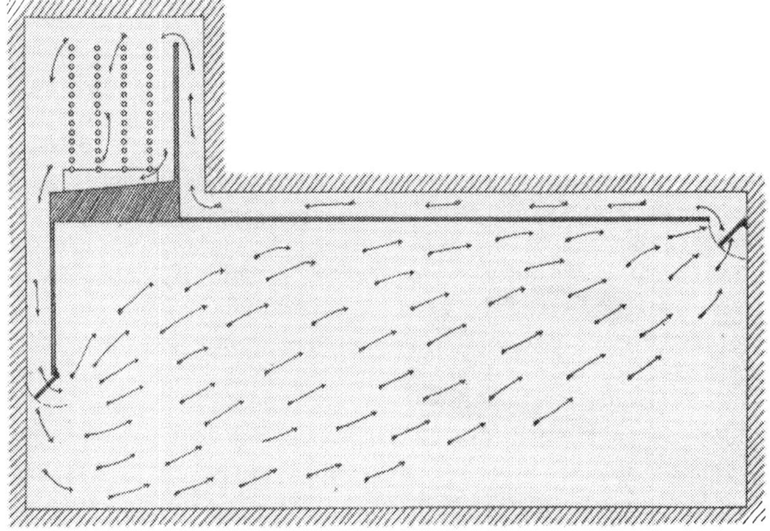

FIG. 10.—THE ST. CLAIR PIPE LOFT SYSTEM.

storage house consists of several floors of storage the pipe loft may be placed at the top and the rooms below all cooled from one pipe loft, but a much better method is to have an independent coil room for each room, and circulate the air through separate air ducts. This prevents contamination from foreign odors when different products are stored in different rooms. The circulation is more vigorous and effective with the St. Clair system than with any pipe-in-the-room system, depending on the law that the higher the column of air the stronger the draft, in the same manner that a tall chimney gives a stronger draft than a short one. The effect of this is to produce a good circulation of air with a comparatively small variation of temperature. The St. Clair system is also better because by suitable trap doors on the air ducts, the pipes may be shut off from the room, when the temperature is such outside as not to require the circulating of the refrigerant. The necessity of keeping the air of a storage room from contact with the frosted pipes when the refrigerant is shut off will be considered in connection with the forced or fan circulation system, to be described further on.

ARGUMENTS FOR IMPROVED SYSTEMS OF AIR CIRCULATION.

We have seen how rooms for the storage of perishable products are cooled by natural or gravity circulation or by direct radiation. Reasons have been given why each succeeding method was superior to the former one. It is very easy to see that where a room is cooled by direct piping, or by any system of gravity air circulation, the goods within such a room cannot all be exposed to the same conditions. Goods piled at the floor and near coils where the air circulates direct from coils are certainly exposed to a much colder air and stronger circulation than those farthest from coils and near the ceiling of room. Gravity air circulation, as its name indicates, depends on a difference in weight, and therefore a difference in temperature of the air in different parts of the room, for its existence, and there must, therefore, be varying temperatures in different parts of the rooms. The difference in temperature will range generally from 2° to 5° F., with the best arrangements here described. The greater the difference the stronger the circulation, usually. With a difference in the temperature of the air in different parts of the room goes a variation of other conditions; especially as to dryness and purity of the air.

Many cold storage warehouses, equipped in many different ways, even some of them cooled by natural ice, are producing results satisfactory to their owners; to use a familiar phrase, "are having good results." This is not at all surprising, when it is considered that a result which is satisfactory to one man would not be satisfactory to another; but it is very confusing to an interested person who undertakes to investigate the various cold storage houses of his acquaintance, with a view to ascertaining which system is best suited to his needs. The variety of opinion expressed depends largely on the individual prejudice of the person giving the opinion. The investigator, if not fairly well posted on the subject himself, usually is so confused that he takes the advice of his most intimate acquaintance, and adopts some old time system which has been found reliable. This means, in a majority of cases, that he is adopting some out-of-date ideas for a new house, which should embody all the latest improvements. Should the investigator be a fair minded man and well informed on the subject, new improvements, with logic and practical results behind them, are adopted, after due consideration. Results are, of course, the final test, but it is very necessary that a person should have actual and not fancied results, and unless new ideas for improvement are investigated and adopted, cold storage men will get "behind the procession," the same as in other mechanical and scientific lines. When a new system or device can show results equal to or better than the older ones, costs no more to install and operate, and, further, is based on scientific principles and common sense, that system is the one to adopt. It will surely demonstrate its superiority in the long run. There are many in the business who still think that direct expansion piping placed directly in the room is the acme of perfection and cannot be improved upon. Argument for improved systems in such a case is useless.

A comparison of the methods of heating our best public buildings in former years, with those in use at the present time, will show us the past and present, or rather the past and future of cooling the best cold storage houses. In years gone by, the best and most costly structures were heated directly by stoves, later by hot-air furnaces, and lastly by the indirect or fan system. A stove for heating a room may be compared with direct piping for the cooling of a storage room. We all know the disadvantages of a stove for house heating—too much direct radiation, and a poor distribution of heat. The same may be said of a room cooled by direct piping, only it is the refrigeration that is poorly distributed. Cooling a room by the pipe loft system is about the same as heating a room with a furnace, with the disadvantages common to both. The advantages of handling the air of a cold storage room by means of a fan are likewise comparable with the advantages to be had from a well designed forced system of heating. The best heating work is now done by means of fans, and the best cold storage work of the future will be done by means of fans. To prove the advantages of the fan system of heating, it is not necessary that people should suffocate and die in a building heated by stoves or furnaces; neither is the fact that goods do not completely spoil or decay rapidly in a room cooled by direct piping any evidence that the fan or forced circulation system is not superior by far to the pipe-in-the-room or any gravity method. Unquestionably the fan system of heating gives a control of temperature, humidity and purity of air, not obtained in any other way. The forced circulation system of cooling also gives a control of temperature, humidity and purity of air in a cold storage room, not to be had otherwise.

PROS AND CONS OF FORCED CIRCULATION.

The chief, and in fact the only objection known to have been urged against forced circulation for cold storage rooms is a fancied notion that it will lead to a greater drying out or shrinkage in weight of goods which are placed in storage for preservation than if a system of gravity air circulation or pipes in the rooms were used. The author has searched long and faithfully for the origin of this old tradition, but has never been able to discover that it was founded on fact. At least none of the most modern houses employing the fan system, so far as known, have ever had complaints from excessive evaporation. The worst shrunken goods which ever came to the author's notice were some eggs from a house cooled by direct expansion piping placed directly in the room. It is probable that the claim that goods evaporate or lose weight more in a room cooled by the fan system is wholly a matter of theory, based, no doubt, on the assumption that the air is circulated at a much higher velocity. It is well known that a movement of the air aids evaporation. Every intelligent housewife knows that linen hung in the open air to dry will be freed of moisture quicker when a moderately strong breeze is blowing than when the air is still. The same principle applies to the goods stored in a refrigerated room, but evaporation from the goods in storage is dependent not only on the movement of air in the room, but on the humidity or dryness as well. If the humidity is properly regulated no harm will result from a very thorough circulation of air, even at a brisk speed. It may have happened in the early days of fan circulation, that the air was rapidly circulated with little or no distribution, and the goods exposed directly to the blast of air where it was blown into the room were excessively evaporated; but in the numerous houses designed by the author, and using one or the other of the two systems of air circulation described further on, no such trouble has been experienced. If the humidity of the air is at the correct point, and the circulation of air well distributed throughout the room, and not too strong, no excessive or damaging evaporation will occur, and where trouble from this cause has been experienced it will be found in every case that no systematic control of humidity has been attempted. It is as easy to control humidity as it is to control temperature, if proper means are provided, and we go about it in the right way. Absorbents and ventilation are both useful for this purpose, but this feature of cold storage is not under consideration here, and is treated on elsewhere under the proper heading.

With a positive and well distributed circulation of air, a storage room may be maintained at a humidity which would be dangerous if only a sluggish gravity circulation of air were in operation. A brisk movement of air in all parts of the room quickly removes the moisture and impurities from the vicinity of the goods, and carries them to the cooling coils, where they are, for the most part, condensed or frozen on the pipe surfaces. This should explain how goods may be carried in good condition and with very little shrinkage in a room where a well designed system of forced circulation is employed.

Three of the houses designed by the author are used exclusively for the storage of cheese. It is well known that cheese loses weight very rapidly in cold storage, and the problem heretofore has been to carry the cheese reasonably free from mold, and with as little evaporation as possible. Cheese has been stored in the houses referred to for three months, with very little mold, and with no shrinkage from marked weights, and the proprietors assert that there is less shrinkage, even on "long-carry" goods, than there was with the overhead ice system which they formerly had in service. This is a sufficient proof of the value of forced circulation for the cold storage of cheese. The same applies equally to other classes of goods. With a room equipped with any of the gravity systems of air circulation, already described, the circulation of air cannot be regulated, because it depends on the temperature of the refrigerant (generally brine or ammonia) circulating through the pipe coils. As the temperature of the refrigerant is regulated to suit outside weather conditions (lower in warm weather, and higher in cold weather), the velocity of air circulation is not constant, being least in the cold weather of fall and winter, when most needed. With a good system of forced circulation installed, the chief problem of the cold storage man is to employ a proper degree of humidity. (See chapter on "Humidity.") Our discussion now brings us to a consideration of the various methods of mechanical air circulation in use. The weak as well as the strong points of the various systems which have been put in operation will be considered, regardless of where or by whom originated.

UNDESIRABLE FORCED CIRCULATION.

The simplest, and probably the most unscientific, form of mechanical air circulation in cold storage rooms is the small electric fan. These fans are usually of the four or six-bladed disk type, of from twelve to eighteen inches in diameter, attached directly to the shaft of a 1/8 or 1/4-horse power electric motor. The electric current for operating is usually obtained from the socket for an incandescent electric lamp. Electric fans are usually placed on the floor in the end of an alleyway, or in an opening in the piled goods, and are used for creating a flow of air from one extremity of the room toward the other. If the circulation is strong enough, these fans tend to create a uniform temperature in the room; but, as the air from the fan will follow a path of least resistance, the circulation resulting from their use is largely confined to the alleyways and openings in the piles of stored goods—it does not penetrate through and behind the goods where it would be most useful. The use of this type of fan in cold storage rooms is of doubtful utility, and is liable at times to lead to a positive harm by causing a condensation of moisture on the goods in storage, as a result of the warm upper stratum of air coming in contact with the cold goods near the floor of the room. In some cases electric fans have been used to propel the air from the cooling pipes, for which purpose they are placed in an opening in a screen or mantle covering the pipes, forcing the cooled air outwardly into the room. This is a first step toward scientific forced circulation, and is useful as far as it goes. In many cases the electric fan is useful only as a "talking point," as it is likely to impress a person, who is not familiar with cold storage work, with the cooling power of the refrigerating apparatus, to stand for a few seconds in the breeze created by one of these high-speed fans. Their use has been adopted to an extent not at all warranted by the results to be obtained, and they will no doubt be gradually discontinued as the fallacy of the idea becomes apparent. Those who use electric fans as above described, by so doing admit the superiority of forced circulation over the gravity system, and also admit that their rooms are in bad condition, and that some mechanical means of agitating or circulating the air is necessary. Instead of such a poor makeshift it seems that they will eventually come to the idea of installing a scientific system of forced circulation.

Having proved a circulation necessary, it is evident that a method which will cause the circulation to be continuous, and at the same velocity, regardless of outside weather conditions, etc., must be better than depending on natural circulation, which varies greatly with the varying conditions and appliances which produce circulation as we have already seen. It follows further, then, that the system which will produce a circulation which is continuous, and at the same velocity, and besides is uniformly distributed to all parts of the room, must be the most nearly perfect way of handling the air for cold storage rooms. Any of the methods of gravity air circulation in which the pipes are placed in the room or otherwise, as shown in Figs. 5 to 10, are dependent on the outside weather conditions, temperature of room, temperature of refrigerant in pipes, length of time goods have been in storage, etc., for their operation. A continuous and uniform air circulation

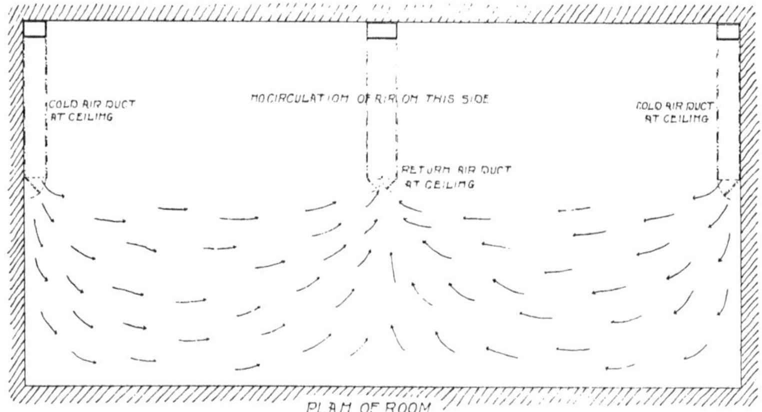

FIG. 11.—PRIMITIVE FORM OF FORCED CIRCULATION.

can only be obtained by the adoption of some mechanical means, and is usually secured by the use of a fan of some kind.

VARIOUS FORMS OF FORCED CIRCULATION.

So far as known to the writer, the systems of forced circulation here described include all of the recognized equipments which have been installed in one or more prominent houses. The patent records show a large number of crude developments which have in most instances been abandoned without having been put into practical use. A system which has been installed in several large houses in the United States, and to some extent abroad, is what may be termed a primitive form of forced circulation. This term fully expresses just what the system is, as no method could be applied in a more crude way. It consists simply in placing the refrigerating pipes outside the storage room, and using a fan to propel the air to and from the room. Fig. 11 shows a floor plan of a room so equipped. The air is forced into the room at each end, and the return air to coil room drawn out in the center as shown. Cold air in this connection is spoken of as being the air from coil room to storage room, and warm air is mentioned as the air from storage room to coil room. These terms are used relatively only, and will be employed in the descriptions contained in this article. It should be understood that in actual practice the difference in temperature between the incoming and outgoing air is very small. In a well designed system this need not be over two or three degrees at the most. The cold air inlets at ends of room are in some cases placed near the floor and in others near the ceiling, but further than this no distribution of air is attempted other than that resulting from the location of the inlet and outlet. Sometimes the ducts are arranged to force the air into the room at the center, and the return air to the coil room is taken out at the ends, or the cold air is allowed to flow from the several openings in a duct running across the center of the room, but no adequate distribution results from this method.

Employing the forced circulation system in this way is very much like the indirect systems of steam heating as at first installed. It is noticeable now that the best steam heating work provides a thorough distribution of the heated air throughout the apartments through a great many small openings rather than forcing a large volume of air into the room at one or two places. It needs no argument or demonstration to show that a room heated or cooled by air forced in at one or two openings must have varying degrees of temperature, humidity and circulation depending on the remoteness or proximity to the direct flow of air from inlet to outlet, for the reason that the air from inlet always seeks the most direct path to the outlet and moves through the area of least resistance, usually through the center alley of room. This is a positive fact and not a theory. The author once visited a large room of the kind above described, and despite the manager's statement that he had tested in every known way and found conditions absolutely uniform, the author for himself saw a temperature variation of two degrees, and this between two thermometers hung in the center alley of room at the same height from floor, and without any extraordinary conditions to cause such a variation. As a matter of fact the real difference in temperature in this room between the coldest and

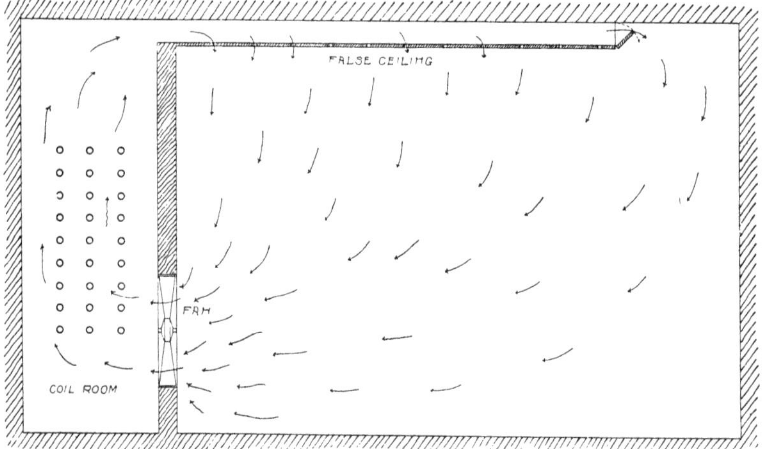

FIG. 12.—A SYSTEM OF FORCED AIR CIRCULATION.

warmest point could not have been less than five or six degrees.

The longitudinal section of a room shown in Fig. 12 illustrates a system of forced air circulation which has been installed to a moderate extent, but has not become as well established as the one first described. A false ceiling is provided for distributing the cold air from cooling coils at the top of the room, but as with the system just described, no collecting ducts are provided for the purpose of uniformly removing the air from the room. The air from coil room comes into the room through narrow, slit-like openings in the false ceiling, and is returned to the cooling coils through and by the disk fan located in the partition between coil room and storage room. It would seem that this is working counter to the natural laws of gravitation, although it may be looked at in another light also. It is often remarked that "cold will naturally drop," but this should not confuse us when studying the means for promoting circulation. If the cold air is admitted to the room at the top, it will of course fall to the floor if allowed to do so; but why admit the cold air at the top of the room if it is wanted at the floor? In a room fitted with direct piping the cold air does not *drop* through the goods in storage, but down over the cooling coils, and *rises* through the goods in

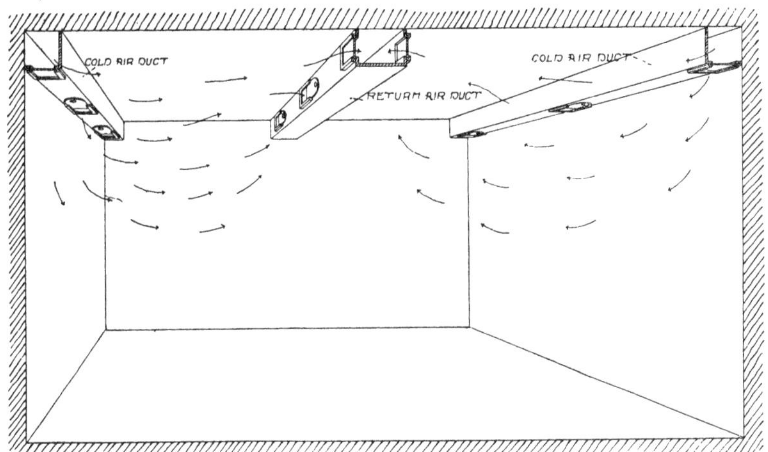

FIG. 13.—COLLECTING AND DISTRIBUTING AIR DUCTS.

storage as it is warmed. It would seem, then, that any method of distributing the cold air at the top of the room is wrong in principle, especially as no means of uniformly drawing off the air at the bottom of the room is provided. When warm goods are placed in a room equipped in this way, the moisture given off as the goods are cooled must be very liable to collect on the cold false ceiling. To provide uniform temperatures and humidity with this system it is necessary to provide a strong blast of air, which is to be avoided, as goods directly in front of the fan may be exposed to too great a drying influence.

The arrangment of collecting and distributing air ducts shown in the cross section of room, Fig. 13, has been installed in a number of houses in America, and, like some of the others, depends on the "cold will naturally drop" theory for its operation. The arrow shows the natural tendency of the air circulation from the cold air ducts on the sides of the room to the warm air collecting duct in the center. In some cases the cold air is distributed in the center and collected at the sides of the room, and where the room is narrow only two ducts are used, as in Fig. 14, a cold air distributing duct on one side of the room and a warm air collecting duct on the opposite side. In every case the ducts are placed at the

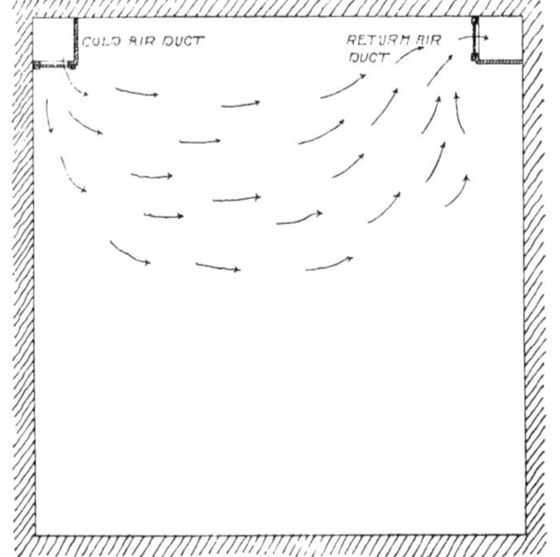

FIG. 14.—SMALL ROOM WITH TWO DUCTS.

ceiling, on the theory that the air from cold air duct will drop and distribute itself along the floor before being drawn back to the coil room through the return duct. The openings provided in the air ducts of this system are usually square openings, fitted with sliding gates to regulate the flow of air into the room and its return to cooler. These gates are placed five or six feet apart, consequently a good distribution of air is not provided, and goods exposed to the rapid flow of air directly in front of the openings will get a much greater volume of circulation than is to be found in any other part of the room. When a room of this kind is filled with goods, preventing the air from falling from the cold air duct to the floor, no circulation of consequence will be obtained near the floor, for the reason that air will travel through path of least resistance, almost directly from feeder duct to return duct, about as shown by the arrows.

A method somewhat similar to the one just described is that in which the cold air distributing ducts are placed at the floor and the warm air return duct is placed at the ceiling, as represented by the cross sections of rooms, Figs. 15 and 16. In narrow rooms only one distributing duct is used, as shown in Fig. 16. In wider rooms two distributing ducts on opposite

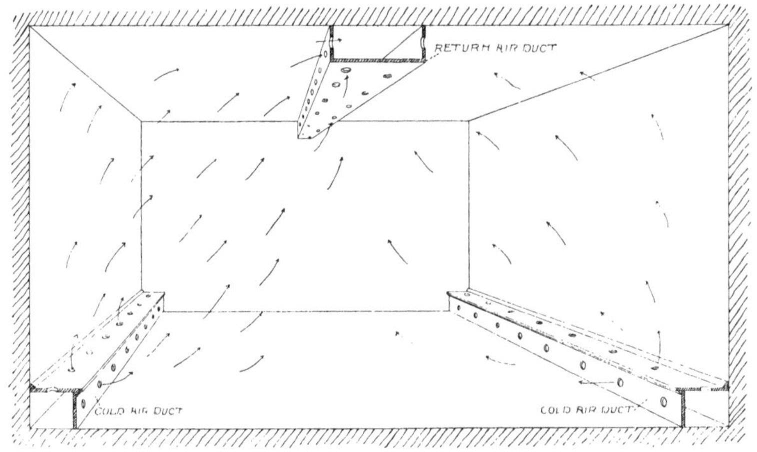

FIG. 15.—ARRANGEMENT OF WARM AND COLD AIR DUCTS.

sides of the room at the floor are used, and one collecting duct at ceiling in center of room. This arrangement has the merit of operating according to the laws of gravity, but still lacks the thorough distribution of cold air and collection of warm air, as shown in the system described further on. It is, however, considerable of an improvement on any of the preceding methods, and the author has demonstrated in actual service that it will produce fairly uniform circulation and temperatures with a comparatively gentle flow of air. This system is to be recommended for goods which do not give off much moisture. It is preferable to use numerous small holes rather than a few large openings in the supply and return ducts.

COOPER SYSTEMS OF AIR CIRCULATION.

The system shown in the cross section of room (Fig. 17) was developed by the author after some experiment and has since been improved by two successive steps, the details of which will be described. It was the old trouble of sluggish circulation, especially during the fall and winter, which impelled the author to experiment for its betterment. As an improvement over the small electric fan already mentioned, an

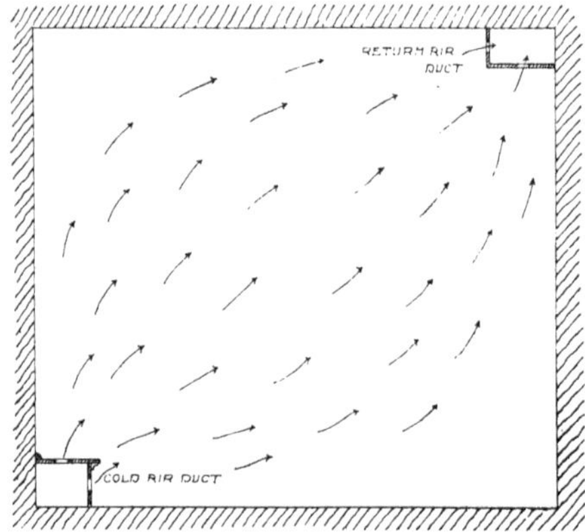

FIG. 16.—ARRANGEMENT OF WARM AND COLD AIR DUCTS.

exhaust fan was fitted up to take air from the cooling apparatus and deliver it to the rear end of the room through a perforated duct. The air was allowed to find its way back to the coils as best it could.

This method was applied to a long narrow room, and certainly was a decided improvement over the sluggish natural circulation which it superseded. Following this, the perforated false ceiling was applied, with distributing cold air ducts on the walls, as shown in Fig. 17. The cold air from coil room was forced into the side ducts and flowed into the room through a great number of small holes in the top, bottom and sides of the cold air ducts. The warm air from the room flowed upward through the small perforations in the false ceiling and through the space between the ceiling of the room and false ceiling and thence to the coil room, where the air was cooled, and caused to repeat the same circuit continuously. The first apparatus was clumsy and the proportions of the various parts not correct, but the efficacy of a forced circulation of air, and a thorough distribution and collection of the incoming and outgoing air of a cold storage room so plainly proven, that a further development of the idea was undertaken.

It was demonstrated by above described experiments that a comparatively small amount of air, well distributed and

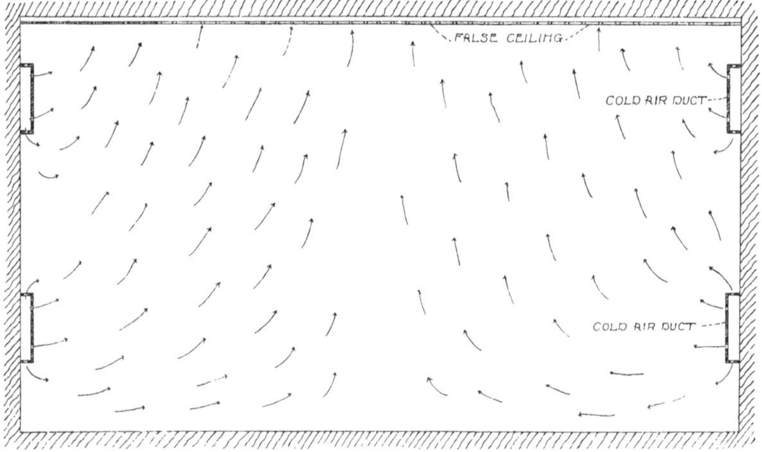

FIG. 17.—COOPER'S FIRST SYSTEM OF AIR CIRCULATION.

uniformly drawn off at the top of the room after flowing upward through the goods in storage, would produce very uniform conditions throughout the entire area of the room. Following up this information, the apparatus was reduced to a more practical form by substituting one broad duct near the floor, as in Fig. 18, for distributing the cold air, in place of the two distributing ducts as used in the apparatus shown in Fig. 17. The top duct of the two did not accomplish any result of consequence, and was considered objectionable, as the air passing from this duct to the false ceiling did not percolate through the goods to any considerable extent, and resulted, practically, in a loss of the work done by the air flowing from the top duct. Two ducts also made the apparatus more complicated. Using the broad single distributing duct near the floor in combination with the false ceiling resulted in a very penetrating and uniform circulation of air, and in practical service it has been found to produce superior results. No practical objections have been urged against it. As shown by the arrows, the air is caused to cover very uniformly the entire cross-sectional area of the room. This was accomplished by perforating the distributing ducts with small holes, and so proportioning them that a larger part of the flow of air is

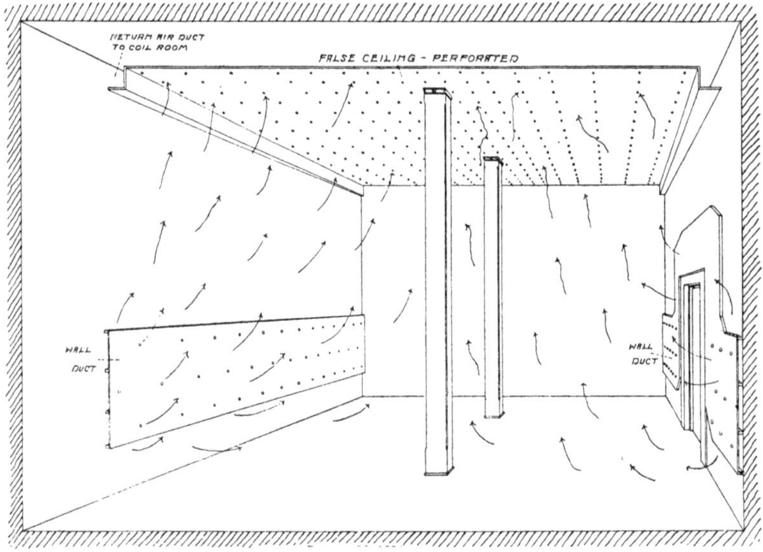

FIG. 18.—COOPER'S IMPROVED SYSTEM OF AIR CIRCULATION.

from the bottom of the ducts. The ducts are also perforated to some extent on sides and top. By piling the goods a few inches off the floor the air from bottom of ducts flows under the goods and out to center of room. This action is also assisted by having the greater number of the perforations in false ceiling in the middle third or quarter of the room, so as to draw the air out from sides of room. As indicated by the arrows, the air moves up from the distributing duct, is drawn into space above false ceiling, and returned to coil room to be cooled.

The system described in the foregoing paragraph is nearly theoretically perfect so far as a uniform circulation of air is concerned, and a more thorough method than any of its predecessors, but it still remained to design the perforated false floor and false ceiling combination (Fig. 19) to produce a system which cannot be improved upon theoretically. Not only is the system theoretically perfect, but its practical application is so simple as to be unobjectionable. As shown clearly by the sketch, the flow of air is directly upward from floor to ceiling,

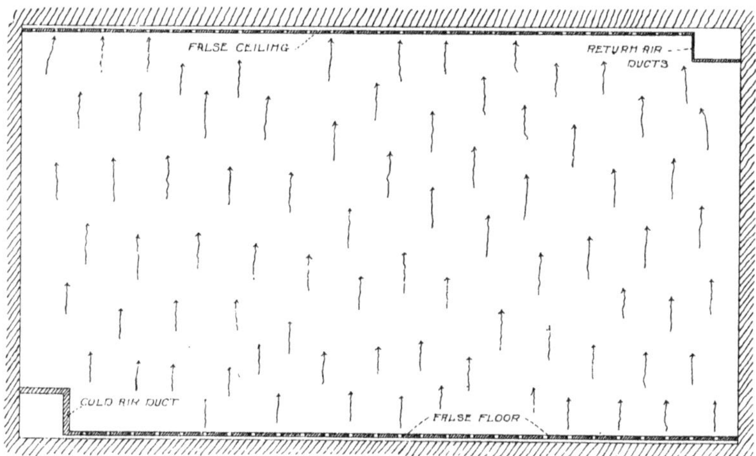

FIG. 19.—COOPER'S LATEST IMPROVED SYSTEM OF AIR CIRCULATION.

consequently all goods piled in such a room are exposed to exactly the same conditions as to circulation, temperature, humidity and purity of the air. In a room equipped with this system, with the parts correctly proportioned, it is entirely safe to pile goods closely, only allowing a fraction of an inch between the packages and at sides of room and placing thin strips beneath the goods to allow air to flow from perforations in false floor. Where, in rooms fitted with direct piping and some of the fan systems as well, a large space must be left at floor and ceiling for a circulation of air, with this system goods may be piled close up to ceiling leaving only half an inch for the air to flow into perforations in false ceilings. As the space occupied in height by false floor and the space underneath is, in most cases, only one and three-fourths inches and that occupied by false ceiling only one and one-fourth inches, it is apparent that much space will be saved by using this system. After a room is filled with goods and cooled down to the correct carrying temperature, no difference in temperature can be noticed in different parts of the room. No blast of air can be felt in any place, a gentle flow from perforations only is noticeable, therefore no particular place has more circulation than another to cause a drying out of the goods. The advantages of this system over any of the others may be summed up as follows:

1. A more equal distribution of air, especially when the room is filled with goods. Goods in center of room are exposed to the same temperature, circulation, etc., as those at sides.

2. Saving in space, as it allows the room to be filled full of goods without leaving large spaces at top and bottom for a circulation of air.

3. Where the air is so perfectly distributed and collected it is not necessary to circulate such a large volume, saving in power and lessening the liability of evaporation of goods.

The objections against this system which have been offered are of no practical consequence. The first one usually mentioned by an inquirer is that the space under false floor is likely to collect litter and become foul. The author admits that this apparent objection for some time kept him from introducing this system to practical service, but when once tried, this was found of no consequence, as the false floor is made in sections, easily handled, and it is as easy to raise these and sweep underneath as to remove the 2x4s or 4x4s generally used to pile goods on. Another objection is the supposedly high initial expense. A contract was awarded for the construction of this system for a fair sized house, in which the cost for air circulating system, including fans and motors, did not exceed

$20 per 1,000 cubic feet. It will be seen that the cost is of very small importance as compared with the practical results obtained and the saving in space effected. Those who are skeptical about the advantages of forced circulation, and of this system in particular, are invited to visit some of the plants designed by the author.

The objections against forced circulation are largely fanciful and are not substantiated when investigated. The idea that goods dry out or evaporate rapidly in a room so equipped, has never been even suggested by the author's experience, and this objection may be dropped without further comment, as this ground has been thrashed over before. It is thought by many that a forced circulation system is unnecessary, expensive to install and costly to keep in operation. It may be admitted that forced circulation is unnecessary in the same sense that refrigeration was unnecessary fifty years ago. People are getting along without it because they do not know or understand its advantages. Many other applications of machinery are not absolutely necessary, but are used for the improved results obtained. If properly designed, the cost of equipping a house with an improved system of forced circulation need not be much greater than with direct piped rooms, for the reason, mainly, that only half or two-thirds as much piping is needed, and because of the saving in main pipes by locating the cooling coils centrally and blowing the air to and from the room with a fan. As to cost of power for operating, this is very small, if using the fans specially designed by the author for this purpose. (Fans for use with air circulating systems should be of special construction. This is considered under the chapter on "Ventilation.") It is customary to install a half horse power motor for handling the air in a room of 15,000 cubic feet. The actual power necessary is from one-quarter to three-eighths of a horse power. As an offset to the cost of operating the fan, may be placed the great saving in space gained by the use of the fan system. In no case is this less than 5 per cent of the space refrigerated, and sometimes will amount to over 10 per cent. Even if all the objections urged against the system were true, this alone is enough to compensate and more besides. When from 5 to 10 per cent may be added to the earning capacity of a storage house without additional cost of operation it means a big increase in the net profit of the business.

Not the least of the advantages of the forced circulation system is, that during cold weather when the ammonia or brine is shut off from circulating through the pipes, their frosted surfaces are not exposed in the storage room. It is comparatively easy to clean the pipes, as they are more accessible than they are in any of the direct piping systems. A still greater advantage may be gained by using a process invented by the author, which consists in placing chloride of calcium above the pipes, so that the brine resulting from a union of the moisture in the air with the calcium will drip down over the pipes. (See chapter on "Uses of Chloride of Calcium.") This prevents the formation of frost on the pipes at all times, and during cold weather, when the refrigerant is shut off, by keeping up the supply of calcium, the moisture and purity of the air are under perfect control.

That the tendency is toward the adoption of forced circulation for the best new work cannot be doubted, even by those who do not advocate these systems. It cannot be expected that they will come into use all at once, but the author feels justified in predicting that ultimately more than half the high grade installations will be done under these systems. The present opposition comes largely from the "old line" people in the business who do not like to see changes and improvements made on methods with which they have "had good results" for so many years.

CHAPTER VIII.

VENTILATION.

NECESSITY FOR VENTILATION OF COLD STORES.

In discussing humidity and circulation, it has been explained how a large portion of the gases of decomposition and impurities of various kinds, which are incident to the presence of perishable products in cold storage, are carried by the moisture existing in the air, and that when this moisture is frozen on the cooling pipes, or absorbed by chemicals, the foul matter is largely rendered harmless. It may now be noted further that even with a good circulation and ample moisture-absorbing capacity, there will still be some impurities and gases, detrimental to the welfare of the stored goods, which have little or no affinity for the water vapor in the air, and consequently accumulate in the storage room. Ventilation is necessary to rid a refrigerator room of these permanent gases.

The subject of ventilation for cold rooms has been very much talked of, but there is really little known about it, so far as any practical information is concerned. Some of the more progressive cold storage managers have given some attention to this part of the business, but many of the largest and best known houses do not ventilate their rooms at all, except perhaps during the winter or spring, when rooms are aired out for the purpose of whitewashing. In some cases the change of air incident to opening and closing of doors, when goods are placed in storage or removed therefrom, is relied on to supply ventilation. This is quite inefficient, because goods are mostly stored during two or three months, and removed from storage likewise, leaving several months when no fresh air of consequence can penetrate to the room, except as the doors may be opened for the purpose of taking the temperature of the room. Furthermore, this kind of ventilation during the warm weather of summer and during a large part of the spring and autumn months is worse than no ventilation at all. Some storage men even take so radical a position on this matter of opening doors during warm weather, as to insist that the door shall not be opened for the purpose of reading the thermometer. A double window is placed in the door of each room, with the thermometer hanging so that it can be read from the outside without opening the door. While the author has not practiced this method, it seems to be a good idea, and it is certainly preferable to ventilating the room through doors which open to the outside air. When doors into rooms open into a corridor, the evil is partly prevented, but opening the door or window of a storage room directly to the outside air when the temperature outside is materially higher will always result in more or less bad effect on the goods, because of the water vapor in the warmer incoming air being condensed on the stored goods.

Another source of ventilation similar in its results to the opening of a door or window is that resulting from the leakage of air directly into the storage room, through the pores and crevices in the walls, around the doors and windows, etc.—leakage of air literally—air that gets in when everything is supposed to be closed. The amount is usually imperceptible, but is enough in some houses to be a serious detriment to the quality of work done. In small houses with large outside exposure and poor insulation this air leakage is considerable, but in the big refrigerators of several hundred thousand cubic feet capacity, and with thorough insulation, it is reduced to practically nothing. The loss of refrigeration caused by air leakage, while of some importance, is of small moment beside the bad effects resulting from the moisture and impurities brought in by the warm air from the outside. The value of prime, tight insulation, as a conserver of refrigeration, aside from a matter of keeping out the warm, moist air, is discussed in the chapter on "Insulation," but a word about windows and doors is properly in line with the present discussion.

WINDOWS AND DOORS.

Rather than consider what might be a good way of placing windows in a cold storage building, their use should be discouraged. Even with four or five separate glass, divided by air spaces and with all joints set in white lead, the loss of refrigeration is large. It is also very difficult to fit insulation around the window frame so as to make a good job; and even if a passable job were practicable, the expense of putting in windows is sufficient to condemn their use. The increased fire exposure is of some consequence, too, and with the low cost of electric light, windows should not be thought of for cold storage work. Barring the small amount of heat given off, the incandescent electric lamp is an ideal device for lighting cold storage rooms, as the air is not vitiated by gases and odors as is the case when using gas, kerosene or candles.

Doors which will shut tight, forming a nearly perfect air seal, with a small amount of pressure, have long been wanted for cold storage rooms. Most of the ordinary bevel doors, either with or without packing on the bevel, will not shut even approximately tight; and in operation nine out of every ten stick and refuse to open except after many persuasive kicks and surges—we all know how it is. The special cold storage doors on the market, the author believes to be far above anything else in this line, and does not hesitate to recommend them to those wanting a door which will prevent air leakage. The prices are very reasonable, considering the excellent material and fine work put into their construction. The slight additional cost over the common door will be quickly saved, by reason of their quick action—opening quickly when the fastener is worked. If a door is built on the job, the chief idea to be considered in its construction, is to build a door which is tight at one point all around. It is absolutely impossible to make a door fit on a long bevel, but the effort is very frequently made. (See chapter on Doors and Windows.)

AIR LEAKAGE.

Having presented the subject of air leakage, we may as well ask how it is caused and why it must be guarded against.

It is amenable to the same law as gravity air circulation, which was explained quite thoroughly in the first part of the chapter on "Air Circulation." When the outside air is very much warmer than that of the storage room, the air in the storage room produces a pressure on the floor and lower part of the room, by reason of its greater weight, and consequently it seeks to escape there. If there are openings near the floor where the air can flow out, and others at the ceiling or upper part of the room, the air will flow in at the top and out at the bottom of the room. Reverse the conditions of temperature, and the direction of flow of air is also reversed. That is, when the air outside is colder than the air of the room, the cold air will flow into the room at the bottom and the comparatively warm air of the room out at the top. This action is nicely illustrated by noting the air currents in a door which is opened into a cold room when the temperature is very warm outside. The warm air rushes in at the top of door and the cold air of room out at the bottom. In cold weather the direction of air flow will be reversed.

Perfect inclosing walls for a cold storage room would be perfectly air tight, as they would be if lined with sheet metal, with soldered joints. The interior conditions would then be under more perfect control. It is hardly necessary to do this (although it has been done in cases of some old time houses), as a practically tight job may be had by using the right materials, well put on. Air leakage may not be exactly ventilation, but it is a kind of ventilation which has given the author some trouble in the past, and does still, consequently the difficulties of operating a house with defective insulation and large outside exposure, and still turning out first-class goods are very thoroughly appreciated.

PRACTICES TO AVOID.

Methods of ventilation which are permissible when applied to the work of supplying fresh air to ordinary structures are generally dangerous when used to ventilate cold storage rooms. The problem in ventilating non-insulated structures is merely the supplying of fresh air from the outside without causing a marked change in the temperature, and without creating strong drafts. Air for the ventilation of refrigerator rooms, during warm weather, must be of very nearly the same temperature and relative humidity as the air of the room to be ventilated, and free from the germs which hasten decay and cause a growth of fungus on the products in storage. If a door or window of a storage room is opened directly to the outside atmosphere, there will be little or no circulation of air into and out of the room when the temperature outside and in is about the same, unless the wind should be favorable. As we cannot ventilate in this way when the air outside is colder than the storage room, on account of freezing the goods, and the introduction of fresh air, which is warmer than the storage room, is not permissible, for reasons already given, the matter reduces itself to not ventilating at all during warm weather (which most houses practice) or of properly cooling and purifying the air before forcing it into the storage room. It will bear repeating that it is positively bad practice to allow air from the outside to get into a cold storage room during the summer months, also during a large portion of the spring and fall months, unless cooled and purified first. The fact that we cannot see the moisture deposited in the form of beads of water, or floating in the air in the form of fog or mist, does not indicate that it is not present. The sling psychrometer, described in discussing humidity, will give an accurate indication of the result of this unscientific method of ventilating.

MEANS FOR AIR HANDLING.

Any natural means of handling air or ventilation is inaccurate and inoperative, or it may be positively harmful, except under favorable conditions. If depending on natural gravity for ventilation it will be guesswork, to a greater or less extent, because depending on conditions which vary with the season, temperature, direction and force of the wind, etc. The late Robert Briggs,* an authority on ventilation, makes a concise statement of the advantages of using fans for ventilation,

*Proceedings Am. Soc. Civil Engineers, May, 1881.

in his "notes on Ventilating and Heating." He says: "It will not be attempted at this time to argue fully the advantages of the method of supplying air for ventilation by impulse through mechanical means—the superiority of forced ventilation, as it is called. This mooted question will be found to have been discussed, argued and combated on all sides in numerous publications, but the conclusion of all is, that if air is wanted in any particular place, at any particular time, it must be put there, not allowed to go. Other methods will give results at certain times or seasons, or under certain conditions. One method will work perfectly with certain differences of internal and external temperature, while another method succeeds only when other differences exist,. * * * No other method than that of impelling air by direct means, with a fan, is equally independent of accidental natural conditions, equally efficient for a desired result, or equally controllable to suit the demands of those who are ventilating."

PLENUM VS. EXHAUST METHODS OF VENTILATING.

There are two general methods, with some modifications, for handling air for ventilation: The plenum or pressure method, in which the fresh air is forced into the room; and the vacuum or exhaust method, in which the foul air is drawn out. The exhaust method is to be avoided for ventilating cold storage rooms, for reasons which we shall see presently. With this method, sometimes the exhaust steam from an engine is utilized to induce a draft of air upward from storage room, by heating the air in a stack or ventilation flue connected at its lower end with the room to be ventilated. In some cases no provision is made for an inflow of fresh air, in which case it will seep in at every crack, crevice and pore (by reason of the partial vacuum created by exhausting the foul air), bringing a load of moisture and germs of disintegration into the storage room. This exhaust steam method is no different in its result than if a fan were placed so as to draw the air out of the storage room under conditions which are otherwise the same as described in connection with the exhaust steam method. Should we provide an inlet for fresh air, through proper absorbents, the same law would be operative, only to a lesser degree, as a partial vacuum must in any case be created before the air from outside would flow into the room, tending to the dangerous air leakage already fully discussed.

The plenum or pressure method is by far the best for our purpose. The air should be forced into the room by a fan, after first properly cooling, drying and purifying it. An outlet for the escape of the foul gases which it is desired to be rid of should be provided near the floor, as these gases, by reason of their greater gravity, tend to accumulate in the lower part of the room. It will be observed that forcing the fresh air in creates a pressure inside the room, and if there is any air leakage, it will be outwardly from the room—exactly the way we want it to go. Having brought our subject to the point where it is found that the best way to ventilate is by the use of fans forcing the air into the storage room, we will determine what type of fan is best adapted to our needs. What is said of fans for ventilation is equally true if they are to be used for forced air circulation, described in chapter on "Air Circulation."

FANS FOR VENTILATION.

It is admitted by a majority of experts on air moving machinery that the disk or propeller wheel type of fan, through which the air moves parallel to the axis of fan, is not efficient or desirable for work where the air has to travel through a series of tortuous air ducts, as in the forced air circulation system for cold storage work, or for ventilation purposes where there is some resistance. Where any resistance of importance is encountered, the disk fan must be driven at a high rate of speed, and at an immense loss of power, to compel it to deliver its full quota of air. Another disadvantage of the disk type is the difficulty of belting to the shaft, or of getting power to the fan in any form, if it is inclosed entirely in an air duct. The disk type will therefore be dismissed, and the well known centrifugal, or peripheral discharge fan taken up.

This type of fan draws the air in at its center parallel to the shaft, and delivers it at right angles to the shaft at the periphery or rim of the fan wheel, the law governing its action being the well understood centrifugal force, which is commonly illustrated when we see the mud fly from a buggy wheel, or the water off a grindstone. The advantage of these fans over the disk type is that the centrifugal action set up by the rotary motion of the fan is utilized to give velocity to the air in its passage over the fan blades. In the selection of a fan for the purpose of forced circulation in the storage room, or for forcing in fresh air for ventilation, it should be noted that a large, slow running fan wheel is very much more economical of power than a small fan running at a high rate of speed, both doing the same amount of work. The loss of refrigeration, too, in a rapidly moving fan, is of consequence, because the air is warmed by impact with the blades. The proportion of power saved by the use of a large fan running at a slow rate of speed rather than a small fan running at a high rate of speed, both delivering the same amount of air, is almost phenomenal, and does not seem at all reasonable at first view. The volume of air delivered by a fan varies very nearly as the speed, while the power required varies about as the cube of the speed. That is, doubling the speed doubles the volume of air, while the power required is increased eight times. We will take a specific case. A 45-inch fan wheel, revolving at a speed of 200 revolutions per minute, delivers, say, 5,000 cubic feet of air per minute, and requires but one-quarter of a horse power to operate it. If the speed is increased to 400 revolutions, the volume of air delivered will be only about 10,000 cubic feet, while the power required to drive it will be raised to two horse power. These figures are theoretical, but within certain limits are approximated in practice.

For use in cold storage work the objection common to nearly all the air moving machinery found listed by the manufacturers is the seemingly unnecessary amount of metal used in its construction. The heavy weight of the fan wheels, and the large diameter of shaft necessitated by such weight, causes much friction on the journals, so that when running at the slow speeds desirable for cold storage work, more power is required to overcome the mechanical friction than is actually required to move the air.

No doubt the high speeds necessary for some work have obliged the manufacturers to make their fans amply strong for the highest speeds, consequently they are not economical for the slower speeds. It would not be appropriate for a person to fan himself with a dinner plate—it would do the work, but would not be economical of power.

Having been unable to find a fan wheel well suited to the requirements of cold storage duty, the author has designed and constructed a line of fan wheels especially for slow speeds, which are amply strong and capable of moderately high speeds, when necessary, but are very much lighter than most fans on the market, and consume proportionately less power in mechanical friction.

TREATING AIR FOR VENTILATION.

So far we have found out what kind of ventilation is not desirable, and have an inkling of what kind would be desirable. The question before us now is to properly treat the air before introducing it into the storage room, so that it may be fresh—*i. e.,* pure oxygen and nitrogen, without excessive moisture, and free from the impurities and germs which may contaminate the product which is being refrigerated.

The free outside air during warm weather, especially in the vicinity of our large cities, contains, among many others, germs which produce the parasitic plant growth which is called mildew or mold. The exhalation from the lungs of the many animals and men who inhabit our cities, and the evaporation from the dust, dirt and decaying matter of various kinds peculiar to the street, render the air a receptacle and conveyor for impurities and germs of many species. The species of germs which concern us are active in proportion to the temperature and humidity of the air. In a warm atmosphere which contains much moisture they take root and grow rapidly, throwing off more germs of their kind, which impregnate the air in an increasing ratio as the humidity and temperature are increased. The humidity of the outside air is not necessarily increased with the temperature, but it is always increased to some extent, and as the temperature of the outside air rises we must necessarily be more and more careful how we treat and handle the air which we are to use for the ventilation of refrigerated rooms.

It is readily understood why it is necessary to cool the air before introducing it into the storage room to at least as low a temperature as that of the room to be ventilated, and some cold storage managers have ventilated on this basis, thinking that this was all that was necessary for successful ventilation. Air cooled only to the temperature of the storage room will be saturated with moisture at that temperature, and will be in condition to develop mold rapidly. An improvement on this manner of handling is to cool the air to be used for ventilation to a few degrees (say five or six) below the temperature of the storage room. The air will then be rendered as dry as that of the storage room. This is a good method of ventilation, and one which the author has practiced, but it is open to criticism, because of the fact that the air is not purified fully at the same time it is cooled and dried. If the air is first cooled to several degrees below the temperature of the room to be ventilated it will be of benefit to the room, if not overdone, but in results will not be equal to a system to be described and illustrated further on in this chapter.

Several houses known to the author ventilate by letting the warm outside air in at a point near to the ceiling, directly over the cooling coils, expecting that the air will be properly cooled and dried before it flows into the room itself. The same objections are applicable to this system as are applicable to any plan of ventilating where the air is cooled only to the temperature of the room to be ventilated, because the air will be at the saturation point, and will therefore raise the humidity of the room, as well as introduce a quantity of germs and impurities.

SIMPLE AIR COOLER FOR VENTILATING.

If we ventilate by simply cooling the air, the simplest and most effective method, as shown in Fig. 1, is to take the air from as high and sheltered a place as is accessible about the building; draw it down over frozen surfaces in the form of brine or ammonia pipes, which may be arranged anywhere along the wall of a room, outside of the storage entirely, if more convenient. An exhaust fan takes the air from the coils

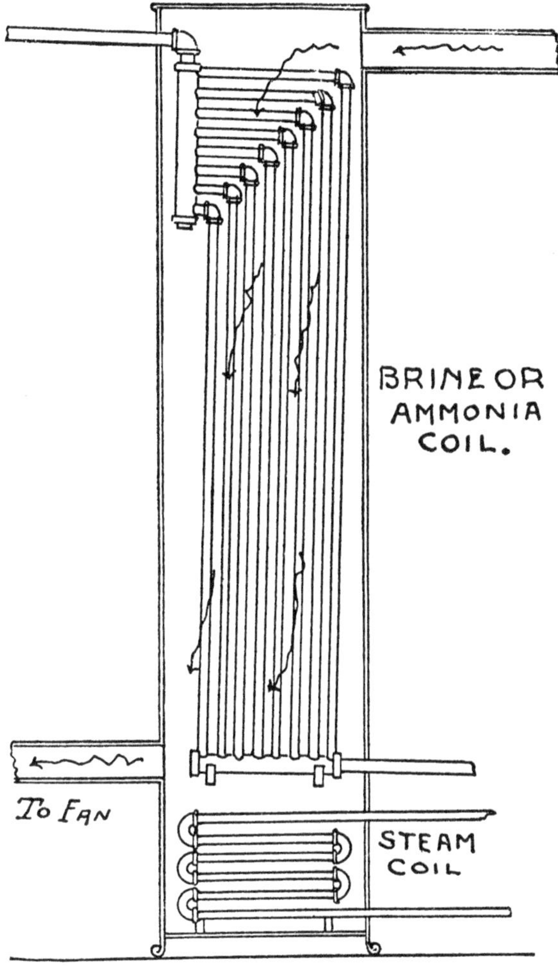

FIG. 1.—BRINE OR AMMONIA COOLER.

in the ventilating flue and forces it into the room to be ventilated, allowing it to escape in the neighborhood of the cooling coils, where it will mix with the air circulation, and flow into

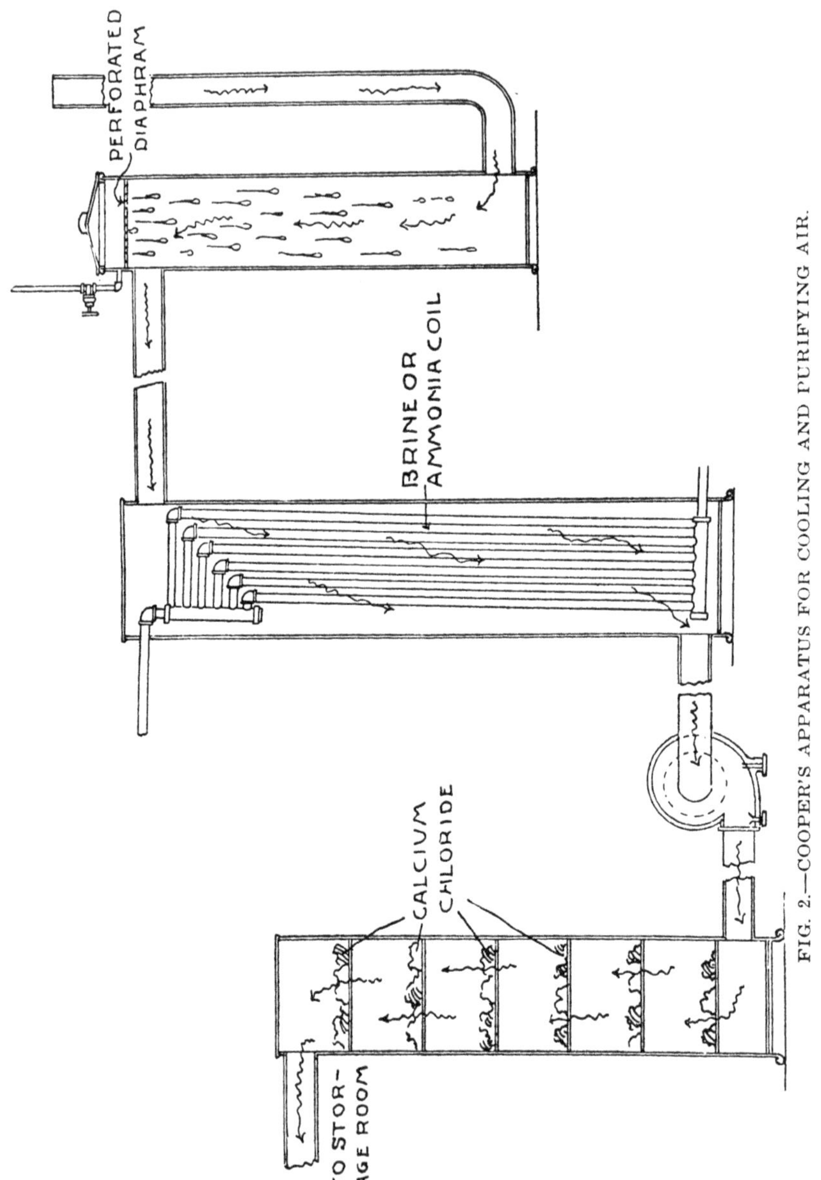

FIG. 2.—COOPER'S APPARATUS FOR COOLING AND PURIFYING AIR.

the room through the regular channel. It is necessary to provide an outlet for the escape of foul air whenever fresh air is forced into the room. This outlet should be near the floor, and of about the same area as the inlet pipe. A steam coil may be provided beneath the cooling coil in ventilating flue, as shown in the sketch for the purpose of melting the frost off the pipes. The casing around the cooling coil should, of course, be insulated moderately, as well as the pipe leading from it to the storage room, wherever exposed to the warm outside air. The size of apparatus necessary for this purpose need not be large as the quantity of air which is generally required for the ventilation of storage rooms is quite small, comparatively.

"Americus"* mentions a method of washing air for ventilation, which seems to have advantages. The idea is to draw or force air through a body of water or brine by immersing the intake pipe so that the air will bubble up through the liquid. This seems quite simple, but when it comes to forcing air through a liquid with a fan it is not so simple, as nothing short of an air pump will drive air through a pipe submerged as above described, unless the opening from pipe is placed quite near the surface of the liquid; in which case the benefit to the air is very small. Experiments conducted by the author along this line were considered failures.

COOPER SYSTEM FOR WARM WEATHER VENTILATION.

Shown in Fig. 2 is what appears as a rather complicated apparatus, but on investigation it proves to be quite simple. There are three parts to this apparatus, as follows:

First.—The air-washing tank, in which the air flows upward against a rain of water from a perforated diaphragm above, as clearly shown in the sketch. This not only cools the air to the temperature of the water, say 55° or 60° F., but it also takes out a large portion of the impurities of various kinds. From the washing tank the air is passed on, in a comparatively pure and cool state, to be still further cooled.

Second.—The cooling tank, in which the air is cooled to several degrees lower temperature than that of the storage

*In *Ice and Refrigeration*, July, 1898.

room. This removes the moisture which holds in suspension the few impurities which may have passed the washing tank, the moisture being deposited on the frozen surfaces within the cooler.

Third.—The drying box, into which the air from the cooler is passed, and which contains chloride of calcium. This chemical is a well known absorber of moisture, what is technically known as a deliquescent substance. If moisture of any account passes the cooler it is surely stopped in the dryer, which "makes assurance doubly sure," so far as delivering a pure dry air is concerned. It would be a hardy germ, indeed, that would not succumb to the washing, cooling and drying processes of this system of ventilation, which is as thorough as it well may be theoretically, and practically is very effective.

FREQUENCY AND AMOUNT OF VENTILATION.

The volume of air necessary for ventilating a given size of storage room can only be estimated, and probably no two storage men will agree as to what is a correct quantity. Some say that the introduction of a volume of air equal to that of the room to be ventilated should take place each day; others twice each day; some even take so radical a view of it as to say the oftener the better if the air is properly dried and cooled. This is of course true enough, but foul gases which can be gotten rid of by ventilation accumulate but slowly in a storage room, and it is probable that the introduction of a volume of fresh air, properly treated, equaling that of the storage room, twice each week will be ample for the purpose of keeping the room in good condition, and in most cases once each week may do nearly as well. There is much to be developed yet in the direction of ventilation of refrigerated rooms, more particularly in the way of some method of knowing when a room requires ventilating. Perhaps some bright chemist will in time make investigations and ascertain what the gases are which we must dispose of, and indicate some simple method of determining their presence, and in what proportion.

All that has been said about ventilation so far applies only to the ventilation of cold storage rooms when the air without is warmer than the air of the storage room. We will now give our attention to another kind of ventilation that is applicable when the air without is at about the same temperature as the storage room, or at some degree lower. This will be designated as cold weather ventilation, as this term seems to express its function perfectly.

COLD WEATHER VENTILATION.

It has long been a well understood fact that products held at about 30° F. or higher are more liable to be injured in cold storage during the cool or cold weather of fall and winter than during a long carry through the heated term. Much has been said and written about why the old style overhead ice cold storages give such poor results during fall and winter, the reason assigned being lack of circulation, as the meltage of ice ceases when the cool weather comes. This is true; further, the large body of ice becomes an evaporating surface, and the dirt and impurities which are found in all natural ice, to a greater or less extent, have accumulated on the top of this ice, and the evaporation which takes place carries gases from this miscellaneous matter into the air of the storage room, with consequent bad results. In some houses this may be avoided by closing the trap doors covering circulation flues, but it is seldom done, and in many houses it is impossible.

Now are we who cool our storage rooms with brine or ammonia pipes very much better off in this one respect than those who have these much despised overhead ice cold storages? Our rooms are cooled by frozen surfaces, on which accumulates the evaporation from the goods in store, which, as we have already plainly seen, contains much foul matter and impurities. Precisely as in the ice cold storages, the cooling surfaces, which absorb moisture during warm weather, become evaporating surfaces, and give back to the air of the room a considerable portion of the various impurities and germs which have been accumulated during the warm weather of summer. To make this point more plain it may be considered thus: During the period when the outside air is considerably warmer than the air of the storage room it is necessary to keep some refrigerant at work cooling the air within. This is usually done by circulating brine or ammonia through pipes and the air of the room is circulated in contact with the pipes. When the outside temperature is high, more of the refrigerant must be circulated, or its temperature must be lowered; as the weather turns cooler in the fall, less refrigerant, or the same amount at a higher temperature, must be circulated, and when the air without reaches the temperature of the room, the circulation of refrigerant must be discontinued altogether. When this is done the moisture on the cooling pipes begins to evaporate. This evaporation added to that which is given off by the goods themselves soon causes the air to be saturated with very impure and poisonous vapors which cause the goods to deteriorate very rapidly.

DISPOSAL OF MOISTURE.

The influence which the temperature of the refrigerant flowing in the cooling pipes has on the condition of a storage room may be better understood by taking a specific case: A room with a temperature of 33° F. and a humidity of 70 per cent has a dew point (temperature at which the air precipitates moisture) of 25° F. Therefore any cold surface (as a pipe surface), having a temperature of 25° F. or lower, will attract moisture when exposed to the air of the room. If the pipe surfaces are heavily coated with frost, as they usually are as cold weather approaches, the frost acts as an insulator, and the refrigerant flowing in pipes must be at a considerably lower temperature than the air of the room, or no moisture is attracted. We have all noted how the accumulation of moisture on pipe coils is slower and slower as the thickness increases, until finally a limit is reached where no more frost will form; yet owing to the largely increased surface the room can be kept at its normal temperature. If pipes are badly loaded with frost, sometimes no absorption of moisture will take place when the refrigerant flowing in the coils is 10° or 15° below the temperature of the room. The surface exposed to the air of the room, whether in the form of frost or otherwise, must be at or below the temperature of the dew point, or no moisture will be absorbed. The value of suitable moisture-absorbing surfaces as the cool weather of fall and winter approaches cannot be overestimated, as many have found to their sorrow that two weeks stay in cold storage under bad conditions in cold weather will do more harm to eggs in particular than four months during hot weather.

The remedy for this trouble is found in keeping the air of the room from coming in contact with the poisonous frost which has been accumulated on the pipes during their period of duty during warm weather; or what is still a better way is not to allow the frost to accumulate on the pipes at all, by using a device, described elsewhere under head of "Absorbents." How to keep the air from contact with the frost on pipes is not an easy matter, and in case of piping suspended directly in the room it is an impossibility.

With a system of screens arranged around coils, as described in the first part of the chapter on "Air Circulation," trap doors may be very easily fitted to the openings and the air circulation shut off in this way; but the simplest and best way is to equip the rooms with forced circulation, and locate the pipes outside of the room entirely. Then it is only a matter of shutting off the circulation over coils, or allowing it to continue through a by-pass, or if the process described in the chapter on "Uses of Chloride of Calcium" is used, the circulation may be allowed to continue over coils. It seems quite clear, from what has been written, why a storage room gets foul quickly during cool weather, and also that the bad conditions may be bettered by cold weather ventilation. The harm resulting from the foul evaporation from frost on cooling pipes may be obviated by not allowing contact between it and the air of room, but the evaporation from the products themselves must be taken up by other means when cooling surfaces are no longer operative.

HINTS ON COLD WEATHER VENTILATION.

By carefully observing conditions a storage room may nearly always be kept in prime condition during cold weather by no other means than the introduction of fresh outside air at as frequent intervals as right conditions of temperature and humidity will permit. It is quite safe to force in plenty of air which has about the same temperature and humidity as the room to be ventilated. There are few impurities in the clear, crisp air of a bright fall day, and many such are available for our purpose in the latitude of Minnesota and New York, and a somewhat smaller number, perhaps, in the latitude of Iowa or Ohio. It is only a matter of handling the free air of heaven understandingly. One's impressions, however, will hardly do in judging what air is good to use for ventilating purposes. If you have a bright, clear day, or, what is still better, a clear cold night, which has the appearance of being what you want, get out your sling psychrometer and set all guesswork aside. It is frequently possible to fill your storage rooms with fine, pure air at a temperature about the same as that of the room, as early as the latter part of October, if you are watching for the opportunity. Provide a good big fan wheel, which will handle a large volume of air in a short time, and when conditions are right blow your rooms full of it. Repeat this whenever the weather conditions will permit.

We may now consider cold weather ventilation under another condition, viz.: When it is colder outside than inside the storage room. Whenever the outside air is 8° or 10° below that of the storage room it is always perfectly safe to introduce it into the storage room, after it has been first warmed to the temperature of the room to be ventilated. That is, it is safe so far as introducing moisture or impurities is concerned. If we should ventilate in this way continuously our humidity would be lowered to a point where the goods might suffer from evaporation. It is necessary, therefore, that observation of the humidity of the room so ventilated be taken, so that this kind of ventilation may not be overdone.

The method of getting air into the rooms under these last two systems of ventilation is of no special moment, except that it be under control, and we have already noted that the only good way of handling air was by the use of fans, preferably large and of light weight, and running at a slow speed. Where the forced circulation is installed, it is sometimes practicable to so connect the fans used for this purpose, that cold weather ventilation may be handled by them; but a separate fan is much better and while it seems more complicated it is really simpler to operate, because handled independently. When using an independent fan or when using the forced circulation fan for ventilating, the fresh air mixes with the circulation and is well distributed by it to various parts of the room.

The ventilation of cold storage rooms is not a matter which can be safely left to such help as may be at hand, and if good results are to be secured "the boss" should see to it himself. Cold weather ventilation, especially, must be handled carefully and scientifically or trouble may result instead of benefit. No absolute rules can be given for handling ventilation because of widely varying conditions, but if what has been written is read and studied carefully the subject can be taken up intelligently and followed out to its legitimate conclusion.

CHAPTER IX.

HUMIDITY.

IMPORTANCE OF ASCERTAINING HUMIDITY IN COLD STORES.

Up to about the year 1898 the subject of humidity of cold storage rooms was given very little or no attention by cold storage operators, and no successful means of testing humidity had been found for the requirements of refrigerated rooms. About the year above referred to the author secured a sling psychrometer, such as is used at stations of the United States Weather Bureau, and made some tests. This instrument is now in general use for the purpose, and is well adapted for obtaining the humidity of cold storage rooms. More attention has been given to the subject each year, and as it costs practically nothing and requires very little time, all houses should make tests to know how they stand in this important respect.

The humidity of a cold storage room under ordinary conditions depends on the season to a moderate extent, and the condition of the room, as regards ventilation, in some cases. In late fall or winter, especially, if air is taken directly into the room from the outside, the humidity will be low. As cool weather approaches, the tendency is for the humidity to rise, and unless kept down by ventilation or by the use of absorbents, serious consequences are sure to follow.

To enable us to thoroughly understand the meaning of relative humidity, as it is called, we will study a few extracts from "Instructions to Voluntary Observers."* Humidity is considered on a decimal scale, with 100 the saturation point of the air, at which it will hold no more water vapor and 0 the point at which air contains no moisture whatever. The vari-

*Issued by the United States Weather Bureau, Washington, D. C.

ous percentages between these points is a degree of humidity relative to these two extremes, or relative humidity. The quotations below are not contained in the recent issue of instructions, but are from the issue of 1892, which is now superceded by that of 1897.

WATER VAPOR IN AIR.

The air contains vapor of water, transparent and colorless like its other gaseous components. It only becomes visible on condensing to fog or cloud, which is only water in a fine state of division. The amount is very variable at different times, even in the vicinity of the ocean. The amount of moisture that can exist as vapor in the air depends on the temperature. There is a certain pressure of vapor, corresponding to every temperature, which cannot be exceeded; beyond this there is condensation. This temperature is called the temperature of saturation for the pressure. When the temperature of the air diminishes until the saturation temperature for the vapor contained is reached, any further fall causes a condensation of moisture. The temperature at which this occurs is called the dew point temperature of the air at that time. The less the quantity of moisture the air contains, the lower will be the temperature of the dew point. For different saturation temperatures, the weight of vapor, in grains, contained in a cubic foot of air is as follows:

Temperature of Saturation, Degrees F.	Weight in a Cubic Foot, Grains.
0	0.56
10	0.87
20	1.32
30	1.96
40	2.85
50	4.08
60	5.74
70	7.98
80	10.93
90	14.79
100	19.77

The air is never perfectly saturated, not even when rain is falling; neither is it ever perfectly dry at any place. Relative humidity expresses relative amount of moisture in the air only as long as the temperature of the air remains constant. For this reason relative humidity is an imperfect datum. At a low temperature even a high relative humidity represents a very small amount of vapor actually in the air, while a low relative humidity at a high temperature represents a great deal.

The most important law relating to above concise statements, and one which, if carefully noted and applied, will make all work in humidity easily understood, is best expressed thus: *The capacity of air for moisture is increased with its temperature.* Strictly speaking, air has no capacity for moisture, the water vapor being simply diffused through the air after the nature of a mechanical mixture. For all practical purposes, we may regard it as being absorbed by the air, and it is usually so treated.

At a temperature of 40° F., air will hold in suspension more water vapor than at any lower temperature (see table); and when the difference is as much as 10° F., the difference in the amount of moisture the air will hold is very considerable. To illustrate: Air which is saturated with moisture at 30° F., when raised in temperature to 40° F., then holds but 68 per cent of its total capacity.

INSTRUMENTS FOR DETERMINING HUMIDITY.

There are two kinds of instruments in use for determining humidity-hygrometers and psychrometers. The hygrometer depends on the expansion and contraction of some substance, as a human hair, in the presence of more or less moisture in the air. The hair used is fastened at one end, the other end passing around a pulley, to which is fastened a pointer, which moves over a graduated arc as the hair changes its length The scale reads from 0 to 100. The chief advantage of these instruments is that results are obtained at once, the reading corresponding to the percentage of saturation or relative humidity; but these instruments are affected by changes of temperature, and shocks or vibration materially affect the reading. Further, they are more expensive in first cost, and not so convenient to use, as they must hang for some time in the room to be tested, while with the sling psychrometer, described in another paragraph, an observer can pass from room to room, getting observations in less than two minutes in each room, needing but one instrument and making all observations at practically the same time.

A psychrometer is simply two thermometers mounted on a frame; the bulb of one being covered with muslin so as to retain a film of water surrounding it. The working of this instrument depends on a law which may be roughly expressed, as "evaporation carries off heat." The evaporation of water from the bulb incased in muslin, known as the wet bulb, cools it somewhat, depending on how dry the air surrounding it may be. The difference between the reading of the wet bulb thermometer and the reading of the dry bulb thermometer, when compared with reference to a prepared table, gives the relative humidity of the air at the time of making the observation. Psychrometers are of two kinds, stationary and sling.

The stationary psychrometer is essentially like the sling psychrometer, both depending on the same principle. The sling instrument is more compact and provided with a handle for whirling, while the stationary instrument is intended to be fastened against the wall, or on a post, the muslin covering the wet bulb being connected by a porous wick with a reservoir of water, to keep the supply of water continuous. This is essential, as it takes some little time to obtain a correct reading with this pattern of instrument. For this reason it is open to the same objections as the hygrometer. Also, after short use the muslin covering the wet bulb, and the wick feeding water to it, become clogged with solid matter and fungous growth affecting its accuracy. At any temperature below 32° F. this instrument is useless, as the water will freeze in the wick supplying the muslin on the wet bulb, and the muslin becomes dry in consequence.

For practical, accurate and quick results at any temperature there is no instrument so reliable and convenient as the sling psychrometer, preferably of the pattern known as Prof. Marvin's improved psychrometer, shown in Fig. 1. This is a standard Weather Bureau instrument, and when used in connection with the tables of humidity published by the bureau, any needed results may be obtained with a fair degree of accuracy. The sling psychrometer, as illustrated, consists of a pair of thermometers mounted on an aluminum plate, one higher than the other, the lower having its bulb covered with a small sack of muslin. At the top, the frame or plate supporting the thermometers is provided with a handle for whirling, this handle being connected by links to the plate, and provided with a swivel to allow of a smooth rotary motion. The bulb of the lower thermometer is wet at the time of making an observation, the muslin serving to retain a film of

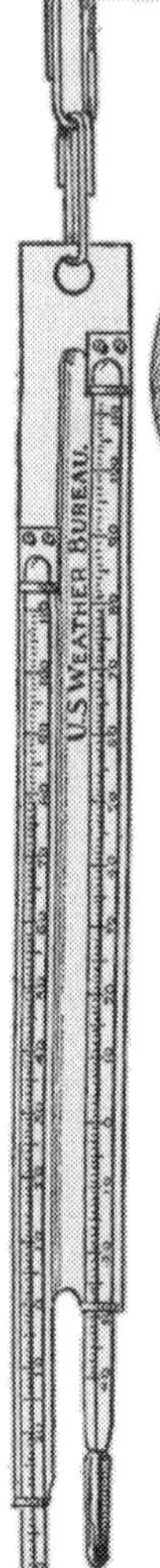

FIG. 1.—SLING PSYCHROMETER.

water, surrounding and in contact with what is known as the wet bulb of the psychrometer. The muslin should be renewed from time to time, as the meshes between the threads will gradually fill with solid matter left by the evaporation of the water and the natural accumulation of dust from the air. The muslin in this condition will neither absorb nor evaporate the water readily.

HOW TO USE THE SLING PSYCHROMETER.

To make an observation dip the muslin-covered bulb in a small cup or other wide-mouthed receptacle containing water. Whirl the thermometer for ten or fifteen seconds, then dip the wet bulb of the psychrometer into the water again. Whirl again for ten or fifteen seconds, stop and read quickly, reading the wet bulb first. Repeat once or twice, noting the reading each time. When two successive readings of the wet bulb agree very nearly, the lowest point has been reached. Dip the wet bulb only after the first whirling, as this is done only to make sure that the muslin is thoroughly saturated with water. If the water used is of nearly the same temperature as the room, correct readings are sooner obtained. If the psychrometer and water are at a much higher temperature than the air of the room, it will take a proportionately longer time to reach a correct reading, but the accuracy will not be impaired, if sufficient time is allowed for the mercury to settle. It is very important that the muslin-covered bulb should not become dry in the least; it should be saturated with water during the full time of observation. There will be no difficulty in getting accurate readings down to 29° F., as indicated by the dry bulb. At about this temperature, and with the wet bulb at about 27° F., ice will form on the wet bulb and cause the psychrometer to become somewhat erratic in its behavior. Readings below 30° F. are, therefore, very difficult to obtain, and it is only after repeated trials that results may be obtained in some cases. By dipping the instrument in water at a temperature near the freezing point and then rapidly whirling it results may usually be obtained. A stationary hygrometer is entirely inoperative at any temperature below 32° F., as the water in the fountain and wick will freeze solid. The sling psychrometer, according to Prof. Marvin, its originator, is supposed to be as accurate when the wet bulb is covered with ice as when covered with water, but this is not borne out by the author's personal experience. There is something to be desired in the way of further information on this point.

It is difficult to describe the proper movements for whirling the sling psychrometer, a little practice being the best instructor. The handle is held in a horizontal position, the frame mounting the thermometers revolving around the pivot, after the manner of the weapon with which David slew Goliath, and from which our moisture-tester gets the easy part of its name. A high rate of speed is unnecessary, a natural, easy motion of the forearm or wrist being all that is required. When stopping the psychrometer the arm should follow the thermometer from the highest point of the circle of rotation, whereby the radius of the path of the psychrometer is increased, and the momentum overcome. The stopping can be accomplished in a single revolution, after a little practice. The psychrometer will come to rest very nicely by simply allowing the arm to stand still, but the final revolution will be quite irregular and jerky.

In making observations in a storage room, the psychrometer should be held as far from the body as convenient, and toward the direction from which the circulation comes—the observer standing to the leeward as it were. In some cases it is necessary, or advisable, to step slowly back and forth a few steps, and the observer should turn his head from the direction of the psychrometer so that his breath will not affect the reading. In reading a thermometer, read as quickly as possible, and do not allow the breath to strike the bulb. It is a common practice with the author to hold his breath while reading a thermometer. It is unnecessary to caution against allowing the psychrometer to strike any object while whirling. In case it should, the observer will have $5 worth of experience, but no psychrometer.

The following short table needs no explanation further than has been already given. It will cover most cases in cold storage observations. It was not intended for cold storage

TABLE OF RELATIVE HUMIDITY, PER CENT.

Dry Ther.	Difference between dry and wet thermometers ($t-t'$).												Dry Ther.
	0°.5	1°.0	1°.5	2°.0	2°.5	3°.0	3°.5	4°.0	4°.5	5°.0	5°.5	6°.0	
25	94	87	81	74	68	62	56	50	44	38	32	26	25
26	94	88	81	75	69	63	57	51	45	40	34	28	26
27	94	88	82	76	70	64	59	53	47	42	36	30	27
28	94	88	82	76	71	65	60	54	49	43	38	33	28
29	94	89	83	77	72	66	61	56	50	45	40	35	29
30	94	89	84	78	73	67	62	57	52	47	41	36	30
31	95	89	84	79	74	68	63	58	53	48	43	38	31
32	95	90	84	79	74	69	64	59	54	50	45	40	32
33	95	90	85	80	75	70	65	60	56	51	47	42	33
34	95	91	86	81	75	72	67	62	57	53	48	44	34
35	95	91	86	82	76	73	69	65	59	54	50	45	35
36	96	91	86	82	77	73	70	66	61	56	51	47	36
37	96	91	87	82	78	74	70	66	62	57	52	48	37
38	96	92	87	83	79	75	71	67	63	58	54	50	38
39	96	92	88	83	79	75	72	68	63	59	55	52	39
40	96	92	88	84	80	76	72	68	64	60	56	53	40
41	96	92	88	84	80	76	72	69	65	61	57	54	41
42	96	92	88	84	81	77	73	69	65	62	58	55	42
43	96	92	88	85	81	77	74	70	66	63	59	56	43
44	96	92	88	85	81	78	74	70	67	63	60	57	44
45	96	92	89	85	82	78	75	71	67	64	61	58	45
46	96	93	89	85	82	79	75	72	68	65	61	58	46
47	96	93	89	86	83	79	76	72	69	66	62	59	47
48	96	93	89	86	83	79	76	73	69	66	63	60	48
49	97	93	90	86	83	80	76	73	70	67	63	60	49

work, being a part of the regular humidity tables published by the Weather Bureau. The full set of tables can be had by addressing the chief of the Weather Bureau, Department of Agriculture, Washington, D. C. They are published in pamphlet form, along with tables giving dew point temperatures. Observers must work out the small fractions for themselves, if they think necessary, but results within the limits covered by the table are near enough for practical purposes.

It is of no use to test for moisture unless having the means to control it, any more than a thermometer would be of use unless the means of regulating temperature were at hand. Humidity can be controlled by ventilation, already discussed, and the use of absorbents, which are considered in the following chapter.

CHAPTER X.

ABSORBENTS.

USE OF ABSORBENTS IN COLD STORAGE.

The use of absorbents in cold storage rooms has been common since the industry was in its infancy; their use originating, no doubt, from an appreciation of the fact that the air of a storage room quickly became too moist and impure to do the work of preservation perfectly. When absorbents and ventilation are applied to refrigerated rooms they practically have one duty in common—that of purifying the air. Ventilation purifies by furnishing pure air which displaces the foul air; absorbents by attracting the moisture, and with it the impurities of the storage room. But while ventilation is largely for the purpose of forcing out the permanent gases or impurities which have little affinity for moisture, absorbents are for the purpose of taking up the moisture and the germs and impurities which are absorbed by it.

Active absorbents can be made to perform duty in absorbing the moisture which is usually condensed on the cooling coils, as illustrated in one style of the antiquated overhead ice cold storages, called Prof. Nyce's system. In this system the ice is supported above a water-tight sheet iron floor which forms the ceiling of the storage room, the air of the room being cooled merely by contact with this cold metal surface, which is cooled by the ice above. The moisture given off by the goods in storage, and that resulting from air leakage was taken up by an absorbent, chloride of calcium being the chemical mostly in use for this purpose. It was applied by suspending it in pans at the ceiling of the room, or in some cases on the floor under the goods. Prof. Nyce's system gave good results years ago in competition with the Jackson, Dexter,

McCray, Stevens, etc., systems of overhead ice cold storage; and which low temperatures, and the improved systems of air circulation now in use have rendered obsolete to a greater or less extent. Mention is made of this system not as recommending it, but to show the possibilities of absorbents in drying and purifying storage rooms.

LIME.

The two chemical absorbents in general use for taking up moisture and the impurities from cold storage rooms are chloride of calcium and lime (either unslaked or air-slaked, or in the form of whitewash). (See chapter on "Keeping Cold Stores Clean.") Occasionally waste bittern from salt works is used, but the active principle of bittern is chloride of calcium. Ordinary quicklime has the property of absorbing moisture and impure gases from the air, and is used in very much the same way as chloride of calcium; that is, it is placed around the room on trays or pans. Lime, however, has very little capacity for moisture as compared with chloride of calcium, and when exposed to the air it will simply air-slake, which means that it will absorb moisture enough from the air to disintegrate into the form of a powder. Lime in this form is known as air-slaked lime, and is used to a large extent in storage rooms. Air-slaked lime as it comes from the lime house will absorb very little moisture, but it gives off minute particles of lime which have a good effect in preventing the growth of fungus, which we have already fully discussed. Air-slaked lime is usually applied by spreading on the floor of the room, between the 2x4s (which are used at the bottom of each pile of goods), to the depth of an inch or more. This must necessarily be done when the goods are piled, and consequently its efficiency is very low when the cool weather of fall comes. This defect has been overcome by scattering fresh air-slaked lime through the rooms so as to create a cloud of lime dust, but this is objected to because it musses up the packages. A better way of using lime is in the lump form (quicklime) which can be placed around the top of the room in trays or pans and renewed from time to time through the season.

CHLORIDE OF CALCIUM.

Chloride of calcium is the most vigorous absorbent (or drier, as it is called) which we are discussing. It is the same salt of the metal calcium as common salt (chloride of sodium) is of the metal sodium. Both have a strong affinity for water, but chloride of calcium is much the more energetic of the two. Where, in moist air, common salt simply attracts enough moisture to become damp, chloride of calcium will absorb enough water to lose its solid form entirely, uniting with the moisture of the air to form a solution or brine. The strong affinity of this salt for water has been utilized for the purpose of drying and purifying refrigerated rooms, and in this capacity has been a general favorite for years. The most primitive method of applying it is to place it in a simple iron pan, allowing the brine to run off into a pail as fast as formed. A better way is to support the calcium on a screen of galvanized wire, with a galvanized pan below for catching the brine. This allows of a free circulation of air around the calcium. This apparatus should be suspended near the ceiling of the room, one end slightly higher, to allow the brine to run off into a galvanized iron pail, supported at the low end of the pan. Galvanized iron is specified because black iron rusts badly when exposed to the air. (In the chapter on "Uses of Chloride of Calcium" a complete description of the uses of this material and illustrations of methods of applying are given.)

Do not in any method of using chloride of calcium evaporate the water from the brine and use the salt over again. The impurities will stay in the salt to a large extent, which is quite harmful, and the calcium has at least lost its value as a purifier, to a large extent. The quantity of calcium necessary depends on the conditions under which it is to be used, but in any case it is safe to use much more than the author saw in use in one eastern house. A room about 30x50 and about fourteen feet high had the refrigerant shut off, and the room was in rather bad condition as to moisture, etc. In each end of the room a pail was placed, on which rested a wire screen, with perhaps ten or fifteen pounds of chloride of calcium on it.

Electric fans were playing on the calcium, which was doing its best, but it seemed "like trying to dip the sea dry with a clam shell." This room should have had at least two drums (about 1,200 pounds) at work in it to do it justice.

CHAPTER XI.

USES OF CHLORIDE OF CALCIUM.

CALCIUM CHLORIDE AS AN ABSORBENT.

Chloride of calcium is a substance which is known in chemistry as a deliquescent salt, which term means that it will become liquid by the absorption of moisture from the air. It is obtained as a by-product in the preparation of ammonia from ammonium chloride and lime; in the preparation of potassium chlorate from calcium chlorate and potassium chloride; in the ammonia-soda or Solvay process, and in the manufacture of carbon dioxide or carbonic acid gas. The greater portion of the commercial product comes from the waste bittern from the salt works, and the Solvay process for the manufacture of soda.

The capacity of chloride of calcium for water depends largely on the temperature at which the solution from which it is prepared is evaporated, and to the presence of a greater or less percentage of impurities (chloride of magnesium, chloride of sodium, gypsum, sulphates, etc.), some of which possess comparatively little or no value as absorbents. Commercial chloride of calcium, as generally prepared, holds about 25 per cent of water, and it will absorb in addition to this, when exposed under average conditions in cold storage rooms, somewhere from one-half to nearly its own weight of water, depending on humidity of the air, temperature, method of applying, etc. It is the most active moisture absorber, or drier—as it is sometimes called—in common use, and because of its low price ($10 to $15 per ton), it has come into general use for many purposes. In general character, common salt (chloride of sodium) and chloride of calcium are similar, both having strong affinity for moisture.

It is a well known fact that cold storage rooms are purified to a large extent by extracting the water vapor which is held in suspension by the air contained in the rooms. The water vapor contains a greater part of the foul gases, germs of decay, etc., which are given off by the goods, or introduced into the rooms by admitting impure moist air from the outside. The water vapor laden with these impurities is disposed of in mechanically refrigerated cold storage rooms by being frozen on the cooling pipes. Because of the strong

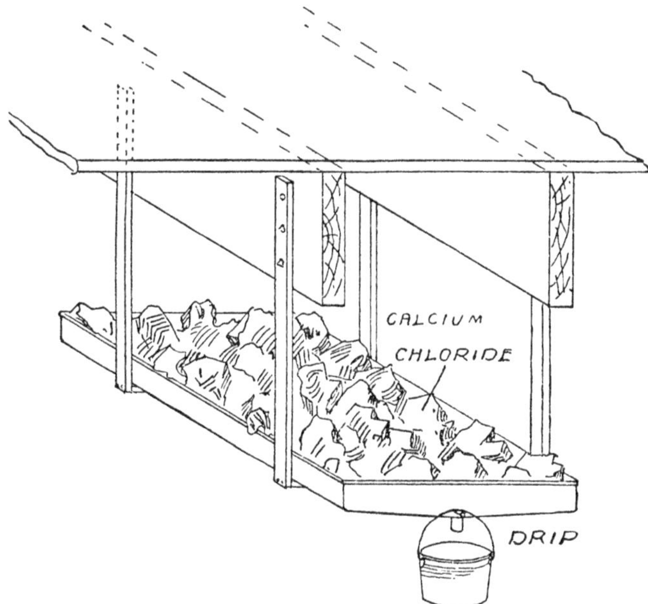

FIG. 1.—CALCIUM SUPPORTED NEAR CEILING.

affinity of chloride of calcium for moisture, it can be utilized to accomplish the same duty in moisture absorbing and purification which can be accomplished by the refrigerating pipes. It has been in use for years for this purpose; the natural ice cold storage houses having used it largely before the advent of the refrigerating machine. When used in a room cooled by air circulated directly from the ice, it is of very little service except during very cold weather, because such a room is held at a positive humidity by the air circulating continually in contact with the moist surface of the melting ice.

The possibilities in the use of chloride of calcium for moisture aborbing are well illustrated in the system of overhead ice cold storage originated by Professor Nyce. (See chapter on "Absorbents.")

The success of this system depends on chloride of calcium as its only agent for moisture absorbing and purification, and proves conclusively its value for the purpose, and those who

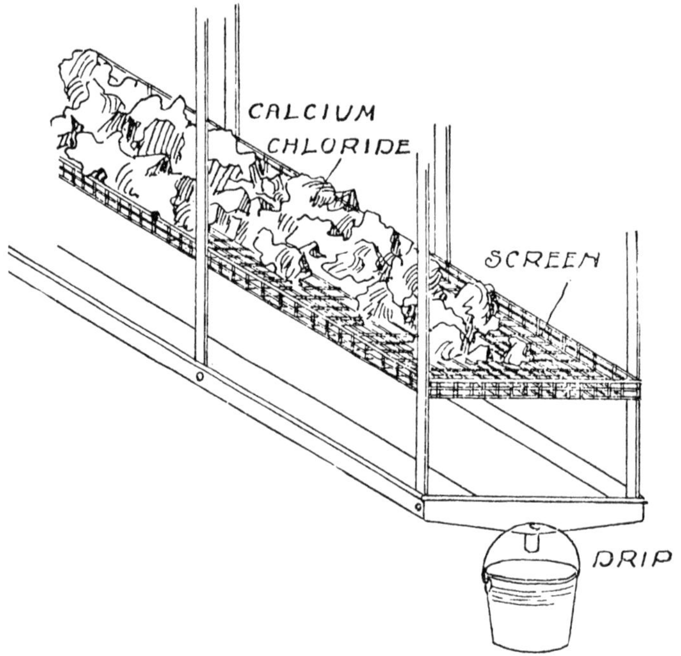

FIG. 2.—METHOD OF CONSTRUCTION OF CALCIUM PANS.

are operating mechanically refrigerated houses can take some ideas from this old system which will assist them through the cold weather of fall and winter, when they are obliged to discontinue the flow of refrigerant through the cooling pipes. When this becomes necessary, the frost on pipes must be promptly cleaned off (which is at times impossible, owing to the stock of stored goods in the room), or the frost will throw off water vapor which is laden with impurities which have been absorbed from the air of the room. The result is easy to foresee. The air becomes moist and foul, and goods stored in such an atmosphere deteriorate very rapidly. The remedy for such a state of things is to expose to the air of the storage rooms a large quantity of chloride of calcium; or, what is better still, this condition can be made impossible by preventing the formation of frost on pipes by the application of chloride of calcium by a process invented by the author, which will be described further on.

DEVICES FOR APPLICATION.

The methods of applying chloride of calcium to the work of moisture absorbing are numerous, but the devices illustrated and described here have been found to do well and will fit almost any case that may come up. Fig. 1 is a cheap, simple way of supporting the calcium near the ceiling of room. It is best to support the calcium near the ceiling, as the space is less valuable and the moistest air is to be found there. The pan or trough of galvanized iron, shown in the sketch, should be inclined toward the outlet, so that the liquid calcium will flow off into a receptacle as fast as formed. The pan is usually suspended over the alley-way between goods, so that it may readily be refilled as required. These pans may be of any size and shape desired, corresponding to the space which they will occupy, but in placing them in the room plenty of space should be left on the sides for the free access of air. The pan shown in Fig. 2 is an improvement on the first, in that the calcium is supported on a wire screen, several inches above the pan below, allowing a free flow of air around the calcium, exposing a greater surface to the action of the air. The liquid dripping from above covers the pan beneath with a film of brine, and the air in contact with this brine will give up its moisture to some extent, resulting in a more dilute brine and, consequently, greater economy in the consumption of the calcium. In other words, a pound of the calcium used in the device shown in Fig. 2 will absorb more moisture than the same quantity used in the device illustrated in Fig. 1,

The general explanation of proper method of using, given in connection with Fig. 1, is equally applicable to Fig. 2. These pans should be constructed of galvanized iron throughout, as they are exposed intermittently to the action of the chloride and the dry air outside when they are out of service; and,

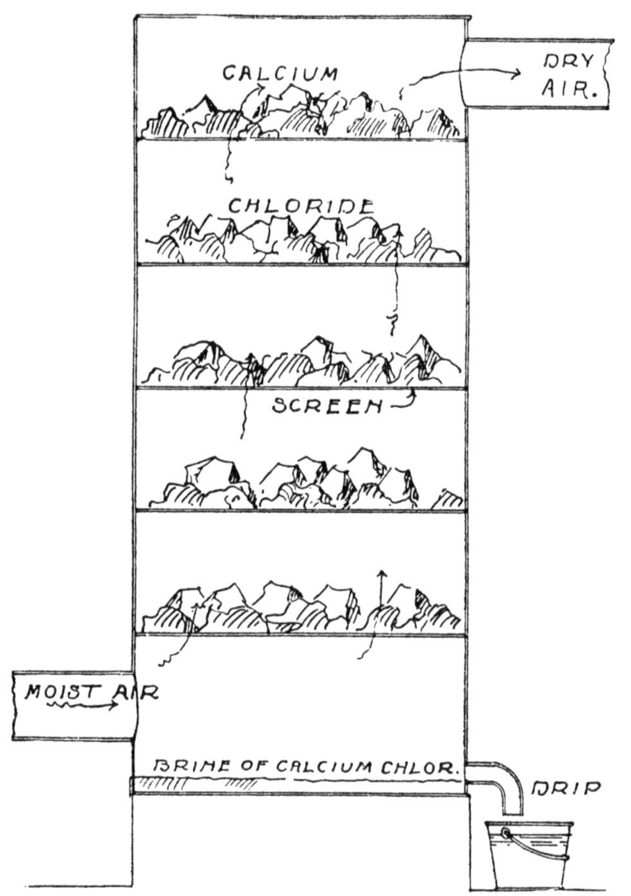

FIG. 3.—ARRANGEMENT FOR DRYING THE AIR IN ROOMS BY USING CHLORIDE OF CALCIUM.

as the calcium will keep them moist a long time, the action of the air in connection with this moisture will cause them to rust badly. Any iron surface continually covered with calcium brine will rust very little—no more, probably not as much, as it would if exposed to the atmosphere under ordinary conditions.

The device shown in Fig. 3 is a more positive and powerful arrangement for drying the air of storage rooms than either of the two described. The chloride is placed in a tank or box on wire screen shelves, as shown, and the air forced or drawn through the box by an exhaust fan, which may be placed on the inlet or outlet end, as may be most convenient. The moist air should be taken from the top of the room to be dried, and conducted to the bottom of box, the dry air to be taken out of the top of box and discharged at the opposite end of the room. In this way the moist air comes first in contact with the liquid calcium, or brine, which lies at the bottom of the box. As the drip from the top shelves drops from one shelf to another, always in contact with the air moving upward, it becomes more and more dilute. It will be seen, therefore, that the air which is moistest comes first in contact with the dilute brine at bottom of tank, and last with the dryest calcium at the top of box. This results in a greater economy in the use of calcium, and gives a more perfect drying effect. The devices shown in Figs. 1 and 2 are much slower in their action, because depending on the ordinary air circulation in the room to bring the air containing the mixture in contact with the calcium.

COOPER CHLORIDE OF CALCIUM PROCESS.

A better method of utilizing chloride of calcium than those described has been designed and thoroughly tested by the author. Claims fully covering this process have been allowed by the patent office at Washington, and it has been put in service in a large number of refrigerating plants. In this system the calcium is made to perform two distinct duties, that of keeping the pipes free of frost during warm weather, and during cold weather, that of maintaining the air of the storage room at the correct degree of humidity, at the same time maintaining it in a pure state. The process is applicable to any of the mechanical systems of refrigeration wherein a refrigerant is circulated through coils of pipe, or to any system where the rooms are cooled by refrigerated metal surfaces. A smaller amount of surface is required to do a given refrigerating duty when the pipes are clean than when the frost is allowed to accumulate on the pipes, and the economy of a device which will keep the refrigerating pipes free of frost at all times will be appreciated by any person familiar with the business, as it is well known that frosted pipes are insulated partly, the degree to which they are insulated depending on the thickness of the coat. We have Mr. E. T. Skinkle's ("The Boy") opinion that this is probably about as the square of the thickness of the frost. Mr. John Levey states: "Perhaps the best system of using chloride of calcium for reducing humidity is one designed by Madison Cooper, in which the calcium is placed in perforated troughs over the cooling pipes in such a manner that the brine formed by absorbing moisture will trickle down over the coils and cut off the frost. This not only reduces the humidity of the air but increases the efficiency of the coils from 15 to 25 per cent., and will result either in a lower temperature in the rooms or a slowing down of the brine pump to maintain the same temperature." The author's process consists simply in placing a quantity of chloride of calcium in proximity to the refrigerating surfaces, so that the brine resulting from a union of the moisture in the air with the calcium will drip over the refrigerating pipes. After passing down over the pipes, the brine falls onto a water tight floor, which is provided with drip connections to the sewer, or the brine may be collected and used as a circulating medium in the system. This effectually and continually disposes of the brine which contains the moisture and impurities from the air of the storage room, therefore contamination from this source is impossible. The apparatus illustrated in Fig. 4 is a simple and effective manner of applying the calcium, although it can be applied in any other manner to produce the desired result; as in case of ceiling coils the calcium may be placed directly on the pipe. The film of brine, covering the pipes, which is produced in this way, practically prevents the formation of frost, and the cooling surfaces of the pipes are, therefore, maintained at their maximum efficiency at all times. The economical advantages of this process are great, the cost of installing the apparatus very small, and the expense for calcium not large.

The disadvantages of the system are very few, if any. The chief one which has been suggested so far is that the chloride of calcium brine trickling over the pipe surfaces would cause the pipes to rust. Rather than rust the pipes, the brine has a cleaning and protective effect, and coils which

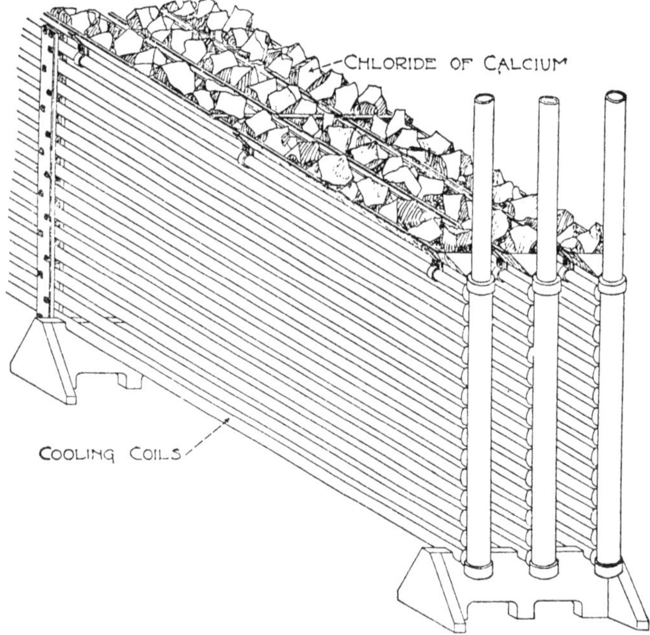

FIG. 4.—COOPER'S CHLORIDE OF CALCIUM PROCESS.

have been equipped with this process show freer of rust after being in service for a few weeks than when first fitted up. It is generally conceded by those who have observed carefully that the most favorable condition for rusting of iron is alternately wetting and drying in the presence of a free circulation of air. When the pipes are coated with a film of brine, no corroding action of consequence will take place, because the air cannot have free access to the surface of the pipes.

The expense for chloride of calcium has also been cited as an objection to the process. When it is considered that it is only necessary to supply about the same weight of the salt as of the frost to be kept off the pipes, it will be seen that expense for this salt is of very small importance. The estimated weight of frost which will accumulate on the pipes during the season in a room of 20,000 cubic feet is about 2,000 pounds. The amount will vary greatly with the season of the year, product stored, and whether room is opened often or not, but above figures will cover average conditions. The cost of calcium as compared with the economy which results from maintaining clean pipes at all times is of small moment, amounting to only a very small percentage of the saving affected by maintaining the refrigerating surfaces at their maximum efficiency at all times.

To show the possibilities of this process, combined with the system of forced air circulation designed by the author and fully described in the chapter on "Air Circulation," the following is quoted from a letter received from a gentleman using these systems. He says:

> A remarkable thing is the small amount of cooling surface required. I put eleven coils, sixteen and one-half feet long, fourteen pipes to the coil, in the coil room, and I am indeed surprised to find that with this system I only need one of these coils, containing 231 feet of 1-inch pipe, brine entering at 14° F. from our ice tank.

This statement refers to the cooling of a room of about 20,000 cubic feet capacity to a temperature of 33° F. This means that a lineal foot of 1-inch pipe is cooling about eighty-five cubic feet of space, with brine at an initial temperature of 14° F.

Naturally, this process, like all others, would have some limitation as to its application; and this limitation is found when a temperature of about 10° F. is reached. It has been used successfully in a room where the temperature was carried at 10° to 12° F., but when tested in a freezer at a temperature of 8° F. the action of the calcium was very slow and the process partly inoperative. At a temperature of 30° F. the action is rapid, and no difficulty was experienced in keeping a coil of sixteen 1-inch brine pipes, one above another, practically free of frost.

In case of slow action of the calcium in cutting frost off pipes when it has already formed, or in low temperature freezers, the action may be hastened by pouring strong chloride of calcium brine over the lumps of calcium in the gutter and allowing it to drip down over the pipes. This starts the action and this may be repeated at intervals if necessary.

Brine circulating pipes can also be kept free from frost by coating them on the outside with a very strong solution of chloride of calcium using a brush dipped in brine for the purpose; the moisture precipitated on them will not freeze and form frost, but will be absorbed by the calcium and drip off the pipes.

This can be done even after the pipes are coated with frost and the frost will soon be absorbed and leave the pipes bare. It is best, however , to coat the pipes with calcium before starting to cool the room, then apply the brine during the season as often as necessary to keep them clean. This plan is especially adapted to small rooms where the coils are placed over water-tight floors or pans to catch the drip.

PREPARING AND HANDLING.

The preparation of chloride of calcium for use is attended with some very disagreeable features, unless a person has had experience and knows the nature of the material to be handled. Some of those who have used calcium have been discouraged from using it again by the hard labor required to put it in shape, and the wetting of floors it causes when carelessly handled. It is also very destructive to leather shoes, and for this reason rubber overshoes or rubber boots should be used when handling this material. For the benefit of those who have never handled this salt, and for those who have experienced difficulty in its preparation, the following directions are given, which if adhered to, will make the preparing of calcium for use as simple a matter as any of the routine work about a cold storage warehouse.

Chloride of calcium in the commercial form comes from the manufacturers in the form of a solid cake, encased in an air tight sheet iron jacket. These jackets are known as drums.

They are simply ordinary black sheet iron of a very light gauge, and are of no value, and when removed from the calcium may as well be thrown away at once. The drums of calcium weigh about 600 to 700 pounds each, and, though heavy, are easily rolled or trucked, and require very little space for storage.

For use, the calcium needs to be broken into lumps, ranging in size from ten pounds downward. This is for convenience in handling and for the purpose of exposing a fair amount of surface to the action of the air. For breaking the calcium select a clear floor space, where nothing can be injured by the moisture, which soon collects on the small pieces which are scattered in breaking. Pound the drum with a

FIG. 5.—BREAKING UP DRUM OF CALCIUM.

sledge hammer, using strong, vigorous blows, working around the drum and do not strike twice in the same place (see Fig. 5), as this tends to pulverize the calcium too much for easy handling and for air drying purposes, though for brine making the finer the calcium is broken the better. After pounding the drum outside thoroughly, stand it on end and take off the top of the drum by prying it out with an old ax or chisel. It is then an easy matter to cut down the side with an ax, when the sheet iron jacket may be easily removed. Any large pieces needing further breaking may be reduced in size without much trouble by striking on the flat side. It is a very simple and easy matter to break the calcium in this way. An active man will prepare and place a drum in an hour or so.

The calcium begins absorbing moisture from the air very quickly, especially in warm, humid weather, and for this reason when a drum is once broken into, it should be disposed of as quickly as possible. The small pieces which fly about when the cake is being broken should be swept up promptly to prevent making a muss; some dry sawdust, scattered over the place where the cake was broken, will be found useful in taking up the moisture which accumulates. As before stated, chloride of calcium is of a similar character to common salt, and aside from the disagreeable property of making everything damp with which it comes in contact, and keeping it so for some time, is entirely harmless.

CHLORIDE OF CALCIUM BRINE.

A non-congealable liquid is used in refrigeration as a secondary or circulating medium for absorbing the cooling effect of an expanding gas, and applying it directly to the work to be done. This non-congealable liquid has been in the past usually a solution of common salt in water; but of late chloride of calcium has come into use quite generally for this purpose. Probably the chief reason why it has not come into general use before to the entire exclusion of common salt brine, are: That it is, or has been, much more expensive in first cost; that it is more difficult to prepare and handle the solution, and also that it cannot be obtained everywhere like common salt. Chloride of calcium possesses positive advantages over common salt for brine making. It is now used by many of the leading engineers in the business, and where once adopted, has not, in a single instance known to the writer, been discarded for common salt. As the use of the so-called brine coolers have made the brine circulating system more desirable, and the brine system is now in favor for most purposes, the proper understanding of chloride of calcium and its use should be a part of the information possessed by every engineer connected with the business.

Those who have written on the subject of refrigerating machinery and refrigeration, have had very little to say regarding the merits of the two different salts for brine purposes. Most of the information formerly available relates to common salt brine, which is a sort of tacit recommendation for its use; but brine and brine making in a general way have until recently been given very little attention by writers on refrigeration. In connection with some investigations bearing on the process for preventing frost on refrigerating pipes already described, the author has collected all the available information on the general subject of chloride of calcium, and all facts obtainable show that calcium brine has important advantages over that made from common salt.

The manufacturers or venders of chloride of calcium claim that it is a better conveyor of refrigeration and that "it does

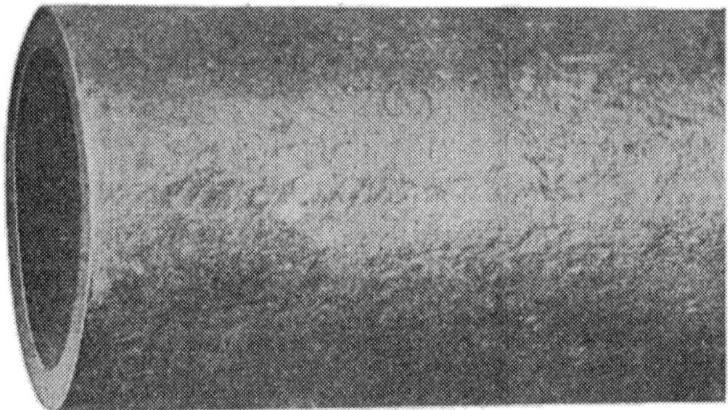

FIG. 6.—PIPE USED FOR FOUR YEARS WITH CALCIUM CHLORIDE.

not eat up the pipes like salt." These claims are, roughly speaking, true, and if the reasons why had been given, the claims would have more weight with engineers. The author's reason why chloride of calcium brine will not rust refrigerating pipes has already been given in connection with the explanation why calcium brine trickling over the pipes in the frost preventing process will not rust the pipes. Probably ordinary salt brine will not corrode the pipes very much more on the inside, but wherever it has access to the exterior of the pipes in contact with air, as from a leaky joint, the corrosion and deterioration are much more rapid than where calcium brine is used. It is probable that the impurities encountered in common salt are responsible to a great extent for the peculiar rotting action which it has in some cases on cast or wrought iron or steel. Calcium also contains damaging impurities at times. Figs. 6, 7 and 8 illustrate the "pitting" or corrosion of pipe when using salt brine, and freedom from same when chloride of calcium brine is employed. The surfaces of pipes moistened by common salt brine, are, owing to varying conditions causing a tendency to dry at one season of the year and become moist at another, subject to the action so favorable for the corrosion of the metal. Calcium brine will not, under any conditions to be met with in cold storage rooms, give up enough water to lose its liquid form, so will not allow of a drying out on the pipes except after a considerable length of time has elapsed.

Without the aid of chloride of calcium the present perfect types of brine coolers would not have been possible. Now the

FIG. 7.—INTERIOR OF A UNION USED IN SALT BRINE FOR THREE YEARS.

brine cooler is recognized as a feature of nearly all up-to-date cold storage plants, and in many ice factories the brine for freezing is cooled in a brine cooler. The saving of space, low cost, and perfection of interchange of temperature between the ammonia and the brine, makes the brine cooler an ideal device. Operating engineers appreciate the saving to them in care of looking after a large number of expansion valves scattered throughout the plant.

Obviously calcium brine has a great advantage over common salt brine at temperatures below zero F. Common salt brine at its maximum density will freeze at about 7° below zero F., while calcium brine can be made which will not freeze at 50° below zero F. It will be seen that where a temperature of zero F. or lower is required in cold storage rooms with brine circulation, calcium brine only can be used. For a given minimum brine temperature a less dense brine of calcium can be used than of common salt, giving more conducting power per pound. The advantages of this are that a given weight of calcium brine can be made to convey more

FIG. 8.—PIPE USED FOR FIVE YEARS WITH SALT BRINE.

units of refrigeration than the same weight of salt brine, saving in the weight and amount of brine to be circulated. Chloride of calcium brine has the advantage of not being liable to deposit crystals in the pipes should the temperature drop below normal, and there is practically no danger of freezing if reasonable care is used in its preparation. Reference to the subjoined table shows that calcium brine has an ultimate freezing point of about 54° below zero F. with a 30 per cent solution. A 25 per cent solution is all that is required in almost any work, and for most purposes a 20 per cent solution is amply dense. For ice making, where a brine tem-

PROPERTIES OF SOLUTION OF CHLORIDE OF CALCIUM. (CALCIUM BRINE.)

Degrees on Salometer, 60° F.	Degrees Baumé, 60° F.	Specific Gravity, 60° F.	Per Cent of Chloride of Calcium.	Freezing Point, Degrees Fahrenheit.	Freezing Point, Degrees Celsius.	Ammonia Gauge.	Specific Heat.
						Pressures.	
4	1	1.007	1	+31.10	— .5	46	.996
8	2	1.015	2	+30.38	— .9	45	.988
12	3	1.024	3	+29.48	— 1.4	44	.980
16	4	1.032	4	+28.58	— 1.9	43	.972
22	5.5	1.041	5	+27.68	— 2.4	41.5	.964
26	6.5	1.049	6	+26.60	— 3.0	39.5	.960
32	8	1.058	7	+25.52	— 3.6	38	.936
36	9	1.067	8	+24.26	— 4.3	37	.925
40	10	1.076	9	+22.8	— 5.1	35.5	.911
44	11	1.085	10	+21.3	— 5.9	34	.896
48	12	1.094	11	+19.7	— 6.8	32.5	.890
52	13	1.103	12	+18.1	— 7.7	30.5	.884
58	14.5	1.112	13	+16.3	— 8.7	28	.876
62	15.5	1.121	14	+14.3	— 9.8	26	.868
68	17	1.131	15	+12.2	—11.0	23.5	.860
72	18	1.140	16	+10.0	—12.2	21.5	.854
76	19	1.150	17	+ 7.5	—13.6	20	.849
80	20	1.159	18	+ 4.6	—15.2	18	.844
84	21	1.169	19	+ 1.7	—16.8	15	.839
88	22	1.179	20	— 1.4	—18.6	12.5	.834
92	23	1.189	21	— 4.9	—20.5	10.5	.825
96	24	1.199	22	— 8.6	—22.6	8	.817
100	25	1.209	23	—11.6	—24.8	6	.808
104	26	1.219	24	—17.1	—27.3	4	.799
108	27	1.229	25	—21.8	—29.9	1.5	.790
						Vacuum.	
112	28	1.240	26	—27.0	—32.8	1	.778
116	29	1.250	27	—32.6	—35.9	5	.769
120	30	1.261	28	—39.2	—39.6	8.5	.757
....	31	1.272	29	—46.3	—43.5	12	....
....	32	1.283	30	—54.4	—48.0	15	....
....	33	1.294	31	—52.5	—46.9	10	....
....	34	1.305	32	—39.2	—39.6	4	....
....	35	1.316	33	—25.2	—31.8	1.5	....
....	35.5	1.327	34	— 9.7	—23.2	....	....
....	36.5	1.338	35	+ 2.8	—16.2	....	....
....	37.5	1.349	36	+14.3	— 9.8	....	....

NOTE.—The + sign denotes temperature above zero, the — sign, temperature below zero.

perature of 10° to 20° F. is carried in the tank, a brine ranging from 12 to 18 per cent is all that is required. The brine must of course, be strong enough to prevent ice forming on the expansion coils, so that the temperature of the expanding ammonia must largely regulate the density of the brine. It will be noted from the table that a very strong solution of chloride of calcium has a much higher freezing point than a more dilute brine. A brine containing too much calcium is therefore to be guarded against. The most common test for brine is the salometer, a hydrometer scaled from zero of pure water to 100 per cent or more, which is about the point of a saturated solution of common salt brine. A Baumé hydrometer scale can also be used for ascertaining percentage of calcium. The per cent of calcium given in the table represents the total per cent, and as the commercial fused chloride of calcium already contains about 25 per cent of water, more of this article will be required for a given quantity of water than is stated in the table. The small sub-table of approximate practical proportions of the commercial calcium and water, for brine of a required test, will be found useful in the making of brine. (See page 230).

The preparation of brine, using chloride of calcium, is a simple matter but somewhat slower than where common salt is used, owing to the much smaller surface exposed to the action of the water. It is difficult to break calcium by hand into small grains like salt, therefore it dissolves comparatively slowly. The simplest way is to put the correct proportion of calcium and water in a barrel or barrels, and stir slowly with a piece of gas pipe to facilitate solution. Another method is to put the correct quantity of calcium and water in the brine tank, and start the pumps running. The circulation of water in contact with the crushed calcium is what is necessary. Others use a steam pipe lead directly into the brine tank or other receptacle. This is perhaps the more rapid way, but it is not desirable, from the fact that the solution may not be at the correct degree when completed, because of the indefinite amount of steam necessary to effect a solution. It is best to have the solution amply strong at first, as it can be readily reduced by adding water in sufficient quantity. If the live steam method is used, a good proportion to put into the brine receptacle is six pounds of the calcium to each gallon of water, or a drum to each 100 gallons. This will make a very strong brine which can be diluted as required. In testing brine it is necessary to have the solution at a tem-

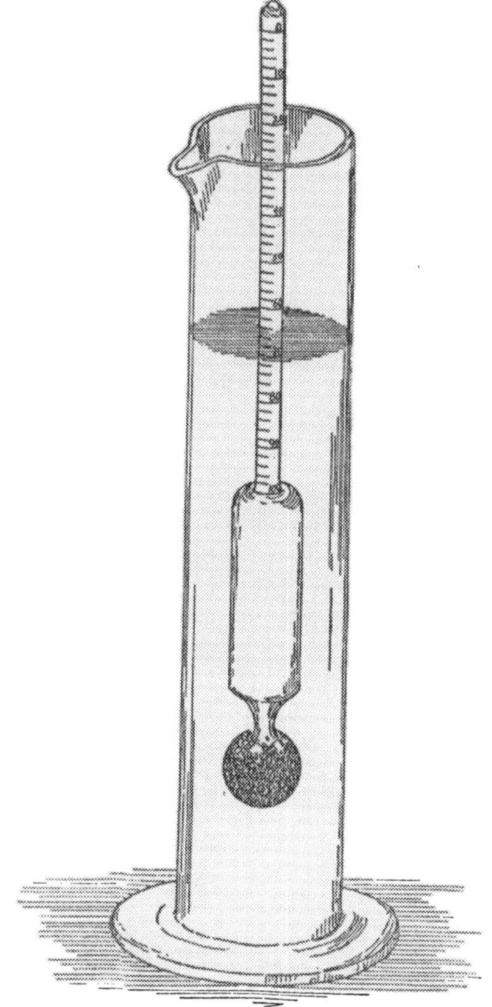

FIG. 9.—HYDROMETER IN GLASS JAR.

perature of 60° F., as any variation from this temperature will cause error in the test. The brine is easily warmed or cooled to the correct degree. A glass hydrometer jar (see Fig. 9) is useful, as supplying a convenient tall vessel, and the scale on the hydrometer can be read more accurately than with a piece of gas pipe with a cap on one end, which some use.

The following table is the one referred to above for the making of calcium brine and will be found of practical value:

PRACTICAL TABLE FOR MAKING CALCIUM BRINE.

Pounds Chloride of Calcium (Commercial fused) to one gallon of water	Degrees Salometer, 60° F.	Degrees Baumé, 60° F.	Freezing Point, Degrees F.
2½	80	20	4
3	88	22	— 2
3½	96	24	— 9
4	104	26	—17
4½	112	28	—27
5	120	30	—39
5½	...	32	—54

A part of the preceding tables and some of the information contained therein has been kindly supplied by the manufacturers of chloride of calcium, and while the tables have been proved inaccurate, they will answer for practical purposes.

CHAPTER XII.

ESTIMATING REFRIGERATING DUTY.

FACTORS TO BE CONSIDERED.

A query which comes up constantly in cold storage practice is the amount of refrigeration required for any given purpose. One of the first questions asked by a prospective purchaser of a refrigerating apparatus is "How much will it cost to run it?" Such a question is not possible of accurate answer without knowing the conditions, and the conditions are in many cases difficult to determine. We will attempt to outline the facts and information needed before calculations are possible which will give the figures so often called for by the above question. Usually no calculations are made in handling the problem, but what is known as "rule of thumb" method is used, commonly so many linear feet of pipe for so many cubic feet of space.

The sources of heat to be taken care of in connection with the refrigeration of cold storage rooms may be roughly considered as follows:

First—Conduction through the walls containing the refrigerated space.

Second—Introduction of heat by means of goods or warm air.

Third—Generation of heat within the refrigerated space.

It is obvious, from a consideration of these conditions, that the volume or cubic capacity of the refrigerated space has no direct bearing on the problem except perhaps to limit the possible requirements; hence the common question as to the cost of operating a given capacity is entirely "beside the mark."

Necessarily there must be some unknown quantities to deal with like opening of doors, workmen employed in the rooms, etc., but we may calculate the greatest number of known quantities obtainable and assume or estimate on the unknown quantities. The refrigerating duty may be considered under the following heads, although they combine partially in making the necessary calculations:

First—Temperature to be maintained in the room.

Second—Temperature of the outside air.

Third—Exposed insulated surface of the room.

Fourth—Character and thickness of insulation employed.

Fifth—Quantity, temperature and kind of goods placed in the room per day.

Sixth—Number of lights and workmen employed and time.

Seventh—Frequency of opening of doors and time standing open.

As this is a practical and not a scientific discussion we will try to eliminate the unimportant and arrive at approximate results, which as a matter of fact, is all that is possible in advance of actual performance. Workmen in the room are to be reckoned with sufficiently to place the goods in storage and remove them therefrom, and this factor may be covered by allowing a small percentage, say 3 per cent to 5 per cent to the duty of cooling or freezing the product stored. About 1,000 B. T. U.'s per hour per person may be allowed. The burning of lights is also another necessary factor, and practically keeps pace with the time of workmen in the room. Modern electric lights make very little heat, and the heat evolved by lights may be omitted or estimated along with the heat from workmen. About 160 B. T. U.'s for a 50 watt p. lamp per hour may be allowed. The generation of heat within the space to be cooled from fermentation of goods may be mentioned, but cannot be specifically discussed as very little is known with reference to this factor as applied to the different goods which are commonly held in cold storage. If electric motors are used for driving fans or pumps it may be considered that all the electrical energy put into the motors is turned into heat and if the motor is located within the refrigerated space this is sometimes worth considering. One electrical watt-hour may be allowed as equal to about 3.2 B. T. U.'s.

The opening of doors is, in some cases, a very serious thing, especially if the room is a retail cooler and has no vestibule between it and the outer air. Also carelessness on the part of workmen in leaving doors open is well known, and as may be readily appreciated, even a rough assumption of an allowance for this factor is almost out of the question.

This brings us then to the five factors which may be handled with some degree of accuracy: Superficial area; outside temperature; room temperature; insulation and goods stored. The difference between outside temperature and inside temperature and the square feet of surface exposed taken with the conductivity of the insulation really make one factor, which may be accurately calculated by taking the average. Say, for instance, the day temperature is 85° F. and the night temperature 55° F. Then assume the inside temperature averages during the twenty-four hours 35° F., and you have a difference between outside and inside temperature of 35° F. Suppose now we assume that the insulation is equal to 15 inches of mill shavings with double boards and paper on each side. Stoddard gives this a value of .03 B. T. U.'s or three one-hundredths of a B. T. U. per hour per degree difference, or applied to our problem 1.05 B. T. U.'s. Call it 1 B. T. U. This would mean that one pound of ice would take care of 142 square feet of surface under the above set of conditions; and applying this to a room say 20x30x12, which has 2,400 sq. ft. outside exposure, and dividing by 142 gives about 17, the pounds of ice required to take up the heat coming through the insulation per hour. Multiply by 24, and we get about 400, the number of pounds of ice required per day. This is, of course, a very small amount for the size of the room, but it is not often that as efficient insulation is used, although it should be. The holding of temperature then with good insulation is a very small matter were it not for the other factors.

This brings us to the cooling or freezing of the goods, which is by far the most important factor in many cases. In determining refrigerating capacity required or amount of refrigerating or pipe surface necessary to do the work, the cooling or freezing service to be handled must be considered as the most important thing. Service means the use to which the room is to be applied. If, for instance, the room is used for the storage of eggs, which mostly go into storage in April during cool weather, the refrigeration required will be to take up the heat which comes through the insulated walls. On the other hand, if the room is to be used for the cooling of freshly killed poultry or freshly killed meat, then the heat which finds its way through the insulation is of minor importance, and the cooling of the product is of greatest importance.

If the weight of the goods to be cooled is known it is a comparatively easy matter to ascertain accurately the amount of refrigeration required for cooling, and in doing this we must know the so-called specific heat of the goods to be cooled.

SPECIFIC HEAT.

Different products or substances of equal weight require different amounts of heat to raise them to a given temperature. Water requires the most heat of any and is, therefore, used as the unit of measurement. The figure which is used to express the heat required to raise the temperature of any given kind of material 1 degree as compared with the amount of heat required to raise an equal weight of water 1 degree, is called the specific heat of that particular material. For instance, eggs and poultry have a specific heat of .80. This would mean that one unit of heat being required to raise 1 pound of water 1 degree, only eight-tenths of a unit would be required to raise 1 pound of eggs or poultry. What is said as applied to raising the temperature applies also to the lowering of the temperature, and the following table covers some of the goods which are carried in cold storage.

The last column showing latent heat of freezing refers to the comparative effort required to freeze the various products, based on the latent heat of freezing water into ice, which equals 142 heat units. It will be noted that the latent heat of freezing the different substances is almost in direct proportion to the amount of water contained in them, as compared with the solids. Another point which will be noted is that fat meat requires much less refrigerating effort than lean meat, for the reason that it contains much less water.

Products	Water Per cent.	Solids. Per cent.	Specific heat above freezing. B. T. U.'s per lb.	Specific heat below freezing. B. T. U.'s per lb.	Latent heat of freezing. B. T. U.'s per lb.
Lean beef	72.00	28.00	0.77	0.41	102
Fat beef	51.00	49.00	0.60	0.34	72
Veal	63.00	37.00	0.70	0.39	90
Fat pork	39.00	61.00	0.51	0.30	55
Eggs	70.00	30.00	0.76	0.40	100
Potatoes	74.00	26.00	0.80	0.42	105
Cabbage	91.00	9.00	0.93	0.48	129
Cream	59.25	40.75	0.68	0.38	84
Milk	87.50	12.50	0.90	0.47	124
Oysters	80.38	19.62	0.84	0.44	114
Fish	78.00	22.00	0.82	0.43	111
Poultry	73.70	26.30	0.80	0.42	....

LATENT HEAT OF FREEZING.

The refrigeration required to reduce the temperature of a given product through any specified range is practically constant, but varies widely for different products as the above table indicates. If the cooling process is not carried below the actual freezing point of the product the amount of refrigeration may be found by multiplying the specific heat of that product by the number of degrees through which the product is to be cooled. If the goods are actually to be frozen the amount of refrigeration must be increased by the latent heat of freezing, and if the product after freezing is to be lowered in temperature still further the refrigeration required must be still further increased by the specific heat of the product below the freezing point multiplied by the number of degrees through which it is cooled below freezing.

As an example we might consider 10,000 pounds of freshly killed poultry to be cooled through a range of 68° F. By referring to the table it will be found that the specific heat of poultry above freezing would be 0.80, and, therefore, the heat to be extracted would be represented as follows:

$$0.80 \times 10{,}000 \times 68 = 544{,}000.$$

544,000 divided by 142 (number of B. T. U.'s per pound of refrigeration) gives us the cooling equal to 3,830 lbs. about. If the poultry is frozen the additional refrigeration will be as follows:

10,000x105 (latent heat of poultry when freezing)= 1,050,000 B. T. U.'s which, divided by 142 gives 7,394 lbs., and if additional cooling to say 0° F. is required, the additional refrigeration would be as follows:

10,000x0.42x32=about 134,400 B. T. U.'s or 946 pounds.

The total refrigeration duty required to cool the 10,000 pounds of poultry through a range of 68° F. freezing it at 32° F., and then chilling it to 0° F. would be as follows:

3,830+7,394+946=12,170 lbs., which divided by 2,000 (number of pounds in a ton) gives us somewhat over 6 tons of refrigeration required for the total work.

It should be noted in this connection that the poultry is figured to be frozen at 32° F. whereas the actual freezing point of poultry would be somewhat lower than this and, therefore, the calculation is not absolutely correct, but near enough for practical purposes.

ROUGH ESTIMATES FOR CUBIC SPACE.

As suggested at the beginning of this discussion estimates based on cubic capacity are necessarily pure "rule-of-thumb" estimates, but they are, of course useful as a guide, and the following figures are given as a rough approximation of the quantity of refrigeration needed for cold storage houses of varying capacity:

For storage houses of 1,000,000 cubic feet or over from 20 to 30 B. T. U.'s per cubic foot per day.

Storage houses of from 250,000 to 1,000,000 cubic feet, 25 to 40 B. T. U.'s per cubic foot per day.

Storage houses of from 50,000 cubic feet to 250,000, 35 to 50 B. T. U.'s per cubic foot per day.

Storage houses of from 15,000 to 50,000 cubic feet per day, 40 to 75 B. T. U.'s per cubic foot per day.

Cooling boxes or rooms of from 1,000 to 10,000 cubic feet, 60 to 100 B. T. U.'s per cubic foot per day.

Refrigerators or coolers of less than 1,000 cubic feet, from 100 to 500 B. T. U.'s per day.

Rooms which are used for chilling or cooling of such goods as meats, etc., should have an allowance of 50 to 100 per cent additional, and for the actual freezing of goods the amount should be multiplied by two or three.

To find the refrigeration required for a building of average insulation under average conditions an approximate method may be used, by multiplying the exposed surface by a factor which depends upon the temperature at which the building is to be carried and upon its size. Temperatures ranging from zero to 32° F., for surfaces of less than 5,000 square feet it may vary from 2.00 to .20. For buildings having from 5,000 to 20,000 sq. ft. of surface the factor of 1.5 to .15 may be used, and for surfaces of buildings from 20,000 sq. ft. upward from 1.2 to .12 respectively. These figures, it must be noted, are for average insulation. Good insulation should not require more than one-half the above figures. Average insulation is altogether too poor in quality or not enough of it in thickness, but insulation has been discussed elsewhere.

It is customary to allow in packing house work one ton of refrigeration for the cooling of ten 750 pound cattle, or thirty-five 350 pound hogs, and a rough estimate for small plants is a ton of refrigeration for from five to seven beeves and the same amount for from fifteen to twenty-five hogs. Another rough estimate is a ton of refrigeration for from 3,000 to 5,000 pounds of meat to be cooled. It must be noted in this connection, however, that all these rough estimates give a large surplus of capacity on account of small plants, and the actual requirements are very much less.

For average conditions and with fair insulation and for plants of 25 to 50 tons and larger for general cold storage warehouses one ton of refrigeration in 24 hours will maintain the following capacities and temperatures:

12,000 cu. ft. of general storage space at 32°-35°F. temperature.
8,000 cu. ft. of egg storage space at 28°-30°F. temperature.
5,000 cu. ft. of butter storage space at 10°-14°F. temperature.
3,000 cu. ft. of game and poultry storage space at 10°-18°F. temperature.
2,000 cu. ft. of game and pultry storage space at 0°F. temp.

Modern Refrigerating Machinery states that:

"In breweries one ton of refrigeration will maintain 8,000 cu. ft. of general storage space at 30° to 36° F., or will cool 40 barrels of beer-wort from 70° to 40° F.

In abattoirs and packing houses one ton of refrigeration will maintain

10,000 cu. ft. of curing space at 35°-40°F.
3,000 cu. ft. of freezing space at 20°F.
1,500 cu. ft. of freezing space at 0°F.

If the number of animals is known, independent of the heat losses through walls and exposed surfaces, one ton of refrigeration is required to properly chill

7-10 beeves, each weighing about 700 lbs., and surrounding space.
20-25 hogs, each weighing about 250 lbs., and surrounding space.
50-60 calves, each weighing about 90 lbs., and surrounding space.
70-75 sheep, each weighing about 75 lbs., and surrounding space.

CHAPTER XIII.

EGGS.

IMPORTANT FACTORS TO BE CONSIDERED.

Eggs are the most important goods now taken care of by cold storage methods, both as regards aggregate value and benefits to the community. They are also among the most difficult products to successfully refrigerate. In 1898 the author estimated the total value of eggs under refrigeration for safe keeping at about $20,000,000 annually for the United States alone. Statistics show that the consumption of eggs doubles every five to ten years. Therefore the value of eggs annually cold stored in the United States at this time (1913) cannot be very far from $70,000,000. Appreciating the importance of the industry and the lack of accurate information available, the author, some years ago, in the interest of a better understanding and dissemination of knowledge on the cold storage of eggs, communicated with quite a large number of individuals and companies, requesting that they give full answers to a printed list of questions sent them. The result was most gratifying; nearly one-half of those written to acknowledged receipt of the inquiry, and more than one-half of this number gave fairly complete replies to the questions submitted. Considering the fact that the inquiries were regarded by some as being of a rather personal nature, the proportion of managers sending replies in full was large. Several gentlemen were frank enough to say that personal considerations prevented them from giving any information; others gave guarded or partial replies. In the main, however, storage men have shown themselves willing to give information and exchange ideas.

The list of inquiries sent out covers the subject quite thoroughly, and divides it into six different parts, namely, temperature, humidity, air circulation, ventilation, absorbents and packages, with three separate questions relating to each. To the data furnished by others is added information from the author's experience and practice with such explanation of underlying laws as may seem necessary to a clear understanding of the principles of successful egg refrigeration. It is hoped that those who are new to the business may obtain valuable information from these collected data, and that those with experience may derive some benefit in the way of a review, and possibly pick up some new ideas as well.

TEMPERATURE.

Questions regarding the correct temperature of egg rooms have been asked repeatedly of storage men who have been in the business long enough to be looked to for advice, the same person, perhaps, giving a different answer from time to time, as his ideas change. There is no temperature on which a large majority of persons can agree as being right, and as giving superior results to any other. The claims made by the advocates of different temperatures will be considered, to determine, if possible, what degree is giving the best results in actual practice.

The three questions relating to temperatures were written to draw out opinion as to the right temperature, the lowest safe temperature, and what deleterious effect, if any, the egg sustained at low temperatures, which did not actually congeal the egg meat. The three temperature queries were:

First.—At what temperature do you hold your rooms for long period egg storage?

Second.—What temperature do you regard as the lowest limit at which eggs may be safely stored?

Third.—What effect have you noticed on eggs held at a lower temperature?

All the replies received contained answers relative to temperature, and by a very small majority 32° F. is the favorite temperature for long period egg storage. Some few, 33° F. and 34° F., with a few scattering ones up to 40° F. Under the freezing point, none recommended a temperature lower than 28° F., and for a very obvious reason, this being near to the actual freezing temperatrue of the albumen of a fresh egg. A very respectable minority say a temperature ranging from 30° F. to 31° F. is giving them prime results; and several recommend 30° F. straight, and say they should go no lower. In recent years there has been a decided tendency among storage men to get the temperature down near the safety limit, but many houses are so poorly equipped that they are unable to maintain a uniformly low temperature below 33° F., without danger of freezing eggs where they are exposed

FIG. 1.—VIEW IN EGG ROOM SHOWING METHOD OF PILING THE EGGS. NOTE PERFORATED FLOOR AND CEILING.

to the flow of cold air from coils. A house must be nicely equipped to maintain low temperatures with safety. More houses would use temperatures under 32° F. were they able to without danger to the eggs. A very successful eastern house issued a pamphlet in 1892. At that time they maintained a temperature of from 32° to 34° F. in their rooms. In sending out this little book during the winter of 1897-98 a postscript was added, as follows: "This pamphlet was published in 1892, when our plant was started. Since that time all first class cold storage houses have lowered their temperatures materially." No better illustration than this can be cited to show the tendency of the times. These people now use a temperature of 30° F. for eggs.

Most of the replies received contained answers to the second question, and the greater portion state this as being about 2° F. lower than that recommended for long period storage It is presumed that these two degrees are allowed as leeway, or margin of safety, for temperature fluctuations. Some state that eggs cannot be safely held below 32° F., but give no reason why, while two or three say a temperature of 27° F. will do no harm to eggs in cases. One reply states that eggs held in

FIG. 2.—EGG ROOM—FALSE FLOOR AND FALSE CEILING SYSTEM OF COOLING.

cut straw at 25° F. for three months showed no bad symptoms. It has never been made clear how the package can be any protection against temperature, when the temperature has been continuously maintained for a length of time sufficient to allow the heat to escape; and eggs will positively freeze at 25° F., as proven by experiments mentioned in another paragraph.

The answers to the third question were few in number, but cover a wide range. The scarcity of data on this point indicates that few have experimented with eggs at temperatures ranging from 25° F. to 30° F. Some say: "dark spot, denoting germ killed"; others, "white gets thin"; others, "eggs will decay more quickly"; or, "they will not 'stand up' as long when removed from storage." It is also claimed that "yolk is hardened or 'cooked' when temperature goes below 32° F." Some answers state a liability of freezing if eggs are held in storage at a temperature below 32° F. for any length of time.

As far as possible, we will dig out reasons for the claims made by advocates of both high and low temperatures, both having equal consideration. Taking 29° F. or 30° F. and 38°F. or 40° F., as representing the lowest and highest of general practice, we will see what is claimed by each; and also the faults of the other fellow's way of doing it, as they see it. Those who are holding their egg rooms at 40° F. say it is economical, that the eggs will keep well, that the consistency of the egg meat is more nearly like that of a fresh egg after being in storage six months, than if held at a lower temperature. As against a low temperature they say: A temperature of 30° F. is expensive to maintain; the yolk of the egg becomes hard and the white thin, after being in store for a long hold; and that when the eggs are taken from storage in warm weather it will require a longer time to get through the sweat than if held in storage at a somewhat higher temperature, resulting in more harm to the eggs. Some claim that the keeping qualities are impaired by holding at a temperature as low as 30° F., and others note a dark spot, or clot, which forms in the vicinity of the germ, when eggs are held below 33° F. Against this formidable array of claims, the low temperature men have some equally strong ones, although fewer in number. They say: "There is very much less mildew, or must, at 30° F. than at temperatures above 32° F.; the amount of shrinkage or evaporation from the egg is less; an egg can be held sweet and reasonably full at this temperature from six to eight months." This last claim is a broad one, and comparatively few houses are turning out eggs answering to this description.

The following, relating to high temperatures, is quoted from a letter written by one of the best posted men in the business, who has spent much money and time on experiments, and studied the question for years. He says: "A temperature of 40° F. is very good for three months' holding, but if they run over that, it is more than likely the eggs will commence to cover with a white film, which grows the longer they stand, and finally makes a musty egg." This gentleman advocates a temperature of 30° F. for long period holding. It should be noted that the high temperature men ignore entirely the effect of high temperatures on the growth of this fungus, spoken of as a white film. The worst thing about most storage eggs is the taste caused by this growth (usually called mildew or mold), which results in what is commonly called a musty egg. To enable us to understand the validity of these claims made by the 30° F. people, it will be necessary for us to ascertain the conditions which are favorable, and also the conditions which are unfavorable for the propagation of this growth of fungus, which has given storage men so much trouble, ever since cold storage was first used for the preservation of eggs.

Heat and moisture are the two conditions leading to its rank growth, and the opposite—dryness and cold—will retard or stop the growth entirely. In moist, tropical countries many species of this parasite grow, while in the cold, dry regions of the north its existence is limited to a single variety. The causes leading to a growth of the fungus on the outside of an egg are not far to seek. It feeds on the moisture and products of decomposition which are being constantly given off by an egg, from the time it is first dropped until its disintegration, unless immersed in a liquid, or otherwise sealed from contact with the air. This evaporation not only takes moisture from the egg, but carries with it the putrid elements from the egg tissue, resulting from a partial decomposition of the outer surface of the egg meat. Conditions of excessive moisture and the presence of decaying animal or vegetable matter, together with a moderate degree of heat, are essential to the formation of fungus of the species which are found growing on eggs in cold storage. As the heat and moisture are increased, the growth of fungus will be proportionate. Furthermore, we all understand that heat hastens decomposition, and the partial decomposition of an egg results in a growth of the fungus, as before explained, when conditions of temperature and humidity are favorable. If the temperature is low, this growth is slow; for instance, if eggs are held at a temperature of 30° F. in an atmosphere of given humidity, the growth of fungus is less rapid than if held at any temperature higher, with the same per cent of humidity. As our subject merges into humidity here, the reader is referred to what is said in regard to this under the head of "Humidity."

Returning to the objections urged against low temperatures, we will see what damage is claimed from the use of a temperature of 29° to 30° F. The objections are: Liability of freezing; germ is killed; white becomes thin; yolk is hardened, and eggs will not keep as long when removed from storage. Some interesting results are obtained from experiments made by the author. Half-rotten or "sour" eggs freeze at temperatures just a trifle under 32°. Fresh eggs freeze at 26° to 27° F. In testing eggs which had been held in storage for several months, it was noted that the freezing point had been reduced from 1° to 2° F. An egg which is leaky will freeze at 2° to 3° higher temperature than one which is sound, probably owing to the evaporation from the uncovered albumen resulting in a lower temperature. The freezing point of eggs, as above, is understood as being the degree at which they begin to form ice crystals inside. Of the replies received touching on the freezing point of eggs, nearly all agree with above experiments. The "dead germ" theory the author has never been able to locate in fact, having never seen anything of the kind in eggs held as low as 28° to 29° F. for several weeks' time; nor in eggs held at 30° F., or a trifle under, through the season. As only two or three mention having noted this result, it would seem that some local conditions, and not low temperature, were responsible.

The matter of the white becoming thin when eggs are held at low temperatures has some bearing; in fact, any egg held at a cold storage temperature for a long carry will show this fault, to a certain extent, especially if cooled quickly when stored, or warmed suddenly when removed from storage. It is the author's opinion that a difference of 4° to 6° F. in carrying temperature will not be noticeable in its effect on the albumen of an egg; and as to the effect of a low temperature on the egg yolk, it has been demostrated that any temperature, which will not actually congeal the albumen, will not harm the yolk of an egg. There is a slight tendency, in this case, to a similar effect to that produced by a low temperature on cheese; that is, causes it to become "short" or crumbly.

In regard to a low temperature egg not keeping as long when removed from storage, it has been the experience of the author that no difference was noted between the eggs put out from storage and the current receipts of fresh eggs, so far as any complaint or objection was concerned, the eggs being shipped in all directions, in all weathers and subject to many different conditions. A test was also made, by placing three dozen of eggs, which had been carried in storage at a temperature of 28° to 30° F. for five months, in a case along with three dozen fresh eggs. After three weeks no pronounced change was noted in either, both showing considerable evaporation as a result of exposure to the dry fall atmosphere. They were exposed to the temperature of the receiving room, fluctuating from 50° F. to 80° F. The eggs from storage went through a "sweat," while the fresh were not subjected to any such trial. As most eggs are consumed inside of three weeks after being removed from storage, this would seem like a good practical test of the vitality of a low temperature egg. A mere matter of economy between holding a room at 40° F. and from 29° to 30° F., while readily appreciated and admitted, seems of very small importance, when a positive advantage can be obtained by carrying eggs at the lower temperature; and a difference of from 4° to 5° F. would be scarcely worth considering.

An advantage of low temperature, not yet mentioned, is the increased stiffness, or thickness, of the white of the egg while in storage, holding the yolk in more perfect suspension. When eggs are held at a temperature of 36° F., or above, for any period longer than four months, the yolk has a decided tendency to rise and stick to the shell, causing rotten eggs, known as "spots." It is usually understood that the yolk settles; but, being of a fatty composition, it is lighter than the albumen, and rises instead. If the albumen is maintained in a heavy consistency, the yolk is retarded from rising, and held in a more central position. It was long a practice with storage men to turn eggs at least once during the season, to prevent the above trouble, and some recommend it even now; but the practice has been generally abandoned with the advent of low temperatures for egg storing.

It should be noted that what is said above applies to conditions as they were some twelve or fifteen years ago, but the same general ideas prevail at this time with reference to the storage of eggs. Fifteen years ago or more the author was the first to advocate a storage temperature of 30° F. for eggs. Now very few eggs are stored at a temperature above 31° F., and the most of the big city houses carry them at temperatures ranging from 28° F. to 30° F. The author recommends 29° F. as being the best temperature everything considered for general long period egg storage purposes; 28° F., is perfectly safe if using an improved system like the Cooper false floor and false ceiling system, described in the chapter on "Air Circulation," but it is not safe with direct piped rooms, as eggs may freeze near the cooling pipes, whereas the temperature will be considerably higher in the center of the room near the ceiling. Eggs will not freeze at 28° F., but they will freeze at 27° F. and possibly at 27½° F. If, therefore, the egg room does not go below 28° F. it may be relied upon that good, sound eggs will not freeze.

When eggs are put in cold storage they should not be cooled rapidly. The effect on the egg tissues is bad—they should have time to rearrange themselves to the changed temperature. This is especially true where eggs are placed in storage in extreme warm weather. Sudden warming is also detrimental to the welfare of an egg, for a similar reason to above. The most noticeable effect of either is a thinned albumen. If this process of cooling and warming could be accomplished slowly (which is not always practicable commercially), a well kept storage egg would come out of storage with nearly the same vitality it had when fresh.

HUMIDITY.

Information on the subject of humidity, as applied to the cold storage of eggs, is very meager. Not more than a dozen of the replies received in answer to the list of inquiries sent out contained information on the three queries under the head of humidity. Considering the amount of talk we have all heard, with dry air as a subject, this scarcity of knowledge is rather surprising. Those who have had experience with cold storage work and the products handled are well aware that an essential for good results in egg refrigeration is a dry atmosphere in the egg room; but just how dry, very few are able to give even an approximate estimate. Very likely if a cold storage man is asked in regard to it, he will reply that an egg room should be "neither too moist nor too dry." What this "happy medium" is, that will not shrink or evaporate the eggs badly, and yet keep down the growth of fungus to a minimum, is what all are striving for, and very few have the means of knowing when this point is reached. A few years ago a prominent commission man, in conversation with the author, speaking of storage eggs, said: "You storage men are between the devil and the deep sea. You always shrink 'em or stink 'em"; meaning that eggs which were held long in storage would show either a considerable evaporation or a radical "musty" flavor. To some extent this is true, but with a penetrating circulation, careful ventilation and a judicious use of absorbents (all of which are considered under their proper heads) eggs can be, and are, turned out of storage without this strong, foreign flavor, and with little evaporation or shrinkage.

The questions relating to humidity were written with a full understanding of the scarcity of information on the subject, and were designed to locate, if possible, those who were making tests of air moisture, and get opinions on the correct humidity for a given temperature. The following are the queries:

First.—What tests, if any, have you made of the dryness or humidity of your egg rooms?

Second.—What per cent of air moisture do you find gives the best results at the temperature you use?

Third.—What instrument do you use for testing air moisture?

The first and third questions are practically the same, the latter being written simply to make the query plainer and indicate whether an instrument or some other test was used for determining air moisture. Four houses reporting were using the dry and wet bulb thermometers; the others were using hygrometers of French or German make.

The answers to the second question varied greatly; some also giving actual testing humidity of their rooms and their opinion of a correct degree as well. From 70 to 80 per cent of humidity is the test of nearly all reporting, and of the rooms tested by the author, nearly all show a similar humidity, with one occasionally going as high as 85 per cent, and some as low as 65 per cent. Two answers recommended a humidity of 65 per cent, and one a humidity of 60 per cent, with a temperature of 30° to 32° F. Others hold that their testing humidity of 70 to 80 per cent is correct.

Under the head of "Temperature," it is stated that: "If eggs are held at a temperature of 30° F. in an atmosphere of a given humidity, the growth of fungus is less rapid than if held at any temperature higher with the same per cent of humidity." By referring to the table on page 168 we see that a cubic foot of air, when saturated at a temperature of 40° F., contains 2.85 grains of water vapor, while at 30° F. it contains but 1.96 grains, or only about two thirds as much as at 40° F.

The same hold true with any relative humidity, the same as when the air is saturated. Take, for instance, air at a temperature of 40° F., with a humidity of 75 per cent, then a cubic foot of air holds 2.14 grains of water vapor per cubic foot; and at a temperature of 30° F., with the same relative humidity, it would hold but 1.47 grains. This great difference in the amount of moisture contained in the air at different temperatures, and still having the same relative humidity, has as radical an effect on the growth of fungus as does the difference in temperature. This is no mere theory, as the writer has demonstrated it, to his own satisfaction, at least, during several seasons' observation. If it is hoped to keep down the growth of fungus in a temperature of 40° F. by maintaining an atmosphere with a lower relative humidity, the result is a badly evaporated egg, which loses its vitality and value very rapidly when held in storage for a term exceeding three or four months; the white becomes thin and watery, with a strong tendency to develop "spot" rotten eggs. As the fullness or absence of evaporation is of only secondary consideration to their sweetness, when eggs are tested by buyers, it is necessary to prevent this trouble if the eggs turned out from storage are to be considered first-class.

From the foregoing it seems clear that to turn out sweet eggs at a temperature of 40° F. it is necessary to maintain a lower relative humidity than at any temperature lower, and the result cannot fail to be as described. The author has already given a summary of the replies to the questions relating to humidity, which are few in number, and not very complete. A little is better than nothing, however, and by comparing his own data with the results obtained by others, and paying careful attention to their opinions, the following table of correct humidity for a given temperature in egg rooms has been compiled. There are no data on the subject in

RELATIVE HUMIDITY FOR A GIVEN TEMPERATURE IN EGG ROOMS

Temperature In Degrees F.	Relative Humidity Per Cent.
28	85
29	83
30	80
31	79
32	75
33	74
34	70
35	68
36	66
37	64
38	61
39	59
40	56

print, so far as known, and no claim for absolute accuracy is made in presenting this first effort in that direction, but as the figures are taken from actual results, no great mistake can be made by depending on them. The percentages of humidity given are modified, to some extent, by the intensity and distribution of the air circulation employed.

CIRCULATION.

A thorough and penetrating circulation of air must be maintained in a cold storage room for eggs if good results are to be insured, and the importance of this condition is quite generally appreciated. It is also a fact that a strong, searching circulation will do much to counteract defects in a cooling apparatus, or wrong conditions in the egg room in some other particular.

The reason why a thorough and well distributed circulation of air in an egg room will give superior results over a sluggish or partial circulation may not be readily apparent. A circulation of air is of benefit in combination with moisture absorbing capacity in the form of frozen surfaces or deliquescent chemicals. Stirring up the air merely, as with an electric motor fan, without provision for extracting the moisture, is of doubtful utility, and may, in some instances, prove positively detrimental, as it is liable to cause condensation of moisture on the goods, or walls of storage room, instead of its correct resting place—the cooling coils and absorbents. Let us see how the circulation of air in a storage room operates to benefit its condition.

Under head of "Temperature," we have seen that the evaporation from an egg contains the putrid elements resulting from a partial decomposition of the egg tissues, and that the air of a storage room carries them in suspension. It is probable that these foul elements are partly in the form of gases absorbed in the moisture thrown off from the egg; and if, therefore, this moisture is promptly frozen on the cooling pipes, or taken up by absorbents, the poisonous gases and products of decomposition are very largely rendered harmless. This is also true of the germs which produce mold and hasten decay, which are ever present in the atmosphere of a storage room, being carried to a considerable extent by the water vapor in the air, along with the foul matter of various kinds referred to. If the vapor laden air surrounding an egg is not removed and fresh air supplied in its place, the air in the immediate vicinity of the egg becomes partly charged with elements which will produce a growth of fungus on its exterior, affecting and flavoring the interior—the flavor varying in intensity, depending on how thoroughly impregnated with fungus-producing vapor the air in which the egg is kept may be. In short, then, circulation is of value because it assists in purifying the air. It should be kept up so that the air may be constantly undergoing a purifying process to free it from the effluvium which is always being thrown off by the eggs, even at very low temperatures.

The questions bearing upon circulation contained in the list of inquiries sent out by the author are as follows:

First.—In piping your rooms what provision was made for air circulation?

Second.—What difference in temperature do you notice in different parts of the same room?

Third.—Do you use a fan or any kind of mechanical device for maintaining a circulation of air in the rooms?

More answers were received on this subject than on the subject of humidity, but not exceeding one-third contained practical replies to all three inquiries. Several of the answers confounded circulation with ventilation, as before alluded to. The first question, in particular, was badly neglected, indicating, no doubt, that no provision was made for circulation in a majority of cases. The common device in use for causing air to circulate more rapidly over the cooling coils, when they are placed directly in the room, is some form of screen, mantle, apron, false ceiling or partition. Many of these have been put up after the house has been in operation for some time, and are very crude affairs, applied in all conceivable combinations with the pipe coils. In some cases canvas curtains, or a thin wooden screen, have been suspended under ceiling coils with a slant to cause the cold air to flow off to one side, and with surprising improvement to the room, considering the simplicity of the device. Forced circulation with a complete system of distributing air-ducts is coming into general use, as the merits of this way of producing circulation are better understood and appreciated.

The second question was answered more generally, but that some of the answers were mere guesses, or statements made without testing, is very evident, as they state that no difference was noticed in different parts of the same room. With open piping or gravity air circulation, this is an impossibility—it is only possible with a perfectly designed forced circulation system. In contrast to this claim some answers state a difference in temperature of as high as 4° to 5° F., but most answers show a difference of 1° to 2° F.; a few of ½° to 1° F.; and, still others, as before stated, none at all. A marked variation of temperature in different parts of a room, while in most cases caused by defective circulation, is due sometimes partly to location of room as to outside exposure, proximity to freezing rooms, character of the insulating walls, etc. An egg room placed over a low temperature freezing room will show more variation between floor and ceiling than when located over another egg room, conditions being otherwise the same. Where this arrangement occurs, and the egg rooms are operated on a natural gravity air circulation system, eggs may be frozen near the floor, when a thermometer hanging at the height of a person's eyes would read 30° F. or above. Even with the very best insulation, the result of this very common arrangement is a defective circulation and more or less variation in temperature between floor and ceiling.

In reply to the third question, about a dozen state that they are using some form of mechanical forced circulation. The advantages of this method are discussed quite fully elsewhere in this book. About double this number are using the small electric fans. These also have been treated in the discussion of mechanical air circulation in another chapter.

As air circulation is a somewhat neglected subject, and comparatively few have experimented enough to have positive opinions, based upon practical experience, regarding the merits of different devices and methods, some of the more prominent and successful ones are illustrated and discussed elsewhere in this book. (See chapter on "Air Circulation.")

VENTILATION.

The introduction of a large volume of fresh air is not essential for the purpose of purifying rooms in which eggs are stored, because the accumulation of permanent gases in an egg room is quite slow, comparatively (as in rooms where well ripened fruit is stored); but a small supply of fresh air continuously, or at regular intervals, is of much benefit.

The questions referring to ventilation contained in the letter of inquiry sent out by the author are as follows:

First.—What plan do you pursue in ventilating egg rooms?

Second.—Under what circumstances and how often do you ventilate?

Third.—How often do you consider it advisable to make a complete change of air?

Outside of a bare dozen, the replies on this much-talked-of subject were of no value whatever for our purpose. Most of those answering do not ventilate; many others get their ventilation through the opening of doors; some ventilate through an elevator shaft, by opening doors at top and bottom, etc. Only three or four were properly cooling and drying the air before introducing it into the egg rooms. One successful storage manager says that: "It is trouble enough to take microbes, bacteria, moisture, etc., out of one batch of air" (meaning the air in his rooms at the beginning of the season), without adding to his troubles by sending in more air loaded down with the same mischief makers. As pointed out in the chapter on "Ventilation," unless the air to be used for purifying the rooms is itself first cooled and purified, this man's idea is perfectly correct. Ventilated egg rooms will, however, turn out eggs which are in every way better than from rooms not ventilated, other conditions being equal. Eggs from ventilated rooms are clearer and stronger bodied (albumen thicker) than from non-ventilated rooms.

No accurate data have yet been established regarding the volume of fresh air which is advisable to use for ventilating egg rooms, but it is a simple and inexpensive matter to supply enough, and too much cannot be used if it is first properly dried and purified and brought to about the same temperature as that of the storage room. Ordinarily it is unnecessary to ventilate egg rooms until filled with goods and closed for the season. After a short time (two to four weeks) begin ventilating, as the accumulation of gases commences at once as soon as the rooms are permanently filled and closed. Ventilate in small quantities and for several hours at a time once or twice a week, rather than in large quantities less often.

For a discussion of the principles involved and mechanical details of this subject see chapter on "Ventilation."

ABSORBENTS.

The letter of inquiry sent out by the author contained three questions referring to absorbents, written with an idea of ascertaining the coating used for the walls of a storage to the greatest extent; what absorbent was the favorite, and in what manner applied. The questions are as follows:

First.—Do you use an absorbent or purifier in your egg rooms?

Second.—In what way do you use or apply them?

Third.—Do you paint or whitewash? What kind and how often applied?

The most common wall coating in use for egg rooms is plain, every-day whitewash, in various proportions of lime and salt. Several recommend one part of lime and one of salt. This makes a very good whitewash, giving a firm, hard surface, but unless some method of blowing warm, dry air through the rooms is feasible, it will dry very slowly, which is likely to cause it to have a mottled appearance instead of the pure white which gives a storage room such an attractive appearance. A better proportion for ordinary cold storage work is three parts of lime and one of salt. This mixture will dry faster, and will give a white surface which will not easily rub or flake off. There are many formulas for good whitewash, some of them so complicated as to be impracticable; but plain lime and salt, with perhaps the addition of a little Portland cement, will be good enough for our purpose. See chapter on "Keeping Cold Stores Clean" for details of whitewash making, etc.; also chapter on "Uses of Chloride of Calcium" for application of his material, also chapter on "Absorbents."

STORAGE PACKAGES.

Eggs are continually giving off moisture from the time they are first dropped by the hen until they disintegrate, unless sealed from contact with the air, and we can therefore never hope to keep them in cold storage for several months without their losing some weight by evaporation. To prove that eggs must evaporate, the following experiment was tried by the author in his early experience: An ordinary 30-dozen egg case was lined with tin, with all joints carefully soldered. The eggs were then placed in the fillers in the tin lined case in the usual way, and an air-tight tin cover soldered on, forming a hermetically sealed package. After about sixty days' stay in an ordinary refrigerator the tins were unsoldered. The result noted was peculiar and startling. The inside of the tins was dripping wet, and very foul smelling, and the eggs were all rotten. This same experiment was tried by a friend, working independently and without knowledge of the author's experiment. He used an ordinary fruit jar, with screw top fitting onto a rubber ring. His results were similar. In addition this gentleman packed some eggs in flour in a fruit jar, otherwise under the same conditions as the other experiment. The eggs packed in this way were all found to be in good condition when the jar was opened, as the moist evaporation from the eggs had been taken up by the flour. These experiments prove beyond a doubt that an egg must evaporate continually, and they prove further that the eggs must be surrounded by some medium which will absorb this evaporation.

In the chapter on "Air Circulation" it is explained how the air is best circulated so as to remove the moisture and impure gases from the vicinity of the goods. This must be done, otherwise the fillers and package containing the eggs would shortly be in as bad condition as the fillers in the experiment just mentioned. The theory and explanation of the other conditions in the storage room necessary for successful egg refrigeration have also been taken up under the various heads. We will now look into the requirements of the package containing the eggs while in cold storage.

The questions contained in the letter of inquiry relating to the egg package are as follows:

First.—What egg package have you found to turn out the sweetest eggs?

Second.—Have you used any kind of ventilated egg case, and with what results?

Third.—Have you ever used open trays or racks, and with what results?

As many different people have experimented with different packages, hoping to get something which would turn out perfectly sweet eggs, with little evaporation, the replies received to the questions relating to packages are interesting, and many contained information valuable as data. The favorite package is the ordinary 30-dozen egg case, made of whitewood, using medium weight hard calendared fillers. The term whitewood is usually meant to include either poplar, cottonwood or basswood, but two or three other varieties of wood, not so well known, are designated as whitewood. Basswood is by some not placed in the whitewood list, but the best authority known to the author says that basswood is as properly a whitewood as poplar or southern whitewood. Poplar and cottonwood are most in use for storage purposes, and many insist that basswood is objectionable because of its liability to ferment or sour and cause tainted or musty eggs. All kinds of cases have been in storage in the house operated by the author, and if all were thoroughly dry, no difference could be noted in the carrying qualities of the different kinds of whitewood, and the preference has been for well seasoned basswood cases. It may be that basswood is more likely to sour and affect the eggs than poplar or cottonwood, but it is always advisable to get stock for egg cases in the fall and have them nailed up during the winter, allowing two or three months for the cases to season and dry out before the opening of the egg storing term. Some have dry kilns for cases, but a naturally seasoned case is to be preferred, as then it has a chance to deodorize as well as dry out. In some localities other woods are used for egg cases. Ash, maple, hemlock and spruce have been used for storage cases, generally because they are cheaper than whitewood in that locality. Any strong scented wood like pine will not do because of the flavor imparted to the eggs.

The pasteboard frames and the horizontal dividing or separating pasteboard pieces which form for each egg an individual cell in the case are usually spoken of as fillers. For years only one grade of these was made—those of ordinary strawboard. When moistened by the evaporation from the eggs this material has a peculiar rank odor, which was taken up to some extent by the eggs if they were allowed to remain in the fillers for several months. Much of the flavor resulting from a growth of fungus has been laid to the fillers, and much of the flavor resulting from fillers has been laid to a growth of fungus or must, but there is no question about strawboard fillers not being a perfect material for cold storage use. Many kinds of fillers have been tried, and many ideas suggested for the improvement of cold storage eggs. A whitewood pulp filler made its appearance some years ago, but did not come into general use. After being in storage a few months, it absorbed moisture to such an extent as to become very soft, and they were objectionable on this account. A good manila odorless is now on the market which is giving good satisfaction where tried. Ordinary strawboard fillers have been coated with various preparations, shellac, paraffine, whitewash, etc. Any substance in the nature of waterproofing might better be left off for the reason, as we have seen, that eggs must evaporate, and a waterproof filler would hold the moisture and not allow it to escape into the air of the room. It is essential to the well being of an egg that it should evaporate, as proven by the experiments in hermetically sealing, before described. Many have gone to the expense of transferring the eggs into dry fillers in the middle of the season. One season of this was enough for the author. A better way is to decrease the humidity of the room as the fillers become more and more loaded with moisture. The humidity may be decreased by the use of absorbents or by ventilation, as already discussed in their proper places. Fillers made of thin wood have been used in years gone by with fair success, but their manufacture has now been entirely discontinued. They were made of maple, shaved very thin, and were a prime filler so far as odor was concerned, but in cold storage the frames warp badly, and the time and eggs wasted in getting the eggs out of the fillers was a serious item against their use. As a shipping filler they were also a failure because of the excessive breakage. Some years ago an eastern company began the manufacture of what is known as the odorless fillers. These fillers are light brown or buff in color, and from the best information the author can obtain, are composed largely of scrap paper stock, with some long fibre like manila added for strength. In the manufacture the stock is treated to a thorough washing and deodorizing process, and the result is a filler with very little odor. Eggs put up in these so-called odorless fillers and subjected to the same conditions as a similar grade of eggs packed in common strawboard fillers have come out of cold storage in better condition in a good many cases. A ventilated filler made by a well known creamery supply house, has been suggested as an ideal filler for cold storage, but they are so poor mechanically that they are not to be thought of. The material cut away to form the air circulation space weakens the structure of the filler to such an extent as to make it dangerous as a shipping filler. Whatever filler is used, it should fit the cases, not crowding in, nor still so loose as to shake. If this point is looked after much breakage and consequent poor results from storage in the cold room may be avoided.

Many styles of ventilated egg cases have been placed on the market in years past, but very few or none survive the test of time. A ventilated case, made by having the sides cut an inch narrower than the ends, has come into use, especially in one large eastern city. Making the sides narrower forms a space of half an inch on both sides of case at top and bottom, for the ready access of air to the interior of the case. This case is of very simple construction, and efficient in allowing a free circulation of air into the case. Others, however, prefer a case with sides in two pieces, claiming that the cracks will allow enough air circulation. Still others prefer the shaved or veneered cases with solid sides and bottom, claiming that this kind of a case will prevent excessive evaporation from the eggs. As pointed out elsewhere, humidity and circulation have much to do with the evaporation from eggs; in fact, are of much more importance than the package, although the package necessarily has much to do with it. A tight package will allow of less evaporation than an open one. In a very dry room with a vigorous circulation a moderately tight package is the thing, but in a comparatively moist room with poor circulation the more open the package the better.

An appreciation of the poor circulation and damp air of the overhead ice systems has caused many of their operators to resort to the use of open trays or racks for the storage of eggs. Very palatable eggs have been turned out in this way, but the use of trays in any ammonia or brine cooled room would lead to very excessive shrinkage of the eggs and consequent heavy loss in candling. On a commercial scale, too, the storing of eggs in trays is hardly practicable, as it increases the risk of breakage immensely, and the eggs must be transferred from the cases when received at the storage house, and back into cases again when shipped, involving much labor, and perhaps loss of valuable time at some stages of the market. In any but a very moist room, eggs stored in open trays, in bulk, will lose much from evaporation, and the loss will be proportinately higher than on an equal grade of eggs stored in ordinary cases and fillers. The advantage of trays, if any, for some houses, is that contamination from fillers is avoided, and about 40 per cent more eggs can be stored in a given space. The eggs are, however, more liable to must as a result of moisture condensing on their surface with change of temperature, or on the introduction of warm goods into the storage room.

The material used for forming a cushion in the case on top and bottom of the fillers to protect the eggs from contact with the case, and so that they will carry in shipping, is generally either excelsior, which is finely shaved wood, usually basswood, or the chips made in the manufacture of corks, known as cork shavings. The big cold storages have in the past recommended cork in preference to the best excelsior. Here again comes a question of dryness. If the excelsior has been in stock for a year and stored in a dry place it is to be preferred to cork shavings, otherwise cork is the best, because we know cork is always dry. Cork makes a very poor cushion as compared to excelsior; it is liable to shift in the case, leaving one side without protection. As a matter of cost, too, cork is much more expensive than excelsior. If people want cork in their cases they can have it by paying the price, but dry, seasoned, fine basswood excelsior is better, for reasons stated. The best houses are now recommending dry excelsior in place of cork on account of the excessive breakage when a cushion of cork is used.

Eggs have been packed in oats for years, but the practice has gradually fallen off, as eggs stored in cases from the best cold storage houses have been improved in quality from year to year. Oats, if dry, will absorb moisture from the eggs quite rapidly, and are objectionable on this account. If the oats are not dry the germs of mold are developed rapidly, and as the moisture is given off by the eggs the mold will grow, causing the eggs to become "musty." Therefore the main difficulty in using oats as packing for eggs in cold storage is to have them at the correct degree of dryness. It is almost impossible to have them in the same condition at all times. Oats have also been used in cases inside the fillers; that is, the layers of eggs are first put into the filler; then the oats are sifted into the spaces around the eggs flush with the top of the filler. This is repeated through the whole case; all the space in the case not occupied by the eggs being filled with oats, excepting the small space taken by the fillers themselves, the object being, of course, to prevent the "filler taste."

At intervals we read of some method of preserving eggs, which is said to be sure to supersede ordinary cold storage for the good keeping of eggs. A scheme was tried on a large scale somewhere across the water, in which the eggs were suspended in racks in a cold room—the racks being turned at regular intervals by automatic machinery to keep the eggs from spoiling; that is, to keep the yolk from attaching to the shell. A low temperature will prevent this, as pointed out in this chapter under the head of "Temperature," and why a man should waste good energy inventing such a machine is passing all comprehension. The quantity of various chemical preparations manufactured and sold for egg pickling or preserving is even now quite large, but the high class stock now turned out by the best equipped cold storage houses has made any other method of preserving eggs at the present day almost entirely obsolete.

HINTS.

There is a long string of "don'ts" in regard to packing, handling and storing eggs which might be put down, but the author will be content with a few of the simpler and most useful ones. To start with, don't store very dirty, stained, cracked, small or bad appearing eggs of any description. Have your grade as uniform as possible. The culled eggs will usually bring within two cents of the market price, and it pays better to let them go at a loss rather than try to store them. Don't use fillers and cases the second time; they are more likely to cause musty eggs than the new ones. Don't ship eggs in cold cars or place eggs which are intended for storing in ice boxes. In shipping eggs from the producing section to the storage house in refrigerator cars, no ice should be put in the bunkers, because if the eggs are cooled down and arrive at their destination during warm or humid weather they will collect moisture or "sweat," and an incipient growth of mold will result. Don't use heavy strawboard fillers for storing eggs. If "the best way to improve on a good thing is to have more of it," then the best way to improve on a poor thing is to have less of it; and if strawboard fillers are objectionable, then the thinner they are the better, because less of the material is present to flavor the eggs. Further, the thin board fillers are more porous, and allow of a freer circulation of air around the eggs.

The grade known as "medium" fillers are best for cold storage purposes. As already stated, odorless fillers are better than any strawboard fillers. Don't use freshly cut excelsior. It should be stored in a dry place at least six months. Use no other kind but basswood or whitewood. Don't store your cases, fillers or excelsior in a basement or any damp place. Don't run warm goods into a room containing goods already cooled when it can be avoided. For this reason very large

FIG. 3.—VIEW IN EGG TESTING ROOM.

rooms are not to be desired. A small room may be quickly filled with goods and closed until goods begin to go out in the fall. If a large room is used it may require several weeks to fill completely, during which time the fluctuation of temperature is at times excessive, causing condensation on the goods, which will propagate must quickly.

To illustrate: We will suppose the egg room partly filled with goods cooled to a temperature of 30° F. Several cars of eggs at a temperature of, say, 70° F. are run into the same room. The new arrivals, in cooling to the low temperature, give off large quantities of vapor from cases, fillers and the eggs themselves, the vapor condensing, of course, on any object in the room which is below the dew point of the air from which the warm goods came. This may seem like a finely spun theory, but the author has had some experience which amply justifies this explanation. That the moist vapor given off by the warm goods does not show in the form of beads of water, or fog, or steam, is no proof that it does not exist. If the extremes of temperature are as great as 25° F. condensation will occur on nine days in ten during the egg storing season. The goods already in storage are raised in temperature materially

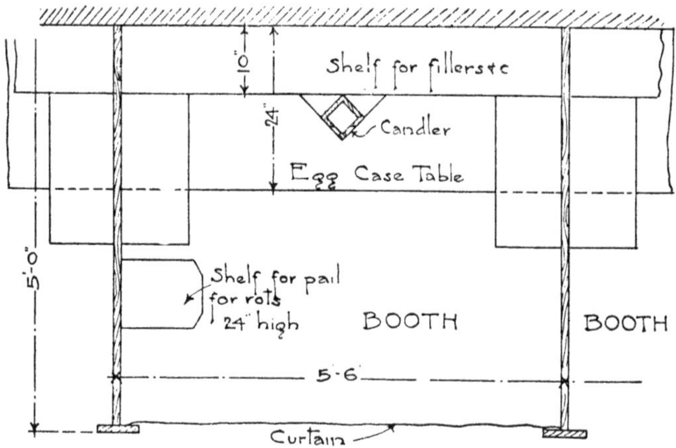

FIG. 4.—PLAN OF EGG TESTING BOOTH.

by placing in warm goods, which is harmful to some degree. The logical deduction from above seems to indicate that warm goods should not be placed in a room with goods which have been reduced to the carrying temperature. A separate room should be provided for this purpose near the receiving room in which the goods coming in warm may be cooled to very nearly the temperature of permanent storage room. This is a refinement which small houses cannot afford, and which most of the larger ones do not have.

If you wish to progress compare your results with those of others. Don't say: "My eggs are as good as fresh;" test carefully from time to time through the season and compare quality with those from other houses.

It should be positively understood that merely theoretical knowledge on this subject is of only limited assistance; and those who undertake new work are advised to put a man in charge who has had experience with the product which it is proposed to handle in storage, as well as acquaintance with the mechanical details of the plant.

CONSTRUCTION OF EGG CANDLING ROOM.

The construction of rooms to be used for the testing or candling of eggs has not reached a stage where it may be stated that there is any design which might be called standard or that is generally approved by the trade. Nearly every pro-

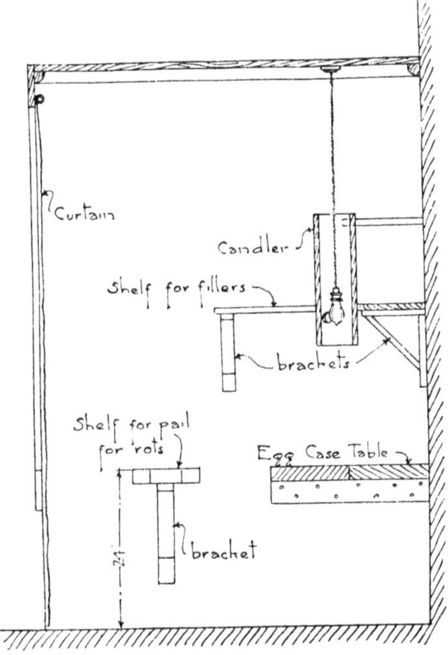

FIG. 5.—SECTION OF EGG TESTING BOOTH.

prietor has his individual ideas on this subject, and, therefore, nearly every different plant is fitted up in a different way. The booth system, which consists of individual stalls or separate small rooms for each person, is coming into general favor. The advantage of this system is that each person works by himself, and, therefore, better work is possible, and each operator must stand on his own individual merits. In other words, the system allows of a closer inspection and a closer systematizing of the important operation of candling eggs.

The number of booths necessary depends upon the volume of business to be handled and any number of booths may be arranged in a large room, which may be called the "Candling Room." (See Fig. 10.) Candling is simply a misnomer in this

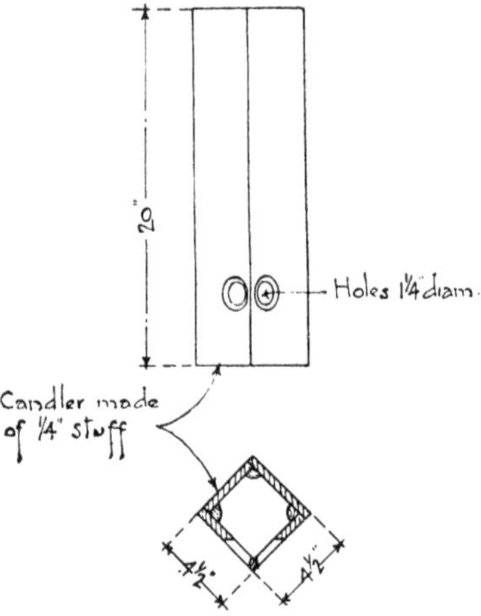

FIG. 6.—DETAIL OF CANDLER.

connection and originated from the fact that a candle was originally used for testing eggs. Very few candles are now in use for this purpose and the electric light is in general favor.

The construction of the booth is subject to some modifications; the one shown in accompanying view, plan, section and elevation, also detail of candling box or candler (see Figs. 3, 4, 5, 6 and 7), has been proved by practical service to be economical of space and, withal, convenient. As shown, sufficient shelf room is provided for fillers above a plank table on which rest cases which contain the eggs to be tested and for the different grades into which the eggs are classed. With the size booth shown, sufficient room is provided for five cases on the table. The booth may be constructed of a single thickness of boarding on three sides and top, the fourth side being closed by a curtain of heavy, dark colored denim, or any suitable material. This curtain should be hung on a wire or rod with rings so as to slide easily, and it may or may not be divided in the center. In Fig. 3, which is reproduced from a photograph, is shown the booth system in service. The white candling boxes

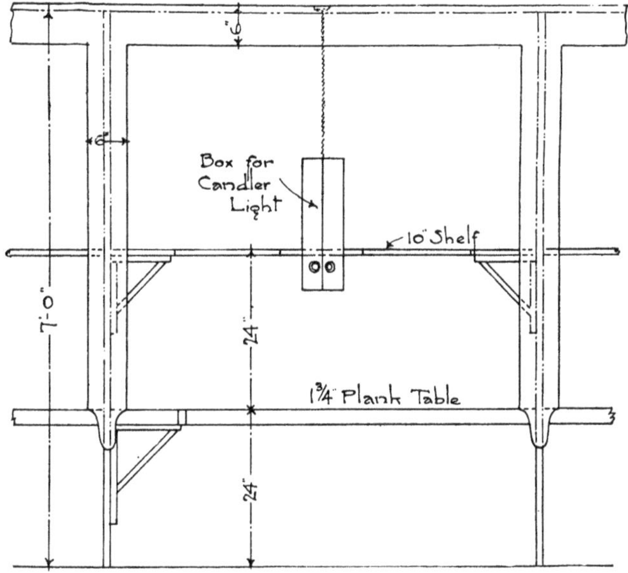

FIG. 7.—FRONT ELEVATION OF EGG TESTING BOOTH.

show plainly in the center of the booths. Cork shavings used as a cushion at top and bottom of egg cases are shown in the box between booths. The barrels are intended to receive the litter of various kinds, such as old newspapers, which accompany country packed eggs. The pail for rots is shown above the barrel.

The candling box or candler which contains the electric light or lamp may be constructed of ordinary egg case material one-quarter of an inch thick. One-half-inch quarter round is nailed into three corners of this box, inside, to strengthen it. The fourth corner is pierced with two holes placed close together, as shown in the detail. The holes should be one and one-quarter inches in diameter. The bottom of the candler is left open so that light from the electric lamp may be thrown into the cases on the table below. The top of the candler may be partly closed by a piece of cardboard, or otherwise, in case too much light is reflected to the ceiling so as to make the candling room too light for close candling. The circulation of air, however, through the candling box should not be entirely shut off, for the reason that it will cause rapid destruction of the electric lamps by overheating.

FIG. 8.—EGG CANDLING SCENE, MAIN RECEIVING ROOM FLOOR.

Sometimes the candler or candling box is made of tin or sheet metal about four inches in diameter and six to ten inches in height. In one or both sides of this cylinder a single oblong or oval hole is provided about one and one-fourth inches in its shortest dimension and about one and one-half to two inches in its greatest dimension, or separate holes may be provided as suggested in connection with the box above described. Very good candlers or egg testers as they are sometimes called are to be had on the market ready made and usable either with electric light, kerosene lamp or candle.

The details of candler or tester are subject to many modifications to suit individual ideas. The question of candling two eggs at once or singly has been much discussed among professional egg candlers. Many prefer the candler with two holes, and still others insist that one hole is sufficient. The author has personally used the booth and candler described and believes it to be equal to anything which has come to his knowledge. All dimensions are given on the diagrams so that the construction of the booth system as above described will be simple for those who desire to test the practicability of the scheme as applied to their local requirements.

The room in which the booths are arranged, or what may be called the candling room proper (see Fig. 10), should be of

FIG. 9.—REFRIGERATED EGG CANDLING ROOM.

sufficient size to accommodate the handling of goods in and out of the room and allow space for empty cases, fillers, etc., and it should also be large enough to provide storage space for one or two days' packing. The room should be insulated in any fairly substantial manner, and means for maintaining same at about 55° to 60° F. in warm weather should be provided. In other words, we should provide a cold storage room for candling eggs. Eggs coming in from the country during warm weather may be placed at once in this room and should be reduced to a temperature of about 55° to 60° F. before being candled.

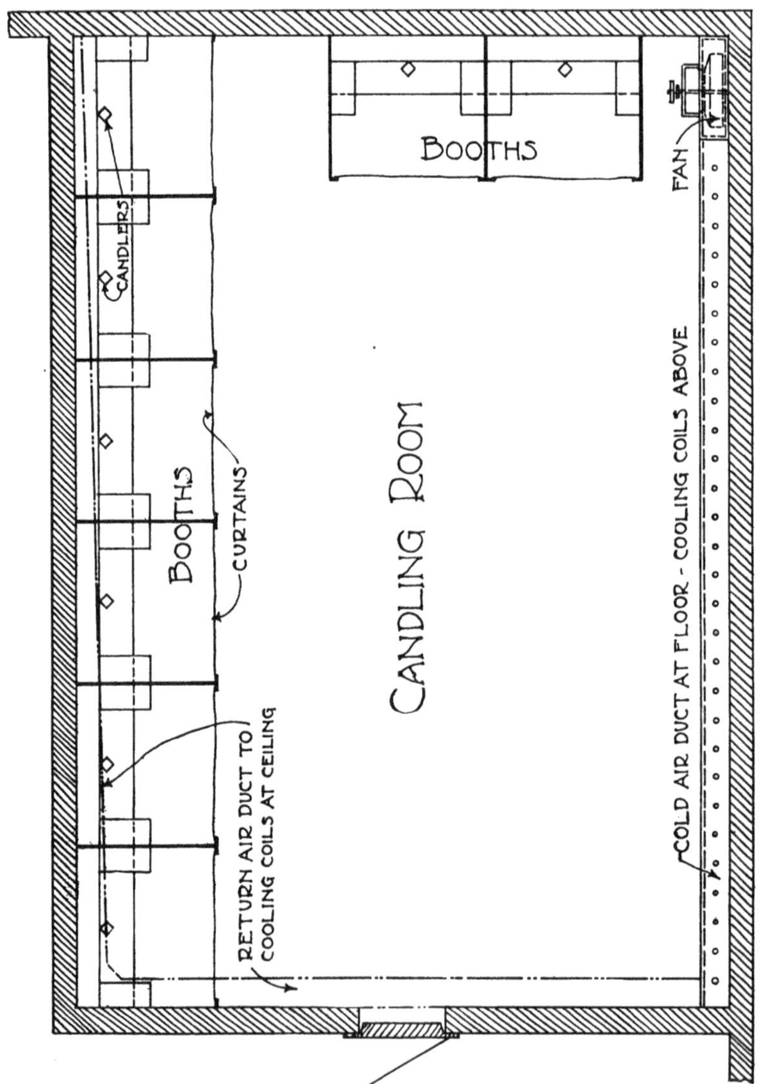

FIG. 10.—PLAN OF CANDLING ROOM.

It is fully understood among practical egg shippers that when eggs come in from the country in hot weather they often appear to be in very much worse condition than they really are. After being cooled to about 60° F. the eggs may be candled and judged for their actual quality. A refrigerated candling room is also a great benefit in stopping further immediate deterioration when the eggs come in heated. The more progressive modern egg houses which are being erected at the present time, where cold storage is an adjunct, have refrigerated candling room facilities. The value of these arrangements will be at once appreciated by those who have had experience in candling eggs during the heated term.

The candling room may be refrigerated in any suitable way, but a fan system of air circulation is generally preferred about on the lines shown in plan of candling room. In this way the cold air from coils is distributed along the floor on the opposite side of the room from the operators and the warm air is taken out over their heads, above the booths. It is advisable to provide openings in the top of the booth for a circulation of air; these openings to have hinged doors so as to regulate the circulation.

The cooling pipes may be placed in a box or bunker near ceiling of the room and a drip pan provided underneath. This will avoid all dripping or spattering on the goods or cases, and, by distributing and drawing off the air as suggested, uniform temperatures are obtained and strong drafts are prevented. If the room is reasonably high, fairly good results may be obtained by placing the cooling pipes on the side walls near the ceiling and providing drip gutters beneath, or the room may be cooled by natural ice by proper arrangement of openings from the source of ice supply. If the refrigerated candling room is once put in service for use during hot weather, its advantages will be so apparent that the operator will wonder how he was ever able to get along without it.

FREEZING EGGS IN BULK.

For ordinary commercial purposes eggs which are to be frozen in bulk, in the form of egg meat, after having been removed from the shell, are best handled in a hermetically sealed package. Tin is better than glass or crockery, as the liability of breakage in handling is less, less danger of bursting in freezing, less space required in storage and less weight to handle. The only weak point of the tin package is its liability to rust if a cheap grade of tin is used, especially when the white and yolk are frozen separately. The white may be discolored by the rusting of the tin. This may be reduced to a minimum by using a good grade of tin.

Some of the large packers pump the air out of the package before soldering, with the object in view of preventing contamination by the impure imprisoned air. Good results may be had, however, by soldering tight after the egg meat is frozen solid, as the small amount of air trapped in the tins contains little moisture and impurities, and is partly sterilized by exposure to the low temperature of the freezing room. Soldering or otherwise sealing the package is not absolutely necessary for successful results, but it makes a more practical package to handle, and prevents evaporation from the surface of the egg meat, which evaporation makes a leathery "skin," which may necessarily be a waste. The author is one of the pioneers in successful egg freezing, and the standard package at first in use was the ordinary lard can holding about twenty-five pounds. These cans were provided with slip covers with no pretense of making them air tight, and very successful results were obtained. It is advisable, however, to protect the surface of the egg meat in some way.

A very good package is what is known as the Record package or butter tub, and consists of an inner tin package and an outer thin wood shell with an air space between the two. These packages are fitted with covers which are pretty nearly air-tight when carefully closed.

To prevent the yolks from becoming solid as if cooked, which prevents their proper melting or dissolving when thawed, they must be effectually broken, and thoroughly mixed with the white. This is sometimes done by placing the egg meat in a churn and churning vigorously for a few minutes, but this has the disadvantage of not surely breaking all the yolks, or if they are broken, the mass may become frothy from the beating up of the whites. The method used by the author was to dump the eggs after removing from the shell on a wire screen of about 5/8 or 1/2 inch mesh, and scrape the yolks through with a wooden paddle or scraper. The screen should be of tinned or galvanized wire, and should be arranged in the bottom of a basin about four inches deep. This screen bottom basin should be fitted to a pail or utensil of convenient size. The pail may be provided with a spout about 1 1/2 inches in diameter to facilitate pouring into the permanent storing packages. Before pouring the egg meat into the permanent storing pack-

FIG. 11.—EGG-BREAKING OUTFIT.

(A tin pan, 10 by 10 by 2 inches, having a tin-bound wire screen, 1/2-inch mesh, fitting firmly over the top. In the middle, at opposite sides, are uprights 3 inches high, each having a slot into which is slipped a piece of tinned boiler steel. The slot binds at the bottom to hold the steel strip firm. It is sharpened on its upper edge. This knife gives a clean crack in the egg shell and can readily be replaced after a bad egg. The sherbet cups are of smooth glass. They withstand steam sterilization. A larger container than a sherbet cup can not safely be used. Neither can an opaque vessel be used. Glass is necessary for good grading.)

age, stir thoroughly from the bottom, as the yolk has a tendency to remain on top, being lighter than the white. Forcing through the screen will break every yolk without fail, and if stirred carefully, the white and yolk will be mixed together, and when thawed no lumpiness or specks will be present.

The United States Department of Agriculture, Bureau of Chemistry, represented by Dr. Mary E. Pennington, has issued a bulletin with suggestions on the preparation of frozen eggs. A suitable apparatus for egg breaking and separation which is recommended is shown in the illustration.

A refrigerated breaking room is also recommended, held at a temperature of not higher than 65° F. This room should be free from foreign odors and have plenty of light and be supplied with fresh air. It is also suggested that all utensils and apparatus in the room be of metal or other material permitting an easy cleaning and sterilization, and that eggs as soon as broken should be promptly placed in the freezer.

In freezing, fill the package only about two-thirds full at first. When frozen solid there will be some expansion of the egg meat, causing it to bulge or hump up in the center. After freezing solid two-thirds full, complete filling, and when all frozen there will be very little hump in the center of the package. Do not fill the package completely, but leave from ½ to 1 inch at the top, depending on the size of package. When the filled package is frozen solid, solder on the cover if the package is to be hermetically sealed, or if an ordinary package is used, without sealing, proceed as follows: If the eggs are separated (yolk and white to be frozen separately), reserve some of the white to apply to the tops of cans after freezing. Pour on about half an inch of the white on the yellow, and allow to freeze, then put parchment paper circles on the top of both white and yolk, pasting or sticking it down carefully with the egg white. The effect of this treatment is to make the top of the package fairly air-tight and protect the egg meat from the air, and the half inch of white on top of yolk will prevent the leathery "skin" already referred to. Some packers reserve a little of the white to cover eggs, frozen yolks and whites together. A parchment circle stuck on top of package with a little of the white, will prevent largely the formation of the leathery "skin," which is caused by the drying out of the surface of the egg meat. After the tins are filled, frozen and capped as above, the slip covers are put on, and they are stacked up in the permanent storage room. Sometimes the tin cans are placed in wood crates to facilitate handling, but more space is required in storage.

In freezing the white and yolk separately, which is very desirable for the better class of trade, it is advisable to keep the white and yolk about even in quantity and this may be done if the people who do the work are skillful, and the eggs are of good quality. In fact, the white will naturally run a little ahead of the yolk. It is better that the yolk have a little of the white mixed with it, as it is easier to thaw out smooth without lumps. It is better to keep whites and yolks of even quantity, as then it is easier to sell an equal amount of each. The white should not be sold separately except at a much higher price.

The author has demonstrated by actual trial that a temperature of 20° F. is best for freezing and storing egg meat in bulk. It has been recommended that eggs be frozen at 18° F. or 20° F. and stored at a somewhat higher temperature, say 25° to 28° F. It has also been recommended that zero or thereabouts was better than any of the higher temperatures. At temperatures of 25° F. and above, the white of the egg softens and becomes gummy, and deteriorates rapidly in quality. The damage is especially noticeable when white and yolk are frozen separately. When frozen at 10° F. and lower a greater expansion of the egg meat takes place, and when thawed it is watery, and not as useful for all purposes as the stock frozen at somewhat higher temperature. One of the best bakers in Boston informed the author that he could use the separated eggs frozen and held at about 20° F. for any purpose for which a perfectly fresh egg was adapted. In putting up eggs for freezing they should be placed in a cold room (not necessarily a freezing room) immediately when removed from the shell, as fermentation begins soon in warm weather, and loss of quality results. If the eggs are broken out of the shell at the storage house, remove them to the freezing room every hour. If they are made ready at a distance, provide a refrigerated room for temporary cooling, and see that all broken stock is in the freezer every night. If the frozen stock is thawed by setting in a tank of cold water, better results are to be had than when allowed to thaw in a warm room. The egg meat should be used up at once when thawed, as fermentation commences soon, and the stock soon becomes useless.

In estimating quantity of frozen or bulk eggs they are sometimes figured on a basis of ninety eggs to the gallon and at the rate of one and one-fourth pounds to the dozen, but this, of course, is only for estimating purposes as there is a considerable variation in different sizes of eggs and at different seasons of the year.

CHAPTER XIV.

BUTTER

TEMPERATURE.

In the early days of butter refrigeration it was thought that temperatures of from 35° to 40° F., such as might be had by cooling with ice only, were sufficiently low. These temperatures kept the butter in a reasonably solid state, and for a storage period of two or three months gave good results in preserving the flavor and preventing deterioration. The tendency has been steadily toward lower temperatures until now zero and below is thought by many to be most suitable for butter storage. This is by no means a general impression, and the majority of produce men still believe that any temperature below 20° F. is sufficiently low for ordinary commercial results. It may be stated that the average of opinions on the subject at this time favors 12° to 15° F., and as there are only a few results of accurate tests available at this time to prove any particular temperature as best for varying conditions and purposes, the present status of the matter is presented in some detail for the consideration of the reader. The author has maintained for some time that any temperature below 15° F. was low enough for periods of three to six months, which covers the average time for which three-fourths of the butter is cold stored. If the butter is stored in suitable packages, and is well made to begin with, no important good can be accomplished by storing at lower temperatures. On the other hand if the butter is in packages not suitable, of inferior packing and grade, and it is desired to store for long periods, or it becomes necessary to carry from one year to another, temperatures of from 10° above zero to below zero Fahrenheit may produce improved results. It is certain that 20° F. (above zero) and perhaps somewhat lower will retain the desirable butter flavors better than from 35° to 40° F., so it appears reasonable that 10° F. above zero will retain flavors better than 20° F. or thereabouts. For a number of years the author has recommended a temperature of from 10° to 14° F. for butter storage, and sees no reason at this time to change.

FREEZING BUTTER.

We hear nowadays about freezing butter for holding in storage. This commonly refers to any temperature below the freezing point of water (32° F.). Some houses have recommended and practiced "freezing" the butter at zero or thereabouts for a few days, and then storing permanently in a temperature of from 10° to 20° F. above zero. Butter does not freeze in the ordinarily accepted sense of the term. It is of an oily nature, and simply gets harder and harder as the temperature is reduced. The freezing point of butter, if it may be so called, is from 92° F. to 96° F. as determined by test. (See "Specific Heat of Butter" further on.) The freezing point of a substance, as ordinarily understood, means the temperature at which it changes from a liquid to a solid, and butter therefore freezes at many degrees above the freezing point of water. The talk about rupture of fat globules in butter by freezing, therefore, is not well applied. Butter does not freeze at any cold storage temperature, but simply becomes harder and denser as the temperature is reduced. It will, however, probably be ultimately shown by repeated tests that storing butter at an extremely low temperature will cause a "shortness" or rupture of the grain, but this theory is advanced by the author on his own responsibility.

PROTECTION FROM THE AIR.

The successful holding of butter in cold storage depends as largely on the protecting of the product from air contact as in maintaining a low temperature in the storage room. Possibly with extreme low temperatures of zero or thereabouts, protection from the air will be of less consequence, but this point cannot at present be overlooked if best results are desired.

Butter being composed largely of an oil or fat, is susceptible of becoming rancid or "air-struck" when exposed to the air for a considerable time; the higher the temperature the quicker the butter becomes rancid. It is reasonable to suppose, therefore, that the lower the temperature the longer butter may be held in contact with the air without becoming rancid. In other words as the temperature of a butter storage room is held lower, the less the necessity of care in protecting the butter from the air of storage room, but it is in any case desirable that the package should be as air-tight as possible. It cannot be known at time of storing how long the butter will be held, and the nearer air-tight a package is, the longer will it keep the butter in good flavor and condition. Butter packed under direction of the United States Government for export and use in warm climates is put up in hermetically sealed cans, and some of our "boys in blue" bear witness to the palatability of same, even when carried under insufficient and inferior methods of refrigeration. Another means of canning is the method formerly in use for packing butter for shipment to California. The butter was made up in rolls and packed in tight casks which were afterwards headed up and all spaces between rolls and at sides and ends of casks were filled with brine or "pickle" as it is called. As the refrigerating means were formerly inadequate, this method was necessary in order that the butter might be carried through to destination in palatable condition. Firkins (kegs holding about 100 lbs. of butter) were much in use at one time, especially for shipment to foreign countries. These wooden packages were thoroughly soaked with brine, packed solid and nearly full. The head was put in and when the butter was cooled, the space resulting from shrinkage of the butter was filled with pickle composed of salt, saltpetre, and sugar. Attempts have also been made to cover the butter in ordinary tubs with brine pickle after the tubs were placed in storage, in order to protect the butter from the air, but the muss and slop resulting made this scheme impracticable. These methods of packing butter are mentioned as representative of the former practices in use to prevent the butter becoming air-struck and rancid. At this time very little butter is stored under any of these methods, owing to the expense of packing and impracticability of the packages for the retailer. Butter stored immersed in pickle also has a soaked appearance, where it comes in contact with the pickle, which is objectionable.

BUTTER FLAVOR AND AROMA.

It was at one time thought that flavor and aroma of butter were due to the food upon which cattle were fed. During the "full grass" months of May, June and July, this was especially noticeable, and at this time cows give milk which makes a fine quality of butter. The bacteriologist has changed our ideas on this matter, and by the use of a "culture," nearly as fine an aroma and flavor may be produced in midwinter as on full grass. By pasteurization and ripening the cream by the use of a culture of suitable bacteria, fine flavored butter may be made at all seasons of the year. One of the chief objects accomplished in cold storing butter is to retain the flavors and aroma which are produced by the ripening or souring bacteria of milk and cream. Loss of these is *prima facie* evidence that butter is no longer fresh. Low temperature and protection from the air will accomplish the desired results.

PREPARING BUTTER FOR COLD STORAGE.

Butter intended for cold storage purposes should have the buttermilk thoroughly removed by washing and working moderately in water. The working should not be carried too far so as to spoil the grain of the butter, but as much of the buttermilk as practicable should be worked out and a moderate amount of pure water and salt incorporated in its place. The butter should be well salted so that the water content shall be in the form of strong brine. Butter containing a large portion of moisture keeps best in cold storage. Butter made by the old deep setting process or by raising the cream by setting in cold water, keeps much better than the best separator butter. No doubt some will be somewhat surprised to learn this. The reason is that more of the casein is left in the butter by the centrifugal separator and this causes a fermentation which deteriorates the butter more rapidly. The author has seen two lots of butter placed in cold storage at the same time and stored in the same room—one lot was fancy separator creamery butter worth at that time about 18c per lb.; the other lot was a second grade gathered-cream creamery, worth 14c per lb. When removed from storage four or five months later the 14c butter sold the best on the open market. The gathered-cream butter, as most of my readers are aware, is made from cream skimmed by the farmer and collected and churned at a creamery, and the cream in this case was secured by setting or gravity as it is sometimes called. The resulting product is always inferior in flavor when first made. The case above is mentioned to show the comparative keeping quality of centrifugal separator and gravity raised cream butter when placed in cold storage. Other things being equal, butter from gravity cream is better than separator butter.

PROCESS BUTTER.

"Process" or renovated butter, which is made from a miscellaneous lot of dairy butter melted, purified, regranulated and flavored by the use of a bacteria culture, has comparatively poor keeping qualities in cold storage and therefore very little is stored. The most common way is for the process operator to store the original package or by repacking into barrels. If barrels are used they should be soaked well with brine and then lined with parchment paper before packing in the butter. Don't store rancid butter for processing—select only that which is fresh and reasonably sweet. Butter which is slightly sour from presence of buttermilk is not as good for cold storage, but in processing this largely disappears, and butter which is sour from this cause may be stored to advantage if fresh. In fact it is difficult to get the medium grade dairy butter which is largely used for processing, during the months of June and July, which does not have more or less this sour character. The chief point of importance to guard against in selecting butter to be cold stored for future processing is rancidity. Butter which has once become even slightly rancid will deteriorate more rapidly in cold storage and is unfit for making anything but low grades of process butter. Store in the original package if possible, providing it is in good condition, as repacking breaks the grain and injures the keeping quality. If it is necessary to repack, pack solidly without leaving air holes.

USE OF JARS FOR BUTTER STORAGE.

For a limited local trade a good grade of dairy butter in small jars is very desirable. Select good flavored, even colored butter for storage, and turn everything else into "packing stock" or low grades. Remove all miscellaneous cloth and paper coverings, replacing by cloth or parchment paper circles or caps and spread on evenly a fine grade of dairy salt to a thickness of one-eighth of an inch, or sufficient to cover the surface of the butter thoroughly. Over this tie a cover of light colored manila wrapping paper, and you have a package which is practically air tight. It is also in good shape for sale when removed from storage. The jars may be piled one upon another to a height of three or four feet. Racks are best for piling jar butter with shelves at intervals of three or four feet. In piling in an ordinary room without racks, there is great danger of a collapse of piles of jars and the result may be imagined. Jars are undesirable for shipping, hard to handle, and liable to be broken, but they make a fine package for cold storage, and are desirable for retailing. For storage in a small way, for local consumption use jars.

STORING BUTTER IN TUBS.

Tubs of various sizes larger at the top are the standard butter packages, and by far the greater portion of the butter made in the United States is handled in tubs containing about sixty pounds. The best material for tubs is white ash, but some markets, notably Boston, prefer tubs made from white spruce. The covers of tubs should be of the same material as the staves and bottom, or of some sweet hard wood. The soft woods, particularly pine, may impart a foreign flavor to the butter. The following directions for soaking tubs and preparing them for packing are given by P. M. Paulson:*

In packing butter it is first necessary to properly prepare the package; this I do by soaking the tub and then placing in a tank of

*In *New York Produce Review.*

brine so that the tubs are held completely in the brine for about 12 to 14 hours. The liners I also place in brine for about the same length of time. When butter is worked sufficiently and ready for packing, I line the tubs. If I am alone to pack, I line five or six tubs at a time; if my helper has time to help me, we line enough tubs to hold what butter we have in a working. The liners we place in smoothly in the tubs in a way so that the top edge of the liner can be turned down over the edge of the tub about ½ inch. Next I put the bottom circle in position. If I am packing alone, I take five or six pounds of butter (not more) and put in each tub that I have lined; I then press it firmly together with the packer, seeing that there are no holes left in the butter and also that it is pressed firmly against the edge of the tub. I repeat this operation until tub is filled and enough more so that there is from one to two pounds on top. When this has been pressed firmly down, I take a string, wet it, and cut the butter off level with the tub; next I take the paper lining and turn it back over the edge of the tub and on to the butter, neatly and with care, being careful not to tear the paper, and smooth it down. Then I place the cloth circle on the tub; this should be large enough to reach to the outside edge of the tub. Then I take a little water with my hand and moisten the cloth, next sprinkle a little salt on, and rub it lightly with my hand, so that it is even all over. In placing the cover on, care must be taken to get it on properly; if it don't go on easily I place my knee on the cover and tap the edge lightly with a hammer until I get the cover on; it is better to hammer on the edge of the cover than to hit the staves on the tub, as it keeps the butter in better shape. In placing the tins, I place the first one on over the end of the cover rim; this will prevent the rim from tearing off if it should by accident get caught; the second tin I place directly across from the first one, the third and fourth at equal distance between first and second. I always try to place the tins so that they will reach down into the top hoop on the tub; last I drive a ⅝d. nail in the lower end of tin; the end on the cover I have always found does very well with one nail. I always use a tin that has one nail in each end; they are the most convenient to use. Wire tub fasteners should not be used, the trade does not like them. Before I place butter in refrigerator I always see that the tubs are perfectly clean.

The liners mentioned by Mr. Paulson are of parchment paper and come ready cut to proper size for tub used. A pint of brine in the bottom of the tub when starting to pack, is desirable, as it fills all cavities and the pores of the wood. In packing keep the butter pressed down in the center first and then at the sides so as not to leave openings in the butter which may later become air spaces by evaporation of the brine and cause the butter to sooner become "air-struck."

OLEOMARGARINE AND BUTTERINE IN STORAGE.

Oleomargarine and butterine are of a similar nature and resemble butter, but are much more easily preservable by refrigeration, and may be kept for long periods in fine condition. The reason is that they contain very little casein or other substance liable to fermentation and decay, being composed almost wholly of fats and oils which do not spoil quickly, even at ordinary temperatures. A temperature somewhat higher than that recommended for butter is generally used for butterine and oleomargarine. Temperatures of from 20° to 30° F., are in common use for the storage of these products.

LADLE BUTTER.

"Ladle" butter is butter reworked, resalted and repacked, so as to put it in marketable condition and give it a uniform grade. Much of this is butter of good quality, but lacking in uniformity of color, salt, and package. The ladler takes the miscellaneous "farmers," "dairy," "store" or "packing stock" butter, and by rehandling turns out a butter which is improved commercially to an extent which has in the past made the business profitable. The ladler makes his profit in intelligent grading and in the increase of weight by resalting, washing, and reworking. "Ladling" has now been largely superseded by "processing." Very little ladle butter is placed in cold storage at the present time. Those who have had experience, know that "ladles" do not keep well in cold storage. The reworking incorporates thoroughly throughout the mass any rancidity or bad flavor present in any part of the butter, and the result is that after standing a comparatively short time "ladles" are off flavored and take on a "ladley" taste and odor, even when carried in low temperatures. As in processing, butter intended for ladling is cold stored as original butter and rehandled as wanted by the trade. Directions given for the handling of original butter apply equally when used for processing or ladling.

CREAMERY BUTTER.

Creamery butter is so well known as not to need much description. At the present time nearly all creamery butter is made from cream which is separated from the milk by a centrifugal machine known to the trade as a separator. Separator butter has poorer keeping qualities than butter made from cream raised by setting the milk in cold water or what is called the gravity process, for reasons already stated, but for the ordinary commercial storage term of three to six months, keeps well enough for practical purposes when held at temperatures of 10° to 14° F. The sixty-pound tub is the package generally used, particularly by the retailer, but much butter after having been stored in large tubs is tempered to soften it slightly and then repacked into smaller packages; the one pound print, wrapped in paraffine or parchment paper being a favorite. Butter which is to be "printed" before sale should be stored in as large, air-tight and well soaked or impervious packages as possible. Some dealers use firkins or butter carriers holding 100 to 200 pounds. Do not try to store butter in prints for any length of time as the grain is somewhat broken in printing and its keeping qualities therefore impaired. For the same reason do not store in small packages which are not impervious to air and moisture. The directions for packing previously given apply especially to creamery butter. In some cases a covering of paste salt (salt which is ground fine) is used. This is mixed with water and is put on as a paste, which hardens on drying, forming an air tight crust over the top of the butter. The butter cannot well be examined without mussing or destroying this paste salt covering, and it is not used to any extent except for cold storage purposes. The Australian butter box, a rectangular and nearly cubical package, holding about 60 lbs., is coming into quite general use. Butter from these boxes cuts up with little waste when printed in one pound bricks, and the shape of the package makes it very economical of storage space.

FISHY FLAVOR IN BUTTER.

The development of a peculiar flavor in butter which is stored under refrigeration has long been under discussion and for a time the cold storage house was blamed for this trouble. Afterwards the trouble was attributed to the use of inferior salt containing lime. More recent investigations prove beyond a question that neither cold storage nor impure salt is the chief cause of fishy flavor in butter.

A circular by L. A. Rogers, Chief of the Dairy Division, Department of Agriculture, Washington, D. C., goes to show that fishy flavor is most often caused by the development of a certain acid produced in the ripening of cream. It is also suggested that other causes might produce fishy flavor and that butter sometimes described as fishy was merely oily flavored, or otherwise off in flavor, the fishy flavor being a peculiar oily taste suggesting salted fish.

During the past three years the Dairy Division has made a large number of lots of experimental butter and in no case has fishy flavor developed. The reason ascribed is that all the butter has been made from pasteurized cream without ripening in the regular way, and by the addition of a so-called starter or bacteria for the development of flavor. The subsequent ripening of the pasteurized sweet cream treated with starter improves the flavor of fresh butter without adding acid in sufficient quantity to cause fishy flavor. The pasteurizing of cream which has already soured in the ordinary way will not prevent the development of fishy flavor in butter stored for long periods under refrigeration.

It is suggested in addition to the findings of the Dairy Division as above, that fishy flavor in butter is a comparatively recent trouble and it has developed only since the centrifugal separator has come into general use. It is suggested that the fishy flavor referred to is caused by a fermentation caused largely by the presence of an excess of the curd or cheese elements in the butter made from separator cream. Butter made from cream raised in the old fashioned way contains very little of the cheese element.

MOLD IN BUTTER PACKAGES.

Mold in butter packages has given much trouble, both in cold storage and in the regular cooling rooms when held for temporary storage. This may be caused by improper soaking of the tubs or a badly constructed refrigerator or cooling room at the creamery, or the empty tubs may have been stored in a damp place such as a cellar or basement at the creamery. A growth of mold once started is quite likely to continue to grow and may in a short time affect and flavor the butter. A growth of mold may be prevented by storing the empty packages in a dry place; providing a good refrigerator with suitable air circulation at the creamery; and by care and attention in packing the butter, as already outlined. Instead of using water for soaking the tubs, use brine. Water promotes mold—brine destroys it. Salt is cheap. Use it in connection with your butter packages, and mold will not trouble you. Use parchment paper liners and use brine for moistening at time of packing. See chapter on "Creamery and Dairy Refrigeration" for information regarding suitable facilities for cooling rooms, etc., in connection with creameries.

SPECIFIC HEAT OF BUTTER.

The following regarding the specific heat of butter* by G. H. King, Agricultural Physicist, University of Wisconsin, is reproduced here for the valuable scientific information it contains:

It would be a very difficult, if not an impossible, task to determine the true specific heat of the butter fat of commerce, making corrections for the elements of latent heat, for the reason that butter is so complex a product, and the butter fat itself varies so much in composition with the season and with the stage of the lactation period, and even with the individuality of the animal producing the butter.

I have made an approximate determination of the specific heat of butter fat between 100° C. and 0° C., and find it to be .5494.

This result was obtained by taking ordinary butter, melting it and boiling until all water was driven off, and skimming to remove solids not fat, and then filtering hot.

There was then placed into a pocket in a block of ice 200 grams of the clear butter fat at a temperature of 100° C., and brought quickly to 0° C., when the butter fat and ice melted were weighed. Calculating the specific heat from the amount of ice melted, the result found was .5494.

Butter fat, leaving the other ingredients out of consideration, is largely a solution of tripalmitin and tristearin in triolein, or, in commercial language, butter fat is a solution of palmitin and stearin in olein. But in addition to these three fats there are also found varying amounts of five others, viz., butyrin, myristin, caproin, caprylin and caprin.

The pure triolein, or olein of vegetable fats and oils, becomes solid only at a temperature as low as 21° F. The tripalmitin, or palmitin of vegetable and animal fats, occurs in three isometric or allotropic forms, with melting points as high as 115°, 142° and 144° F., respectively, while the tristearin, or animal and vegetable stearin, also occurs in three forms, which remain solid, when pure, until a temperature of 124°, 148° and 157° F., respectively, is reached.

The temperature at which butter becomes solid, or semi-solid, varies with the relative amounts of the three chief fats which happen to be present in the sample. It is stated that ordinarily butter be-

*From *Ice and Refrigeration*, June, 1901, page 278.

comes fluid or melts at between 92° and 96° F., which should be understood that below these temperatures the olein is no longer able to hold all of the palmitin and stearin in solution. Pure lard melts at 78° to 87° F., and its composition is given as 62 per cent olein and 38 per cent of palmatin and stearin. Butter fat, in the spring, from fresh cows on green grass has a composition near 50 per cent of olein, 30 per cent of stearin and 20 per cent of palmitin; but later in the period of lactation, and in the fall when the feeds are drier, its composition may change to 30 per cent olein, 50 per cent stearin and 20 per cent palmitin.

It seems likely from these observations that the amount of heat necessary to be applied to butter, in raising it from freezing to its melting point, and to be withdrawn from it in cooling it from its melting point down to freezing, will not be very far from the amount which would be required to make a corresponding change in temperature of water, pound for pound.

HUMIDITY, CIRCULATION, VENTILATION.

Humidity, air circulation and ventilation have been given comparatively little attention as applied to the storage of butter. At the low temperatures at which butter is generally stored the air contains so little moisture as to be amply dry to prevent mold, and nothing further is thought of it. In fact, most butter storage rooms are dryer than necessary, and it is difficult to prevent the butter drying out from this cause. It is only necessary to have a butter room dry enough to prevent mold on packages, as the goods are supposed to be sealed from air contact. What this humidity should be there are no records to show, but moisture does not trouble the general run of storage rooms for butter. A circulation of air in butter storage rooms is of no great consequence, as sufficient air circulation for purification of the air is usually present. Most butter storage rooms are equipped with direct piping, but some are provided with air circulation by means of fans, when a quicker cooling is possible. Ventilation of butter storage rooms is advisable at regular intervals, using the apparatus described in the chapter on "Ventilation." Gases from the oxidizing of butter fat and odors from the wooden packages accumulate in the storage room unless disposed of by ventilation.

CHAPTER XV.

CHEESE.

THE DESIRABILITY OF COLD STORAGE FOR CHEESE.

The cold storage of cheese on an extensive scale is of comparatively recent date. Formerly it was considered sufficient to store cheese in an ordinary cellar or basement room, perhaps cooled by ice bunkers, but about thirty years ago cheese were first placed in cold storage, both with the old overhead ice method of cooling and the first ammonia refrigerated houses. The author remembers distinctly when as a boy he visited the old St. John's Park Depot in New York, which was then cooled with one of the first ammonia systems to be put in commercial cold storage service and was used quite largely for cheese storing. The success of the early experiments in keeping cheese in cold storage was such as to extend the practice, and at the present time practically all cheese which are to be held for consumption at some future time are placed in cold storage for preservation. In fact the advantages of low temperature have been so thoroughly appreciated that it has led the various experiment stations to conduct some very extensive experiments in what is called "the cold curing of cheese."

As a matter of fact, cheese "ripen," "cure" or mature at any low temperature at which they may be safely stored. Cheese is one of the products that improve with age. It is not at its best when first made; in fact it is unpalatable and unhealthful when new or "green." It requires "curing" in order to make it a suitable article for food. Under ordinary conditions the curing process goes on regardless of temperature, but the action is much slower as the temperature is lower. The results of experiments which are here given prove conclusively that a much better quality of matured cheese results when the cheese are placed in cold storage soon after being made. It seems that the low temperature prevents the development of bad flavors and deleterious gases which injure the flavor and texture of the cheese, while at the same time it allows the rennet which is used in the manufacture of cheese to fulfill its mission of curing or developing. The experiments which are described in detail further on need no additional explanation.

The result of these experiments seem to prove the advisability of establishing centralizing stations, which are in reality cold storage plants, for the receiving of cheese when first made. A plant of this character may be built at any convenient railroad point and the cheese from a number of different factories hauled thereto at frequent and regular intervals. They are then placed under suitable temperature, and other conditions, and are ready for immediate shipment at any time. The advantage of this method over the old factory system of allowing the cheese to remain on the curing room shelves for a time is that the flavor is improved, shrinkage reduced to a minmum and the cheese are protected from exposure to hot weather, which is one of the worst things that the cheese manufacturer has to contend with. Appreciating this difficulty, the sub-earth duct system has been adopted by some of the more progressive factories. This is simply an air duct running underneath the ground through which circulates air which is introduced into the curing room. In passing below the surface of the earth, the temperature of the air is reduced to 60° to 65° F. and the temperature of the curing room is therefore modified during extremely warm weather. It was found that this system in many cases had the disadvantage of causing cheese to mold badly and no doubt it will be abandoned in favor of the cold storage or cold curing method. There is an advantage in having cheese brought to a central cold storage or curing station in that it is easier for buyers to inspect and brings the cheese all into a market center as it were. There is no reason why they should be out of possession of the salesman any more with this system than they would under the old method. Co-operation and consolidation will enable cheese manufacturers to realize much better prices for their product, owing to improved quality, if they will but adopt the cold storage system instead of the old-time curing room method. A large part of the expense of a central cold storage station would be paid by the saving effected at the factory in not being obliged to provide for shelves and curing room space.

The best refrigerating system for use in connection with cold curing will depend upon the section where located and local conditions to a large extent. In cooling with air circulating in direct contact with ice, a temperature below 40° F. cannot be depended upon and as experiments demonstrate that cheese stored at a temperature of 30° to 35° F. are of a better flavor and texture, it is evident that some system which will produce a lower temperature would be advisable. In addition, the humidity of a room cooled directly from the ice is very high (in other words, very moist). It has been demonstrated that the relative humidity of such a room when used for the cold curing of cheese would be at times somewhat above 90 per cent. These conditions are very favorable for the growth of mold. The direct ice system therefore is not advisable for the reason that sufficiently low temperatures and regulation of humidity cannot be obtained. The Cooper brine system, cooling with ice and salt, described elsewhere in this book, is recommended as a system which will control temperatures, and in connection with the Cooper chloride of calcium process, also described elsewhere, the humidity of the room may be regulated to any desired degree. In situations where natural ice cannot be obtained cheaply, the use of refrigerating machinery is advisable and the temperature and humidity can thereby be controlled in the same way as with the Cooper brine system.

THE COLD CURING OF CHEESE.*

The prevalent opinion among cheese dealers has always been that low temperatures, varying from 35° to 50° F., or thereabouts, resulted in the production of an inferior quality of cheese, in comparison with that from 60° to 70° F. No care-

*Extracts from Bulletin No. 49, Bureau of Animal Industry, U. S. Agr. Dept., giving results of experiments conducted under the directions of Henry E. Alvord, Chief of Dairy Division. More detailed information may be obtained by consulting same.

fully controlled experiments bearing on this problem have been recorded earlier than those undertaken by Babcock and Russell at the Wisconsin Agricultural Experiment Station, and described in the fourteenth (1897) annual report of that station. The results of those tests showed that cheese placed at refrigerator temperatures (45° to 50° F.), directly from the press, was of superior quality as to flavor and also as to texture, and that such cheese was wholly free from any bitter or other undesirable taints.

In connection with their studies on the influence which galactase and rennet extract exert on the progress of cheese ripening, the same investigators later employed still lower temperatures 25° to 30° F.). Cheese were kept at these excessively low curing temperatures for a period of eighteen months. The quality of these cheese, cured as they were below the freezing point throughout their whole history, was exceptionally fine, and emphasized still more than the previous experiments did the fact that the ripening of cheese can go on at much lower temperatures than has heretofore been considered possible.

These results led to an extended series of experiments, in which cheese made on a commercial scale was cured at a range of temperature from below freezing (15° F.) to 60° F.—a point which common practice has now accepted as the best obtainable temperature that can be secured without the use of artificial refrigeration.*

In these experiments (consisting of five series made at intervals throughout a period of two years) 138 cheeses were used for which 30,000 pounds of milk were required. These experiments were upon a scale which represented commercial conditions, and therefore obviated the objection which is often urged in commercial practice against the application of results derived simply from laboratory experiments.

The Ontario Agricultural College began experiments on the cold curing of cheese in April, 1901. As a result of these tests the conclusion was drawn that the cheese cured at low tem-

*No doubt the term " artificial refrigeration" as here used means cooling by any means other than natural earth or air temperature, and not the general accepted meaning, viz., refrigerating machinery.

peratures (37.8° F.) was much superior to that cured in ordinary curing rooms (average temperature during season 63.8° F.). Mr. R. M. Ballantyne, a prominent cheese expert, said of this cheese that "they (the merchants) universally expressed surprise at the condition of the cheese that was put into cold storage at the earliest period (that is, directly from the press), as they expected to find the cheese still curdy and probably with a bitter flavor."* If this experiment is borne out by other experts, it would appear as if the best way to handle hot-weather cheese would be to ship it to the cold storage directly after making, and this would certainly mean a great revolution to the trade.

A considerable number of experiments have also been made at other stations (Dominion government tests and New York State and Iowa experiment stations), where somewhat lower temperatures were used than those which are normally employed for ripening. The results obtained all show an improvement in quality that becomes more marked as the temperature is reduced.

In order that a much larger experiment might be instituted, covering the different types of cheese as represented by eastern as well as western manufacture, Drs. Babcock and Russell, of the Wisconsin Station, presented this matter for consideration to the Dairy Division of the Bureau of Animal Industry. As a result of this proposal the officers of the New York Agricultural Experiment Station were also consulted and plans perfected for the co-operative experiments conducted simultaneously in Wisconsin and New York.† It should be noted that it was so late in the season of 1902 when the arrangements for this work were completed that it was impossible to obtain favorable conditions in all respects.

In addition to the influence which a range in temperature exerts on the quality of cheese, as determined by flavor and texture scores, instructions were also issued to secure data regarding the loss in weight which the different lots of cheese suffered

---

*Bulletin No. 121, Ontario Agricultural College, June, 1902.

†The eastern experiments are not given here as the results differ in detail only, general conclusions being the same in both series of experiments.

at the different temperatures. The commercial quality of the product was to be determined by a jury of experts who were thoroughly in touch with the demands of the market. Although the effect of coating cheese with paraffin soon after being taken from the hoop was not at first proposed as a part of this work, it was finally included.

The reasons for selecting 40°, 50°, and 60° F. as the temperatures to be used in these experiments are fully given on a later page. It may be assumed that the advantages of a cool and even temperature in curing Cheddar cheese have been already established in preference to a warm temperature or to very variable conditions which frequently include periods above 70° F. and sometimes much higher. As already stated, 60° F. or thereabouts is regarded as the lowest temperature practicable without artificial refrigeration; this may therefore be taken as fairly representative of what may be called a "cool" temperature for curing cheese. And rooms held at 40° and 50° F. were selected as representative of a "cold" temperature for curing, or comparatively so. It is thus hoped to emphasize by these experiments the distinction between cool curing and cold curing.

The cheese for these experiments were purchased by the United States Department of Agriculture, which also paid all expenses of transportation and storage and for the experts who made the periodical examinations. The two experiment stations selected the cheese, arranged all details of storage and examination, supervised the work throughout, performed the chemical and other incidental scientific work, kept the records, and reported results.

Each of the reports, prepared by the two experiment stations participating in this work, treats the same general subject and similar lines of experiment and observation from its own point of view. The reports therefore differ in many respects, and yet they may be easily compared upon all essential points. Both support the same general conclusions as to the advantages of curing cheese at low temperatures, summarized as follows:

1.—The loss of moisture is less at low temperatures, and therefore there is more cheese to sell.

2.—The commercial quality of cheese cured at low temperatures is better, resulting in giving cheese a higher market value.

3.—Cheese can be held a long time at low temperatures without impairment of quality.

4.—By utilizing the combination of paraffining cheese and curing it at low temperatures the greatest economy is effected.

THE WESTERN EXPERIMENTS.*

For the purposes of this experiment Chicago would naturally have been chosen as a curing station, but it was found difficult to make arrangements for the range of temperatures desired. Suitable arrangements, however, were made at the cold-storage warehouse of the Roach & Seeber Co., Waterloo, Wis., where rooms were fitted up and the desired temperatures secured.

As Wisconsin is the leading cheese-producing state of the west, the bulk of the product selected for experiment was of the type of cheese manufactured in this state. In order, however, to cover more thoroughly the cheese-producing territory of the west samples were also secured from a number of the neighboring states. In this way all types of American cheese were obtained, ranging from the firm, typical Cheddar suitable for export, to the soft, open-bodied, moist cheese, intended for early consumption. For convenience we may group these various lots of cheese under three different types, as follows:

I.—Close-bodied, firm, long-keeping type, suitable for export trade (typical Cheddar).

II.—Sweet-cured type.

III.—Soft, open-bodied, quick-curing type, suitable for early consumption.

Type I represents the class of cheese that is especially manufactured in Wisconsin, while, as a rule, type III represents the kind of cheese that is chiefly made in Michigan. The representatives of the sweet-curd type were taken from Iowa and Illinois, although this class is made to some extent in all sections.

---

*Conducted by S. M. Babcock and H. L. Russell, assisted by U. S. Baer, all of the Wisconsin Agricultural Experiment Station.

In having the cheese made at these various factories directions were given for the use of a uniform amount of rennet and salt. Color was left optional for each maker to follow his customary practice. The use of 3½ ounces of Hansen's rennet extract and 2½ pounds of salt per 1,000 pounds of milk was recommended in each case with the exception of the smaller cheese (dairies and 10-pound prints), which were salted at the rate of 2¼ pounds per 1,000 pounds of milk. The cheese was made from September 26 to October 4. The condition of the milk was influenced in several instances by the fact that severe frosts had occurred in some sections, which injured the quality of the product. This was particularly true in the case of the Alma cheese, which was in consequence somewhat tainted. The milk from which the Iowa cheese was made was also reported as of inferior quality. The Michigan goods were too high in acid, and were cooked low, making a soft cheese which was quick-curing and which kept poorly.

Where it was necessary to secure cheese from such a wide range of territory it was manifestly impossible to expect that the curing could be carried out as satisfactorily as if it had been done at or near the factories. The varying period of transit to which the cheese was subjected with no especial temperature control, affected, of course, the initial stages of curing, but the conditions of the experiment prevented the carrying out of immediate installation of the cheese in the cold curing rooms, especially in the case of those made outside of Wisconsin, although the shipments were made in October, when the temperature range was moderate.

TEMPERATURES AT WHICH THE CHEESE WAS CURED.

The cheese was weighed and put in the respective rooms as soon as received at Waterloo. It was stored in boxes during the curing, as is the custom in the handling of cold-storage goods. The temperatures at which it was desired to hold the cheese for curing were 40°, 50°, and 60° F. These points were selected for the following reasons: In our previous experiments we had found that the character of the cheese cured at the lower temperatures (40° and 50°) was much better than that produced at 60° F. Perhaps it would have been better for the purpose of the experiment if the cold-cured cheese could have been compared with the same make of cheese cured under the widely variable conditions which prevail in most factories, where often the maximum temperature is in the neighborhood of 80° F. and the fluctuation is 20° or more; but we have made this comparison with the very best conditions that obtain in factories provided with subearth ducts and other means of temperature control. In such cases a temperature of 60° F. can be maintained with a fair degree of constancy. The experiments, therefore, compare the cold-curing process with that of the best prevailing conditions.

The temperatures actually maintained varied only slightly from the chosen points, and in the two colder rooms were remarkably uniform. The 60° room was subject to somewhat wider fluctuations, but was much more uniform than is obtained in summer where no artificial refrigeration is practiced.

DETAILS OF SCORING THE CHEESE.

It would have been advisable to have the cheese examined a considerable number of times by the commercial judges, but it was impossible to carry out this test so frequently. The tests were therefore arranged to come at those periods which would give the judges the most accurate idea of the character of the cheese held at the different temperatures.

As a jury of commercial experts, representing the different markets, the following gentlemen were selected: C. A. White, of Fond du Lac, resident representative in Wisconsin of a leading dairy produce house of New York; T. B. Millar, of London, Ontario, a cheese expert and large buyer for the export trade, and John Kirkpatrick, a member of a leading produce firm of Chicago.

For the jury trials representative cheese were taken from storage and shipped by refrigerator service to Chicago, where they were submitted to a thorough examination by the commercial judges. The first of these commercial scorings was made when it was found that the 60° product was ready for market. This test was made on January 6, 1903. Another test was made on March 23, when the cheese was about 7 months old.

It might at first thought seem preferable to have had the cheese sold in the open market and thus secured a strict commercial valuation on the product, but, as everyone knows, a considerable variation in quality may exist without an appreciable difference being made in the market price. Then, too, the inevitable fluctuations in the market price would render comparisons at different periods untrustworthy. To obviate these difficulties the cheese was scored on the basis of a standard price (13 cents). The fact that but few of the cheese reached this standard should not be interpreted as indicating a poorer quality than the average market product, for the cheese was adjudged by the jury to be superior in quality; but the price was in part determined by the market appearance of the goods, which was somewhat inferior because of the fact that they had been box-cured and had received practically no care in curing, as the curing station was located at a distance from Madison.

The scores of the commercial jury were supplemented by a series of scores made by Mr. Baer which covered the entire history of the cheese from the time it was received until its final disposition. In this study it was possible to follow more closely the course of the ripening.

SHRINKAGE OF CHEESE IN WEIGHT WHEN CURED AT DIFFERENT TEMPERATURES.

The losses in weight which cheese undergoes in the curing process is a matter of such practical importance that it is advisable when possible to accumulate data relating to it. This is all the more important in this connection because no studies have yet been reported on cold-cured cheese, and it was therefore deemed advisable to keep a record of the losses in weight so that the shrinkage at these lower temperatures might be compared with those which normally obtain at the best temperatures now employed. The average shrinkage under existing curing conditions in the majority of factories results in a loss of 5 to 7 per cent for the first thirty days, with a gradually diminishing rate for longer curing periods. This results in a heavy tax to the producer, and any factor which reduces these losses increases thereby the total receipts from the milk produced.

There are a number of factors which modify the rate at which a cheese loses its water content during the course of ripening. The following factors are known to exert a more or less marked influence, although it is impossible to arrange them in order of their relative importance, as they are always interdependent:

1.—Temperature of curing room.
2.—Relative humidity of air in curing room.
3.—Size and form of cheese.
4.—Moisture content of the cheese.
5.—Protection to external surface of the cheese.

The influence of temperature is closely connected with the relative humidity of the curing room; but, in addition to the effect which the higher temperatures exert on this factor, it should be observed that water evaporates more rapidly at a high than at a low temperature, even though the relative humidity remains the same. The more potent influence of temperature is, however, the effect which varying degrees of heat exert on the relative humidity of the atmosphere. A fall of 20° F. from ordinary air temperatures practically doubles the relative humidity, provided the point of saturation is not passed. As the average relative humidity of the air is generally over 50 per cent, it therefore follows, in cold-curing rooms supplied with outside air, the temperature of which is from 30° to 40° F. higher in summer than the inside temperatures, that the air of these rooms is practically saturated, thus greatly reducing the loss of moisture from the cheese.*

So far as the cheese itself is concerned, the moisture of the room may be materially altered by the way in which the cheese is handled during the curing process. If the cheese is shelf-cured, as is the custom in most factories, the surrounding air more nearly approximates the average relative humidity of the entire room than is the case where the goods are box-cured. In

*Conclusions so positive as these are not warranted. Temperature and humidity are not necessarily closely related. Water evaporates more rapidly at high temperature because the capacity of air for moisture is increased with its temperature, but it does not necessarily follow that the humidity is increased as the temperature is reduced, and a room in which the air is nearly saturated with moisture seldom exists. If it did it would be a bad place to store cheese because mold would grow rapidly. See chapter on "Humidity."

the latter case the air is more nearly saturated, as is shown by the greater liability to mold and rind-rot.

This point is well shown in a series of observations on the relative humidity of the air in a box containing a cheese placed directly therein from the press.

A factor which is frequently overlooked is the varying moisture content of the cheese. The more moisture there is left in the cheese the more rapid the evaporation. The varying moisture content of different types of cheese is determined by the temperature at which the curds are cooked, the time of exposure, and the acidity of the curd. A cheese in which the acidity is developed is materially drier than a sweet-curd cheese. Salt also has a tendency to diminish the water content. In the foregoing cases the cause of this diminution in moisture is due to the shrinking of the curd particles under the influence of these factors. An increase in fat lessens the drying of the curd. Much loss of moisture can also be prevented by coating the cheese with paraffin, a practice which is now coming into very general use for the prevention of mold and to lessen shrinkage in weight.

EXPERIMENTS IN SHRINKAGE OF COLD-CURED CHEESE.

In these experiments the first careful weighings were made when the cheese was received at the cold-storage plant in Waterloo. The cheese was shipped from the factories directly after it was removed from the press, but was in every case several days upon the road. In no instance was the interval between making and installing in cold-curing rooms less than five days, and it ranged from this up to seventeen days with one lot from Michigan, which was delayed in transit. During this period, which was in early October, the cheese was subjected to varying conditions of temperature and exposure. In a few cases boxes were broken, and in other instances the cheese was delayed at points of transfer. It was impossible to obviate these difficulties, as the cheese was purchased at distant points in order to secure representation from a wide range of territory and from different types of cheese. This variation in initial drying changed, of course, the rate of loss when cheese was placed in cold-curing rooms, so that this factor must be taken into consideration in studying the data presented below.

The losses reported here cover those only which took place in the cheese after it had reached the cold-curing rooms, but careful records have been kept for the entire curing period; and these data, we believe, are of sufficient importance to warrant full consideration in this connection.

DETAILS OF WEIGHING.

The cheese was all weighed on counter scales, weighing accurately to fractions of an ounce. In order to check the accuracy of the weights, each cheese was weighed separately and the weight recorded; then the whole lot was weighed collectively. As these weights agreed within a few ounces, they show the accuracy of the weighings. For practical purposes it is desirable to know the losses which occur for stated periods. It was, however, impracticable for all of the cheese to be weighed at exactly the same intervals, as it was put in storage at different dates, but it was designed to secure at least three weighings for the first month of storage, two weighings for the second, and at about monthly intervals thereafter. If these data are charted, it is possible to deduce an estimated loss for any stated period, and in doing so we have selected the following intervals as being those concerning which data would be most frequently desired. For this purpose ten, twenty, thirty, sixty, ninety, etc., days have been selected.

CONDITIONS UNDER WHICH THE CHEESE WAS STORED.

In this work the attempt was made to hold the cheese at 40°, 50°, and 60° F. The actual temperatures secured averaged 36.8°, 46.9°, and 58.5° F. The variation in temperature in the two lower rooms was practically negligible, as it was only 2° to 2½°. The tempreature of the 60° room oscillated somewhat more (4° F.), but was very much more uniform than ordinary factory curing rooms.

Hygrometric data were not secured during the whole period, as it was at first thought that a saturated atmosphere would prevail where the cheese was box-cured, but during the course of the experiments it was noted that the 50° cheese was not molding as much as was that at 40° to 60°. This fact could only be explained by the assumption that a less humid atmosphere was present in the case of the 50° room.*

DISCUSSION OF RESULTS.

As there are several factors which affect the rate of shrinkage which the cheese suffers in curing, it will be desirable to discuss the data collected under several heads. The conditions of the experiment were such as to temperature that an especially favorable opportunity was had for the study of the influence which this factor exerts on the cheese. It is, of course, necessary in a study of this sort to have the cheese uniform in size. The moisture contents of the cheese cannot, of course, be made alike, but in this study the cheese of the same type have been grouped together—that is, as firm Cheddars suitable for export, and softer, moister cheese intended for home trade.

INFLUENCE OF TEMPERATURE ON SHRINKAGE.

To study the rate of loss of Cheddar cheese when kept at different temperatures, 129 flats were selected from nine different lots of cheese made by six different makers. These were exposed at three different temperatures, which averaged, respectively, 36.8°, 46.9°, and 58.5° F. The results obtained were calculated upon the number of cheese which were subjected to stated weighings. During the experiments much more data were collected on the lower temperatures than on the 60° lot. This was regarded necessary, as up to this time we have no published data on cheese cured at so low a temperature.

For purposes of convenience the different lots of cheese were divided into three types, depending upon their character.

I.—Firm-bodied cheese (export type), of Wisconsin.

II.—Sweet-curd type, as represented by the Iowa and Illinois makes.

III.—A very moist, soft type, suitable for home trade (Michigan).

The general conclusions arrived at were:

*See previous remarks on temperature and relative humidity.

1.—The losses sustained by the different lots were very much less at 40° F. than at either of the other two temperatures. For a ninety-day period the losses of the 40° cheese ranged from 1 to 1.4 per cent, while the 50° and 60° product shrunk from 3.4 to 4.5 per cent for the same time. In other words, by the use of the lower temperature for curing practically two-thirds of the losses which occurred at the temperatures of 50° and 60° F. were prevented. If these results are compared with what happens under ordinary factory conditions, the loss at these low temperatures for a period of ninety days (the minimum curing period recommended) will not be more than one-fourth of that which obtains under average factory conditions when the cheese are held for a period of about twenty days. The saving for any such factory making 500 pounds of cheese daily would amount to at least 15 pounds of cheese (or $1.50) per day as an average for the season, and considerably more than this for cheese made during hot weather. This saving in itself would go far toward meeting the extra expense of lower temperature curing, even if the product were no better than that cured at higher temperatures.

2.—The differences between the cheese cured at 50° and 60° F. are not so marked as between 50° and 40° F. It is quite probable, as before mentioned, that the 50° room was somewhat drier than the 60° (as shown by the lessened mold growth), and hence the rate of loss was abnormally increased in this room.*

3.—If the firm Wisconsin type is compared with the softer variety, as shown in types II and III, it appears that the losses are considerably less, especially at the higher temperatures, although this difference is not so observable at 40° F.

4.—The data referred to above showed a marked saving in losses where the cheese was cold cured, but in these experiments it must be remembered that the cheese was subjected to higher temperatures during transit, and hence dried out somewhat more than would have occurred if put in storage as soon as re-

*The reason why evaporation is less at the lower temperatures is not necessarily owing to higher relative humidity, but to the lesser capacity of the air for moisture at low temperatures, and the fact that mold naturally grows much more slowly at low temperatures.

moved from the press; also, that this cheese was box-cured, and therefore under conditions which prevented rapid evaporation. Under other conditions the losses would have been greater than represented here, and the difference in the rate of loss between the different lots wider than reported above. This would still further increase the saving.

It must be remembered that the entire loss in weight during the curing of cheese is not due to evaporation. A cheese in curing is constantly breathing out carbon dioxide the same as any living organism, due to the development of microorganisms (bacterial growth within the cheese as well as molds on surface). Aside from these biological factors, it has been shown by Van Slyke and Hart* that profound proteolytic decompositions also give rise to an appreciable amount of CO_2. With cheese at 60° F., in which external mold growth was suppressed, they found a loss of approximately one-fourth of 1 per cent in ninety days. In our cold-cured cheese, copious mold development occurred, and hence the losses of carbon from the cheese due to this growth would be considerably greater than if no such growth occurred. With the nearly uniform rate of shrinkage shown in these cold-cured cheese, regardless of size, it is quite problematical whether this loss in weight may not be chiefly due to the operation of the foregoing factors. If this is so, we may consider such losses as absolutely unavoidable under normal conditions, for the action of microorganisms which can not be suppressed will inevitably result in the production of some volatile products.†

At the temperatures of 50° and 60° F., where the relative humidity was below saturation, the factor of evaporation is apparent and is inversely related to the size of the cheese. From a practical point of view, it is worth noting that the losses in both sizes of cheese cured at 60° F. are approximately 50 per cent more than they are in the cheese ripened at 50° F.

---

*Bulletin No. 231, New York State Agricultural Experiment Station, p. 36.

---

†This interesting deduction is supported by the tests by the author and others on the keeping of eggs in sealed packages. See chapter on "Eggs in Cold Storage."

INFLUENCE OF PARAFFINING CHEESE ON SHRINKAGE DURING CURING.

Within the last few years the custom of coating the cheese with an impervious layer has been suggested, with the object mainly of preventing the development of mold. For this purpose paraffin has been found to be the most suitable agent. The application of such a layer to the cheese not only prevents the growth of mold spores by excluding the air, but materially retards the rate at which the cheese loses its moisture. Paraffined cheese then dries out much more slowly than the untreated product, and the application of this method is of particular service in the handling of the smaller types of cheese, which have a relatively larger superficial area exposed to the air.

In the paraffined cheese at 40° F. the losses were reduced practically to a minimum, as was also the case with the unparaffined at this temperature. As evaporation would certainly be lessened in the paraffined lot, the uniformity of loss between these and the unparaffined still further substantiates the view advanced earlier, that these losses are not so much due to shrinkage from evaporation as they are to metabolic activities of organisms and possibly chemical transformations within the cheese.

EFFECT OF TEMPERATURE ON QUALITY OF CHEESE.

Originally it was planned to have the cheese judged by commercial experts, but it was found impossible to arrange for a sufficiently large number of such tests to closely follow the progressive changes which occurred in the course of the ripening of the cheese. Hence, in addition to the examinations made by the jury of commercial experts, the cheese was carefully scored at Waterloo by Mr. Baer at frequent intervals.

COURSE OF RIPENING IN TYPE I.

Type I was represented by four different lots of Wisconsin cheese. All of them were well-cooked, firm-bodied, slow-ripening cheese that may be regarded as typical Cheddars. In one case the milk from which the cheese was made was evidently tainted, as the cheese was slightly off at the outset.

The results of these periodical scores by Mr. Baer show that good cheese was produced at all temperatures in the first three lots. Naturally that cured at 60° F. developed more rapidly than the goods cured at lower temperatures, but it should be noticed that even at this temperature some of the firm-textured

FIG. I.—THREE CHEESE SECTIONS—TYPE I.
Cheese at top cured at 40°, in middle at 50°, and at bottom at 60°.

cheese went off in five months. At 50° and 40° F. the cheese was about six weeks to two months behind the 60° in development, but in time it reached as high as the 60° lot, and generally of a better quality, and kept this maximum condition much longer than the 60°. This enhanced keeping quality was more pronounced at 40° than at 50° F.

In the lot made from tainted milk the imperfect condition

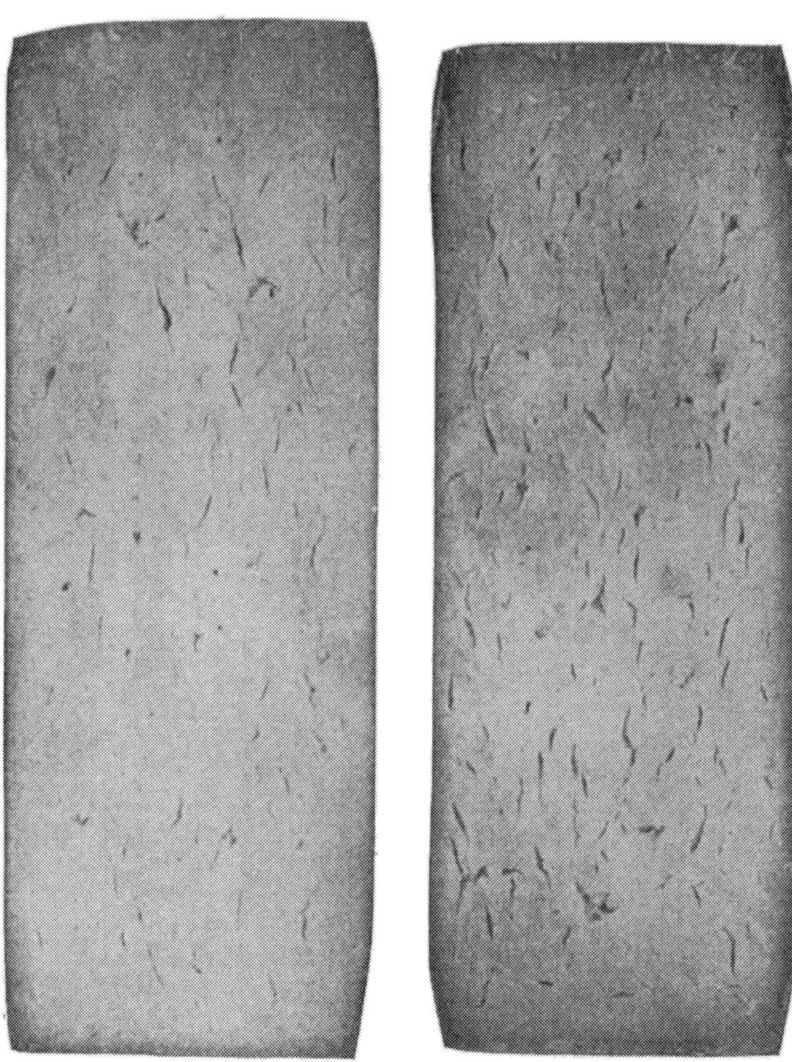

FIG. 2.—TWO VERTICAL CHEESE SECTIONS—TYPE I.
Cheese cured at 40° on left and cheese cured at 60° on right.

was pronounced at all temperatures, but was more prominent at 60° than below.

In studying the scores by Mr. Baer, it is possible to combine the numerical scores of the four different lots of Wisconsin cheese belonging to the same type and so obtain a set of aver-

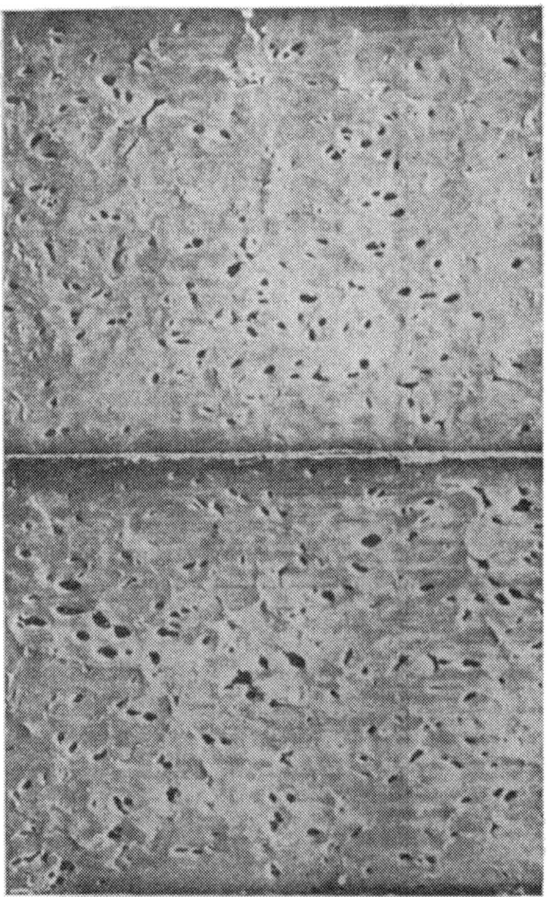

FIG. 3.—TWO CHEESE SECTIONS—TYPE II.
Cheese cured at 40° on top, cheese cured at 60° on bottom.

ages, as to flavor, texture, and price, which indicate clearly the progress of the curing of these various lots at the different temperatures.

The variation in flavor observed at the different temperatures is more marked than any other characteristic. It appears that at the higher temperatures the flavor is more developed during the earlier ripening stages, but as the cheese increases in age the quality of the flavor at the higher temperatures deteriorates more rapidly than in the cold-cured goods. At the end of five months the 40° was still improving, and even at this time was higher than at any period with the 50° and 60°. At the end of eight months the cold-cured cheese was still of excellent quality, and showed no signs of deterioration.

The texture of the cheese followed quite closely a development similar to that noted above. In the earlier stages the 60° had the highest score, but it reached its maximum in three months, while the 50° and 40° continued to improve up to the end of the test, and was higher in the 40° at this time than at any time in the 60°.

The beneficial effect of cold-curing on this firm type of cheese is strikingly apparent from the above data. Not only was this cold-cured cheese free from any bitterness or taint incident to the curing process, but it was much improved in texture, as is evident from Fig. 1, which shows the appearance of cheese made from the same vat, but cured at approximately 40°, 50°, and 60° F. When the cheese is cold cured the body is much closer, as the curd particles are subject to more pronounced shrinkage at higher temperatures, which causes the formation of these irregular, ragged cracks. This is perhaps rendered more obvious in cheese cured at 40° and 60° F., as shown in Figs. 2 and 3. When it is remembered that the results ordinarily obtained in factory curing are not anything like as satisfactory as those shown in the cheese cured at 60° F., the improvement in quality, as shown by the texture of the cheese cured by the cold-curing process over that now in vogue, is emphazised still more.

The 50° cheese stands intermediate between the distinctively cold-cured product and that obtained under best present conditions without artificial refrigeration. Emphasis has already been laid upon the fact that a considerable improvement in quality is to be expected where a slight diminution in temperature is secured over that found in the best type of factory curing now in vogue. This system of "cool-curing"—that is, the use of a temperature from 52° to 58° F., as recently advocated by the Canadian authorities*—stands midway between the cold-curing process and the system now most frequently in use. The benefits to be gained by this system are evident from the Canadian experiments, in which 480 pairs of cheese were cured, one of each lot being kept at 52° to 58° F., while the other was ripened in an ordinary curing room (61° to 70°). Quoting Mr. Ruddick's paper, he says that "in every case the cool-cured (cheese) has been pronounced the best in quality."

From the experiments detailed above it appears that further improvement in quality is possible if the curing temperature is still further reduced (40° to 50° F.). It must be remembered in this comparison that the highest temperature we employed is much lower than the average factory curing room. The difference in quality between cold-cured and ordinary-cured cheese would be much greater than that represented in this work.

The cheese of this type at 60° F. ripened rapidly and showed an excellent quality in all lots but one, which was tainted from the beginning, but they all passed their prime in three months and showed marked deterioration by the end of five months.

With this type of cheese it must be remembered that the quality of the flavor produced at low temperatures is quite different from that found at 60° F. Cold-cured cheese possesses a very mild but perfectly clean flavor, together with a solid, waxy texture.

COURSE OF RIPENING IN TYPE II.

The cheese in Type II is not so uniform in its make-up as that of Type I. but it represents that type of American product in which less acid is developed than is found in the normal Cheddar cheese. This cheese is more open in texture and contains a considerable number of mechanical and small Swiss holes, as shown in Fig. 3. The cheese was somewhat low in

---

*J. A. Ruddick in paper presented at the Ontario Dairymen's Association, January, 1903.

flavor, due in all probability to the milk and method of manufacture, and not to the curing, as this defect was quite as apparent at the lower temperatures as at 60° F.

The Iowa cheese was found to be of only fair quality, but at all ages was better at 40° F. than at other temperatures, although the difference is considerably less than it was with the firmer Wisconsin type of cheese.

The Illinois cheese was quite similar to the Iowa lot, but the texture of this cheese at 60° F. was considerably more impaired than that obtained at the lower temperatures.

COURSE OF RIPENING IN TYPE III.

Type III represents the softer make of cheese intended for home trade, and one which cures more quickly, and therefore does not keep as long as the firmer Cheddar type. This type is represented by four different lots of Michigan cheese made at the same factory. They were not of standard quality, but were too acid. The first three lots were materially delayed in transit and consequently had undergone considerable change before being cold-cured. From the detailed data it is evident that lot four was the best, and in this lot the 40° and 50° were both better than the 60°.

In this case the flavor of the four lots was poor, only once exceeding 40 points. While the 60° scored higher at one time than the cheese at the other two temperatures, the 40° cheese at five months equaled the flavor of the higher temperature cheese at this time.

The difference in price of this cheese at three months was inconsequential, and from this date the cheese at all temperatures fell off rapidly in value.

All four lots of these Michigan goods were more or less delayed in transit, although lot four was no more so than some of the cheese in the other types. But with this moist, quick-curing cheese it was much more susceptible to temperature influences, and hence was materially impaired before being put in storage. This condition, taken in connection with the inferior make (high acid), renders this part of the experiment unsatisfactory.

In the first test the jury consisted of Messrs. White, Millar, and Kirkpatrick. In the second test, made when the cheese was five months old, one of the judges (Kirkpatrick) was unfortunately unable to assist. It is therefore impossible to compare with each other the average scores secured in these two tests, as the judgment of the different members of the jury naturally is not uniform. In comparing, therefore, the course of ripening in the three and five months' tests, it will be necessary to correct the averages given by eliminating the score of the judge who was absent in the second test.

For purposes of study, however, the two tests can be considered independently and the influence of the different temperatures on the character of the cheese determined.

RESULTS OF FIRST JURY TRIAL.

When the cheese had been cured for three months, the sample cheese which had been tested previously at monthly intervals by Mr. Baer, was shipped by refrigerator service to Chicago and submitted to the jury for examination.

Type I. In the four lots of cheese which comprised this group the 50° product was higher in flavor twice, the 40° once, and once the 40° and 50° were alike. In no case, even at this age, when the 60° cheese was at its best (as shown by the serial examinations made by Mr. Baer), did this cheese reach as fine a flavor as at the lower temperatures.

In texture the 40° lot was ahead twice, once the 50° and 60° were alike, and once the 60° was the highest.

As to price, in no case did the 60° equal the value set upon the cheese cured at the lower temperatures, although the difference given by the judges was slight. It must be remembered that the price assigned by the commercial jury was influenced materially by the fact that there is considerable difference in quality, even among the best types of cheese, without a corresponding difference in price. In the majority of cases, when the cheese scored within one or two points of perfect, the price was cut from a quarter to a half cent below the market standard (13 cents), simply because the appearance of the cheese on the surface (mold, etc.) warranted this reduction from a purely commercial point of view. The judges were free to admit that intrinsically the cold-cured cheese was of much better quality than is usually obtained in the market. This cheese was box-cured and received no especial care throughout the experiment; consequently the exterior appearance of the same had been impaired. With proper control this condition could have been entirely obviated, as we have been able to show repeatedly where cheese was cold-cured under our direct supervision.

Type II. In this type, in which less acid was developed, little or no difference was observed in the Iowa goods; but in the Illinois cheese the 40° product had a better flavor and texture than the cheese cured at 50° or 60° F. Fig. 4 shows the appearance of the Illinois cheese cured at the three temperatures when three months old.

Type III. This type is represented by four different lots from the same factory. All of the lots were highly acid and were of somewhat inferior make. Then, too, the earlier lots were delayed in transit from the factory to the curing station, so that the results of the experiment should not be considered as necessarily typical of the cold-curing process. In this group of four tests the 50° goods were ahead twice on flavor, the 60° once, and once the 40° and 60° were alike. In texture the 50° was the highest three times out of four.

GENERAL SUMMARY OF THE FIRST (THREE MONTHS) TEST.

The cheese was examined at this date by the commercial judges, as it was thought that the highest temperature cheese (60°) had reached its maximum condition. It was naturally expected that the 60° product at this time would rank higher in quality than the cold-cured goods.

From this it appears that the 50° cheese was superior in flavor and texture, not only on the basis of the total scores, but also as to the number of times they ranked highest or equal to the cheese cured at either of the other temperatures. This test was made before the 40° goods were marketable, but even at this time this cheese compared favorably with the 60° product.

RESULTS OF SECOND JURY TRIAL.

The second commercial scoring was made at the end of five months, at which time it was thought that the cold-cured goods could best be judged from a market point of view. The results of this scoring follow:

FIG. 4.—THREE CHEESE SECTIONS—ILLINOIS CHEESE.
Cheese at top cured at 40°, in middle at 50°, and at bottom at 60°.

Type I. In the four lots tested of this firm-bodied cheese, the 40° was highest in flavor three times and the 60° once. Averaging the total scores shows that the 40° cheese scored 2.8 points higher than the 60°, and even the 50° was 1.6 points above the cheese held at what has been considered ideal curing conditions.

In texture the 40° was highest twice, while in the other cases the scores were equal. Numerically, the average texture of the 40° was nearly a point above the 60°. At this age the 60° goods began to show signs of deterioration, while the cold-cured goods kept much better.

Type II. In this test one lot of the 60° goods (Iowa) was mislaid in transit, and hence was not tested, but in this case the 40° was 2 points above the 50° in flavor, and 1 point on texture. In the Illinois cheese but little difference was observed.

Type III. In this softer cheese, twice the 40° scored highest in flavor, the 50° and 60° once each. On texture the 40° scored highest twice, the 50° once, and the 50° and 60° tied once.

GENERAL SUMMARY OF SECOND (FIVE MONTHS) TEST.

In this test the average score, as well as the number of times any lot has scored the highest, shows that the 40° cheese was superior to those at either of the other temperatures, while at this age the 60° cheese showed that it had passed its prime.

COMPARISON OF THE FIRST AND SECOND JURY TRIALS AS INDICATING THE KEEPING QUALITY OF THE CHEESE.

It is important to compare the scores of the commercial judges made at the first and second jury trials, as in this way it is possible to study the keeping quality of the cheese cured at different temperatures. Unfortunately one of the judges could not be present at the second test. Therefore the judgment of the other two has been used in comparing the data of the two tests.

Type I. With reference to flavor, type I showed its better keeping qualities, inasmuch as it held its own at 40° F., while at 50° F. the cheese had deteriorated 2 points and at 60° F 2.9 points. The texture improved at all temperatures as the age increased, but was much more pronounced (over a point) at 40° than at 50° or 60° F. This improvement in flavor and texture is also reflected in the enhancement in commercial value. The 40° gained 0.2 cent per pound in three to five months, while the 50° fell off 0.1 cent and the 60° 0.2 cent per pound. Thus in all ways the advantage of cold curing is evident on this firm, solid type of the Wisconsin cheese.

Type II. In this type, in which less acid was developed than in the typical Cheddar type, the deterioration in flavor was less at 40° F. than at either 50° or 60° F. In texture, however, all scored lower at five months, the data showing a wider difference at 40° F. than at the other two temperatures. In price, however, the cheese was considered to be worth 0.2 cent per pound more at 40°, while the 60° cheese had depreciated 0.7 cent.

Type III. In the softer Michigan make, in which more rapid deterioration would be expected, the falling off in flavor was 2 points at 60° F. as against 1.1 points at 40° F. In texture the 40° improved 0.4 point, while the other two depreciated 0.8 and 0.3 point, respectively. In price, all these goods were of less value at five months than at three, but they had depreciated 0.5 cent at 60° and only 0.1 at 40° F.

Summarizing the above, there can be no question but that the keeping quality of all of these various types of American cheese is improved by curing them at these lower temperatures. This is more evident with the firm, solid Wisconsin type of Cheddar than with the softer, quick-curing goods; but even these can be held with less deterioration at these temperatures than is possible under present curing conditions.

SUMMARY OF EFFECT OF TEMPERATURE ON QUALITY.

As the three different types of cheese represented in these experiments varied so much in character, it will be fairer to state the conclusions with relation to each separately. The scores on these lots of cheese were made separately by our own cheese expert throughout the whole curing period, and also at stated intervals by the commercial judges.

Type I. At 60° F. flavor developed more rapidly than at lower temperatures, but the maximum score at this temperature, as indicated by Baer, was equaled or exceeded by the maximum score at 50° or 40° F. In the scoring made by the commercial jury the 50° averaged 0.6 point higher than the 60°, when cheese was three months old. When five months old, the 40°

was 2.8 points higher than the 60°, and the 50° 1.6 points higher.

In texture the course of development was quite the same, the judges scoring the 50° ahead at three months, but at five months the 40° averaged nearly a point higher than the 60°.

Type II. In this low-acid cheese the course for ripening followed the same rule as in the above type, although this cheese was inferior in quality to the preceding type.

Type III. The results on this quick-curing type of cheese were affected by the delay in transit, which permitted of a considerable degree of ripening before the cheese was put in the curing rooms. In this type of cheese the improvement was less marked, but when the enhanced keeping quality is considered, the cold-curing process was found to be advantageous even under these advanced conditions.

INFLUENCE OF PARAFFINING ON QUALITY OF CHEESE.

With the use of lower temperatures for curing, a higher degree of saturation of the atmosphere is always found, which greatly promotes the development of mold, and this growth injures the salability, though not the quality, of the cheese, and hence many attempts have been made to overcome the difficulty.*

The most efficient method yet proposed is to coat the surface of the cheese with an impervious layer, which, by excluding oxygen, prevents development of molds. For this purpose the cheese are immersed in a bath of melted paraffin, which, upon cooling, adheres closely to the surface. While this effectually accomplishes the desired end, it is a question of importance whether the quality of the cheese so treated is affected prejudicially or not. It is possible to conceive that the retention of all volatile decomposition products within the cheese might injure the flavor of the product.

In these cheese-curing experiments it was thought advisable to institute a series of trials to determine what influence paraffining had on the quality, as shown by the flavor and

*The statement that the lower the temperature the higher the relative humidity cannot be allowed to stand in the light of present information. Further, mold is checked by the lower temperature. See chapter on "Humidity."

texture scores. For this purpose the cheese which was used in the experiments on shrinkage (La Crosse lot) was scored by Mr. Baer, and was also submitted to the experts for scoring at the regular periods.

It is evident that the difference between the same lot of cheese when paraffined or unparaffined is very slight. If the course of curing is considered, as is shown by the scores of Mr. Baer, which were taken when the cheese was one, two, three and five months old, it is apparent that the application of paraffin has not injured either the flavor or the texture of the cheese. It will be further noted that in the "daisies" the unparaffined cheese was, with one exception (60°), better at the beginning; but throughout the remainder of the curing and to the end of the experiment the paraffined improved much more rapidly, and without exception was as good or better than the unparaffined.

With the prints the difference in scores was practically negligible.

This same cheese was scored by the commercial experts when it was three and five months old, and it should be noted that the opinions of these experts coincided quitely closely with those of Mr. Baer.

It would be unsafe from these limited experiments to draw any general conclusions, but so far as they go these trials show that no injurious effect was observed on either the flavor or the texture of the paraffined cheese.

GENERAL SUMMARY.

The purpose of the experiments detailed above was to test the value of low temperatures for the curing of cheese made under widely different but commercial conditions. To accomplish this purpose, it was deemed advisable to purchase the product from a wide range of territory. This condition rendered it impossible to install the cheese in the curing rooms immediately after it was taken from the press, and hence the full effect of the process is not so evident as would have been the case if the cheese had not had any preliminary curing.

Naturally a comparison of the cold-curing process would be made with the conditions most frequently found in factories, but in these studies the low temperature cured product has been compared with cheese ripened at about 60° F.—a temperature which has hitherto been considered as the best for the ripening of Cheddar cheese.

EFFECT ON SHRINKAGE.

When cheese is cold-cured, the losses due to shrinkage in weight are greatly reduced over what occurs under ordinary factory conditions.

1.—Influence of temperature.—Cheese cured at 40° F. decreased in weight in ninety days from 1 to 1.4 per cent, while that cured at 50° and 60° F. lost fully three times as much. This saving would be still further increased if comparison were made between the results of cold curing and existing factory conditions. Under prevailing factory practice cheese is sold at a much earlier date than is advisable with cold-cured goods, but the loss under present conditions, for even as brief a curing period as twenty days, is fully four times as great as has occurred in these experiments in a ninety-day period (the minimum curing period recommended) under cold-curing conditions (40° F.). This saving in a factory making 500 pounds of cheese daily would average not less than 15 pounds of cheese per day for the entire season, or considerably more than this if only summer-made cheese were cold cured.*

2.—*Influence of type of cheese.*—In these experiments different types of cheese were used, ranging from the firm, typical Cheddar to the soft, moist, quick-curing cheese made for the home trade. The losses with the firmer type were considerably reduced in comparison with the other, but the conditions to which the softer types of cheese were subjected were not as favorable (because of initial delays), and hence the losses with these types can not be relied upon with such definiteness. As this cheese was exceedingly moist, the total losses from the press were undoubtedly greater than here reported.

---

*It seems to the author that undue strèss is being laid on the great benefit to be derived from a saving in evaporation or shrinkage in weight. If this loss is saved to the manufacturer the retailer or consumer is the sufferer, because moisture has no value as food, and the loss of moisture is practically all that evaporation means. More importance should be given to the improved quality, because the saving in weight comes out of the retailer or consumer.

3.—*Influence of size of cheese.*—The size of package exerts a marked effect on the rates of loss. At ordinary temperatures, the smaller the cheese the more rapidly it drys out. This difference in loss diminishes as the temperature is lowered, and in our experiments at 40° F. was practically independent of the size. This condition, however, was undoubtedly attributable to the relative humidity of the curing room, which at 40° F. was 100 per cent.

4.—*Influence of paraffin.*—By coating the cheese with melted paraffin the losses at 60° were reduced more than one-half; at 50° the saving was somewhat less, and at 40° the losses observed on the paraffined cheese of both sizes used were slightly in excess of those noted on the uncoated cheese.*

5.—As some loss occurs even in a saturated atmosphere, where evaporation is presumed not to take place, it implies that the shrinkage in weight of cheese under these conditions is not wholly due to dessication, but is possibly affected by the production of volatile products that are formed by processes inherent in the curing of cheese.

EFFECT ON QUALITY.

6.—The three types of cheese before referred to can scarcely be compared closely with each other, as they were so different in their make-up and subject to somewhat different conditions during transit. By far the most satisfactory portion of the experiment is that which relates to Type I, in which the best quality of cheese was represented. With these firm, typical Cheddars the influence of temperature on curing could best be studied. This cheese was also placed in storage nearer the press than any of the other types, and hence the test as to the effect of the curing temperature was more satisfactory. In this type the 60° cheese was of excellent quality and naturally developed faster than the cold-cured goods, but in time it was surpassed by the cheese at the lower temperatures (50° and 40°), and, when the keeping quality of the latter was taken in-

*Retailers of cheese in England have in some cases made strong objection to the paraffining of cheese for the reason that they suffer much greater loss from shrinkage when cutting up the cheese for retailing. From these experiments it seems that the cold-curing of cheese has much more to do with preventing loss of weight than paraffining.

to consideration, it was found to be superior in every way to that cured at 60° F. Even when the condition of the milk was not entirely perfect, the quality of the cold-cured cheese was better, although the original taint was not removed.

With the sweet-curd (type II) and the soft home-trade cheese (type III) the effect of the disturbing influences previously noted rendered it impossible to obtain as satisfactory results, but, even under these adverse conditions, the 40° and 50° cheese generally ranked better than the 60°, and, when keeping quality was taken into consideration, was materially better.

This same cheese was also scored independently by commercial experts when three and five months old. The results obtained conform very closely to those mentioned above, and indicate the superiority of the cold-cured product (either at 50° or 40°) in comparison with the cheese cured at 60° F. This improvement in quality reflects itself also in the commercial values which were placed upon the cheese cured at different temperatures, both by our own expert and also by the commercial judges.

In this low-temperature-cured cheese the flavor was remarkably mild but clean, and was free from all trace of bitterness or other taint. The texture was fine and silky and the body close.

7.—*Keeping quality.*—The keeping quality of the cold-cured cheese far excels that of the cheese ripened at higher temperatures. The better types of cheese cured at 40° F. were at the end of eight months still in their prime, while the 60° cheese had long since greatly deteriorated.

8.—*Effect of paraffining on quality.*—Portions of two lots of cheese were paraffined as they came from the press, but were otherwise handled the same as the unparaffined cheese. The results obtained showed that paraffining did not prejudicially affect their quality at any temperature. As paraffining greatly reduced the shrinkage, the beneficial effect of the system is obvious. The rapid introduction of the method in commercial practice further attests its value.

9.—The production of a thoroughly broken-down Cheddar cheese of mild, delicate flavor and perfect texture meets a demand which is impossible to satisfy with cheese cured at high temperatures. Without any question, if the general market can be supplied with this mild, well-ripened cheese, consumption will be greatly stimulated, not only by increasing the amount used by present consumers, but by largely extending the use of this valuable and nutritious article of food.

10.—The improvement in quality of cold-cured cheese, the enhanced keeping quality, and the material saving in shrinkage due to lessened evaporation are sufficient to warrant a considerable expenditure on the part of cheese producers in installing cold-curing stations.

The principle of increasing cost of equipment to lessen cost of production or augment gross earnings is recognized as a sound financial method by all large enterprises, and, while the expense involved is considerably more than is incurred under existing conditions, yet the advantages enumerated more than compensate for such expense where carried out under proper conditions.

11.—This system is particularly applicable where the product of a number of factories can be handled at one point, and such consolidated curing stations must be established before the cold-curing process can be economically introduced. Such stations are now successfully used in a number of localities. The greatest advantage will undoubtedly accrue from the use of this system of curing with summer-made cheese, but the process is equally applicable to cheese made at any season of the year.

AUTHOR'S CONCLUDING REMARKS.

The foregoing report of the result of experiments by the Wisconsin Station demonstrates fully the desirability of low temperatures for cheese storing, and for the curing of cheese by placing it in a low temperature as soon as manufactured. The experiments do not, however, include temperatures of from 30° to 32° F., which are now considered best for long period storage of cheese. It is desirable that the best temperatures for the most successful storing of cheese should be determined and additional experiments should be made for this purpose. It is also practicable to extend the experiments so as to include foreign makes as well as the various types of American cheese.

The cheese business is now practically all handled through cold storage, and temperatures ranging from 30° to 40° F. are in use. The use of cold storage for the curing of cheese is, therefore, not in an experimental stage, and it is to be regretted that the experiments of the Department of Agriculture did not include temperatures of 30° F. and 35° F. as representing the commercial practice of the times, and a still lower range to determine the possibilities in this direction.

The initial quality of cheese has much to do with what is best for it in the way of temperature while curing or cold storing, but nothing positive may be said on this point at the present time, as no results of experiments are at hand as a guide. The author recommends that a good average clean-flavored make of American cheese be first placed in a temperature of about 40° F. After being in storage for a month or two reduce the temperature gradually so that at the end of two or three months the temperature reaches 30° F., which is recommended for a permanent storage temperature. This temperature is somewhat lower than is generally considered best, but if handled as suggested better results may be had than at any higher temperature.

CHAPTER XVI.

CREAMERY AND DAIRY REFRIGERATION.

NECESSITY FOR REFRIGERATION.

It has been estimated that the total amount of butter produced in the United States is about 2,000,000,000 pounds each year. It is probable that the amount is in excess of this, rather than less. The consumption of butter is rapidly increasing and the average quality of same is likewise being improved, but it is probable that not more than one-half of the butter made reaches the consumer in prime condition. The most important reason for this is, no doubt, that refrigeration is not employed to a sufficient extent, or where employed, not intelligently or scientifically applied.

Though nowadays not of the same importance to the dairy as it was before the centrifugal creamers were invented, yet in our climate ice or refrigerating machinery is indispensable to the production of fine butter. To fully control the process, the butter maker must be able to heat and cool the cream at will, and the butter often requires a cooling which cannot be effected without ice or a refrigerating machine. Every creamery and dairy not provided with a machine should, therefore, have an ice house, and a refrigerator or cooling room should always be constructed.

Refrigeration is absolutely necessary to the proper manufacture of butter, and is likewise necessary to the proper keeping or preserving of same after it is made. Refrigeration is applied in the manufacture of butter to the manipulation and proper tempering of the raw materials and the keeping of the butter when made at a low temperature to prevent deterioration. Considering the great importance of refrigeration as applied to creamery products, comparatively little attention has been given to this branch of the business. It is not meant by this that those who are operating creameries have not given careful thought to this matter, but that the refrigerating engineers and the makers of refrigerating machinery have not studied its application to creamery and dairy service as fully as they might.

As in all other branches of the refrigeration of perishable food products, the United States is in advance of other countries in the preservation, by cold, of milk, butter and cheese. Until a comparatively recent day, however, the most progressive of the dairy companies often cooled their cans of milk by immersing them in a bath of cracked ice. This process was not only cumbersome, in that it necessitated the repeated handling of the heavy cans, but the cans themselves were thus injured. The ice and water were scattered over the premises, which rendered cleanliness very difficult. A dairy establishment refrigerated artificially presents a neater appearance. The milk as it is brought in from the country is first tested for quality. It is then placed in a large tank, from which it passes through three sets of fine strainers, which remove all small particles of dirt or dust that may have gotten into the cans. It then passes through a series of pipes, which are submerged in a large brine tank. The tank contains the ammonia expansion coils, by which the brine is kept at the required temperature. After passing through these coils, the milk is drawn off into cans, which, in turn, are stored in a large refrigerator, kept at a temperature of about 35°.

Denmark and Sweden, in Europe, have made the greatest advance in the refrigeration of the products of the dairy, machinery being extensively employed for that purpose. The creameries and butter factories of Belgium and Holland are also becoming more modern in this respect year by year. A late innovation in the dairy industry in northern Germany and Denmark is the process of freezing milk into blocks, and shipping it abroad as milk ice, mostly to England. The required machinery agitates the milk during the freezing process, so that when ready for the market the substance of the frozen milk is uniform throughout.

The pasteurization of the milk, which is now becoming quite general in the larger creameries in this country, as it is in European countries, notably Denmark and Belgium, calls for additional demands on refrigerating apparatus, as it is found essential to reduce the temperature after pasteurization as rapidly as possible.

ICE VERSUS REFRIGERATING MACHINE.

There is at the present time considerable controversy between those who advocate the use of ice for creamery or dairy refrigeration and those who recommend refrigerating machinery. There should be no quarrel between these two different methods, as each one has its proper sphere, and there are cases where the selection of either one or the other would be a matter largely of individual opinion. Where natural ice can be stored cheaply at, say, a cost of $1.00 a ton or less, and where the quantity of milk to be handled would not exceed 10,000 or 15,000 pounds per day, the use of natural ice is usually to be preferred to installing refrigerating machinery. On the other hand, where the quantity of milk to be handled is large and ice is comparatively expensive, a refrigerating machine can profitably be employed.

The advantages and disadvantages of the mechanical systems over ice have been quite fully investigated by Prof. Oscar Erf, late of the College of Agriculture, University of Illinois. His deductions, however, were based on conditions which do not apply in states of about the same latitude as New York, Michigan, Wisconsin and Minnesota, and even south of these latitudes there are places where natural ice can be housed at much less cost than 90 cents per ton, which he has taken as a basis. His investigation seems to have been conducted from an intelligent and fair-minded standpoint, and his results are useful to creamery men, if proper allowances are made for the difference in latitude and other working conditions. Prof. Erf gives his results in detail, but we will only consider his summary of the disadvantages of mechanical refrigeration, as follows:*

*From Ice and Refrigeration, June, 1902.

1.—Large capital invested.
2.—Necessitates daily or continual operation, unless provided with large storage tanks.
3.—Operating expenses for labor, coal, oil, ammonia and repairs.
4.—Excessive dryness in such refrigerators, often causing a great shrinkage in the products.
5.—Great risks for accidents that might happen, such as breakage on machines and the delay of repairs.
6.—Expense of pumping water for condensing ammonia.

The advantages offsetting these disadvantages by using machinery for refrigeration, as compared with the use of natural ice:

1.—No risks to run in securing cold whenever needed.
2.—Practically no variation in cost of producing cold from year to year.
3.—The refrigerator is under better control.
4.—Any temperature may be practically obtained above zero.
5.—Atmosphere is dryer in refrigerator; hence butter is less susceptible to mold.
6.—Less disagreeable labor, such as the handling of ice.
7.—Cold room can be kept cleaner.
8.—Does away with the impurities imbedded in river and pond ice.
9.—Provides for a more perfect method of cream ripening, which results in a better product.
10.—Secures economy of space in the cool room, which lessens the radiating surface for same amount of refrigeration.

The disadvantages as set forth are sufficiently plain to all who have had experience with refrigerating machinery. The advantages which are cited are more or less true, especially as applied to the ordinary application of ice as generally used in creameries. Should the Cooper brine system be used, as described further on, there would be:

1.—Absolutely no risk to run in securing cold whenever needed.
2.—Any temperature may be practically obtained down to 15° F.
3.—The refrigeration would be under fully as good control and a more uniform temperature could be obtained than by the use of refrigerating machinery.
4.—The moisture in the atmosphere of the cold room could be carried at any temperature desired and under as good control as with the mechanical system.
5.—The amount of disagreeable labor required, should an ice crusher and ice elevator be used, would be very small indeed.
6.—The cold room can be kept as clean as with any system.
7.—Impurities in the ice would have no influence on the air of the room for the reason that the air does not come in contact with the ice.
8.—As perfect results can be had in the ripening of cream.
9.—The economy of space in the cold room would be as great as with any system.

In other words, the Cooper brine system will produce any results which can be had with refrigerating machinery down to a temperature of 10°, or even 5° F., and besides this, it is absolutely sure against a breakdown.

THE CREAMERY REFRIGERATOR.

A cooling room for maintaining the butter at a low temperature after being made, is admitted to be absolutely necessary in every creamery, and it cannot be dispensed with, except in cases where butter is loaded into a refrigerator car each day. Even then the butter will handle much better and arrive on the market in much better condition if it is hardened so that it

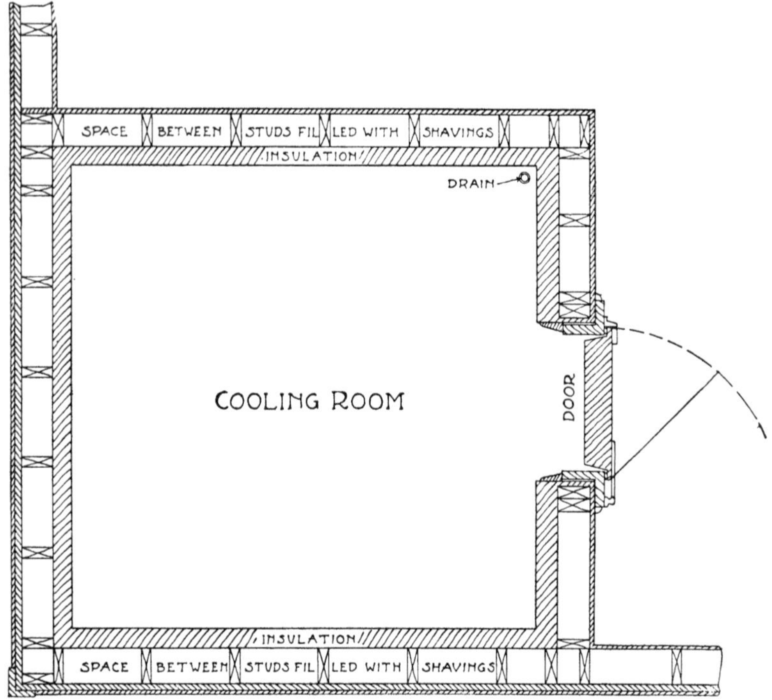

FIG. 1—PLAN SMALL COOLING ROOM.

will carry without shaking or slopping in the tub, to say nothing of the advantages of always having it at a low temperature until consumed. Butter is practically at its best when first made, and the nearer it can be retained in this condition until consumed, the better satisfaction it will give the customer and the greater will be the ultimate gain to the creamery man. When butter is to be shipped frequently, a small cooling room, constructed with ice chambers above the storage room, essentially as outlilned in Figs. 1 and 2, should be built in every creamery not provided with mechanical refrigeration. It is

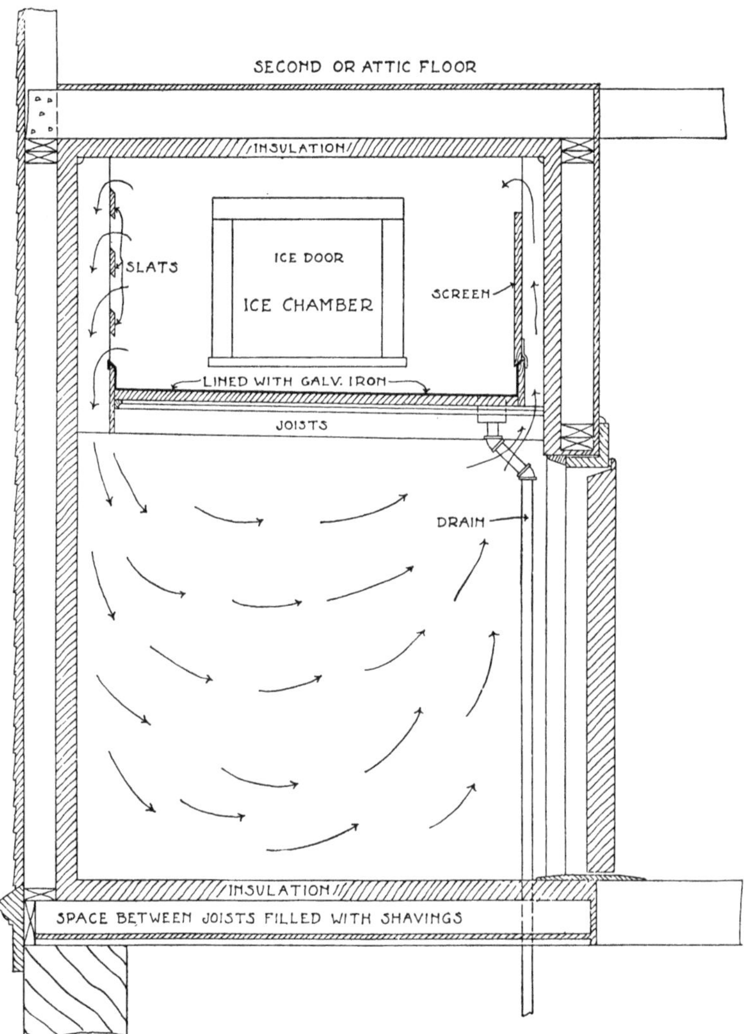

FIG. 2—SECTION SMALL COOLING ROOM.

much better to place the ice over the room than it is to put the ice in a rack at one end or one side of the room. A lower temperature will be obtained and a dryer atmosphere will result, owing to the circulation of air, as indicated by the arrows. A room of this kind should be well built, and a few dollars extra spent in the insulation of same will be saved in a short time in the saving in the quantity of ice required. The temperature to be obtained in the room also depends on good insulation. If the insulation is thorough, a temperature of 36° to 40° F. may be depended upon. Of course, at the time when warm butter is placed in the room, the temperature will naturally rise to quite an extent. The construction of a room of this kind can be adapted to suit local conditions and the nature of the materials which can be most readily obtained. The detailed description which follows, of a room constructed on the plan as laid down in the preceding paragraphs, will be of considerable value and interest to owners of creameries.

The floor joists, ceiling joists and side wall studding should all be filled with mill shavings, sawdust, tan bark, cut straw or any similar material. This material, however, must be dry and protected on the outside and inside by the best grades of insulating paper (not the ordinary rosin sized or common building papers). Care should be taken that in all corners the paper is thoroughly lapped to make an absolutely air-tight surface, so as to prevent a circulation of outside air into the space which is filled with the insulating or packing material. It is best to double-board the outside of the room and put the insulating paper between. On the inside of the studding and on the top of the floor joists and on the bottom of the ceiling joists use matched boarding. Covering this interior surface should be placed a much better grade of insulating material than the filling between studs and joists. This may be of hair felt, sheet cork, granulated cork, rock fiber felt, mineral wool, Cabot quilt, or any of the best grades of insulating materials. If there is any liability of trouble from rats or mice, they can be kept out of a room of this kind by using an inch or two of rock fiber felt or mineral wool on the outside of all walls. Rats or mice cannot work in either of these materials. The rock fiber felt spoken of is practically a mineral wool made up in the form of sheets or boards. The materials indicated may be used to a thickness of 2, 3 or 4 inches, depending upon the amount of money the owner is willing to spend and cost of refrigeration. These various materials must, of course, be put on between battens or cleats of sufficient thickness to flush up even with the layers of insulating material. If hair felt, sheet cork or rock fiber felt is used, the different thicknesses should be separated by a good grade of insulating paper. The interior of the room should be lined with matched stuff, preferably of poplar, spruce or hemlock. (The chapter on "Insulation" may be of interest in this connection.) If it is desired to wash out a refrigerating or cooling room of this kind from time to time, the interior finish may be of shellac or hard oil, preferably shellac, or, the inside surface may be coated with whitewash, which may be renewed from time to time. (See chapter on "Keeping Cold Stores Clean.") The joists for supporting the ice should be of fairly strong material, depending on the size of the room and should be pitched slightly toward the drain end of the ice floor. The joists for supporting the ice are not carried into the insulation, but rest on ribbands of 2x4s spiked onto the outside of the insulation. A batten should be set in the insulation for receiving these ribbands. The pan or floor under the ice consists simply of two thicknesses of dressed and matched stuff with a covering or lining of No. 20 galvanized iron. A loose rack of wood should be placed on this iron floor to prevent its wearing or getting punctured in handling the ice thereon. The galvanized iron should be turned up on the sides 4 to 6 inches. The circulation of air is provided for by placing a tight board screen on one side of the ice space which is carried up to near the ceiling. The other or opposite side of the ice space has cleats or slats which keep the ice in place and allow a circulation of air. This screen and the slats mentioned are, of course, fastened to 2x4s or to 2x6s, which form the open spaces for the circulation of air on the two sides of the ice chamber. The screen and cleats should be beveled on the top and bottom so that any dripping will be on the galvanized iron pan and not into the air flues and then down into the room. These various parts are illustrated in Fig. 2, but are not shown in detail. For filling ice into the ice chamber, a door may be provided at any point, but should not be on the side where the air flows down from the ice room to the storage room or up from the storage room into the ice chamber. This ice door may be at the top, and the room can be filled from the floor above if convenient. Both the ice door and the door for entering the room are preferably one of the special doors which are on the market and which cost but very little more than the home-made door, and are superior in every way.

The above cooling room is intended to be filled from an independent ice house, which should be located as near the cooling room as convenient.

COMBINATION ICE HOUSE AND REFRIGERATOR.

The following description,* by W. G. Newton, will be of interest:

The ice house is in the opposite end of the building from the boiler room and the ice is put right in on the ground floor and the refrigerator is built next to it and holes cut through next to the floor for the cold air to enter and same at top for warm air to go out. The ceiling over the ice needs to be from four to six feet higher than it is in the refrigerator, then if the outlets for the warm air are right up close to the ceiling (not having so much as a piece of molding between the top of the hole in the side and the ceiling overhead), the dampness will all go off with the warm air up over the ice, leaving your refrigerator dry and sweet.

As to the expense of building, it is not much, as a room 20x20 or 20x30 feet at most will hold ice enough to cool most of the creamery refrigerators if they have some ice stored elsewhere for other uses. All that is necessary is to have the walls of the ice house properly insulated with sawdust and air spaces and then the yearly renewal of sawdust in which to pack the ice is saved.

Not only creameries, but several large meat markets here have this kind of an ice house and refrigerator combined, and they are giving the best of satisfaction. There is a patent on them. The days of building refrigerators with ice overhead have gone by in this section of the country.

This appears like a fine arrangement to save labor and produce the lowest possible temperature with ice. The ice room must be as well finished and insulated as the refrigerator and no sawdust or packing material used on the ice. It would be advisable to build the ice room more than four to six feet higher than the refrigerator, and ten or fifteen feet would be better. An ice room of say 20x30x20 feet dimensions should be sufficient for an ordinary creamery, but this depends on what is to

---

*From *New York Produce Review.*

be done with the milk. The storage of ice in a building, as is well known, tends to cause it to decay and deteriorate rapidly, and this is the only real objection to the plan, as the ice room would be in bad condition long before the refrigerator. A well-built house on a stone or brick foundation would be almost a necessity for the purpose.

In practice, in some exterme cases, it has been found advisable to fill into the ice house as many pounds of ice as there are pounds of milk to be treated, or to harvest 100 cubic feet of ice for each cow furnishing milk to the dairy or creamery. Less than one-half this quantity may be ample in many cases, so much depends on the treatment to which the milk and manufactured product is subjected. Where pasteurization is practiced, much more ice is required, especially where no well water is available. During the winter, ice or snow may be used, which is simply hauled together in a heap near the creamery, so that no ice is taken from the ice house until April or May.

Where separators are used, no ice is needed for raising the cream, but the latter needs cooling either as it runs from the separator or after the ripening, before churning.

Ice is also needed in the hot summer months to cool the butter before or between the workings, and for keeping it firm in texture before it is shipped, so that it may leave in the very best condition for standing exposure to heat while in transit to its destination. Butter in prints is sometimes shipped in cases with an ice box filled with crushed ice in the center.

The amount of ice required for these various purposes varies according to local conditions, and cannot be definitely stated, though it may be calculated approximately. The chapters on "Harvesting, Handling and Storing Ice" and "Ice Houses" give methods of handling ice and details for construction of ice houses of various capacities.

TRANSPORTATION OF MILK AND CREAM.

In the transportation of milk and cream, baggage cars, refrigerator cars, and cars especially constructed for the purpose are employed. The railroads adopt the style of car best suited to their individual requirements. In the case of light shipments and short hauls, superannuated baggage cars appear to meet every requirement and are generally moved in conjunction with local passenger trains. In the case of long hauls, however, refrigerator or special milk cars are used. These cars are plentifully supplied with ice during the warm summer months and, in extremely cold weather, are often steam-heated to prevent the milk from freezing. Cleaning generally takes place after each run, the cars being either swept or washed by means of a hose. Trains making long hauls are usually composed entirely of refrigerator or special milk cars and are operated on about passenger schedule time, the actual running time being as fast as fifty miles an hour. The capacity of a large milk car is 325 ten-gallon cans.

Nearly all railroads which handle a large milk traffic have well-built covered receiving and shipping stations along their lines, nearly all of them with an ice house connected in which natural ice in sufficient quantity is stored during the winter.

Shipping stations are equipped with large cooling vats in which cans of milk are placed immediately after being delivered by the farmers. These vats are filled with water and ice, the milk is stirred and cooled down to 40° F. within forty minutes from the time it is received, and kept in ice water until the train arrives, when it is loaded direct from the vats into a refrigerator car.

COOPER BRINE SYSTEM.

The application of the Cooper brine system to the refrigeration of a creamery cooling room and freezing room is shown in Figs. 3 and 4. They also show the arrangement of ice crushing and ice handling apparatus, which will deliver crushed ice to any convenient point in the creamery workroom for the cooling of cream, butter, shipping, or other purposes. Fig. 3 shows the plan view of one end of the creamery with ice house adjoining. The refrigerated space in the creamery consists of a cooling room with a capacity of about one carload. The butter freezing room has a somewhat larger capacity. The relative size of these rooms can, of course, be changed to suit any conditions. The cooling room and freezing room are both entered from the vestibule and not from the workroom. This prevents the access of warm air into the rooms, which is very important, especially in the freezing room. If it is desired, the cream cooling vats may be placed in a cooling room of this kind, but as planned, it is intended that the cream should be cooled with crushed ice or cold well water. There are a number of different ideas on arrangements of this kind, but with the apparatus shown any arrangement can be provided to suit the ideas of the owner or local requirements.

The Cooper brine system, patented by the author, which referigerates the cooling room and freezing room, is described fully in the chapter on "Refrigeration from Ice."

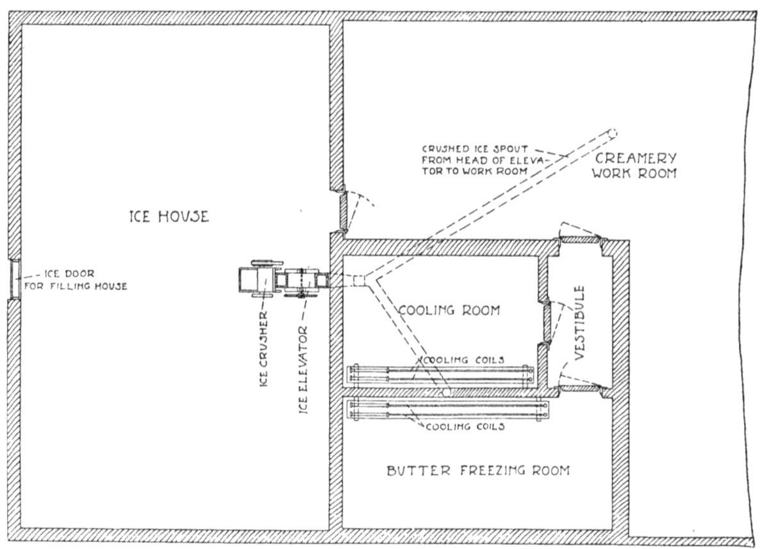

FIG. 3—PLAN COOPER BRINE SYSTEM FOR CREAMERY.

ICE HANDLING MACHINERY.

The ice crusher and ice elevator which are clearly shown in Fig. 4 are quite simple in their operation, and as labor saving machines are the best form of apparatus which has been applied to the handling of ice. It only requires that the ice be broken into irregular pieces of 20 to 30 pounds or thereabouts, and fed into the ice crusher. The crusher breaks the ice into small pieces the size of hens' eggs or smaller, and it drops into the bucket elevator where it is raised to a point sufficiently high to allow of its being spouted to a convenient point in the creamery and to the flexible spout which is used to feed ice into the tanks containing primary coils of the gravity brine

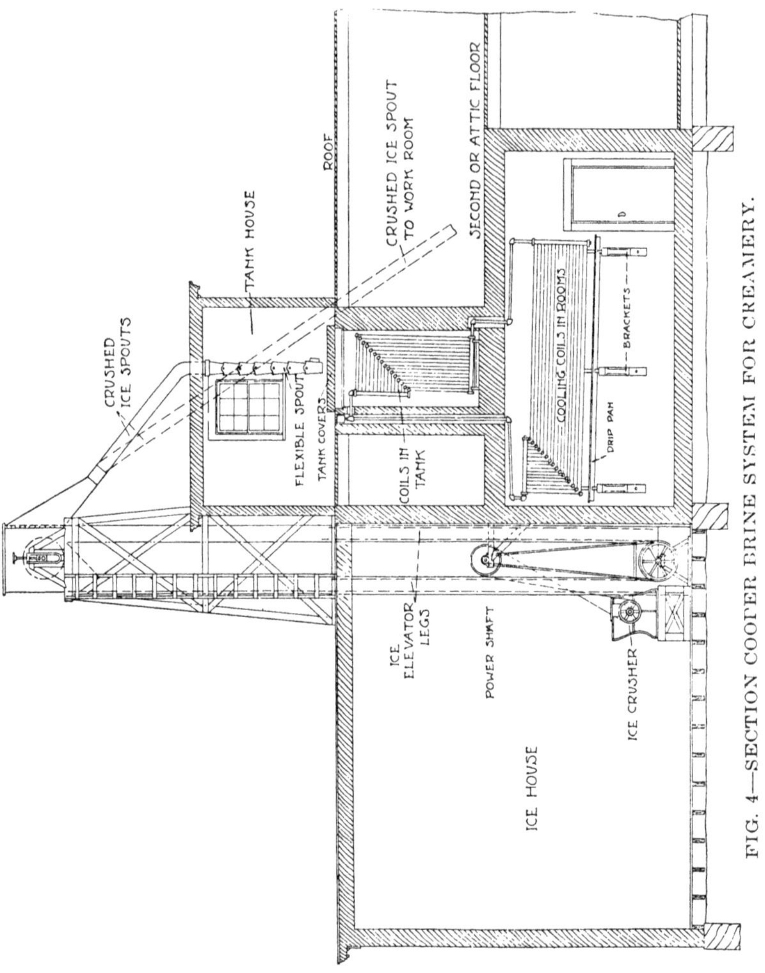

FIG. 4—SECTION COOPER BRINE SYSTEM FOR CREAMERY.

system. Four or five tons of ice may be handled with an apparatus of this kind in half an hour. As recommended in connection with the "Model Creamery Ice House" described in the chapter on "Ice Houses," the ice should not be covered with sawdust or packing material of any kind, and where the ice is clean in the ice house, the labor of handling same to the crusher and delivering to any convenient point in the creamery is very little as compared with the old-fashioned method. This outfit and apparatus is not recommended for the average small creamery, but where several tons of ice are to be handled each day, or where it is desired to store a certain portion of the product of the creamery and carry it for several months, as good results may be obtained with the Cooper brine system as with a refrigerating machine. The expense of installing, while it is considerably more than any of the old-style refrigerators, is less than for a good refrigerating machine.

COOLING OF MILK AND CREAM.*

The following is a portion of an address by Loudon M. Douglas, read before the Cold Storage and Ice Association at Islington, England:

The main object in view in cooling fresh milk for immediate consumption is to arrest the development of the spores which produce bacteria, and which, in their turn, destroy the milk—that is to say, the milk becomes sour. It will be understood, however, that the bacteria referred to are those which are always found in the milk produced, even under proper hygienic conditions. Heat is the essential condition for their development, and, in the absence of that condition, they will remain inactive. Of pathogenic bacteria we need not speak here.

Properly speaking, under good hygienic conditions, it should only be necessary to cool the milk before sending it out, and this is practiced by some retailers. It may, however, be considered an advantage to previously pasteurize the milk at a high temperature, and then cool it down. It is not easy to say which way is the better. In any case both methods are in use.

In cooling town's milk direct from a temperature of, say, 68° to 38° F., all that is necessary is a small refrigerating machine connected to a circular cooler. The cold brine from the machine is circulated through the flutings of the cooler and the milk run over. When it reaches the bottom and escapes, the temperature will be about 38° F., the balance of the heat units having been absorbed by the brine. But by means of a similar small machine a very large cooling effect may be produced by having a large insulated store tank fitted with agitating gear and filled with either water or non-freezable brine.

It is obvious that, by working the small machine for a lengthened number of hours upon this store, the heat will be extracted and utilized to cool in turn a very large quantity of milk in a short time, a quantity quite beyond the power of a small machine to deal with directly. Thus, by intelligent working a small machine costing a comparatively small sum can be made to perform a large amount of work.

---

* Extracted from Ice and Refrigeration, June, 1904.

In the case where milk is previously pasteurized the procedure is different, and can not be better demonstrated than by referring to a large dairy where the work is carried out. The dairy in question handles 1,000 gallons of fresh milk per day, all of which is distributed either directly to consumers or to shops for such distribution, and, in considering the question of refrigeration, it was stipulated that the cooling of the quantity named should be performed in one hour, and that there should also be provision made for cooling a churn store, a cream store, and a butter store of certain dimensions. Now, the machine necessary to cool the 1,000 gallons from 68° to 45° F. in one hour, equal to the elimination of 230,000 B.T.U., would be a very large one, whereas the B.T.U. to be eliminated from the accessory stores would be comparatively few. Obviously, therefore, if a large machine had been installed it would have been much of its time idle. The problem, therefore, was to find a machine which would perform the whole work during working hours. This was done by providing storage tanks for 1,000 gallons of brine, and the heat is extracted from this during a series of hours. Obviously all this brine when cooled down is available, and it is only necessary to run it through a large capillary cooler while the milk to be cooled is run over the outside. The heat of the milk is transferred to the brine, and thus the cooling is accomplished with great rapidity. The pasteurized milk is first of all cooled with ordinary water from the town's supply to 68° F., and from that temperature is lowered through 23° to 45° F. The machine used is capable of eliminating 45,000 B.T.U.s per hour. But the total number of B.T.U.s to be elimated are altogether 355,000, taking into account the accessory work to be done; thus, if 355,000 is divided by the output of the machine, viz., 45,000, you get the number of hours' work necessary, viz., eight, or an ordinary working day. There is a margin of 5,000 B.T.U.s allowed for contingencies.

The creamery system has now become well established throughout Europe, and feeding stations to main creameries are recognized as essential to economical working. The process which is usually carried out in these places is as follows:

The milk is brought by the farmers to the creamery, sampled, weighed, pasteurized, and separated. When the cream leaves the separator it may be at a temperature of from 170° to 180° F., and is therefore immediately run over a circular capillary cooler, through which water is circulated, and reduced to about 65° F. It is then run over another cooler, through which brine is circulated, and cooled to about 45° F., being caught in churns, and in this state taken to the main creamery to be ripened and made into butter. The separated milk is treated in very much the same way. A large surface water cooler reduces the temperature to 68° F., and the milk is then run over a small cooler and reduced to 48° F., at which temperature it is returned to the farmer.

Some actual tests of a machine (at Ballinorig) might be appropriately recorded here:

1.—100 gallons of brine were cooled from 40° F. to 27° F. in one hour (condensing water 57° F.), or equal to the elimination of 13,000 B.T.U. per hour.

2.—100 gallons of brine were cooled from 27° F. to 17° F. in one hour (cooling water, 58° F.) =10,000 B.T.U. per hour.

3.—100 gallons of brine were cooled from 45° F. to 31° F. in one hour (cooling water, 57° F.) =14,000 B.T.U. per hour.

These tests bring out very strongly the fact that at comparatively high temperatures cooling is effected at a much more rapid rate than at the lower range of temperatures, and the amount of energy consumed is greater at lower or ice-making temperatures than at the higher, and this must be borne in mind in specifying the duty of the machine. The machine in question is one of the very smallest made, but the same result is obtained with machines of all sizes.

Perhaps the greatest interest is attached to the application of refrigeration to a central or main creamery, for in such a place all the important applications can be put into effect. These may be classified thus: A, cooling cream from separator; B, cooling separated milk; C, cooling ripened cream; D, cooling water for washing butter; E, cooling a butter store.

As in the auxiliary creamery, the cream is first of all cooled with water to about 68° F., so in the main dairy. The cream is brought down to a temperature of 48° F. by passing it over a circular capillary cooler, and is then run into the ripening vats. Here the process of ripening rapidly increases the temperature again, and in about eighteen or twenty hours it is at about 65° F. At such a temperature it would be ruinous to churn, inasmuch as the texture of the butter would be oily and bad, and there would also be an excessive loss of butter fat in the buttermilk. The perfect churning temperature (in summer) may be anything between 48° and 52° F., and to attain this it is obvious that the temperature of the cream must be lowered some 13° to 17° F. The most economical arrangement by which this can be accomplished is by having the cream-ripening vats sufficiently high up in the creamery to enable the cream to run over a capillary cooler, then flow into the churn. Such an arrangement is simple and works well. By proper and intelligent adjustment of the appliances, the cream can be reduced in temperature to 48° F. precisely, if wanted.

Separated milk in the main creamery is treated in the same way as in the auxiliary, viz., first of all passed over a large circular cooler, in which water takes up the heat from the milk. It is then passed over a small cooler in which brine is the cooling medium, and delivered to the farmer at 48° F.

Cold water in a creamery is very desirable. The average temperature of well water in the British Isles is 52° F., but that is not considered to be low enough for washing purposes; besides, if it were, well water is not always available. Hence, provision has to be made for cooling water to a very low temperature. This is done in a separate tank, usually placed in a sufficiently elevated position to command the butter worker and churn. An insulated tank of, say, one to 500 gallons capacity, is fixed on the wall with brackets, or on a platform, and in this is fixed a direct expansion or brine coil connected to the machine. The cooling is more quickly produced if a small agitator is placed in the tank, as by that means the water is more quickly brought in contact with the cooling surface. Water at from 45° to 58° F. seems to be generally preferred.

John A. Ruddick, Dairy and Cold Storage Commissioner of Canada, in a paper presented at the Chicago meeting of the American Society of Refrigerating Engineers has the following to say about

THE REFRIGERATION OF MILK.

Housewives and dairymaids have, from time immemorial, employed a measure of refrigeration for milk when they placed it in various receptacles, in cool cellars, for the purpose of securing a maximum amount of cream or to keep it sweet as long as possible. It is only within recent years that actual refrigeration has been used in the preservation and handling of milk. Absolutely pure milk, that is, milk free from germs of fermentation, or as it exists in the cow's udder, will keep indefinitely at any temperature if protected from infection, but if any of the members of this society were brought up on farms, as your humble servant was, they will know how impracticable it is to procure milk without more or less, generally more, impurities finding entrance into it. If the multiplication of these germs which are thus introduced is not checked in some manner most profound changes soon take place in the milk.

I should be the last person to decry the efforts which are being made all over Christendom to obtain cleaner and more sanitary milk, because I know the need thereof, but I would emphasize the importance of cooling in that connection, because I believe it to be probably the most potent factor in preserving milk in a sweet and wholesome condition, and one that has not been given the prominence which it deserves. The process of pasteurization, very often looked upon as a heating process, is half refrigeration, because the heating without immediate and rapid cooling would, in most cases, be worse than useless. Refrigeration will not remove impurities from the milk, but it does have the effect of checking the multiplication of bacteria. It is of the utmost importance that the cooling of milk should be proceeded with as quickly as possible after it is drawn from the cow. Milk which is cooled immediately, say to 60° F., will keep longer and be in better condition than if it is allowed to remain at a temperature of 70 to 80 degrees for several hours and then afterwards cooled to 40. I use these figures more to illustrate my meaning than to record actual experience. The refrigerating engineer who is called upon to design or erect a milk-cooling plant should provide for quick cooling with as little exposure to the air as possible.

Some years ago an attempt was made to ship milk long distances in a frozen condition. Milk was sent from Scandinavia to Great Britain, covering a journey of two or three days and it was predicted that it would be possible to ship it by this method across the Atlantic. The scheme has apparently not been commercially successful, because we have heard nothing about it of late years. One of the objections to the freezing of milk is the formation of flocculent particles of albumen or casein compounds which are not readily dissolved when the milk is thawed. It also has the effect of collecting the fat globules into small lumps of fat.

It may be said, therefore, that for practical purposes a temperature of 40° F. or under is low enough for the preservation of milk, and that its preservation can only be a matter of days under ordinary commercial conditions.

CHAPTER XVII.

APPLES.

INTRODUCTION.

Cold storage in its present partly improved state is for the most part a growth of the past quarter century, and it is susceptible of much more development. The first so-called cold storage was by means of ice placed within the space to be cooled, which is very primitive and inferior. In what follows in this chapter, it must be understood that improved methods of storage by refrigeration are referred to, wherein temperature, and conditions of humidity, air circulation, and ventilation are under control.

If it had been predicted in 1880 that in 1910 practically all apples wanted for consumption during winter and spring would be stored immediately after picking in artificially refrigerated rooms or what is popularly known as cold storage, the prognosticator would doubtless have been called visionary. Even as late as 1890 comparatively few apples were stored under refrigeration. At that time it was thought (erroneously, as will be demonstrated farther on) that apples were such a low priced product that they could not afford payment of cold storage charges, and therefore, the greater bulk of the crop was stored in basements, cellars or frost-proof storage of some kind. Such structures may be partially cooled by letting in outside air whenever natural temperatures are suitable, but a control of temperature in this way is only partial.

The value of cold storage as compared with cellar or frost-proof storage has been demonstrated by some experiments conducted at the cold storage plant of the Michigan Agricultural College. The building originally had some rooms cooled with ice, and later some of these rooms were remodeled and cooled by Cooper brine system and chloride of calcium process. A part of the basement was without refrigeration, and left so especially to test the value of cold storage for apples, as compared with frost-proof or cellar storage. The following gives essentially the results of the experiments.

The average temperature of the cold storage room, September to May, was nearly 35° F.; that of the cellar 42°. By January 6, 100 per cent of the Kieffer pears in the cellar had rotted; during the same time three per cent rotted in the cold storage room. By May 22, 100 per cent of Baldwin apples stored in the cellar rotted as compared with two per cent in the storage room. Between the same dates, the results on Spys were twenty-one per cent for the storage and 100 for the cellar; on Baldwins, thirteen in storage, 100 in cellar. These figures not only show the advantage of a cold storage over a cellar, but they show what influence a small margin of 7° F. has on the keeping of apples.

At the present time the apple storage business is almost wholly conducted in comparatively large warehouses. For the most part these are located in big cities, and the apples are shipped to them during the picking season; but there are also many large cold storage houses in the apple growing sections run by firms or individuals; only a few of them are owned and operated by co-operative fruit growers' organizations. For various reasons, some of which are discussed further on in this chapter, the natural development of the storage of apples in artificially cooled structures is in the direction of storage by the producer. By this is meant that the producer will in future largely store his own products, either as an individual in his own house, or as a member of a fruit growers' or fruit shippers' organization. This is the present tendency of the business, and in future it will doubtless be mostly handled in that way, in preference to storing in the big city houses, where the most of the business is now done. The apple grower, therefore, will do well to keep posted on the situation, and study the various means of refrigeration, with an idea of selecting the system best adapted to his circumstances and location, when the time comes for him to install his own cold storage plant.

At the present time most apple growers have all their money invested in planting and have not the surplus money to put into a cold storage plant. Even co-operative concerns are difficult to promote, as no one grower will invest more than a small amount. As the business becomes older and growers accumulate a surplus they will, doubtless, put in plants of their own to a large extent. This problem is more fully discussed further on in this chapter.

HISTORICAL.

To Professor Benjamin M. Nyce (sometimes styled Reverend) is due the credit of giving the first impetus to commercial apple storage. Prof. Nyce was known as a preacher, teacher and chemist, and his home was in Decatur county, Indiana. In 1856, being in poor health, he was advised by his physicians to eat plenty of fruit during the entire year. At that time this seemed impossible, as cold storage was unknown in its present form and perfection, but Prof. Nyce, being a man of resource, undertook to improve on the then common method of storage, which we now call direct ice cooling, and which is discussed at some length in the chapter on "Refrigeration from Ice." The limitations of direct ice cooling were fully understood by Prof. Nyce, and he thought it possible to develop a cold storage system in which the humidity of the air could be controlled. Temperature he did not attempt to control any further than to get the lowest temperature which ice melting naturally would give him. Humidity he controlled by placing in the room lumps of chloride of calcium for taking up surplus air moisture and the moisture resulting from the emanation or evaporation from the fruit in storage. This system was patented in 1858, and is described in some detail in the chapter above referred to.

After considerable experimental work and not a little loss of money and damage to goods in storage, it was demonstrated that apples must be stored by themselves, as should other products. By using a tremendous thickness of insulation (two to three feet) temperatures as low as 34° F. were obtained with this system. Houses were built at Cleveland, Ohio,* In-

*This house was visited by the author during the winter of 1912-1913 and it was still in operation for the storage of apples.

dianapolis, Ind., Covington, Ky., and other points. It is stated that the house at Covington made a profit of $16,000 on apples sold in May and June, 1866, and that 4,000 bushels stored in the Cleveland house in 1870 yielded a profit of $7,200. Publication of these enormous profits was followed by active demand for the right to build under the Nyce patent, and quite a number of houses were constructed, mostly by the inventor, as he would not sell the right to use his patents. In this way he suffered financial loss, and as other and improved methods came into use the Nyce System soon fell into neglect. The results obtainable with this system under favorable conditions were so good that it led to all sorts of unreasonable claims, and much radical and flimsy construction was indulged in, doubtless with the idea that the wonderful system, as it was considered, would do almost anything. We can, however, without question give Prof. Nyce due credit for being the pioneer in the successful cold storage of apples, and, as a man of enterprise, scientific attainments and ability, he is entitled to much credit.

The utilization of a freezing mixture consisting of ice and salt was employed as early as 1865, and this development of cold storage has had an important influence even on present methods. As at first employed ice and salt was placed in V-shaped galvanized iron tanks suspended from the ceiling and filled from above, and in some cases the walls of the room were covered with flat tanks in a similar manner. Temperatures low enough to freeze large quantities of poultry, game, etc., were secured, and prior to the introduction of mechanical refrigeration quite a number of plants were equipped in this way. For a time this scheme was carefully guarded as a secret, but it soon became known and was utilized in quite a number of places, and assisted in the general development of the business. The ice and salt method of cooling is at present in use in a large number of houses, and will doubtless continue for many years to come, as it is a very simple and efficient method, and as applied with the Cooper brine system, described elsewhere in this book, results are obtainable which are not in any way inferior to those obtained by mechanical refrigeration.

Mechanical refrigeration was introduced in this country between 1860 and 1865, but it was fully ten years later, or after 1875, before modern systems of cold storage by this means were introduced and some years later before these were put on a practical basis. The principles of mechanical refrigeration, as at present applied, are discussed in one of the opening chapters of this book. It may be said that the greater part of the apple storage business now is done in houses wherein temperatures are maintained by artificial means.

COLD STORAGE AND COMMERCIAL APPLE GROWING.

The growing of fruit, and especially apples, as a business distinct from general farm operations, has made striking progress during the past half century. Early in the nineteenth century commercial orcharding was unknown, and the growing of apples and other fruits was conducted in a small way about the farm home for family use, or possible for sale in nearby cities or towns. As the cities have grown and the country developed, and transportation facilities have been extended and improved, fruit growing has assumed a commercial aspect. During the past twenty years especially, apple growing has developed into one of the leading agricultural industries.

The free use of fruit as a staple article of food, and here we may remark again, apples especially, has been accompanied by a higher standard of living, and doubtless the future will see less flesh foods eaten and more vegetable foods. By the use of improved cold storage the natural apple consuming season is extended to late winter and early spring, and the crop, instead of being thrown on the market mostly at picking time, is in demand during a period twice as long as formerly. This means a greatly increased average price to the producer, and a lower average price to the consumer, for the reason that waste is largely eliminated. These great advantages are all creditable to modern cold storage. With modern refrigerated transportation, commercial orcharding in the Mississippi Valley and in the North Pacific Coast region has developed largely, and the product of these regions is now a regular factor in the Eastern market. Where formerly the orchards of Michigan and western New York were ample to supply the demand in the large eastern cities like New York, Boston, and Philadelphia, now the crop of the Western apple grower is in demand as well as the product from the increased orchards in Michigan and in New York. Cold storage and refrigerated transportation have not been merely an assistance in this development, but an absolute necessity.

With the developing markets and larger demand and with the improvement of distribution from large wholesale centers, has come the possibility of control of markets by distributors and buyers. Nearly every year complaints are heard that apple buyers create their own market prices. Cold storage in the producing sections, so that growers may hold their own crops, will doubtless solve the problem. Whether these houses should be constructed and operated by individual growers, or on a co-operative basis as a community proposition, is a matter which cannot be determined except by a canvass of the local situation and conditions.

There seems to be no question but that certain abuses in the buying and selling of apples have been practiced, and the growers have been placed under great disadvantage in many cases. A community or co-operative or an individual cold storage plant may be the solution of the problem, and there is no question but what this would be the correct thing for those who have money to hold their crops and market them themselves. If a community could be induced to combine and put up a plant and employ an experienced salesman to market their goods they would doubtless find it most profitable indeed. This means quite a heavy investment in a plant and necessarily considerable of an extra investment tied up in storage, but doubtless local banks could easily take care of this latter problem by using warehouse receipts issued on goods in storage as collateral for the basis of short time loans.

The practice is to stamp goods as received with a lot number, using a rubber stamp. One or more of these lot numbers may be combined in a warehouse receipt, but it is customary to issue warehouse receipts on a carload so as to facilitate the business of taking up warehouse receipts and shipping in round lots. However, large dealers sometimes take out warehouse receipts on 500 barrels to 1,000 barrels. It may be noted in this connection that it is necessary in order to issue warehouse receipts that the apples be packed in a uniform and permanent package. Open barrels, crates or boxes would hardly be permissible, and certainly apples stored in bulk would never do at all.

By handling the business on the basis of warehouse receipts a large amount of money is not necessary, as it is customary for banks to advance about three-quarters the value of goods when placed in storage, taking the warehouse receipt, which is endorsed over to the bank as security. The putting up of a cold storage plant necessarily means somewhat of an investment, and some figures have been given on the cost of a cold storage plant per barrel. These range from $1.00 per barrel up to $3.00 per barrel. The cost depends entirely on the following:

First: Type of building; whether frame, brick, stone, concrete, etc.

Second: Location; whether in the South, where the picking season is comparatively warm, or whether in the North, where the picking season is comparatively cold. Where the fruit goes into storage at high temperature a much heavier equipment is necessary.

Third: Whether the plant will be used for summer storage of other goods, such as eggs, butter, cheese, etc. A much heavier and more expensive insulation and a much heavier refrigerating equipment is necessary if this is to be done.

Fourth: System of refrigeration. Where natural ice may be stored at reasonable cost the Cooper brine system using ice and salt for cooling is best adapted. In certain southern locations the mechanical or chemical systems are necessary, and these are much more expensive.

Fifth: Variety of apples to be stored. If the main crop consists of winter apples coming in at a comparatively late date only a small amount of refrigeration is needed to cool the fruit as received, whereas if the bulk of the crop consists of summer or fall varieties much more refrigeration is needed and a much heavier refrigerating equipment is required for this reason.

Sixth: Size and capacity of cold storage plant. Necessarily if the plant is a small one the cost per barrel is much higher than in a large plant. For instance, a small plant in service in New York State with a capacity of 500 barrels actually cost about $2,000, which is about $4.00 per barrel. A cold storage plant with a capacity of 10,000 to 20,000 barrels would, of course, cost very much less in proportion, as the cost of building and the cost of insulating would be less, for the outside exposure is less and cubic capacity very much greater in proportion to the material required for construction.

So far as the building itself is concerned this may be of cheap frame construction and cheaply insulated in many localities. In the North, insulation which would answer for a frost-proof building would also ordinarily answer for a cold storage building for the storage of winter varieties of apples largely. In some cases larger packing space is required. It is often necessary to store a large number of empty packages and, therefore, as much space or possibly more may be required in connection with a cold storage building which is not under refrigeration, as is actually employed for cold storage purposes. These things are of course subject to local influences and requirements. It may be stated roughly that a plant of even very large capacity and where very little space is required for handling purposes, storage of empty packages, etc., cannot certainly be built for $1.00 per barrel of storage capacity. This figure might have been correct some years ago, but with advancing cost of materials and labor it is doubtful if a suitably equipped cold storage plant of any capacity could be built today at a less cost than $1.50 per barrel. It may be estimated roughly that a cold storage plant of from 5,000 to 10,000 barrels would cost in the neighborhood of $2.25 to $3.00 per barrel of cold storage capacity, and a plant of from 10,000 to 20,000 barrrels would cost from $1.50 to $2.25 per barrel of cold storage capacity. Necessarily these figures are only estimates, and local conditions might make the cost greater, but it is improbable that a plant could be built for less, although perhaps in some places, as in Virginia, with comparatively cheap lumber and labor, these costs might be reduced somewhat. The only way to get an accurate cost is to take a given locality with its material and labor costs and the requirements as to size and shape of the building and make an estimate based thereon.

The question of whether individual plants are best or larger houses located at some central point where it might be on the railroad, is open for discussion, but the personal preference of the author would be for individual plants. In most cases it is difficult to get the best and most influential growers into a co-operative concern, but it is sometimes possible to get practically all interested to combine to the extent of forming a selling organization. If each large grower had his own cold storage house his fruit could be picked and packed under the best conditions and circumstances. It is even possible to pick hurriedly, putting the apples in barrels without heading up permanently, and then sorting and regrading at leisure, thus getting a very careful grading. This is not possible where fruit is delivered to a central or community storehouse. Picking time is always a rush time and it is of great importance that the work be carried forward quickly and grading and packing must necessarily be done hurriedly under these conditions, especially if the grower does his own grading and packing. This is one of the most important reasons why each individual apple grower should have his own cold storage plant, and by those who are now handling orchards as a commercial undertaking this fact will be understood at once.

It may readily be seen from the figures given above that with two or three favorable seasons a grower could pay for a cold storage plant out of profits which could reasonably be expected. The fruit business, like other lines, necessarily has its "off years," but these are compensated for by extra good years, so that we may consider that it would be possible to pay the entire cost of a cold storage plant out of the profits of three to five average years. It is, of course, assumed that large growers will, to an extent, interest themselves in markets and work up a trade or brand by careful and conscientious packing and not put their brand on any fruit which is inferior. One such grower in the State of New York is known to the author who has made enough profit to pay for his cold storage plant in three years' operation.

ADVANTAGES OF LOCAL COLD STORE.

J. W. Wellington, in "Purdue Agriculturist," bearing on the suggestion that growers own their own cold storage houses, suggests the following:

Cold storage has a direct bearing on the marketing of fruits. Apples put into cold storage will keep long beyond their natural season and the commission men make the most of the fact. They buy fruit at moderate prices and hold until the prices advance. It is speculation, but at the same time, a form which rarely fails. With an apple like our friend, the Grimes, cold storage is necessary and consequently a large part of the Grimes raised in Indiana do go into storage to be sold later at much increased prices. The probability is this—when Indiana becomes the fruit state, that it surely will in a very few years,—the fruit growers will own, co-operatively, their own storage houses and thus secure what is of right their own.

The Year Book of the Department of Agriculture for 1903 contains the following comments with relation to the subject as discussed above:

"In handling the apple for cold storage the ideal is reached when the fruit can be taken directly from the tree to the warehouse. So far as the fruit is concerned, a similar condition is approached when it is shipped to a distant warehouse in refrigerator cars, or the ideal is attained in those sections or seasons in which the picking and handling of the crop occur in cool weather. It may not be practicable for the apple dealer who is located in a distant city to store his fruit in warehouses situated near the orchards, nor is the local warehouse advisable in sections where there are inadequate facilities for transporting the fruit to distant markets during the winter. As a general rule, it is to the mutual interest of the owner and the warehouseman that the fruit be stored where it can be watched throughout the season by the owner, as the warehouseman is responsible only for the proper management of the building and its contents, and not for the ultimate condition of the fruit."

The local cold storage warehouse is especially favorable to the apple grower who stores his own fruit and who is not located near a large city warehouse. It is also adapted to apple dealers in cities who have permanent representatives near the orchards. In those sections in which the fruit is likely to ripen in warm weather, like the warmer apple regions of the Mississippi and the Allegheny Mountain districts, the grower is frequently forced to sell his apples in the local market or to a dealer at a low price. If the weather is unusually warm the fruit is likely to arrive in the markets in bad condition, and the apple trade soon becomes demoralized. On the other hand, if the fruit is shipped to a distant storage house and the packing, shipping or handling is delayed, its storage quality has been seriously impaired before it reaches the warehouse.

A system of warehouses located in the orchards and managed by growers, or operated by companies in nearby towns, would reduce some of the difficulties with which the growers in the warmer apple belts have to contend, and would thereby give greater stability to the industry in those sections. There can be no question, from the standpoint of the keeping of the fruit, of the advantage of a warehouse located near the orchard, but its usefulness to the business as a whole depends not on the keeping quality of the fruit alone, but on the larger question of its adaptability to the present requirements of the apple trade.

Experiments by the Michigan Agricultural College, referred to in the introduction to this chapter, have demonstrated that the maturity of fruit and time elapsing from picking until stored, determines largely the possible life of apples in storage. In these experiments it was demonstrated that only 21 per cent of Spys stored immediately after picking rotted, as compared with 49 per cent left in a barn ten days before storing. In Spys fully matured and well colored, but perfectly firm, 18 per cent rotted up to May 22d, as contrasted with 62 per cent taken from the same trees two weeks later.

UNITED STATES GOVERNMENT EXPERIMENTS.

The following is largely extracted from Bulletin No. 48, Bureau of Plant Industry, U. S. Department of Agriculture, by G. Harold Powell, Assistant Pomologist in Charge of Field Investigation, and S. H. Fulton, Assistant in Pomology:

THE FUNCTION OF THE COLD STORAGE WAREHOUSE.

There is a good deal of misapprehension as to the function of the cold storage house in the preservation of fruits. This condition leads to frequent misunderstandings between the warehouseman and the fruit storer, though they might be avoided and the condition of the fruit storage business improved if there was a clearer definition of the influence on fruit preservation of cultural conditions, of the commercial methods of handling, and of the methods of storage.

A fruit is a living organism in which the life processes go forward more slowly in low temperatures, but do not cease even in the lowest temperatures in which the fruit may be safely stored. When the fruit naturally reaches the end of its life it dies from old age. It may be killed prematurely by rots, usually caused by fungi which lodge on the fruit before it is packed, or sometimes afterwards. The cold storage house is designed to arrest the ripening processes in a temperature that will not injure the fruit in other respects and thereby prolong its life history. It is designed also to retard the development of the diseases with which the fruit is afflicted, but it cannot prevent the slow growth of some of them. It follows that the behavior of different apples or lots of apples in a storage room is largely dependent on their condition when they enter the room. If they are in a dissimilar condition of ripeness, or have been grown or handled differently, or vary in other respects, these differences may be expected to appear as the fruit ripens slowly in the low temperature. If the fruit is already overripe, the low temperature cannot prevent its deterioration sooner than would be the case with apples of the same variety that were in a less mature condition. If the fruit has been bruised, or is covered with rot spores, the low temperature may retard, but cannot prevent its premature decay. If there are inherent differences in the apples due to the character of the soil, the altitude, and to incidental features of orchard management, or variations due to the methods of picking, packing, and shipping, the low temperature must not be expected to obliterate them, but rather to retard while not preventing their normal development.

In general it is the function of the cold storage warehouse to furnish a uniform temperature of the desired degree of cold through its compartments during the storage season.*

*The experiments so far conducted cover only the influence of temperature in cold storage. Much has yet to be done in determining the best methods of refrigerating which control air circulation, ventilation and humidity. More is promised along these lines.

The warehouse is expected to be managed in other respects so that the deterioration of the fruit or any other injury may not be reasonably attributed to a poorly constructed and installed plant, or to its negligent or improper management. The warehouseman does not insure the fruit against natural deterioration; he holds it in storage as a trustee, and in that relation is bound to use only that degree of care and diligence in the management of the warehouse that a man of ordinary care and prudence would exercise under the circumstances in protecting the goods if they were his private property.

If the temperature of the storage rooms fluctuates unduly from the point to be maintained and causes the fruit to freeze to its injury, or to ripen with abnormal rapidity, or if the management of the rooms or the handling of the fruit in other respects can be shown to have been faulty or negligent, the warehouse has failed to perform its proper function.

OUTLINE OF EXPERIMENTS IN APPLE STORAGE.

An outline of the apple storage experiments of the United States Department of Agriculture is presented here. The following problems were under investigation during two apple seasons:

1.—A comparative test of the keeping quality of a large number of varieties grown in different regions and of the same varieties grown under different conditions and in different localities.

The fruit was stored in closed 50-pound boxes in a temperature of 31° to 32° F. One-half of the fruit in each box was wrapped in paper.

2.—A determination of the influence of various commercial methods of apple handling on the keeping quality of the most important varieties in the leading apple-growing regions of the eastern United States.

Each variety was picked at two different degrees of maturity: First, when nearly grown but only half to two-thirds colored, but about the time when apples are usually picked; second, when the fruit was fully grown and more highly colored, but still hard. In each picking the fruit was separated into two lots, representing the average of the lightest and of the darkest colored or most mature specimens.

Part of the fruit of each series was sent to storage as soon as picked. A duplicate lot was held two weeks in the orchard or in a building, either in piles or protected in packages, before it was sent to storage.

Comparative tests were made to determine the efficiency of different kinds of fruit wrappers on the keeping of the fruit, and observations on the behavior of the fruit in closed and ventilated packages were recorded.

3.—A determination of the influence of various cultural and other conditions of growth on the keeping quality of the fruit.

Comparison was made with the same variety from heavy clay and from sandy soils, from sod, and from cultivated land, from young, rapidly growing trees, and from older trees with more steady habits.

4.—A determination of the behavior of the fruit under the conditions outlined in temperatures of 31° to 32° F., and in 34° to 36° F.

5.—A determination of the behavior of the fruit when removed from storage, and of its value to the consumer.

The fruit used in the investigations was taken from central and eastern Kansas, southwestern and central Missouri, southern and central Illinois, western Michigan, northeastern West Virginia, northern and western Virginia, western North Carolina, central Delaware, southern Maine, central Massachusetts, and from eastern, central, and western New York. A description of each orchard accompanies the data included in the account of the variety test.

It was necessary to duplicate the work in different parts of the country, as the climatic and other conditions and the varieties differ in each section. The work must be repeated for several successive seasons before general conclusions can safely be drawn from it, as the climatic conditions differ each year and thereby affect the results.

FACTORS INFLUENCING THE KEEPING QUALITY OF APPLES.

In recent years there has been a tendency to pick the apple crop relatively earlier in the season than formerly. It is quite generally supposed that the longest keeping apples are not fully developed in size or maturity and that the most highly colored fruit is less able to endure the abuses that arise in picking, packing, and shipping.

Aside from these general impressions, several important economic factors have influenced the picking time. A large proportion of the apple crop is purchased in the orchard by the barrel or by the entire orchard by a comparatively few apple merchants. The fruit may be picked and barreled either by the grower or by the purchaser, but with the growing scarcity of farm hands and other labor it has become necessary to begin picking relatively earlier in the autumn to secure the crop before the fall storms or winter months set in.

The general increase in freight traffic during the past few years has overtaxed the carrying capacity of the railroads as well as their terminal facilities for freight handling, and has influenced the apple dealers to extend the picking and shipping season over the longest possible time, in order to avoid congestion and consequent delays in shipping and in unloading the fruit. The facilities at the warehouses are often inadequate for the quick handling of the fruit from the cars when it is received in unusually large quantities, and this condition has also favored a longer shipping season.

In localities where the entire crop is sometimes ruined by the bitter rot after the fruit is half grown the picking of the apples is often begun early in the season in order to secure the largest amount of perfect fruit.

It is not generally the case, however, that the immature and partly colored fruit has the best keeping quality. On the other hand, an apple that is not overgrown and which has attained full growth and high color, like the lower specimen of York Imperial in Fig. 1, but is still hard and firm when picked, equals the less mature fruit (upper specimen, Fig. 1) in keeping quality and often surpasses it. The mature fruit is superior in flavor and texture; it is more attractive to the purchaser, and therefore of greater money value. It retains its plumpness longer and is less subject to apple scald. If, however, the fruit is not picked until overripe, it is already near the end of its life history, and will deteriorate rapidly unless stored soon after picking in a low temperature.

In the experiments with the Tompkins King and the Sutton apples grown in New York on rapidly growing young trees producing unusually large apples, the fruit that was three-fourths colored kept longer than the fully colored apples from the same trees. Dark red Tompkins King showed 28 per cent of physiological decay in February following the storage. Light, half red Tompkins King from the same trees, picked at the same time, showed 10 per cent of physiological decay in February following the storage. Fig. 2 shows Tompkins King in February at two degrees of maturity in September, 1902, from young, rapidly growing trees. The upper specimen represents fruit that was highly colored but firm when picked; the lower specimen shows fruit one-half to two-thirds colored. The less mature fruit kept in good condition a month longer than the highly colored apple. These apples were overgrown—a condition likely to occur on young trees. Whether the same conditions hold true of other varieties that are overgrown has not been determined.

From older trees, apples that are fully grown, highly colored, and firm when picked have kept as well in all cases (and better in many, as shown in Fig. 1) than immature and under-colored fruit.

A considerable number of later varieties may be picked when they are beginning to mellow, and will keep for months in prime condition provided they are handled with great care and quickly stored after picking in a temperature of 31° to 32° F. Fruit in this ripe state cannot be left in the orchard or in warm freight cars, or in any other condition that will cause it to ripen after picking, without seriously injuring its value. In this ripe condition it should be stored in boxes, and a fruit wrapper will still further protect it.

Apples that are to be stored in a local cold storage house to be distributed to the large markets in cooler weather may be picked much later than fruit requiring ten days or more in transit, but the use of the refrigerator car makes late picking possible where the fruit must be in transit for a considerable

FIG. 1—SCALD ON YORK IMPERIAL APPLES.

time in warm weather in reaching a distant storage house.

While it is not the purpose of this publication to discuss cultural practices in the orchard, some suggestions in relation to the methods of securing more mature and more highly colored fruit may not be without value to the fruit grower.

A large proportion of the poorly colored fruit from old orchards is caused by dense-headed trees and close planting, which prevent the free access of air and sunlight and delay the maturity of the fruit in the fall. The fundamental corrective in such cases lies in judicious pruning, by which means the fruit may be exposed to the sunlight.

In other cases the poor color may be due to a combination of heavy soil, tillage, frequent turning in of nitrogenous cover-crops, spraying, and neglect in pruning. These conditions stimulate the trees to active growth, the foliage increases in health, size, and quantity, and, as the water-holding capacity of the soil is enlarged by the incorporation of the cover-crops and is retained by the tillage, the trees grow late in the fall and the fruit does not properly color before the picking season arrives. It is often possible to overcome the difficulty by severely pruning the top to let in more air and light. If this treatment does not prove efficient, the cover-crops may be withheld, when the fruit will usually mature earlier in the fall, unless the season is wet. As an additional treatment where necessary, the growth of the orchard may be still further checked by seeding it down until the desired condition is attained.

It is not possible to secure a uniform degree of maturity and size when all the apples on a tree are picked at one time, as fruit in different stages of growth is mixed together on the same tree. The apples differ in size and maturity in relation to their position, the upper outer branches producing the large, highly colored and early ripening fruit, while the apples on the side branches and the shaded interior branches ripen later. Greater uniformity in these respects is approached by proper pruning and by other cultural methods, but the greatest uniformity can be attained when, like the peach or the pear, the apple tree is picked over several times, taking the fruit in each picking that approaches the desired standard of size and maturity.

FIG. 2—TOMPKINS KING APPLES. OVERGROWN ON YOUNG TREES

Summer apples, like the Yellow Transparent, Astrachan, and Williams, are usually picked in this manner, and fall varieties, like Twenty Ounce, Oldenburg, and Wealthy, are sometimes treated similarly. In recent years a few growers of winter apples have adopted the plan for the late varieties, with the result that the size, color, and ripeness of a larger proportion of the fruit are more uniform. This method of picking is not usually adapted to the apple merchant who buys the crop of a large number of orchards, and who can not always secure efficient or abundant labor, but for the specialist who is working for the finest trade and who has a storage house nearby or a convenient refrigerator car service to a distant storage house, the plan has much to commend it.

INFLUENCE OF DELAYING THE STORAGE OF THE FRUIT.

The removal of an apple from the tree hastens its ripening. As soon as the growth is stopped by picking, the fruit matures more rapidly than it does when growing on the tree and maturing at the same time. The rapidity of ripening increases as the temperature rises, and it is checked by a low temperature. It appears to vary with the degree of maturity at which the fruit is picked, the less mature apples seeming to reach the end of their life as quickly as or even sooner than the more mature fruit. It varies with the conditions of growth, the abnormally large fruit from young trees or fruit which has been overgrown from other causes, ripening and deteriorating very rapidly. It differs with the nature of the variety, those sorts with a short life history, like the summer and fall varieties, or like the early winter apples, such as Rhode Island *Greening,* Yellow Bellflower, or Grimes *Golden,* progressing more rapidly than the long-keeping varieties like Roxbury, Swaar, or Baldwin.

Any condition in the management of the fruit that causes it to ripen after it is picked brings it just so much nearer the end of its life, whether it is stored in common storage or in cold storage, while treatment that checks the ripening to the greatest possible degree prolongs it.

The keeping quality of a great deal of fruit is seriously injured by delays between the orchard and the storage house.

This is especially true in hot weather and in fruit that comes from sections where the autumn months are usually hot. If the apples are exposed to the sun in piles in the orchard, or are kept in closed buildings where the hot, humid air can not easily be removed from the pile, if transportation is delayed because cars for shipment can not be secured promptly, or if the fruit is detained in transit or at the terminal point in tight cars which soon become charged with hot, moist air, the ripening progresses rapidly and the apples may already be near the point of deterioration or may even have commenced to deteriorate from scald, or mellowness, or decay when the storage house is reached.

On the contrary, the weather may be cool during a similar period of delay and no serious injury result to the keeping quality, or the ripening may be checked in hot weather by shipping the fruit in refrigerator cars to a distant storage house.

The fungous diseases of the fruit, such as the apple scab (*Fusicladium dendriticum*) (Wallr.) (Fckl.) and the pink mold (*Cephalothecium roseum* Cda.) which grows upon the scab, the blue mold (*Penicillium glaucum* Link) which causes the common, soft, brown rot, the black rot (*Sphaeropsis malorum* Pk.) and the bitter rot (*Glaeosporium fructigenum* Berk.), develop very fast if the fruit becomes heated after picking. The conditions already enumerated which cause the fruit to ripen quickly during the delay between the orchard and the storage house are also most favorable to the development of fruit diseases. It is therefore of the greatest importance that the fruit be stored immediately after picking, if the weather is warm, in order to insure it against the unusual development of the fungous rots.

In the fall of 1901, when the weather in western New York was cool, there was no apparent injury from delaying storage of a large number of varieties two weeks and then shipping the fruit to Buffalo, the transit occupying from one to three days. There was also no apparent injury to the fruit from Virginia treated in a similar manner, but in southwestern Missouri, where it was warmer, the apples delayed two weeks before storing were seriously injured in their commercial keeping qualities.

The results accomplished during 1902 have been of the most instructive character. During the latter half of September the temperature in eastern New York averaged about 62° F., with a humidity of 84°. During the first half of October the average temperature was 53° F. and the humidity 80°.

Rhode Island *Greening*, Tompkins King, and Sutton apples picked September 15, 1902, and stored within three days, were firm till the following March, with no rot or scald, but fruit from the same trees not stored till two weeks after picking was badly scalded or decayed by the 1st of January. None of the immediate-stored fruit was scalded or decayed by the 1st of February, but the delayed Sutton and Rhode Island *Greening* apples were soft and mealy, and one-third were scalded at that time, while nearly 40 per cent of the delayed Tompkins King were soft and worthless. The commercial value of these varieties was injured from 40 to 70 per cent by the delay in storage.

Apples of these varieties picked from the same trees on October 5, 1902, and stored immediately, and also some stored two weeks later, were less injured by the delay, as the temperature and humidity were not sufficiently high to cause rapid ripening or the development of the fruit rots.

From the standpoint of the orchardist or apple dealer who can not secure quick transportation to the large storage centers, or who can not obtain refrigerator cars, or who is geographically situated where the weather is usually warm in apple-picking time, the local storage plant in which the fruit can be stored at once and distributed in cool weather offers important advantages. The importance of this phase of the fruit-storage business and its relation to the fruit-growing industry are emphasized as the apple business enlarges.

INFLUENCE OF STORAGE TEMPERATURE.

The investigations indicate that the ripening processes are delayed more in a temperature of 31° to 32° F. than in 35° to 36° F. The apple keeps longer in the lower temperature, it scalds less, the fruit rots and molds are retarded to a greater extent, while the quality, aroma, flavor, and other characteristics of the fruit are fully as good, and when removed from storage it remains in good condition for a longer period.

The impression is quite general that fall varieties and the tender early winter sorts, like Fameuse, Wealthy, and Grimes, are injured in some way by the low temperature, but the investigations of the Department of Agriculture indicate that these varieties behave more satisfactorily in every respect when stored at 31° to 32° F.

If the fruit is intended for storage for a short time only, and it is desired to have it ripen before removing it from the storage house, then a higher temperature may be desirable to hasten the maturity.

The influence of the temperature on the ripening processes appears to depend on the condition of the fruit. Baldwin, Esopus *Spitzenburg,* Roxbury, Jonathan, Lady Sweet, and other long-keeping eastern-grown varieties have been held in prime commercial condition throughout the storage season in a temperature of 35° F., when carefully picked and handled and stored soon after picking; but when the fruit was carelessly handled or the storage was delayed in hot weather, then a temperature of 31° to 32° F. was required to retard the ripening.

It might be safe to use a temperature of 34° to 35° F. in a storage house located near the orchard, in which the fruit may be stored immediately after harvesting, but for general commercial apple handling, a temperature as low as 32° F. is needed to overcome the abuses that usually arise in picking, packing and shipping.

Apples are sometimes frozen in the storage rooms owing to a considerable drop in the temperature or to a poor distribution of the cold air. If the fruit compartment adjoins a freezer room and the insulation is poor, the fruit may be frozen in packages piled close to the freezer wall. Apples placed near the refrigerating pipes or near the cold-air duct where it enters the room may be injured by freezing if the plant is improperly installed or managed; or if the piping or air circulation is faulty, the temperature at the bottom may be lower than that at the top of the room.

No definite investigations have been made by the Department of Agriculture as to the effect of temperatures lower than 31° F. The exact freezing point of apples has not been determined, but it is below this point. It may possibly vary with the composition or condition of the variety. Under the most favorable conditions, apples are sometimes commercially stored at 30° F.* without injury, but 31° F. should be considered a critical temperature below which it is unsafe to store this fruit, except in houses that are properly constructed and in which the temperature is maintained uniform in all parts of the rooms.

The frosting of the fruit does not necessarily injure it. When the apple freezes, the water in the cells is withdrawn and frozen in the intercellular spaces, and if it thaws slowly and the freezing has not been too severe, the cells may regain the water without injury and resume their living function. If the thawing is rapid, the cells may not reabsorb the water with sufficient rapidity, and in this case it remains in the intercellular spaces and is lost by evaporation. In addition, the tissues next to the area of greatest freezing may be forced apart by the formation of ice crystals in the intercellular spaces.

If the freezing is so severe as to withdraw too much of the cell water, the cells may not be able to absorb it and will be killed in the same manner as if dried out in any other way. Occasionally the freezing is so rapid that besides the withdrawal of water the cell contents are disorganized or possibly frozen outright; at any rate, the cell may be directly killed by a sudden change of temperature. It is probable that varieties may differ as to the degree of freezing they will stand without injury, and further, that the same sort may vary in this respect when grown under different conditions or subjected to different treatment.

---

*The author's personal experience is that a temperature of 30° F. is better than any degree above that, and 29° F. is practicable and advisable for long-period storing of the better keeping varieties. To safely store at 29° to 30° F. it is necessary that a thorough forced circulation of air be employed (see chapter on "Air Circulation"), and in cooling the fruit down to the final carrying temperature, the refrigeration must not be applied too suddenly. If, say, the fruit has a temperaure of 60° or 70° F. when placed in storage, a period of two or three weeks should be consumed in reducing to 29° or 30° F. This applies to the better keeping kinds only. Softer varieties must be cooled quickly, as their life is shorter, and too much deterioration will take place during cooling process if handled as suggested above.—Author.

The most characteristic results of injurious freezing are a translucent appearance of the skin of the fruit, a water-logged and springy or spongy condition of the flesh, a forcing apart of the tissues, and a brownish discoloration of the flesh. The browning may result from any cause which results in the death of the cells and is not necessarily characteristic of freezing. It often happens that the skin of the fruit retains its normal brightness after the interior has discolored.

In the practical handling of frozen stock, the temperature should be raised very slowly until the frost is withdrawn. If possible, the fruit should not be moved until it is defrosted, as it discolors quickly wherever a slight bruise occurs, or even where the skin is lightly rubbed. With these precautions observed it is often possible to defrost stock that is quite firmly frozen without apparent injury to it.

INFLUENCE OF A FRUIT WRAPPER.

In the storage investigations under discussion a comparison has been made between wrapped and unwrapped stock on

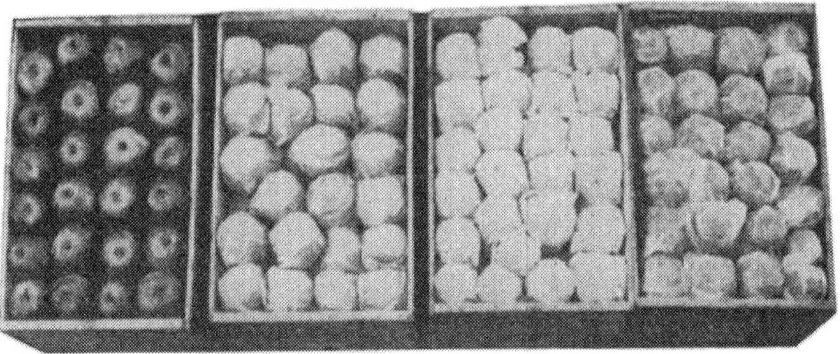

FIG. 3.—APPLES UNWRAPPED AND IN TISSUE, PARCHMENT, AND WAX WRAPPERS.

the keeping quality of the fruit, and the efficiency of different kinds of paper for wrappers—tissue, parchment, waxed or paraffin, and unprinted news—has been tested. A box of unwrapped fruit with packages of fruit wrapped with the kinds of paper mentioned in order above, is shown in Fig. 3.

It has been found that the wrapper may influence the keeping quality in several different ways. It extends the life of the fruit beyond its normal period by retarding the ripening processes. The influence of the wrapper in this regard is apparent especially at the end of the normal storage season of the naked fruit when the flesh begins to grow mealy from overripeness. At this time the wrapped apples may be firm and remain in prime condition for several weeks or even months. The wrapper is especially useful in extending the season of early winter sorts, or in making the long-keeping varieties available for use over a still longer period of time.

The wrapper may be useful in preventing the transfer of rot from one apple to another. If the fungous is capable of growing in the storage temperature, it is not likely that the wrapper retards its growth, but when the spores develop they are confined within the wrapper and their dissemination is difficult or impossible.

The importance of a wrapper in protecting the fruit from decay and in extending its season may be better appreciated by reference to the following table:

AMOUNT OF DECAYED FRUIT APRIL 29 IN BUSHEL PACKAGES

Variety.	News paper wrapped.	Un-wrapped.	Variety.	News paper wrapped.	Un-wrapped.
	Per cent.	Per cent.		Per cent.	Per cent.
Baker	3.7	27.2	Northern Spy	5.6	52.0
Dickenson	6.4	43.0	Wagener	38.0	63.0
McIntosh	7.7	15.0	Wealthy	42.0	60.0
McIntosh (second lot)..	19.7	32.0			

The wrapper protects the apple against bruising and the discoloration that may result from improper packing or rough handling; it checks transpiration, and by the preservation of the attractive appearance and firmness of the fruit adds to its commercial value.

No important difference was noticeable in the efficiency of the different wrappers, except that a mold developed freely on the parchment paper in a temperature of 36° F. This mold grew only to a slight extent in 32° F.

A double wrapper is more efficient in retarding ripening and transpiration than a single wrapper. A good combination consists in a porous news paper next to the fruit, with an impervious wax or paraffin wrapper on the outside. The wrappers vary in cost from 20 cents per thousand for news paper, 9x12 inches, to 70 cents per thousand for the better grades of paraffin.

INFLUENCE OF CULTURAL CONDITIONS.

Preliminary studies have been made on the influence of cultural and other conditions surrounding the growing fruit on its storage quality. Considerable data along this line will be brought out in the comparison of the same variety grown in different sections. It has been observed that the Tompkins King, Hubbardston, and Sutton apples from rank-growing young trees ripen faster than smaller fruit from older slower-growing trees, and therefore reach the end of their life history sooner. From older trees these varieties have kept well till the middle of April, while from young trees the commercial storage limit is sometimes three months shorter.

It has been noticed that Rhode Island *Greening* apples from old trees remain hard longer than the same variety from young trees, but the greener condition of the fruit from the older trees when picked at the same time made it more susceptible to scald. Rhode Island *Greenings* from Mr. Grant G. Hitchings, South Onondaga, N. Y., showed 50 per cent of scald from young trees on April 28, 1903, and 82 per cent in smaller, greener fruit from older trees.

Rhode Island *Greening,* Mann, and Baldwin apples grown on sandy land ripened more rapidly than similar fruit from clay land, where all of the other conditions of growth were similar. Fig. 4 shows the average condition of Baldwin apples on April 28, 1903, grown on sandy and clay soil in the orchard of Mr. W. T. Mann, Barker, Niagara County, N. Y., and stored in a temperature of 32° F. The upper apple was grown on clay; the lower, on sandy soil.

FIG. 4.—BALDWIN APPLES FROM CLAY AND SANDY SOIL.

This fruit was picked in October, 1902, and was stored soon after picking. The fruit from the heavy clay soil was generally smaller and was much less highly colored. Both lots kept well throughout the storage season. The fruit from the sandy land was riper at the end of the storage season, better in quality, and worth more to the dealer and to the consumer.

The subject will require critical study over a period of years before it will be possible to fully understand the influence of various cultural, climatic, and other conditions of growth on the life processes in the fruit.

INFLUENCE OF THE TYPE OF PACKAGE.

The principal storage packages for apples are barrels of about 3 bushels capacity and boxes holding 40 to 50 pounds. The larger the bulk of fruit and the more it is protected from the air the longer it retains the heat after entering the storage room. If the fruit is hot and the variety a quick-ripening sort, it may continue to ripen considerably in the center of the package before the fruit cools in that position. The long-keeping varieties that are harvested and shipped in cooler weather are less likely to show the effect of the type of the package. The smaller package therefore presents distinct advantages for the early, quick-ripening varieties and is most useful in the hottest weather, as the fruit cools down quickly throughout the package and its ripening proceeds uniformly.

There is a wide difference of opinion concerning the comparative value of ventilated and closed packages for apple storage. The chief advantage of the ventilated package appears to lie in the greater rapidity with which its contents cool off, and its value in this respect depends on the amount of ventilation in the package. The contents of an ordinary ventilated apple barrel do not cool much more quickly than the contents of a closed barrel, and the value of the ventilated barrel for the purpose for which it is designed is somewhat doubtful.

Apples in a ventilated package are likely to shrivel if the fruit is stored for any length of time. In the ordinary ventilated apple barrel the exposure is not sufficient to affect the fruit to any extent, but in boxes in which there is much exposure the fruit may be corky or spongy in texture if held until spring.

The size of the package may have an important influence on the length of the storage season. Its influence in this respect is especially marked when the fruit begins to mellow in texture. Barrel stock in this condition needs to be sold to prevent the bruising of the fruit from its own weight, but apples equally ripe may be carried in boxes safely sometimes for several weeks longer.

BEHAVIOR OF THE FRUIT WHEN REMOVED FROM STORAGE.

There is a general impression that cold-storage apples deteriorate quickly after removal from the warehouse. This opinion is founded on the experience of the fruit handler and the consumer, but the impression is not generally applicable to all storage apples. In fact, it is probable that storage apples do not deteriorate more quickly than other apples that are equally ripe and are held in the same outside temperature. If the fruit is overripe when taken from storage—and a good deal of stock is stored until it reaches this condition—it naturally breaks down quickly; but firm stock may be held for weeks and even months after it has been in storage.*

The rapidity of deterioration depends also on the temperature into which the fruit is removed. The following table shows

AMOUNT OF DECAY AFTER REMOVAL FROM STORAGE TO DIFFERENT TEMPERATURES.

Variety.	Date removed from storage (1903).	Date inspected.	Per cent rot			
			44°F.	48°F.	61°F.	67°F.
Baldwin	Jan. 29	Jan. 29	0	0	0	3
		Feb. 10	0	0	3	10
		Feb. 13	0	0	12	14
		Feb. 16	0	0	21	24
		Feb. 20	0	4	23	28
		Mar. 3	5	10	......	......
		Mar. 7	5	15		......
		Mar. 24	20	......		......
		Apr. 6	36	......		......

*This is confirmed by the author's experience, and applies not only to apples, but also to other goods which are cold stored.—Author.

the amount of decay in Baldwin apples from the same barrel after removal and subjection to different temperatures:

Late in the spring the fruit is far advanced in its life and the weather is becoming warmer. All apples similarly ripe, whether in cold storage or not, break down more quickly at this time than in the winter.

In commercial practice the dealer often holds the apples for a rise in price, and finally removes them from the warehouse, not because the market has improved, but for the reason that he finds that a longer storage would result in serious deterioration from fruit rots and overripeness. When a considerable amount of stock is decayed on removal from the warehouse the evidence is conclusive that the apples should have been sold earlier in the season. In the purchase of cold-storage stock the consumer will have little cause to complain of the rapid deterioration of the fruit if he exercises good judgment in the selection of apples that are still sound and firm.

THE IMPORTANCE OF GOOD FRUIT.

Apples do not improve in grade in cold storage. In handling a crop too much care can not be given to grading the fruit properly before it enters the storage house. The contents of many packages are injured by the spread of disease from a few imperfect apples. Rots enter the fruit most easily wherever the skin is bruised or broken, and in the early stages of the rot development it is common to see the diseases manifesting themselves around worm holes or bruises occasioned by rough handling, from nails that protrude through the barrels, or from other causes.

When the crop is light it may pay to store apples that are not of the first grade, but such fruit should be rigidly eliminated from the best stock and stored where it can be removed earlier in the season than the better qualities.

The attractiveness and the value of the best fruit is often injured by careless handling. A bruised spot dies and discolors. Finger marks made by pickers, graders, and packers, and injuries from the shifting of the fruit in transit or from rough handling, become more apparent as the season advances. In fact, all of the investigations of the Department of Agriculture emphasize the fundamental importance of well-grown, carefully handled fruit in successful storage operations.

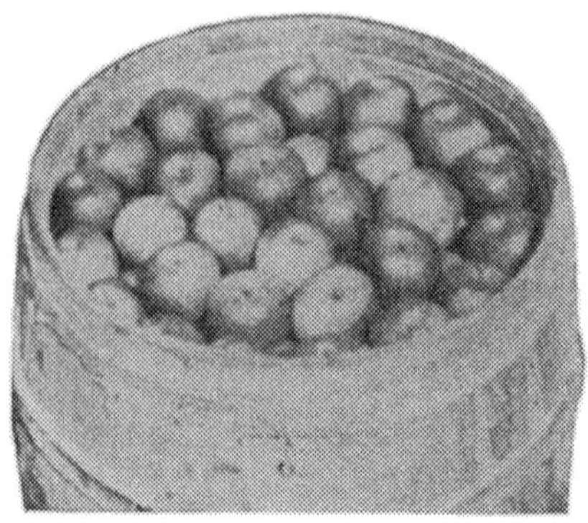

FIG. 5.—WELL PACKED ESOPUS SPITZENBURG APPLES.

FIG. 6.—"SLACK" PACKED NORTHERN SPY APPLES.

Fig. 5 shows a well packed barrel of Esopus *Spitzenburg* apples removed from storage in March, 1903. The fruit was properly packed in the orchard and repacking was not needed when the fruit was sold.

Fig. 6 shows a "slack" packed barrel of Northern Spy apples removed from storage in March, 1903. The fruit was not packed firmly in the orchard. It settled in the barrel, leaving it "slack" when removed from storage. Barrels in this condition need to be repacked. The fruit is easily bruised and it deteriorates more quickly in the storage house and after removal when it is loosely packed.

APPLE SCALD.

When some varieties of apples reach a certain degree of ripeness the part of the fruit grown in the shade often turns brown, not unlike the color of a baked apple. This difficulty does not extend deep into the flesh, but it detracts from the appearance of the fruit and reduces its commercial value. This trouble is commonly called "apple scald." It may appear in fruit held in common or in cold storage.

The exact nature of scald is not well understood, though apple men have many theories by which its appearance is popularly explained. The most common theory gives rise to the name of scald—that is, the brown, cooked appearance is thought to be due to the overheating of the fruit when it is stored, or to a temperature too low for the variety, or to picking the fruit when too ripe; and other matters relating to the growth and handling of the fruit are thought to develop it.

As the scald is an important commercial problem it has been considered from several standpoints in the fruit-storage investigations of the Department. The nature of the scald, the influence of the degree of maturity of the fruit when picked, of commercial method of handling, of fruit wrappers, of different temperatures, and of cultural conditions on its developments are among the problems investigated.

Apple scald is not a contagious disease. According to Dr. A. F. Woods, Pathologist and Physiologist of the Department of Agriculture, it is a physiological disturbance not connected in any way with the action of parasitic or saprophytic organisms such as molds or bacteria. Briefly, it is the mixing of the cell contents or premature death of the cells and their browning by oxidation through the influence of the normal oxidizing ferments of the cell. There are many conditions which influence the development of this trouble. It appears to be closely connected with the changes that occur in ripening after the fruit is picked, and is most injurious in its effects as the fruit approaches the end of its life. Several of the factors that influence it will be discussed. Fig. 7 shows the scald on a Rhode Island *Greening* apple. The cross section shows that the scald is a surface trouble and does not extend into the flesh.

The scald always appears first on the green or less mature side of an apple, and if the fruit is only partly ripe it may spread entirely over it; but the portions grown in the shade and undercolored are first and most seriously affected. The upper specimen in Fig. 1 shows the distribution of scald on an immature York Imperial apple in March, 1903. The apples that are more mature and more highly colored when picked are

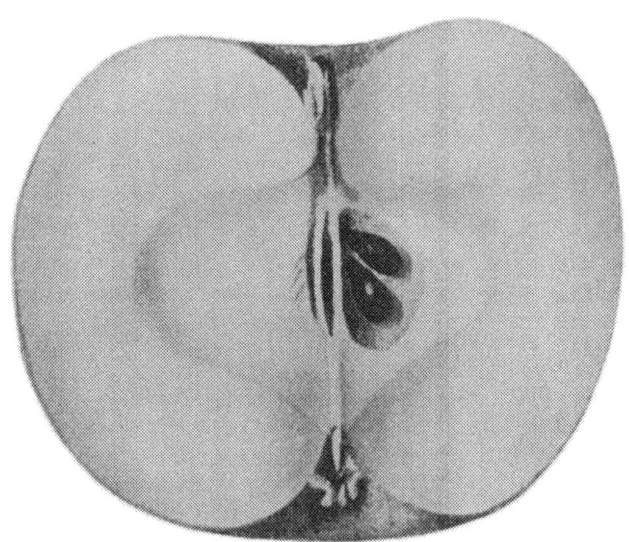

FIG. 7.—SCALD ON RHODE ISLAND GREENING APPLE.

less susceptible to injury, and the side grown in the sunlight may remain entirely free from it. The lower specimen in Fig. 1 (picked from the same tree at the time, October, 1902, when the upper specimen was picked) shows a well-colored York Imperial apple and its freedom from the scald is noticeable. A wall only is shown on the right-hand side of the apple, where the color is not as dark as elsewhere.

When the apple crop is picked before it is mature the fruit is more susceptible to scald than it would have been later in the season. The relative susceptibility of immature and more mature apples is brought out in the table following. The fruit was picked two weeks apart. At the first picking the apples were partly colored, or in the condition in which a large proportion of the commercial apple crop is harvested. At the second picking the fruit was more mature, with better color, but still hard. The differences in ripeness are fairly represented in the fruit in Figs. 1 and 2. The percentages do not represent the relative susceptibility of the different varieties to scald, as the fruit was grown in different States and the observations were made at different times. The percentages show the average amounts of scald in fruit stored at temperatures of 31° to 32° F. and 34° or 36° F.

SCALD ON MATURE AND IMMATURE APPLES.

Variety.	Locality grown.	Mature well colored.	Immature, partly colored.
		Per cent.	Per cent.
Baldwin	New York	3.1	29.2
Ben Davis	Illinois	2.6	15.8
Do	Virginia	13.1	41.6
Rhode Island "Greening"	New York	25.4	43.4
Winesap	Illinois	0.2	31.8
Yellow Newtown	Virginia	2.3	9.4
York Imperial	....do	2.0	18.2
Average		6.9	27.0

In the practical handling of orchards the fundamental corrective of scald lies in practicing those cultural and harvesting methods that develop maturity and a highly colored fruit. These methods have already been briefly discussed. The picking of the fruit when too green, dense-headed trees that shut out the sunlight, heavy soil, a location or season that cause the fruit to mature later than usual and makes it still green at picking time—these are among the conditions that make it particularly susceptible to the development of the scald.

After the fruit is harvested its susceptibility increases as its ripening progresses. Early in the storage season the scald may not appear, but later the same variety may have developed enough to injure its commercial value. The amount of scald at different periods of the season on the same lot of Baldwin apples stored at 32° F. is brought out in the following statement:

AMOUNT OF SCALD AT DIFFERENT PERIODS OF STORAGE SEASON.

	Per cent.
January 29, 1903	0
February 21, 1903	0
March 20, 1903	20
April 21, 1903	23

It should be the aim of the apple storer to remove the fruit from storage before a variety normally begins to scald, and to hold until late in the season only those sorts that do not scald.

INFLUENCE OF TEMPERATURE ON SCALD.

The temperature that checks the ripening to the greatest degree also retards the appearance of the scald. In some of the apple-growing sections it is quite generally believed that bad scalding varieties should be stored in a temperature of 36° to 38° F., and that a temperature as low as 32° F. hastens its development. The investigations of the Department have shown that this impression is not well founded, but on the contrary they indicate that the scald develops more freely in the higher temperature. To illustrate, one lot of York Imperial apples, a variety which is greatly affected by scald, had developed 16.9 per cent of this trouble by January 22, 1902, in a temperature of 36° F., while a similar lot stored in a temperature of 32° F. developed only 3.4 per cent. One lot of Rhode Island *Greening* apples by February 3, 1903, had developed 21 per cent in 32° F., while a similar lot, in 36° F., showed 55 per cent. In the case of the Sutton apple, investigation showed 25 per cent of scald in apples stored at 32°, and 42 per cent where the temperature was kept at 36°.

If the fruit is stored as soon as it is picked, or is shipped in refrigerator cars or in cool weather, and if it has been handled in the most careful manner, the ripening may not proceed much more rapidly and the scald may not develop more freely in the higher than in the lower storage temperature.

When the fruit is removed from the storage house the scald sometimes develops rapidly. Its appearance at this time seems to depend on at least two important conditions—the ripeness of the fruit and the temperature into which it is taken. Late in the storage season the scald is most severe; first, because the fruit is more mature, and, second, for the reason that the warm weather prevailing at that season develops it quickly.*

The development of the scald also seems to be influenced by the amount of humidity in the air. So long as the fruit remains cold and condenses the moisture of the atmosphere upon it the scald is retarded more than in a dry air of the same temperature.

The accompanying table shows the rapidity with which the scald may develop on Baldwin apples when portions of the same barrel are removed to different temperatures. There was no increase in the amount of scald in any of the lots after nine days.

Variety.	Date removed from storage.	Date inspected.	Per cent of scald.			
			44° F.	48° F.	61° F.	67° F.
	1903	1903				
Baldwin	Jan. 29	Jan. 29	0	0	0	0
do	do	Feb. 3	0	6	21	22
do	do	Feb. 4	4	11	21	37
do	do	Feb. 6	4	25	40	63
do	do	Feb. 7	4	25	41	63

*It is suggested that scald develops much more rapidly in case the fruit is allowed to rise in temperature suddenly. When removed from storage, apples, as well as other goods, should not be exposed at once to comparatively high temperatures.—Author.

SCALD DEVELOPED IN DIFFERENT TEMPERATURES WHEN APPLES WERE REMOVED FROM STORAGE.

The upper specimen in Fig. 8 shows the average condition of a lot of Wagener apples in March, 1903, having been picked

FIG. 8.—WAGENER APPLE—SCALD DEVELOPED AFTER REMOVAL FROM STORAGE.

in October, 1902, and stored at a temperature of 32° F. There was no scald on the apples when removed. Forty-eight hours later, after the fruit had been in a temperature of 70° F., the light-colored portion of the apples was badly scalded, as shown in the lower apple. Late in the storage season the fruit is more susceptible to scald, and a high temperature when the fruit is removed from the storage house may develop it quickly.

It should be the aim of the fruit storer not only to remove the fruit before the scald normally appears, but to hold the apples after removal in the lowest possible temperature to prevent its rapid development.

INFLUENCE ON SCALD OF DELAYING THE STORAGE OF THE FRUIT AFTER IT IS PICKED.

The ripening of the fruit between the time of picking and its storage increases its susceptibility to scald.

When the picking and shipping seasons are cool and dry it may be possible to delay the storage of the fruit for some time without injury so far as the predisposition of scald is concerned. In the investigations of 1901-2 in western New York there was no apparent injury from delaying the storage, but the weather conditions at this period were ideal for apple handling.

The scald develops seriously when the storage of the fruit is delayed in hot weather. Detentions in the orchard, in transit in closed cars, in unloading at the terminal, or in the warehouse cause the fruit to ripen quickly and favor the rapid growth of the fruit rots, as they bring the fruit much nearer the end of its life before it enters the storage room. Under these circumstances the fruit may scald badly, mellow early in the season, and rot, and no storage treatment can correct the abuses to which it has been subjected.

The following table brings out the injury that may be caused by delaying the storage of the fruit in hot weather. The mean average temperature between September 15 and 30, 1902, was about 62° F. and the mean average humidity about 84°. Fruit picked from the same trees on October 4, 1902, and stored two weeks later, when the temperature was about 53° F. and the humidity about 80°, was not injured by the delay. The apples referred to were grown in eastern New York and stored in Boston, and these records were taken the following February.

SCALD ON IMMEDIATE- AND DELAYED-STORED APPLES IN FEBRUARY, 1903.

Variety.	Picked Sept. 12, 1902, stored Sept. 15.	Picked Sept. 15, stored Sept. 30.	Picked Oct. 4, stored Oct. 9.	Picked Oct. 5, stored Oct. 19.
	Per cent.	Per cent.	Per cent.	Per cent.
Rhode Island Greening	0	38	(No record)	(No record)
Sutton	0	33	0	0
Tompkins King	0	15	0	0

INFLUENCE OF A FRUIT WRAPPER ON SCALD.

The influence of the various fruit wrappers mentioned has been studied in connection with the scald. Sometimes the wrappers retard it to a slight degree, but often the trouble is as severe or more severe in the wrapped fruit. Whenever the wrapper has been effective in retarding the scald the wax or paraffin paper was most useful.

The following table gives a comparison between wrapped and unwrapped fruit, and emphasizes the fact that for commercial purposes the wrapper should not be looked upon as an effective means of preventing the trouble. The records of each variety are based on 8 to 32 bushels of fruit, one-half of which was wrapped.

SCALD ON WRAPPED AND UNWRAPPED FRUIT.

Variety.	Locality.	Wrapped.	Unwrapped.
		Per cent.	Per cent.
Baldwin	New York	12.4	19.9
Ben Davis	Illinois	5.8	2.8
Do	Virginia	27.1	28.7
Huntsman	Illinois	47.8	40.3
Minkler	....do	22.9	20.1
Rhode Island "Greening"	New York	32.3	37.6
Winesap	Virginia	30.0	47.0
Do	Illinois	17.9	10.2
York Imperial	Virginia	9.6	12.9

VARIETIES MOST SUSCEPTIBLE TO SCALD.

All varieties are not equally susceptible to scald, and there appears to be a wide difference in the amount developed in the same variety grown in different parts of the country. While the light-colored portion of an apple is more susceptible than the more highly-colored part, it does not follow that green or yellow varities are more susceptible than red ones. Of the important commercial sorts used in the investigations of the Department of Agriculture, the varieties named in the subjoined list have proved susceptible. The season when the scald is most likely to appear is given with each kind, though there may be a wide variation from year to year. The time of the appearance of the scald is influenced to an important degree by the method of handling the fruit and by its degree of ripeness.

Arctic, serious midwinter
Arkansas often serious, after midwinter.
Baldwin, often serious, late in season.
Ben Davis, often serious, late in season.
Gilpin, often serious, late in season.
Green Newtown, slight, late in season.
Grimes, serious, early winter.
Huntsman, serious, midwinter.
Lankford, serious, midwinter.
Nero, serious, midwinter.
Paragon, sometimes serious, midwinter.
Ralls, slight, midwinter.
Rhode Island Greening, serious, midwinter.
Smith, Cider, serious, early winter.
Stayman Winesap, sometimes serious, midwinter.
Wagener, serious, midwinter.
White Doctor, serious, midwinter.
White Pippin, slight, late in season.
Willow, slight, late in season.
Winesap, often serious, late in season.
Yellow Newtown, slight, late in season.
York Imperial, serious, midwinter.
York Stripe, slight, late in season.

COMPARISON OF VARIETIES IN COLD STORAGE.

A large number of varieties of apples grown under various conditions were under observation by the Department of Agriculture. It was the purpose of the investigation to determine the keeping quality of the varieties during the commercial apple-storage season, which usually terminates May 1, or shortly afterwards. It was not attempted to carry the varieties longer than the apple-storage season, though many of them when finally taken from the storage house were in prime condition and would have kept well for a longer period.

There is a wide difference in the keeping quality of the same variety when it is grown in different parts of the country. There is a striking variation also in the behavior of the same variety when it is grown in the same locality under different cultural conditions and in different seasons. There may be a permanent difference in the keeping quality of the apples of one region when compared with those of another, but it is not safe to draw general conclusions in this regard until the varieties of each have been under observation during several seasons and have been grown under different cultural conditions. No attempt was made in the investigations to draw comparisons between the keeping quality of the same sort from different places. The behavior of each lot is given in commercial terms rather than in detailed notes, so that the grower or apple handler may know something of the storage value of a variety under the conditions in which it has been observed by the Department of Agriculture. The fruit was stored in bushel boxes in a temperature of 30° to 32° F.

SUMMARY OF U. S. GOVERNMENT EXPERIMENTS.

An apple usually should be fully grown and highly colored when picked, to give it the best keeping and commercial qualities. When harvested in that condition it is less liable to scald, of better quality, more attractive in appearance, and is worth more money than when it is picked in greener condition.

An exception to the statement appears to exist in the case of certain varieties when borne on rapidly growing young trees. Such fruit is likely to be overgrown, and under these conditions the apples may need picking before they reach their highest color and full development.

Uniform color may be secured by pruning to let the sunlight into the tree, by cultural conditions that check the growth of the tree early in the fall, and by picking over the trees several times, taking the apples in each picking that have attained the desired degree of color and size.

Apples should be stored as quickly as possible after picking. The fruit ripens rapidly after it is picked, especially if the weather is hot. The ripening which takes place between the time of picking and storage shortens the life of the fruit in the storage house. The fruit rots multiply rapidly if storage is delayed and the fruit becomes heated. If the weather is cool enough to prevent after-ripening, a delay in the storage of the fruit may not be injurious to its keeping quality.

A temperature of 31° to 32° F. retards the ripening processes more than a higher temperature. This temperature favors the fruit in other respects.

A fruit wrapper retards the ripening of the fruit; it preserves its bright color, checks transpiration and lessens wilting, protects the apple from bruising, and prevents the spread of fungous spores from decayed to perfect fruit. In commercial practice the use of the wrapper may be advisable on the finest grades of fruit that are placed on the market in small packages.

Apples that are to be stored for any length of time should be placed in closed packages. Fruit in ventilated packages is likely to be injured by wilting. Delicate fruit and fruit on which the ripening processes need to be quickly checked should be stored in the smallest practicable commercial package. The fruit cools more rapidly in small packages.

Apples should be in a firm condition when taken from storage, and kept in a low temperature after removal. A high temperature hastens decomposition and develops scald.

The best fruit keeps best in storage. When the crop is light it may pay to store fruit of inferior grade, but in this case the grades should be established when the fruit is picked. The bruising of the fruit leads to premature decay.

The scald is probably caused by a ferment or enzyme which works most rapidly in a high temperature. Fruit picked before it is mature is more susceptible than highly colored, well-developed fruit.

After the fruit is picked its susceptibility to scald increases as the ripening progresses.

The ripening that takes place between the picking of the fruit and its storage makes it more susceptible to scald, and delay in storing the fruit in hot weather is particularly injurious.

The fruit scalds least in a low temperature. On removal from storage late in the season the scald develops quickly, especially when the temperature is high.

It does not appear practicable to treat the fruit with gases or other substances to prevent the scald.

From the practical standpoint the scald may be prevented to the greatest extent by producing highly colored, well-developed fruit, by storing it as soon as it is picked in a temperature of 31° to 32° F., by removing it from storage while it is still free from scald, and by holding it after removal in the coolest possible temperature.

A variety may differ in its keeping quality when grown in different parts of the country. It may vary when grown in the same locality under different cultural conditions. The character of the soil, the age of the trees, the care of the orchard—all of these factors modify the growth of the tree and fruit and may affect the keeping quality of the apples. The character of the season also modifies the keeping power of the fruit.

COMMERCIAL RESULTS FROM THE COLD STORING OF APPLES.

The following outline of various experiments and experimental results in addition to the work of the U. S. Department of Agriculture is of interest as bearing on the actual results already obtained in the storage and transportation of apples. Many of these tests or experiments have been conducted with extreme care and by men of scientific training and attainment. A word of caution, however, must be suggested. Experiments must not be taken too literally. The results apply only to the specific fruit and conditions under which the tests were made. The quality of any variety of apples is extremely variable one season with another, and depends also on the soil on which grown and the climatic conditions. These details must be kept in mind always in considering the practical and scientific aspect of any set or series of experiments, or practical results.

CANADIAN GOVERNMENT EXPERIMENTS.

A report by John A. Ruddick, Dairy and Cold Storage Commissioner of Canada, printed as Bulletin No. 24, and headed "Some Trial Shipments of Cold Storage Apples," gives some very interesting and explicit results secured from the shipment of apples from various Canadian territory to Scotland. The experiments were undertaken to demonstrate the great advantage of cold storage over frost-proof storage, and the educational value of these shipping experiments are most important. The following extracts from Mr. Ruddick's Bulletin No. 24 will prove of general interest:

"The apples used for the experiment were the ordinary commercial packs of different growers, as represented by The Oshawa Fruit Growers, Limited, and The Sparta Co-operative Fruit Growers' Association.

In presenting the results of these trials we have taken each shipment separately, showing the net returns against the total cost, including freight and storage charges and the expenses of members of the staff in looking after the packing and shipment. These costs are necessarily much higher than they would be in a regular commercial transaction, where careful records and notes are not necessary.

It was thought advisable to have one carload of apples held in an ordinary frost- proof storage for the sake of comparison. These apples were from the same orchards and packed by the same persons as the apples stored at Montreal and St. John.

With the exception of lots 1 and 2, the apples were carried in cold storage across the Atlantic, and the two Calgary lots were shipped in refrigerator cars. All the apples carried in cold storage were held at a temperature of 32 to 34 degrees during the whole storage period.

Lot 1.—Apples in barrels stored at Oshawa, Ont., in frost-proof warehouse.

Picked—October 25-30.
Packed—November 22-23.
Stored—November 22-23.
Shipped from Oshawa—February 24.
Shipped from St. John—March 2.

COST.

No. Brls.	Variety.	Purchase price.	Amount.	Cost of Repacking before Shipment.	Total Cost.
26	No. 1 Spy...........	$3.75	$97.50	$1.70	$99.20
20	No. 2 Spy...........	2.75	55.00	1.30	56.30
26	No. 1 Baldwin.......	3.25	84.50	1.70	86.20
20	No. 2 Baldwin.......	2.50	50.00	1.30	51.30
92			$287.00	6.00	$293.00

Sold by Simons, Jacobs & Co., Glasgow, March 15, 1910, ex. ss. *Cassandra* from St. John, N. B.

PROCEEDS.

No. Brls.	Variety.	Average Price Sold for.	Gross Proceeds.	Total Charges.	Net Proceeds.	Net loss per Brl.
		$	$	$	$	$
24	No. 1 Spy........	3.67	88.11	34.08	54.03	1.73
19	No. 2 Spy........	3.40	64.60	26.79	37.81	0.92
24	No. 1 Baldwin....	4.39	105.36	35.04	70.32	0.61
20	No. 2 Baldwin....	3.77	75.42	28.40	47.02	0.21
87			$333.49	$124.31	$209.18	
5 us	ed in repacking.					
92						

A thermograph was placed in the warehouse along with the apples and a continuous record of temperature was obtained. For the first 12 days the temperature was between 40 and 42 degrees. During the following month it averaged about 35 degrees and from that time until the apples were shipped, it varied only between 32 and 34 degrees. The temperature in the car from Oshawa to St. John held steadily at 32 degrees.

These apples were repacked before shipping, the shrinkage being 5 barrels in 92. They were sold at the same time as lot 2, but under separate marks. Although no charge for storage is included in the total cost of this lot, the net loss was greater than in any other lot of the same apples. This was partly due to the loss and expense in repacking and partly to a poor market, but if these apples had been held longer before sale, the loss would in all probability have been greater.

Our cargo inspector at Glasgow, Mr. Jas. Findlay, reporting on this lot, stated: "This mark, while in very fair order as a whole, were mostly slight 'shakes' and the fruit was rather ripe and inclined to give way." He also mentions that the Baldwins showed considerable scald.

Lot 2.—Apples in barrels stored at St. John, N. B., in cold storage.
Picked (Baldwins)—October 25-30.
Picked (Spies)—November 1-5.
Packed (Spies)—November 2-5.
Packed (Baldwins)—November 6-9.
Stored at St. John—November 15.
Shipped to Glasgow—March 2.

COST.

No. Brls.	Variety.	Purchase Price.	Amount.	Cost of opening, examin'g and tighten'g brls. before shipment.	Storage charges.	Total cost.
		$	$	$	$	$
45	No. 1 Spy.....	3.75	168.75	2.92	11.25	182.92
40	No. 2 Spy.....	2.75	110.00	2.60	10.00	122.60
40	No. 1 Baldwin.	3.25	130.00	2.60	10.00	142.60
30	No. 2 Baldwin.	2.50	75.00	1.95	7.50	84.45
155			483.75	10.07	38.75	532.57

Sold by Simons, Jacobs & Co., Glasgow, March 15, 1910, ex ss. *Cassandra* from St. John, N. B.

PROCEEDS.

No. Brls.	Variety.	Average Price Sold for.	Gross Proceeds.	Total Charges.	Net Proceeds.	Net loss per Brl.
		$	$	$	$	$
44	No. 1 Spy....	4.57	201.08	64.68	136.40	1.03
40	No. 2 Spy....	3.60	144.00	56.80	87.20	0.88
40	No. 1 Baldwin	4.66	186.40	58.45	127.95	0.36
30	No. 2 Baldwin	3.90	117.00	42.90	74.10	0.34
154			648.48	222.83	425.65	
1 us	ed in plugging	barrels				
155						

NOTE.—The freight from Oshawa to Glasgow via St. John, N. B., worked out at $1.02 per barrel and the broker's charges for insurance, landing, delivering, etc., at 21 cents per barrel. The usual 5 per cent commission was charged.

This lot was from the same orchards as lot 1. It included two marks, "A" and "B." The A's were carefully packed, to avoid, if possible, the necessity of repacking. The packing of the B's was in accordance with the usual practice and was intended to be temporary, with a view to repacking.

The condition of both marks was found to be so good on March 1st that it was decided to ship them as they were, after "plugging" the slack barrels. Only one barrel was used to plug 154. It was thought that the damage from repacking the B's would amount to more than the possible unevenness of the original temporary pack. It will be observed that this shows a better return than lot 1, after charging the cold storage expenses.

It should be remembered that lot 1 and lot 2 were packed alike, that lot 1 was repacked before shipment, with a shrinkage of 5 barrels in 92, and that lot 2 (in cold storage) was not repacked, one barrel in 155 being used for plugging.

Lots 1 and 2 were carried as ordinary cargo across the Atlantic at a temperature of about 40 degrees.

Mr. Findley reported as follows concerning lot 2:—

"The apples in above steamer shipped by the Department of Agriculture, branded 'Oshawa Fruit Growers' Association.' I found on arrival to be in the following condition:—

"Spy No. 1 and 2 grade, countermarked 'A,' were in good sound condition, almost free from bruise spots. I saw several of the bottoms of barrels of No. 1's and they all were very sound; the color was good, the size even, and they were generally choice. The Baldwins No. 1 and 2 of this mark were also in good condition, free from scald, and of good color and even size.

"Spy No. 1 and 2 countermarked 'B' were also in good condition, but fruit not so even, large and smaller apples being mixed. A trace of bruising was just showing on odd apples throughout the barrels, and coloring was not so even or good. Baldwins No. 1 and 2 were in good condition, an odd apple here and there showing 'brown' in the barrels; otherwise, fruit was clean and of very fair color generally."

SALE OF LOTS 1 AND 2 COMPARED

Variety and Grade.	Lot 1 Frost Proof Storage.			Lot 2 Cold Storage			Difference per Brl. in favor of	
	Storage Period.	Temperature.	Net Loss Per Brl.	Storage Period.	Temp.	Net Loss Per Brl.	Cold st'r'ge after paying storage charge of 25c per bl.	Frost proof st'r'ge
			$		deg.	$	$	$
Spy No. 1......	Nov. 22	About 40 deg. first fortnight; about 34 deg. 2nd fortnight and about 35 deg. for balance of period.	1.73	Nov. 18	32	1.03	.70	..
" " 2......	1909 to		.92	1909 to	"	.88	.04	..
Baldwin No. 1	Feb. 24		.61	Mch. 2	"	.36	.25	..
" " 2	1910.		.21	1910	"	.34	...	.13

In summarizing the results, Commissioner Ruddick states the later-picked apples had, of course, the better color and appearance and kept slightly better. The advantages of quick cold storing after picking were obvious, and this is the greatest lesson to be drawn from the trials. Cold stores should be as near as possible to the place of production, and the fruit should go direct from the orchard to the cold store as soon as possible. If packing is carefully done, repacking is not necessary in cold-stored apples. This means a big saving in expenses and in waste. The season for Greenings can be extended safely several weeks if the apples are well matured on the trees and placed in storage without delay.

In conclusion the report says:

"It is very frequently asserted that apples deteriorate quickly after being removed from cold storage. It would seem to depend entirely on the stage which the ripening process had reached. Apples ripen slowly in cold storage. If they are held until the limit is nearly reached, they naturally deteriorate quickly when removed, but no more quickly than they would if the same stage had been reached in ordinary storage at any temperature. * * * *

NOTE.—The apples ex Oshawa frost-proof storage were shipped on February 24th, and during the six days the car was in transit to St. John the temperature in the car remained steadily at 32 degrees.

"There is evidence in the results of these trials which would go to show that apples which are cold-stored promptly after picking and held at 32-34 degrees for, say five months, then removed to a high temperature for one month, will be in better condition at the end of the sixth month than if they had been exposed to the same high temperature for the first month and then placed in cold storage for the rest of the period. Or, in other words, exposure to a high temperature just after picking, when the life processes are active in the apple, will cause more injury than the same exposure at a later stage."

APPLE COLD STORAGE EXPERIMENT RESULTS BY THE NEW HAMPSHIRE EXPERIMENT STATION.

The following is interesting information as bearing on the handling of apples commercially, by shipping them to the big city houses, paying storage, commission, etc.:

On November 20th, 1899, a number of barrels of apples were shipped to one of the Boston cold storage houses. Beginning with February two barrels were taken out each month until July and examined. The fruit did not receive any extra care and was representative of apples as ordinarily purchased at that time of year on the open market. It was found that the apples could not safely be allowed to remain after April 1st, as they decayed rapidly after that date. The prices at time of shipment ranged between $1.25 and $2.00 and on April 1st they brought $3.50 to $4.25.

On Oct. 27th, 1900, a second shipment of apples was sent to cold storage with the following results. Price when put in storage, $1.25. On April 23rd ten barrels sold for $34.00 Expense, carting, 50c., commission, 8 per cent, $2.72. Net proceeds, $30.78 or $3.08 per barrel. Freight and cold storage charges must be deducted from this amount. The storage rates were 10c. per barrel per month, or for the season ending May 1st, 35 to 50 cents, according to the number of barrels. The freight charges can easily be found out according to the location of the individual.

The greatest care in handling and placing the fruit immediately into cold storage pays for the extra trouble. One must understand that cold storage will simply retard and not prevent entirely the spread of decay. If the fruit is in prime keeping condition on entering, it is likely to come out in proportionately as good condition.

Where apples were placed in brine and cold air storage, the cold air gave the best results.

From an examination of the prices paid in the fall and those paid on April 1st for the past six years, the results show that there has been a sufficient increase to warrant the extra expense of storage in every case and on the average the practice has resulted in good profit.

MR. ROE'S WISCONSIN EXPERIMENT.

J. P. Roe, in *Farm Press,* reports some experience with the storage of apples grown in Wisconsin. He says that the winter apple for long keeping which is adapted to the northwest is yet to be discovered. Of the fall varieties that succeeded in cold storage he reports that the Wealthy, the McIntosh Red, the Fameuse or Snow, and a Russian variety known as the Red Annis, give most satisfactory results. Of the summer varieties the Dutchess of Oldenburg and Yellow Transparent are successful. The Yellow Transparent is ordinarily considered a very soft variety, and not suitable for cold storing, but in Wisconsin this variety is much firmer and a much better keeper than when grown further south. Mr. Roe, however, reports some discoloration in the storage of Yellow Transparents. He reports that for general commercial purposes a red apple is preferable to white, as it is not only more salable, but more durable under cold storage treatment, and the results of a bruise not so conspicuous.

Mr. Roe suggests that if fruit growers are inclined on account of an unusual state of the market to experiment with apples in cold storage, they should go slowly, and only try a few barrels. He reports having stored 50 barrels of Longfields, and that they proved a total loss. As the author has suggested elsewhere in this chapter, the result of a single experiment must not be taken as final, nor as positive data on which to base future action. Repeated experiments under different conditions are necessary in order to be even an approximate guide.

MR. YOUNGERS' NEBRASKA EXPERIMENTS.

Some interesting facts on the cold storage of apples are gathered from the report of Mr. Youngers of the Nebraska Horticultural society, who collected and stored 180 barrels of apples, representing 34 varieties, the fall previous to the Columbian exposition. The following markings were made on a scale of 10 points for a perfect condition, or as nearly so as apples could be at that time of year. These markings were made at the time the apples were taken from cold storage.

	June 15	July 14	Aug. 2	Sept. 2	Oct. 2	Nov. 1
Ben Davis	10	10	10	10	10	10
Winesap	10	10	10	10	10	10
Genet	10	10	10	10	10	10
W. W. Pearmain	10	7	6	6	4	3
Limbertwig	10	10	10	10	10	10
Allen's Choice	10	10	10	10	9	8
Willow Twig	10	10	10	10	10	10
Sweet Russet	10	10	9	9	8	8
Little Red Romanite	10	10	10	10	10	10
Lansingburg	10	10	10	10	10	10
McIntosh Red	9	9	9	9	9	9
Salome	9	9	9	9	7	3
Dominic	9	8	8	8	7	6
Rome Beauty	8	8	8	7	6	5
Iowa Blush	8	8	8	8	7	5

The following varieties retained all of their good qualities up to the time of their last marking, Nov. 1: Ben Davis, Winesap, Genet, Limbertwig, Willow Twig, Little Red Romanite and Lansingburg.

The other varieties which were stored, but which in the percentages showing their condition at the time it was desired to use them, fell below the lowest percentage named in the list given were as follows: Jonathan, G. G. Pippin, Missouri Pippin, Northern Spy, Wallbridge, Yellow Bellflower, Eicke, Price's Sweet, Sheriff, Snow, Fulton, Minkler, English Golden Russet, Roman Stem, Ortley, Milam, Talman Sweet, Perry Russet, Wagener.

All of this fruit was gathered and placed in cold storage during the fall of 1897, most of it during the month of October. Each apple was wrapped first in a sheet of waxed paper, using 9 by 12 inch sheets for small apples and 12 by 12 inch sheets for large ones. Then another covering of common newspaper was added and the apples carefully packed in barrels, filling them up so as to require considerable pressure to get the heads in. They were stored in a cold storage room in South Omaha, and the temperature did not vary over one degree from 36 degrees from the time they were placed in storage until they were removed.

TIME LIMIT FOR APPLE COLD STORAGE.

The extreme practical limit of time possible to carry apples in cold storage has been demonstrated in numerous cases. Ben Davis, for instance, an apple which is known as a remarkable keeper and of poor quality, has been stored for two years in cold storage without decay. The apples were, of course, shrunken or shriveled and discolored, but were sound. Russets have also been carried over from one season to another in an experimental way. These are mere experiments, and apples are not handled on a commercial scale for storage longer than eight or nine months.

When the fruits of spring and summer ripen, apples of the previous season's growth properly go out of the market. They should be considered primarily a fruit for winter's use. Some few fancy varieties or apples of exceptionally good quality are carried over through the spring and into summer mainly for fruit stand and other special trade.

An experiment or practical demonstration was tried in a Buffalo, New York, cold storage house, to demonstrate the time limit of apple storage. It was believed that an apple had a maximum life period beyond which it would succumb to age, whether in or out of storage, and that while cold storage would prolong its life, it could not preserve the fruit beyond a certain limit of time.

The apples used in the test at Buffalo were of several varieties, including Tallman Sweets, Northern Spys, Smith's Cider, Spitzenbergs, Tompkins County Kings, Culverts and the like. Some of these were distinctively what are called soft or fall apples and not expected to keep very long even in cold storage.

The apples were placed in the cold storage warehouse October 4 and 6, 1904. They were taken out during the first week in January, 1906. On opening the boxes in which they had been placed the apples in appearance were found to be all right and in an excellent state of preservation, but some of the varieties lacked flavor, either their own distinctive flavor or any good apple flavor, while others were quite as good as those harvested in the autumn of 1905. They had stood the test for 15 months and looked as though many of them could have lasted another six months.

The above is a good example of so-called experiments by the average warehouseman. It is very difficult for the average person to judge accurately the question of quality. He is too prone to say that a thing is good or bad, and let that settle it. In the above case, for instance, it is stated that, while some of the apples lacked any distinctive apple flavor whatever, yet others were as good fifteen months after storage as the freshly picked apples, or the crop from a year later. The impossibility of any such result need not be questioned. Apples after storing for a year or more would certainly be very much inferior, regardless of how they were kept, and no apples are stored commercially for this length of time.

PACKAGES SUITABLE FOR COLD STORAGE.

A package suitable as a container of apples for cold storage need not necessarily be such as would be convenient for handling. But, practically, it is essential that the packages used for cold storage be such as can be used successfully as a shipping and handling package. By years of trial and demonstration, two kinds of packages have proved commercially successful both for storing and for handling, viz.: The three-bushel barrel and the somewhat more modern one-bushel box.

Much comment, unfavorable to the barrel, has been indulged in by those who think they know what package is most suitable for apples. The box has been lauded as the best package, and this, doubtless, has been owing to the fact that western growers, especially in the North Pacific territory, have used the box exclusively, and have shipped East large quantities of very finely colored and finely selected apples. The use of boxes will not give the fruit a fine appearance, if it is not inherently of excellent quality. The box is a desirable package for many purposes, but has no great advantage over the barrel as a package for the retailer. Practically it is just as easy to sell a three-bushel barrel as it is one-third the quantity in a bushel box. The average consumer has no place to keep even as small a quantity as one bushel of apples, and he buys in a quantity not exceeding one peck and even by the quart or pound.

The three bushel barrel is a package which will remain with us for many years to come, and it should not be condemned nor blamed for those things for which it is not responsible. The fact that those who pack apples, "face" both ends of the barrel with large and fine appearing fruit, is a standing joke, and the old familiar barrel has probably from this reason as much as any, been condemned. The barrel is by no means responsible for the deceit and dishonesty of fruit packers. The great advantage of the barrel over any rectangular package is that its shape makes it very strong, and it thus protects the apples from damage by bruising while being handled. Another great advantage of the barrel is that it is tight, or reasonably so, to an extent which protects the fruit from a direct contact and circulation of the air. This is a big advantage, not only while in cold storage, but while being handled out of storage. The advantage of protecting apples from air in cold storage is mentioned more fully on another page.

The box has none of these advantages. The apples as packed in the box cause the box to spread, and when being trucked or handled the sides are easily dented or sprung, and thus the fruit is more or less damaged. The spreading of the sides and top and bottom of the box, open cracks at the corners, which allow the air to circulate more or less freely in contact with the fruit. This is, of course, offset to some extent by lining with paper, and in case of fancy fruit, where each individual apple is wrapped, as is discussed under the head "Double Wrapping for Apples." The chief advantage of the box is that it shows up the fruit to better advantage; that the package is more portable or easier handled and that it takes very much less space in handling and storage. The saving in space by using a rectangular package is well understood by cold storage warehousemen, and the bushel box for apples as compared with the three bushel barrel, makes a saving of about 50 to 75 per cent. This, of course, in cold storage where space is valuable, is very important.

As a suggestion, the author ventures to offer the following: It is possible to put up a fine grade of apples in some sort of cartons, holding perhaps one dozen or two dozen apples, depending on size, and having these cartons, preferably of paper or cardboard, of some unit size, so they will fit into a rectangular wooden box; possibly a bushel box might be made to serve in this connection. It is, of course, thoroughly appreciated that this plan would make it difficult to show or examine the fruit to advantage, but doubtless the cartons could be so arranged that the covers could be opened easily. The great advantage of some unit package of this kind is that the original package as packed by the fruit grower or by the original packer, would go to the consumer without rehandling, and the scheme has the further great advantage that the cartons suggested would be easily portable, and would make a nice package which the retailer could deliver to the consumer. Fruit growers who desire to work up a fancy private trade direct to the consumer would find some package of this kind a most valuable factor. Portable packages, and packages which will protect the goods are of great assistance as a selling factor in any line of business. The selling of goods in bulk is being more and more done away with, and doubtless this will apply to apples as well as to other lines of goods.

BOX AND BARREL COMPARED.

Prof. S. W. Fletcher, director of Virginia Agricultural Experiment Station, discusses the merits of the box and barrel as follows*:

1. Quantity of fruit—It is probably true that the box is a more convenient quantity of fruit for the "ultimate consum-

*From paper presented before the American Pomological Society.

er," who has recently received so much attention by tariff makers, than the barrel. Over 30 per cent of our population now live in cities, and the percentage of city dwellers is increasing with each census. A majority of the city and town people, constituting the main market for fruits, have no cool cellar in which fruit can be stored. Their storage facilities are limited to the refrigerator. They wish to buy only such a quantity of fruit as will keep, at the ordinary temperature of the house, while it is being used. Under such conditions the box is a more convenient package than the barrel. A large basket of the Climax type, holding about a peck, would be more convenient still, especially for summer and autumn apples.

On the other hand, there is a large demand for apples in bigger bulk,—not only because of the custom of years, but also for the winter supply of those who have a cool cellar, and for export. Certain varieties carry better across the water in barrels, than in boxes, because the latter packages permit the entrance of salt air.

2. Cost of package—On the Pacific Coast, apple boxes cost from six cents to nine cents, knocked down. As three boxes can be packed out of one barrel, at that price the boxes are cheaper than the barrel. In the East we pay from eleven cents to twenty-one cents per box. In Virginia, boxes cost ten cents to twelve cents; in Minneapolis, Minn., fourteen cents; while Mr. Robert Brodie of Montreal states that his boxes cost twenty-one cents. The price of barrels in the East ranges from thirty cents to forty-five cents, with an average of about thirty-five cents. Bought knocked down in carload lots, they have cost certain growers twenty-eight cents to twenty-nine cents. The inferior quality of some eastern-made boxes, as noted previously, should also be considered. The comparative cost of barrels and boxes is a local problem, and each grower will have to get estimates.

3. Grading and packing—The fundamental difference between the two types of packages is here: The box encourages, and almost enforces, honest and uniform grading, while the barrel permits carelessness in this respect. The cost of packing is also an item. Where a very large quantity of fruit

is packed by specially trained men, it costs little if any more for labor to pack in boxes than in barrels. But the small grower, and especially one who has been accustomed to the barrel pack, will find that it costs from one-third to one-half more to pack in boxes than in barrels. It should be noted, also, that very small, or otherwise inferior fruit seldom if ever yields as high returns in the box pack as in the barrel pack. Only the large sizes go well in boxes. It is a question for each grower to decide, whether he can get more by sorting out his fancy and No. 1 stock for boxing, and selling the smaller fruit in barrels, than to sell all in barrels as No. 1's.

Another point to be considered is the shape of the fruit. It is almost imperative that box fruit should be quite regular in shape. Lop-sided and misshapen fruit, like the York, especially from young trees, would not pack well in boxes.

The most important point under this heading, however, is that no one has ever succeeded with the box pack using common stock. Only fancy and No. 1 fruit of the best quality has paid in boxes. By intensive methods, and especially by thinning the young fruit on the trees, many of the best western growers have been able to produce fruit, ninety-five per cent of which is fancy. Practically all of the Hood River fruit is box fruit. I doubt if, on an average, thirty per cent of the apple crop of Virginia, or Ontario, or any other part of the East, is box or fancy fruit. This point must be kept emphatically in mind when the suggestion is made that the box should become the exclusive apple package of the East, as it is now in the West.

4. Quality of fruit.—Of far less importance than the grade of the fruit in the package, in respect to the question before us, is its quality. It is a fact, however, that the box fruit that has commanded the highest prices is mostly of varieties of high quality,—Winesap, Spitzenberg, Newtown. But other varieties, even some of very indifferent quality, have been sold in the box package to great advantage, showing that the style of package and the grade of fruit, rather than its flavor, are the deciding factors. However, the general experience has been that the better the quality of the fruit, the

more apt it is to pay in the box pack. If varieties of inferior quality pay in the box pack, it is because the style of package and the grading outweigh the deficiency in quality.

Experience with the box package in the East.—Having in mind the essential difference between the box and the barrel trade, it does not seem strange that most of the attempts to use the box in the East have not resulted satisfactorily. It is probably near the truth to say that eight out of every ten trials of the apple box in the East have been unsuccessful. A notable example is an experiment by the Field Pomologist of the U. S. Department of Agriculture, Mr. W. A. Taylor, several years ago. He sent abroad during two seasons eight carloads of carefully graded box Baldwin, York, and Newtown, but with indifferent results as compared with barrels. There are many possible reasons for these failures.

1. Custom.—Custom is hard to change,—and the box package is an innovation in the East. As a rule, eastern buyers and grocers do not look with favor upon the box, partly because the profits in repacking and selling a barrel of indifferently packed apples are apt to be greater than in handling three well packed boxes. If the producer could deal with the consumer, it would be different; there is no doubt but that a majority of the consumers would prefer the box, or a smaller package, if the fruit did not cost much more.

2. The market.—A good deal depends upon what a certain market prefers, in the matter of fruit packages, as well as in fruit varieties. West of the Mississippi there is special necessity for caution in this respect. Some buyers want their fruit in boxes, and others prefer barrels, according to the market they expect to reach. The grower who ships should be equally wise.

3. Poor packing and grading.—More failures arise from this cause than from any other. The art of packing boxes is not acquired in an hour. It is work for specially trained men, not for the average farm help. In this respect it differs materially from barrel packing, which may be quite well done by ordinary help. Moreover, the habits of several generations of men who have packed in barrels, using "facers" and "fillers," have descended to the fruit growers of today; and many

of them find it extremely difficult to keep the smaller, poorly colored, or slightly imperfect specimens from gravitating to the bottom of the box. It will take a generation or two, perhaps to breed out that habit. The western man deserves no credit for being more honest in this respect, for, as has been pointed out, honesty was not merely the best policy for him, but the only policy that would pay freight rates.

General conclusions.—The drift is all towards the smaller package. This is in keeping with the trend or the times with respect to other commodities. There is no doubt but that the box package, or at least the smaller type of package, will some time entirely supplant the barrel. The smaller package will not necessarily be made of wood. We can expect the wooden package to be replaced, eventually, by paper, cellulose, or some other cheap material. Even now some very substantial paper boxes are on the market. When speaking of the box type of package, therefore, we refer to the size and shape of package, rather than to the material.

But while the box type of package is the ideal towards which we are rapidly working, it by no means follows that every Eastern fruit grower should begin packing in boxes at once. He should begin only when he is ready; and nine-tenths of the growers are not ready. To be ready for box packing means that the grower can get good boxes about as cheap as barrels, bushel for bushel; that he is able to grow a crop of fruit, preferably of high quality varieties, at least ninety per cent of which is fancy or No. 1; that he is able to command skillful and experienced packers; that he is able to put a large quantity of box fruit on the market, not one year only, but year after year, so as to win a reputation for the brand; and that he ships his fruit to markets that are already familiar with the box pack and take kindly to it. At the present time not one apple grower out of ten, east of the Mississippi, is able to meet these conditions.

With respect to the market, the fruit grower must recognize the different demands of two entirely different types of markets. One of these, the common or general market, will pay a fair price for good or common stock. The other, the special or fancy market, will pay a fancy price for fancy stock. At the present time the box package supplies the special or fancy market almost exclusively, while the barrel package supplies both, but more especially the common or general market. These two classes of markets will always exist, or as long as some people are more successful in accumulating money than others. It goes without saying that the demand for cheap or common fruit, at a fair price, will continue to be very much greater than the demand for fancy fruit at a high price; because there are many more people who are in moderate circumstances than there are people who are able to pay fancy prices for fruit. The proportion of fruit growers who are able to grow fancy fruit is as small as the proportion of consumers who are able to pay fancy prices. Location, soil, and the varieties best adapted thereto may make it more profitable to grow staple varieties for the common market. This cheap fruit—the main supply of the great middle class of people—will be marketed in barrels to best advantage for many years to come.

The successful marketing of apples in boxes depends so much upon skillful grading and packing and upon the possession of a large quantity of fruit so packed, that it seems likely that very little impetus will be given to box packing in the East except through co-operative shipping associations. Here and there an exceptional grower may find it profitable to pack his fancy grade of certain varieties in boxes; but it does not seem probable that box packing will make much headway in the East except through the co-operative shipping association, with its trained business manager and its crews of trained packers.

These conclusions indicate that the eastern fruit grower should be a conservative on the subject of the box apple package. The drift is towards the smaller package—but, at the present time and for many years to come, apple growers who are so situated that they must produce apples for the general or common markets—which means a majority of the growers—will find the barrel more profitable. With the advent of co-operative shipping associations, the box package will become more and more common in the East, and eventually even for the common grades of fruit.

DOUBLE WRAPPING FOR APPLES—BEHAVIOR OF DIFFERENT VARIETIES.

A striking example of the possibilities of cold storage in the preservation of apples was furnished by the work of the Nebraska State Horticultural Society at the Transmississippi exposition of 1898, reported by V. A. Clark. The fruit was gathered and put in cold storage during the fall of 1897, most of it during the month of October, though some not until December. Each apple was wrapped first in a sheet of waxed paper, using 9 by 12 inch sheets for small apples and 12 by 12 inch sheets for large ones. Then another covering of common newspaper was added. This double wrapping made practically an air-tight cell for each apple, thus preventing any spread of decay. The fruit was then carefully packed in barrels, filling them up so as to require considerable pressure to get the heads in. The temperature of the room in which they were stored did not vary over one degree from 36 degrees from the time they were placed in it until they were removed. A number of varieties were still in good condition Nov. 1 of the following year.

To determine how such double wrapping lengthens the period of keeping a few barrels of unwrapped Ben Davis and Winesap apples were placed in the same storage room at the same time and received exactly the same treatment as the others. Seventy per cent of them were decayed when taken out June 1. Those remaining in firm condition were so badly discolored and had lost flavor to such an extent as to render them wholly unfit for either show or market. A few of the same varieties were also wrapped in newspaper only. Of these about 30 per cent were in very poor condition June 1. The fruit which went into cold storage in 1897 was taken out at intervals during the summer and fall of 1898 and at that time was examined, and each variety received a mark, according to the condition in which it was found.

One of the most interesting parts of the report of these experiments is the account of the behavior of the different varieties in cold storage. Some retained all their good qualities up to the close of the exposition, Nov. 1, 1898. These were

Ben Davis, Winesap, Ralls Genet, Limbertwig, Willow Twig, Gilpin and Lansingburg. Although the Salome lost a little in quality, it kept well in storage and on the table. Fruit taken from storage June 1 retained color and firmness for nearly five weeks. Some retained a good outward appearance, but lost in some other quality, as, for instance, the Iowa Blush, the skin of which became so bitter as to render the fruit unfit for use.

On the other hand, some varieties retained their eating qualities, but lost in outward appearance. Such was the Milam, which kept well, but lost in color. There were also numerous other kinds of deterioration. Minkler lost flavor and began to decay, the English Golden Russet and Fulton shriveled, the Roman Stem became mealy and lost flavor. Sheriff and Walbridge discolored so badly as to render them unfit for show or market and they deteriorated rapidly; Fameuse retained color, but many burst and after a few days became mealy, and the Yellow Bellflower went down suddenly.

Moreover, the behavior of varieties having a certain characteristic in common was not always the same in respect to it. The Missouri Pippin, a dark apple, faded in storage, but the Walbridge and Sheriff, also dark apples, came out almost black. Nor did the lighter colored apples fade more than the dark red ones, for Grimes Golden and Yellow Bellflower, both yellow apples, held their color unchanged, while Missouri Pippin, a dark red apple, as has been said, faded.

Too much reliance must not be placed on the results of storing different varieties. So much depends on the condition of the fruit when stored, the soil on which it is grown, and the local characteristics of the variety. It is reported, for instance, that Fameuse retained color, but that many of them burst and after a few days became mealy. That might be characteristic of Nebraska Fameuse, but certainly not be characteristic of northern New York and New England and Canada Fameuse. The above tests are valuable only as they relate specifically to the apples grown in Nebraska and the cultural conditions prevailing.

Another caution is that the results from storing double wrapped fruit must not be taken too literally. The exact kind of paper used, and the exact character of the barrels used, and the system of refrigeration employed in cold storing have not been given in detail, and all these have an important bearing on the final result. An absolutely air-tight covering for the fruit, or an air-tight barrel or package would be fatal to its keeping qualities, and although doubtless many feel that wrapping fruit in waxed paper encases it in an air-tight envelope, this is not by any means the case. Waxed paper, in fact, is quite porous so far as penetration or flow of gas or air through it is concerned.

PILING BARRELS OF APPLES.

Owing to the comparatively heavy weight of a barrel of apples and its somewhat awkward shape, the proper handling and stowing of same in cold storage rooms is difficult and where the rooms are high it is a very laborious task to get the top tiers of apples into place. Figure 9 shows a method which has been worked out in practice, and which has proved to be entirely satisfactory as well as simple and inexpensive.

The device used consists of a tackle consisting of two single blocks as shown. The rope is fastened to the upper block, then run through the lower block, and then through the upper block, and thence to the hands of the operator. The lower block is attached to what is known as a barrel clamp, consisting of two flat hooks for engaging the chimes of a barrel, which are fastened to a chain, the center of which is fastened to the lower block as shown in the cut. The upper block is fastened to a piece of 2-inch pipe supported in any convenient way over the central piling alley. The pipe may be supported by hooks from the ceiling or by blocks of wood from the ceiling against a post. In any case the piping is removable. The upper tackle block is fastened to the pipe only by a loop of rope, which can be slid along on the pipe to the different piles, and untied completely for removal to another support.

With this device a barrel can be raised by one man, as shown, to a position where it can be rolled back to a location on any tier as the apples are built up; 2 x 4's are laid on the floor, one at each end of the barrel directly underneath the head of the barrel, so that no weight comes on the sides or bilge of the barrel. Other 2 x 4's are placed on each tier of barrels up to the 3d, or 4th, or 5th, or possibly the 6th tier according to height of the room. Above that the barrels are piled without 2 x 4's, each one being placed in the notch or space between the two barrels of the tier below as shown in the illustration. It is not advisable to pile in this way more than four tiers in height.

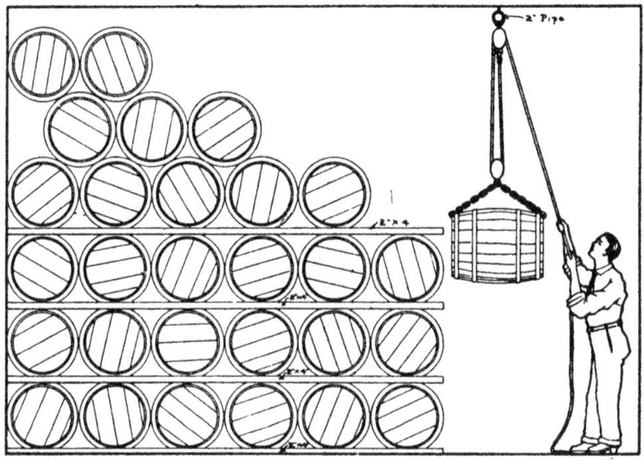

FIG. 9—PILING AND LIFTING APPLES.

While one man can hoist a barrel in this way, where rapid work is wanted, it has been found expedient to have two men on the rope and one on the pile. This apparatus has been found entirely practicable in service, and considering the small expense and ease of application it should work out wherever there are many barrels to pile, either of apples or of any other similar goods.

Apples have been piled in this way in a room 20 feet in height, and we see no reason why they could not be piled in a room higher than this. Of course, there is some danger in handling apples on high piles, as a weak barrel in the bottom may cause trouble. However, by arranging to have the grain of the wood in the heads of the barrels come vertical as much as possible, and by alternating from one side to the other so that the vertical grain would be first on one side of the pile and then on the other, it would seem that with good substantial piling strips, possibly pieces which were 2½ inches or three inches thick and 4 inches or 5 inches wide, that apples could be piled to a height of 30 feet if required.

HANDLING APPLES IN BARRELS AFTER COLD STORING.

The question often comes up as to whether it is detrimental to apples to handle them from one storage room into another after they have been in storage for some months. It is occasionally desirable to do this so as to concentrate the lots remaining on hand into one room or into less space than they would occupy when scattered about the house. It is also sometimes desirable to move apples from the public cold storage into private storage at the end of the storage season.

As a general statement it is damaging to apples to move them after they have been in storage for several months or more. The damage comes from the fact that after being stored, apples become" slack" in the barrels on account of natural shrinkage, and may be badly bruised in handling. Further than this, if the apples are affected with rot to any extent, handling the barrels in "slack" condition, smears the good apples from the decayed ones. In any case, apples after being in storage for four or five months are pretty well matured, and it is damaging to handle them to any extent. If they are of extra good stock and reasonably tight in the barrels, and not affected with rot, they can be moved if carefully handled. Under these conditions, they should not be rolled on the bilge or sides of the barrels, but should be rolled on the chime, or handled carefully on trucks without allowing them to drop or be jarred in any way.

PACKING APPLES.

We are indebted to John A. Ruddick, Cold Storage and Dairy Commissioner of Canada, for the following on packing apples in barrels and boxes:

The barrel in common use in Ontario is made of 30-inch staves, that in use in Nova Scotia of 28-inch staves. The dimensions are: Between heads, 27½ inches; head diameter, 17 inches; middle diameter, 19½ inches. The specifications for a good apple barrel call for a sound stave, 9/16-inch jointing,

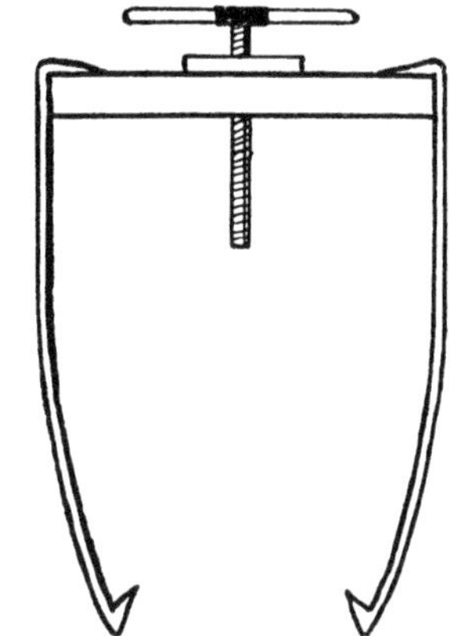

FIG. 10—SCREW PRESS FRAME.

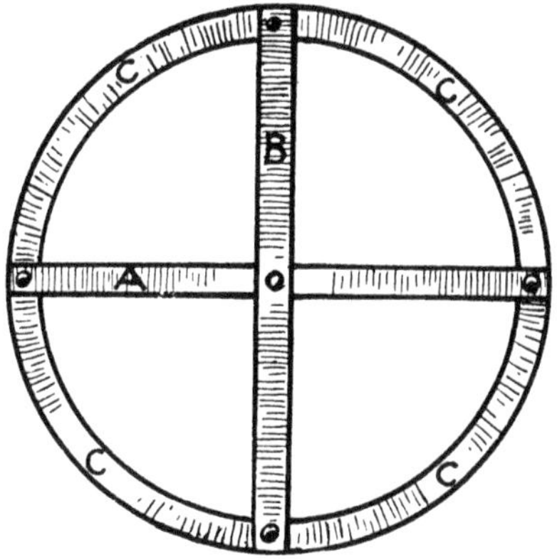

FIG. 11—IRON CIRCLE PRESS HEAD.

cut five to two inches, and averaging four inches in width at the bilge. The head to be not less than ½ inch in thickness, dressed, and to have eight hoops.

When packing the apples in barrels each apple is laid with the stem end down, the stem having been previously cut off with a stemmer. Upon no consideration should a very large or very small apple be used to finish up in the center of the face. If the apples are colored, the second layer should be placed so that the color of the apples will show though between the apples for the first layer. After this second layer is laid the apples may be turned in from the round-bottom baskets in which the graded apples have been placed. Never use any device that will require the apples to fall any distance into their place on the grading table or in the barrel. The presumption is that the grading has been done off the grading table, and that fruit of a perfectly uniform grade is put in each barrel.

Heads cut from heavy paper or light pulp board are very desirable on both ends of the barrel. The patent corrugated heads cannot be recommended. It is doubtful, too, whether there is any advantage in using fancy paper heads.

The pressure will depend somewhat upon the variety. The Spy must be pressed very moderately; the Russets, on the contrary, will stand much heavier pressure. If packed for storage, the pressure need not be as heavy as when packed for export. Slackness in barrels is as often caused by over-pressing as by under-pressing. Over-pressing will break or bruise the skin, inducing decay.

The most efficient and handy form of barrel press is the screw press frame, shown in Fig. 10. To make the pressure equal, an iron circle press head is used, as shown in Fig. 11. The bars A and B are made with an arch and with a shoulder to fit against the iron circle, C. The circle should be fourteen inches in diameter and made of quarter-inch bar iron.

BARRELS VERSUS BOXES.

The question of boxes versus barrels has been discussed in eastern Canada for a number of years. The British Columbia fruit growers use no barrels. A careful analysis of the conditions in eastern Canada would seem to show that neither package possesses all the virtues. The following facts are well established.

1.—The highest priced apples are shipped in boxes.

2.—The box is the only practical package in which an apple can be transported with any reasonable degree of economy in a fit condition for the highest dessert trade.

3.—Only the best grade of apples will pay in boxes.

For the last four or five years a few Canadian shippers have each year experimented with boxes. In only one or two cases have they pronounced it a success. A fairly close inquiry into the conditions under which these experiments were carried on shows that the business was not handled in the best way. Nearly all who experimented with boxes did so with unskilled packers. In many cases the boxes were faced and then the apples were simply rolled in on top of this face, after the manner of barrel packing, and finished in every respect like barrel packing, with no attempt at arranging the apples in tiers. Of course, nothing but failure could be expected from such a style of packing.

The size of the Canadian apple box is 10x11x20 inches, inside measurement. This is obligatory for the export trade. It is recommended that the box should be made with the following specifications: The end pieces not less than 5/8 inch nor more than 3/4 inch thick; the sides not less than 3/8 inch, the top and bottom 1/4 inch thick. These dimensions cannot be changed to any great extent. Dovetailed boxes are not a success with fruit.

Whether the apples should be wrapped or not depends somewhat upon the variety and the grade of fruit. Wrapping has several advantages: 1, it serves as a cushion in the case of delicate fruit; 2, it prevents rot and fungus diseases from spreading from specimen to specimen; 3, it maintains a more even temperature in the fruit, and 4, it has a somewhat more finished appearance when exposed for sale.

Wrapping has also some disadvantages: 1, it adds to the cost of packing, and 2, it prevents rapid cooling in cases where the fruit is not cool at the time of packing.

Lining papers for the boxes are not often used. At the Ontario Horticultural Exhibition for 1906 not more than twenty-five per cent of the boxes shown for prizes were lined. The practice, however, is to be commended. It costs but a trifle and adds greatly to the appearance.

When packing apples in boxes, packing tables are absolutely essential. These should be of two sorts, as it is impossible to get packing and grading done at the same table economically. When the apples are brought to the packing house the first operation is grading them into Fancy, No. 1, No. 2 and Culls, which may be done by help that knows nothing about the practical part of box packing. The grading is best done on tables lined with canvas or burlap.

FIG. 12—PACKING BENCH FOR BOX APPLES.

The basis of rapid box packing is good, even grading. The packer should have before him an even run in point of size, without which it will be impossible for him to do rapid work, or, indeed, do good work. The really skillful packer will take the very slightly smaller apples and use these at the ends of the boxes, the larger always going toward the middle of the box. But this difference in the size of the end and the middle apples is so slight that only the practiced eye of the packer would detect it. The skillful packer will also take advantage of the slight inequalities in shape. If the packer finds that there is a slight slackness in a row of apples which he is packing across the box, he can usually make this perfectly tight by simply turning the specimens one way or the other. Of course, the opposite fault of being somewhat too crowded can be remedied by the same process. It is, perhaps, not equally important to grade to color, yet this adds greatly to the appearance of the finished box. If, then, the packer has the choice, he will put the lighter-colored apples in one box and the highly-colored apples in another. Both boxes may sell equally well, but neither would have sold so well had the apples been mixed in color in each box.

After the packing is completed the cover of the box must be carefully nailed in position. The lining papers are folded neatly at the edge of the top of the box, to allow for the swell, and will then overlap slightly at the center. The staff of the Fruit Division of Canada have been using a bench, illustrated in Fig. 12. This is the style, with some modifications, in general use on the Pacific Coast, and can readily be made by any one handy with tools. The box is set between, and is held firmly by, the clamps shown.

PACKING IN TIERS.

The simplest method of packing boxes is nothing more than the barrel pack practised with boxes. It is needless to say that such a method of packing a box will result in absolute failure. The number of apples in a box can be determined almost instantly by the style of pack, of which there may be a large variety. Some practiced packers claim to distinguish as many as sixty different styles of pack. Familiarity with half a dozen, however, will enable an intelligent person to pack successfully all common varieties. In a general way the size of the apples is indicated by the number of tiers or layers in the box. The box is supposed to be open so that it is eleven inches wide and ten inches deep. If, then, three layers or tiers of apples will fill the box properly, that sized apple is spoken of as a 3-tier apple. In the same way, if five layers or tiers fill the box, the size is said to be 5-tier. The 3-tier apples would be the largest that would be packed, such as the Alexander or overgrown specimens of the King and Spy. These may be so large that only forty-five will go in a box. It is possible to get a 3-tier apple with sixty-three in a box. In the same way, a 4-tier apple usually contains ninety-six specimens, but it may contain as high as 112.

If the apples of one layer are placed in the spaces between the apples of the one below, there would be, say, four

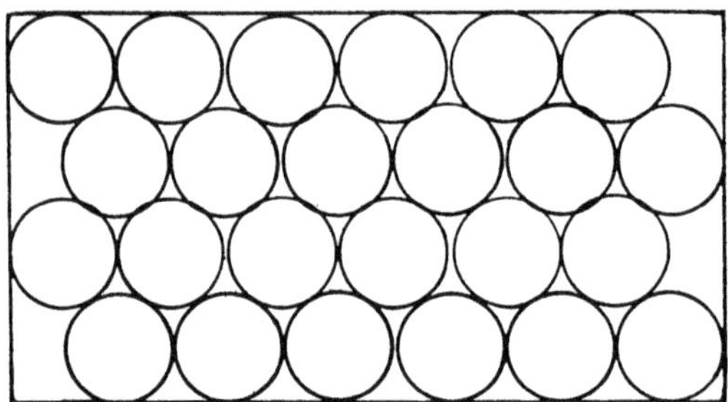

First and Third Layers.

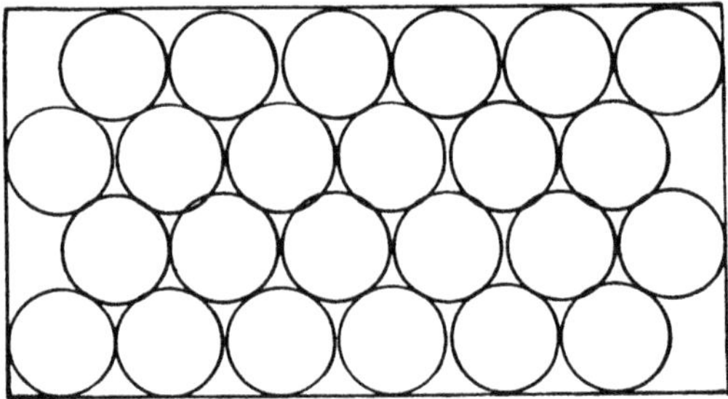

Second and Fourth Layers.

FIG. 13—SHOWING DIAGONAL 2-2 PACK, 3½ TIERS, 4 LAYERS, 96 APPLES.

layers of apples intermediate in size between those that would fill the box in three layers or in four layers if packed directly over each other or straight pack. Such intermediate size would be styled a 3½-tier size. Similarly, the intermediate size between a straight 4-tier and a straight 5-tier would be spoken of as a 4½-tier. From the smallest Fameuse that should be packed, to the largest Kings or Alexanders, there are between thirty-five and forty different sizes, each of which requires a different style of pack. These different styles of packing are really only modifications of two general types. The first is called the "Straight" pack, where every apple but those in the first layer is directly over another. The second is called the "Diagonal" pack, in which no apple is directly over any other which it touches. Usually the apples in the alternate layers are directly over each other, but never in the contiguous layers.

First and Third Layers.

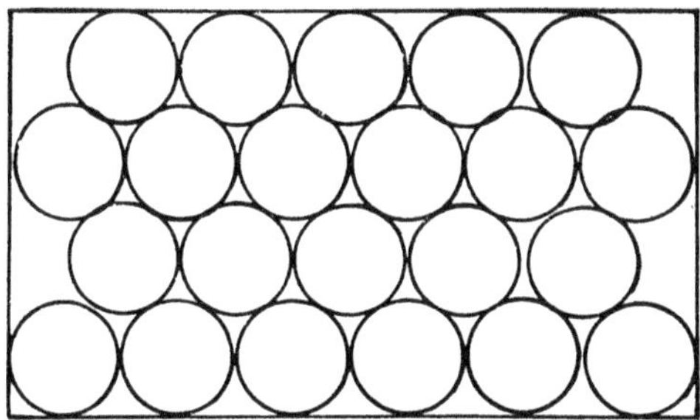

Second and Fourth Layers.

FIG. 14—DIAGONAL 2-2 PACK, 3½ TIERS, 4 LAYERS, 88 APPLES.

Both straight and diagonal packs may be modified in a number of ways. A modification of the diagonal pack in common use is called the "Offset." Place three apples touching each other, but leaving a space about the width of half an apple between one side of the box and the last apple. The next row of three would be placed so as to leave the space on the opposite side. A very useful diagonal pack is made by placing three apples in the first row, one in each corner and one in the middle. The second would then be made with two apples, the third with three, and so on, until the tier is completed. The second layer would be commenced with two apples and alternated with three, as in the first layer. The first and third and fifth layers, and second and fourth would be the same, and directly over each other. By commencing this pack with two apples, instead of three, the box will contain two apples less. With larger apples, the 2-2 pack is used. This is begun by placing an apple in one corner of the box and then dividing the remaining space evenly with another apple. Into these spaces

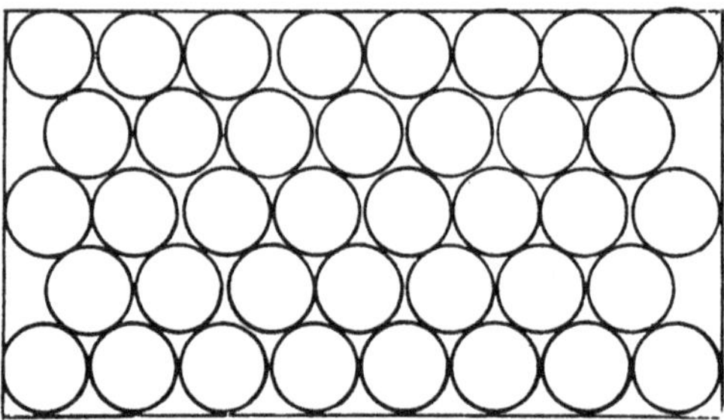

First, Third and Fifth Layers.

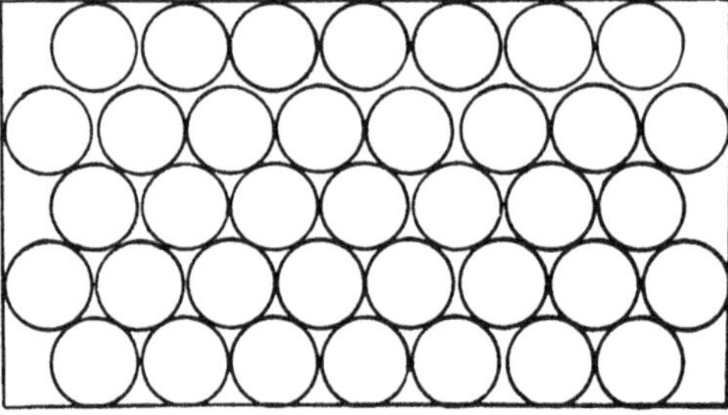

Second and Fourth Layers.

FIG. 15—SHOWING A 3-2 PACK, 4½ TIERS, 5 LAYERS, 188 APPLES. IF LAYERS ARE REVERSED THERE WILL BE 187 APPLES.

are pressed two apples forming the next row. This is continued until the box is filled. Four layers will fill the box, the first being directly over the third, and the second over the fourth.

The art of packing can only be learned by packing. It

requires a deft hand and a trained eye, so that slight differences may be recognized and utilized to fill the box and so tightly packed that the box may be put on end with the lid off and yet no apples fall out.

The accompanying illustrations will give a very good idea of the method of packing apples in boxes and of the appearance when packed.

Fig. 18 shows several different styles of pack. The upper left hand box has five rows, straight pack, for part of the face layer, and four for the remainder. This device is quite unnecessary and seriously mars the look of the pack. The

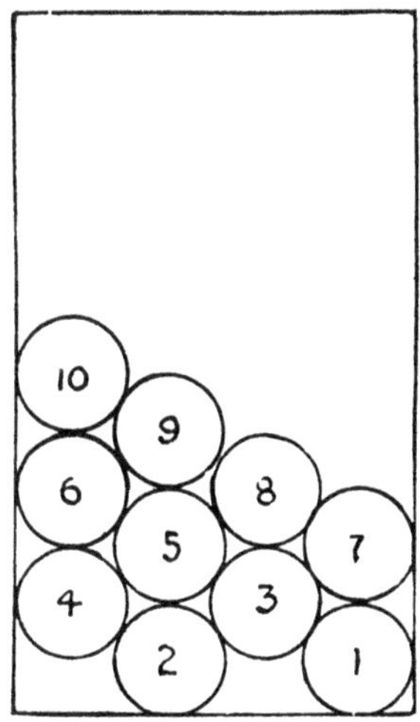

FIG. 16—HOW TO START A 2-2 DIAGONAL PACK.

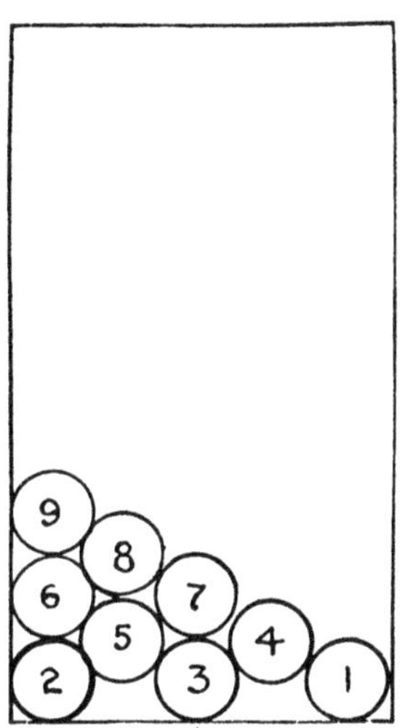

FIG. 17—HOW TO START A 3-2 DIAGONAL PACK.

middle box of the upper row is an "offset" pack, not desirable if any other can be used. When opened on the side the spaces are prominent. The upper right hand box is a 3½-tier, 2-2 pack. The left hand box of the middle row is a slight modification of the same pack for larger apples, there being only eighty-eight instead of ninety-six in the box.

Fig. 19 shows four boxes of Alexanders. The upper right hand box is a straight 3-tier, containing forty-five apples, but shows the defect of having one row smaller than the rest. The right hand lower box is a 3-tier with sixty-three apples. The left hand upper box is a 3-tier with fifty-seven, but defective in grading. The lower left hand box is a very even pack, larger than 3-tier, containing forty-one apples.

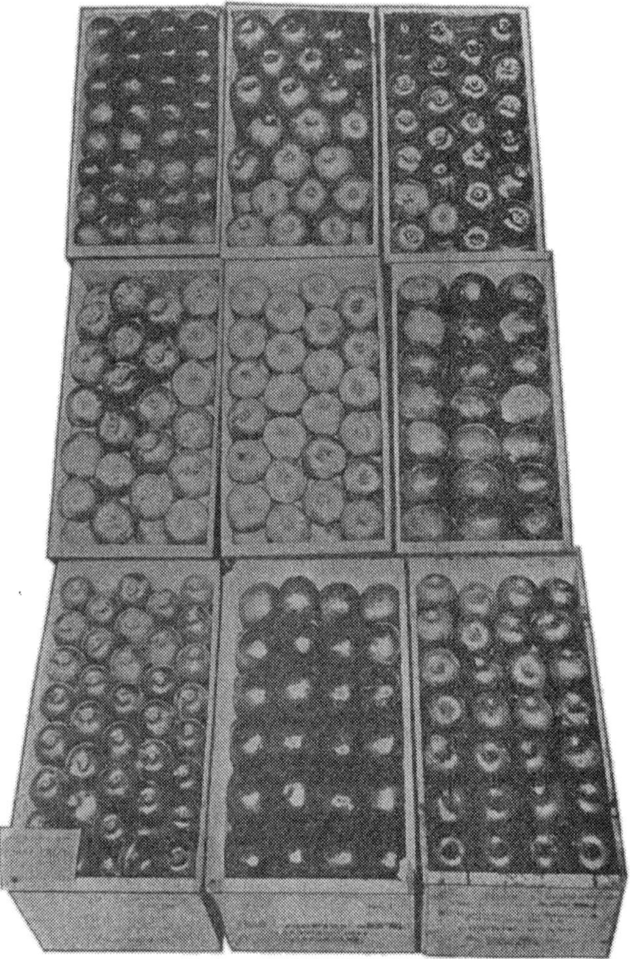

FIG. 18—SHOWING VARIOUS STYLES OF PACKING.

It will be noted that the left hand box has twenty apples in each layer, while the next has eighteen.

Fig. 20 shows how wrapped apples will accommodate themselves somewhat more easily than unwrapped to the different styles of pack, owing to the elasticity of the paper covering.

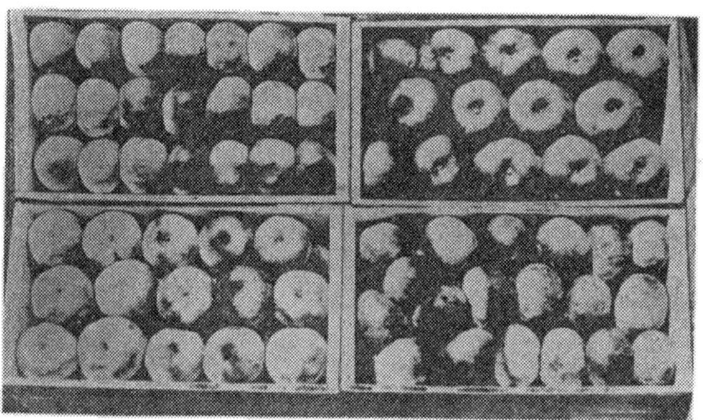

FIG. 19—SHOWING DIFFERENT STYLES OF PACKING ALEXANDERS

FIG. 20—SHOWING WRAPPED APPLES IN BOXES.

THE "SWELL" IN BOX PACKING.

Fig. 21 illustrates a very neat pack for a conical apple, like certain types of Northern Spy and Ben Davis.

Eastern packers have been so long accustomed to the barrel, a rigid package, that it is difficult for them to conceive that the essential difference between box packing and barrel packing lies in the fact that the box is an elastic package. The secret of rapid and good packing is largely in a recognition of the elasticity of the top and bottom, and, to a very slight extent, of the sides of the box. It is understood, of course, that the box remains always the same dimensions, but the apples to be packed are constantly varying in size, and yet the experienced packer has no difficulty in securing an arrangement of the tiers, so that after a certain number of tiers are placed in the box, the box is properly filled without the aid of any extraneous packing material, such as paper shavings, excelsior or pulp pads.

Nevertheless, even the most skillful packer requires for the best packing slight difference in the size and shape of the

FIG. 21—NEAT PACK FOR CONICAL APPLES.

individual apples, differences so slight that they would escape the attention of all but the practiced eye. Small as the differences may be between the specimens of any particular package, this difference in size and shape is very important, and is taken advantage of by the packer to secure the swell in the center of the top and bottom.

This swell serves two purposes. It enables the packer to find a place for the apples slightly larger or smaller than the main run in one or both diameters. There is a careful grading as to size by the eye, so that the smaller specimens are placed at both ends on each tier and the slightly larger one toward the center. This must be done by selecting the proper shaped fruit, because it is not desirable to break the plan of any particular tier; that is, if the packing is begun with the apple stems down, it is desirable that the whole tier should be packed stems down. In that case the flatter apples would be placed near the ends of the box, while the apples that were equal in transverse diameter, but not longer through the axis, would be placed toward the middle. Where this is done consistently, it will be found that when the box is packed ready for covering, the apples at the ends of the box project half an inch or more above the box, while at the middle they would rise about an inch and a half above the sides of the box. This selection and placing of the apples becomes, in the skillful packer, automatic, and he scarcely feels that he is making the selection, so rapidly is it done. Yet if a selection of this sort is not made, there is no possibility of securing a box that will not go slack.

VENTILATION.

There are necessarily conflicting opinions among authorities as to the best way to keep apples, and especially so as to the kind of package, and whether the package should be ventilated or not. Some seem to think that a package with a free circulation of air is necessary, but the majority of experienced fruit packers and fruit storers contend, and their contention is backed by experience, that apples in common with some other goods must be protected from the air of the storage room. It has been demonstrated by long experience that a package like the ordinary apple barrel, which, while being tight, is really of porous wood, and allows just enough penetration of air and escape of gas from the fruit, gives the best results in cold storage. The fruit must not be tightly sealed, or there will be an accumulation of gas and moisture, which will very quickly rot the apples. Should anyone doubt this statement, it would be very easy to make an experiment by putting apples away in a tightly sealed tin package with a soldered top, which will not allow penetration of air nor the escape of gases and moisture from the interior. Place this receptacle in cold storage for two months, and then cut it open and examine the interior. The necessity for ventilation will be at once apparent. The exact quantity of ventilation is, of course, impossible to determine accurately, but as before stated, the ordinary apple barrel thoroughly dry before placing in storage, allows just about the right ventilation. It is common practice among growers who have their own storage facilities at the orchard, to store apples in barrels without heading up the barrels. They are placed on end with the top uncovered. This is permissible only for a few weeks at most and even for that short period the open end should be protected by paper or by the cover laid on loose.

The necessity for ventilation has been thoroughly demonstrated by Professor Fred W. Morse, of the New Hampshire Experiment Station in a bulletin which he is pleased to call "The Respiration of Apples and Its Relation to Their Keeping Quality." We extract below some of the main points demonstrated:

RESPIRATION OF FRUITS.

"The respiration of animals is a well known action and the necessity for it in the living creature is fully appreciated. The fact that plants and parts of plants must also breathe is not so commonly understood. Yet all living cells, whether a part of animal matter or vegetable matter, must have oxygen to keep them alive and they give up carbon dioxide and water as a result of the action of the oxygen on some of their contents. Parts of plants when cut off from the main stem do not die at once, and must continue to breathe. This is true, whether the severed part is a leafy branch, a fruit or a root; but some parts live much longer after removal than others, and the apple continues to breathe for many weeks after it has been picked from the tree.

Respiration, whether in animals or in plants, causes a destruction of matter in the cells much like the destruction of wood in a stove, and the rate at which this destruction goes on can be measured by determining the amount of carbonic acid that is breathed out in a given length of time.

In animals, under usual conditions, the food which they eat makes good the losses produced by respiration. An animal, however, may live without food for some time, during which period it still breathes in oxygen and breathes out carbonic acid and water, but it steadily loses weight and grows thin in flesh because there is a steady destruction of cell material with no food to replace it.

Fruit, after having been picked from the tree, is in the condition of the starving animal. Its cells still keep up respiration with nothing in the way of food to make good the losses produced by the action. Since apples and other fruits have no body heat to maintain, the breathing process is not so active as in animals, and they may last months after being picked from the tree. Yet there is a steady, continuous loss in weight as the weeks go by, although the fruit is sound and firm.

For example, fruit put in cold storage Nov. 13 and weighed at intervals of two months had lost as follows:

January	2,	0.33 per cent.	March	5,	2.34 per cent.
May	6,	3.60 per cent.	July	1,	4.71 per cent.

That the shrinkage in weight is due to respiration and not to simple drying out of the water is shown by the practically constant percentages of water and dry matter, since if the solid material was not destroyed it should gradually increase in proportion while the water would decrease. Results proving this point are here given.

A lot of Baldwin apples were set aside in October and a few of them analyzed at intervals.

October	24,	water, 85.45;	dry matter, 14.55.
October	31,	water, 85.41;	dry matter, 14.59.
November	21,	water, 85.23;	dry matter, 14.77.
November	29,	water, 85.02;	dry matter, 14.98.
December	27,	water, 85.56;	dry matter, 14.44.
April	20,	water, 86.19;	dry matter, 13.81.

Respiration is partly a chemical reaction and in apples, like most chemical reactions in the laboratory, it grows more rapid as the fruit becomes warmer, and is slowed down when the fruit is cooled. If two sets of experiments were carried out as described in a previous paragraph, one set in a refrigerator, and the other in a warm room, it would be easily seen at the end of four or five days that the warm room had caused the larger amount of respiration. Since no exact figures had been obtained showing just how rapidly an apple was changed in composition when stored at an ice cold temperature compared with another apple at 45 degrees and another at summer temperature, it seemed possible to measure the rate by determining the amount of carbonic acid given off by the fruit at different temperatures. The carbonic acid would not show the kind of changes taking place within the cells of the apple; but it would be a measure of the rate at which those changes were progressing since the formation of the carbonic acid must be one of the reactions concerned in them.

It was seen on comparing the average rates of exhalation of carbonic acid at the different temperatures, that in passing from melting ice (32°) to cellar temperatures (45° to 50°) the rate nearly triples, and in passing from the medium temperature to summer temperatures the rate doubles.

Since the breathing out of carbonic acid is an indication of the rate of chemical change within the fruit, it follows that changes of composition must take place from four to six times as fast at summer temperatures as in cold storage and from two to three times as fast in cool cellars as in cold storage.

These increases in rate are in agreement with the laws of chemical action, as the speed of such reaction is found to double and sometimes to triple when the temperature is raised 18 degrees Fahrenheit (10 degrees Centigrade).

There is a practical application of this law to be made to the care of fruit, especially at apple picking time.

It is frequently the case that warm days with temperatures of 70 degrees occur in October and sometimes continue for a considerable period. Fancy apples intended for long keeping in cold storage should be cooled as soon as possible and kept cold. The breathing process is at the expense of cell contents and must weaken the keeping qualities as it goes on. And this destructive action is from four to six times as fast out of cold storage as inside it.

Another fact in connection with the respiration is important. It is not stopped in cold storage, but simply slowed. Apples cannot be kept indefinitely but keep about twice as long in cold storage as in a cool cellar.

AUTHOR'S SUGGESTIONS IN CONNECTION WITH APPLE STORAGE.

In conclusion the author wishes to offer suggestions and an outline of information in connection with picking of fruit at the correct maturity; the proper packing of the fruit in suitable packages; and sundry suggestions with reference to the handling of fruit before placing in storage, while in storage, and when removed therefrom. Hints as to suitable systems of refrigeration and illustrations of a practical nature will also prove of service to those who are interested in operating cold storage houses.

As a general statement, apples for cold storage should be carefully selected or graded. At the same time it is possible to store them for short periods of one to three months without the necessity for careful grading. This is desirable and necessary sometimes to tide over a temporary surplus of fruit at picking time. It is also common practice with some fruit growers, who have their own cold storage, to put the apples in barrels as picked, hurry them into cold storage, and then take their time about grading, and this work is done without removing them from the cold storage room. The great advantage of this method is the ability to pick the apples very rapidly, and thus save time during the picking period, which is an extremely "rushed" time with apple growers. It also enables the growers to pick the apples at the proper stage of maturity, which is most important. As has been stated elsewhere, if apples are placed in barrels without being headed the open ends of the barrels should be protected in some way, if no more than with ordinary building paper. This protects the apples from a drying out or evaporation, which is fully discussed elsewhere in this chapter.

To develop flavor to the greatest extent possible it is necessary that apples be picked on a dry day, and when the sun is shining brightly, and after the dew has disappeared in the morning. Sun on the apples is beneficial while they are hanging on the trees, but after they are picked they should not again be exposed to direct sunshine. The common practice of putting apples in piles in the orchard causes a vast damage each year, which is far greater than is supposed. If apples are picked and placed in cold storage the same day, they will ripen slowly and come out of storage at the proper period in much better condition for use than they were when picked from the trees in the fall. They will be better because they have slowly ripened, and the flavor has been retained and developed to its fullest extent.

Whether it is most desirable to grade in the orchard as many do, or store the apples without grading as above suggested, can only be determined by the individual grower, but if the crop is of considerable magnitude it is suggested that with the cold storage house in close proximity to the orchard, the prompt storage without taking the time to grade and pack carefully, is doubtless the best and most practicable method commercially. Of course, if the apples are graded at the time they are picked, the work is completed once for all, but on the other hand, the apples after remaining for several months in cold storage always show some defective ones, and it is necessary to rehandle and repack in many cases. It would seem, therefore, that quick and rapid work at picking time, and a leisurely and careful grading prior to shipment and while still in cold storage, would be a most desirable method. It is, of course, not suggested that apples should be handled and graded after they have been long in cold storage, but if the grading is done within two or three months from the time the apples are stored, the better keeping or winter varieties can be handled without any damage. The softer or short keeping varieties cannot, of course, be handled in this way to the same advantage, and in fact, the softer cannot be stored for long periods under any circumstances.

CHAPTER XVIII.

PEARS AND PEACHES.*

INFLUENCE OF COLD STORAGE ON THE PEAR INDUSTRY.

Before the advent of the cold storage business the supply of summer pears frequently exceeded the demand. This condition of the markets, which were demoralized in hot, humid seasons, pertained especially to the early varieties, like the Bartlett, which ripen in hot weather and need to be sold in a short time to prevent heavy losses from rapid decay. The introduction of the refrigerator car and of the cold storage warehouse, together with the rapid growth of the canning industry, has done much to improve the pear situation by artificially establishing a well regulated and more uniform supply of fruit throughout a longer period of time. The pear acreage of the country has more than doubled within a decade, and is enlarging the relative importance of cold storage to the pear-growing business, though a large part of the increase especially in California, along the Atlantic coast from New Jersey southward, in Texas, and in the central west, is primarily related to the canning industry.

Pear storage has developed most largely in the east. In New York and Jersey City from 60,000 to 100,000 bushels of summer pears, 30,000 to 60,000 bushels of later varieties, and many cars of California pears are stored annually. In Boston, since 1895 there have been stored each year from 5,000 to 15,000 bushels of early pears, principally Bartlett, and from 7,000 to 20,000 bushels of later varieties, such as Anjou, Bosc, Angouleme (*Duchess*), Seckel and Sheldon. In Buffalo

*Extracts from Bulletin No. 40, Bureau of Plant Industry, United States Department of Agriculture, by G. Harold Powell, Assistant Pomologist in charge of Field Investigations, and S. H. Fulton, Assistant in Pomology.

10,000 bushels are sometimes stored in a single season, and in Philadelphia from 30,000 to 35,000 bushels. While there are no accurate statistics available and the quantity fluctuates from year to year, it is probable that as many as 300,000 bushels are stored in a single year throughout the country at large.

There are many practical difficulties in pear storage. The early-ripening varieties which mature in hot weather, like the Bartlett, often "slump" before they reach the storage house, or are in soft condition, especially if they have been delayed in ordinary freight cars in transit. They may afterwards decay badly in storage, break down quickly on removal, or lose their delicate flavor and aroma. When stored in a large package like the barrel, the fruit, especially of the early varieties, often softens in the center of the package, while the outside layers remain firm and green. Frequently no two shipments from the same orchard act alike, even when stored in adjoining packages in the same room, and the warehouseman and the owner, not always knowing the history of the fruit, are at a loss to understand the difficulty. It has been the aim in the fruit-storage investigations of the Department of Agriculture to determine as far as possible the reasons for some of these storage troubles, and to point out the relation of the results to a more rational storage business.

OUTLINE OF EXPERIMENTS IN PEAR STORAGE.

The investigations in pear storage were of a preliminary nature only. The experiments undertaken have been planned with a view to determining the influence in the storage room of various temperatures, of the character of the storage package, of a fruit wrapper, of the degree of maturity of the fruit when picked, and of other factors in relation to the ripening processes in the storage house, and also to ascertain the behavior of the fruit and its value to the consumer when placed on the market.

The Bartlett and Kieffer pears principally were used in the experiments, but several other kinds were under limited observation. The Bartlett represents the delicate-fleshed, tender pears, ripening in hot weather, which are withdrawn from storage before the weather becomes cool. The Kieffer, on the other hand, is a coarse, hard pear, ripening later in the fall in cooler weather, and in which the normal ripening processes are slower. It is a longer keeper, and like other fall varieties is withdrawn in cool weather.

The Bartlett experiments extended through the season of 1902. The fruit was grown by Mr. F. L. Bradley, Barker, N. Y., in a twelve-year-old orchard on a sandy loam, with a clay subsoil. The orchard is a half mile from Lake Ontario and is 50 feet above the level of the lake. The fruit, which was full grown, but green, was picked early in September, and was packed in tight and ventilated barrels, in 40-pound closed boxes, and in slat bushel crates. Part of the fruit in each lot was wrapped in unprinted news paper, and an equal amount was left unwrapped. Part was forwarded at once by trolley line to the warehouse of the Buffalo Cold Storage Company at Buffalo, N. Y., and a similar quantity was held four days before being stored. The fruit reached the storage house within ten hours after leaving the orchard.*

The Kieffer experiments have extended over two years. In 1901 the fruit was grown by Mr. M. B. Waite, Woodwardsville, Md., in a Norfolk sandy soil, on rapidly growing five-year-old trees, from which the fruit was large, coarse, and of poor quality. It was stored in the cold storage department of the Center Market at Washington, D. C. In 1902 the fruit with which the experiments were made was grown by Mr. S. H. Derby, Woodside, Del., on heavy-bearing ten-year-old trees on sandy soil with a clay subsoil. The fruit was smaller, of finer texture, and of somewhat better quality than that used the previous year. It was stored in the cold storage department of the Reading Terminal Market in Philadelphia, Pa.

The Kieffers were picked at three degrees of maturity: First, when two thirds grown, or before the fruit is usually picked; second, ten days later, or about the time that Kieffers are commonly picked, and third, ten days later, when the fruit was fully grown and still green, but showing a yellowish tinge around the calyx. In each picking, part of the fruit was

---

*Bartlett pears should be picked when the seeds begin to turn brown. This is a sure sign of proper maturity and may be relied upon.—Author.

shipped to storage and was placed in rooms with a temperature of 36° and 32° F. within forty-eight hours. Equal quantities stored in each temperature were wrapped in parchment paper, in unprinted news paper, and were left unwrapped. A duplicate lot of fruit remained in a common storage house ten days in open boxes, when it was packed in a similar manner and sent to storage. This fruit colored considerably during the interval, but was still hard and apparently in good physical condition on entering the storage house. The pears were stored in 40-pound closed boxes and in five-eighths bushel peach baskets. One hundred and fifty bushels were used in the experiments.

INFLUENCE OF THE DEGREE OF MATURITY ON KEEPING QUALITY.

The experiments with the Kieffer pear show that under conditions similar to those in Delaware and eastern Maryland this variety may safely be picked from the same orchard during a period of at least three weeks, or when from two-thirds grown to full size, and that the fruit in all cases may be stored successfully until the holidays, or much longer if there is still a demand for it. It is absolutely essential that the fruit be handled with the greatest care, that it be sent at once to storage after picking, that it be packed carefully to prevent bruising (preferably in small packages, like a bushel box), and that it be stored in a temperature not above 32° F. if it is desired to hold it for any length of time. If stored by the middle of October, the fruit, by the latter part of December, will take on a rich, yellow color when kept in a temperature of 32° F., and earlier if a higher temperature is used. The fruit may be withdrawn during the holidays, and will stand up, *i. e.,* continue in good condition, for ten days or longer if the weather is cool, and will retain its normal quality if the rooms have been properly managed. While the later picked fully grown pears keep well, they are already inferior in quality at the picking time, as the flesh around the center is filled with woody cells, making it of less value either for eating in a fresh state or for culinary purposes. These coarse cells in the Kieffer and some other late varieties do not develop in the early picked fruit to so large an extent. Pears of all kinds need to be picked before they reach maturity and to be ripened in a cool temperature if the best texture and flavor are to be developed. It is a matter of practical judgment to determine the proper picking season, but for cold storage or other purposes the stem should at least cleave easily from the tree before the fruit is ready to pick. Many trees bear fruit differing widely in the degree of maturity at the same time, and in such cases uniformity in the crop can be attained only when the orchard is picked several times, the properly mature specimens being selected in each successive picking. This practice not only secures more uniformity in ripeness, but the fruit is more even and the average size is larger than when all the pears are picked at the same time.

INFLUENCE OF DELAYED STORAGE ON KEEPING QUALITY.

Pears ripen much more rapidly after they are picked than they do in a similar temperature while hanging on the tree. The rapidity of ripening varies with the character of the variety, the maturity of the fruit when picked, the temperature in which it is placed, and the conditions under which it has been grown. If the fruit is left in the orchard in warm weather in piles or in packages, if it is delayed in hot cars or on a railroad siding in transit, or if it is put in packages which retain the heat for a long time, it continues to ripen and is considerably nearer the end of its life history when it reaches the storage house than would otherwise be the case. The influence of delay in reaching the storage house will therefore vary with the season, with the variety, and with the conditions surrounding the fruit at this time. A delay of a few days with a quick-ripening Bartlett in sultry August weather might cause the fruit to soften or even decay before it reached the storage house, though a similar delay in clear, cooler weather would be less hurtful. A delay of a like period in storing the slower ripening Kieffer would be less injurious in cool October weather, though the Kieffer pear, especially from young trees, can sometimes be ruined commercially by not storing it at once after picking.

FIG. 1—KIEFFER PEARS IN MARCH—REDUCED ONE-FIFTH.

From the experiments with the Bartlett and the Kieffer pears, from which these general introductory remarks are deduced, it was found that the Bartlett, if properly packed, kept in prime condition in cold storage for six weeks, provided it was stored within forty-eight hours after picking in a temperature of 32° F.; but that if the fruit did not reach the storage room until four days after it was picked there was a loss of 20 to 30 per cent from softening and decay under exactly similar storage conditions.*

The Kieffers stored within forty-eight hours in a temperature of 32° F. have kept in perfect condition until late winter, although there is little commercial demand for them after the Holidays. The fruit grown by Mr. Waite on young trees in 1901, which was still hard and greenish-yellow when stored ten days after picking, began to discolor and soften at the core in a few days after entering the storage room, though the outside of the pears appeared perfectly normal. After forty to fifty days the flesh was nearly all discolored and softened, and the skin had turned brown. The fruit from the older trees on the Derby farm in 1902, which was smaller and finer in texture, appeared to ripen as much as the Waite pears during the ten days' delay. This fruit, however, did not discolor at the core and decay from the inside outward, but continued to ripen and soften in the storage house and was injured at least 50 per cent in its commercial value by the delay. Fig. 1 shows the condition of the Kieffer pears stored in a temperature of 32° F. as soon as picked and withdrawn in March. Under these conditions the fruit kept well until late in the spring. Fig. 2 shows the condition of fruit picked at the same time and stored in the same temperature ten days after picking, when withdrawn in January. The delay in storage caused the fruit to decay from the core outward.

Fig. 3 shows the influence of immediate and delayed storage on Maryland Kieffer pears. The fruit in the box at the right represents the average condition of pears picked October

---

*Bartlett pears have been successfully stored, at the orchard where grown, in a Cooper brine system plant for periods of from six to ten weeks and even longer, and have been held for the Christmas trade.—Author.

FIG. 2—KIEFFER PEARS IN JANUARY—REDUCED ONE-FIFTH.

21, stored October 22, and withdrawn March 3. Storage temperature 32° F. The fruit was wrapped in parchment paper. It was in prime commercial condition when withdrawn from storage. The fruit in the box at the left represents the average condition of pears picked from the same trees at the same time. It was stored in the same temperature ten days later and withdrawn March 3. All of the fruit had decayed.

The results of the experiments point out clearly the injury that may occur by delaying the storage of the fruit after it is picked, and emphasize the importance of a quick transfer from the orchard to the storage house. If cars are not available for transportation and the fruit can not be kept in a cool place, it is safer on the trees so far as its ultimate keeping is concerned. It is advisable to forward to storage the delicate quick-ripening varieties, like the Bartlett, in refrigerator cars. The common closed freight car in warm weather soon becomes a sweat box and ripens the fruit with unusual rapidity. The results show clearly that the storage house may be responsible in no way for the entire deterioration or even for a large part of the deterioration that may take place while the fruit is in storage, and that the different behavior of two lots from the same orchard may often be due to the conditions that exist during the period that elapsed between the time of picking and of storage.

INFLUENCE OF DIFFERENT TEMPERATURES ON KEEPING QUALITY.

There is no uniformity in practice in the temperatures in which pears are stored. Formerly a temperature of 36° to 40° F. was considered most desirable, as a lower temperature was supposed to discolor the flesh and to injure the quality of the fruit. The pears were also believed to deteriorate much more rapidly when removed to a warmer air. In recent years a number of storage houses have carried the fruit at the standard apple temperatures, i. e., from 30° to 32° F. Large quantities of Bartlett, Angouleme, and Kieffer pears have been stored in 32° and 36° F. in the experiments of the Department. The fruit of all varieties has kept longer in the lower temperature and the flesh has retained its commercial qualities longer after removal from the storage house. Bartlett pears were in prime commercial condition four to five weeks longer, Angouleme two months longer, and Kieffer three months longer in a temperature of 32° F. Figs. 1 and 5 show the condition of Kieffer pears in March, 1902, in 32° and 36° F., the two lots having received similar treatment in all respects except in storage temperatures. The fruit held at 36° F. did not keep well after December 1.

Fig. 4 also shows the influence of 36° and 32° F. storage

FIG. 3.—WRAPPED KIEFFER PEARS, REMOVED FROM STORAGE (32° F.) MARCH 3.

temperature on the keeping of Kieffer pears. The fruit in both packages was picked October 21, and stored October 22. The package at the left represents the average condition of the fruit when withdrawn March 3 from a temperature of 36° F. All of the pears were soft and discolored, and some of them decayed. The fruit in the package at the right, kept in a temperature of 32° F., with bright yellow, firm, and in prime commercial condition.

In the higher temperature the fruit ripens more rapidly, which may be an advantage when it is desirable to color the fruit before it leaves storage; but the fruit in that condition is nearer the end of its life history and breaks down more quickly on removal to a warm atmosphere.

There is a much wider variation in the behavior of pears that have been delayed in storage or that are overripe when they enter the storage room at 32° and 36° F. than in pears stored at once in these temperatures. In the higher temperature the fruit that has been improperly handled ripens and deteriorates more quickly. The lower temperature not only keeps the fruit longer when it is stored at once, but it is even more essential in preventing rapid deterioration in fruit that has been improperly handled.

FIG. 4—WRAPPED KIEFFER PEARS, REMOVED FROM STORAGE (36° AND 32° F.) MARCH 3.

INFLUENCE OF THE TYPE OF PACKAGE ON THE KEEPING QUALITY OF THE PEAR.

Pears are commercially stored in closed barrels, in ventilated barrels, in tight boxes holding a bushel or less, and in various kinds of ventilated crates. The character of the package exerts an important influence on the ripening of the fruit and on its behavior in other respects, both before it enters the storage house and after it is stored, though this fact is not generally recognized by fruit handlers or by warehousemen. The influence of the package on the ripening processes appears to be related primarily to the ease with which the heat is radiated from its contents. The greater the bulk of fruit within a package and the more the air of the storage room is excluded from it the longer the heat is retained. Quick-ripening fruits, like the Bartlett pear, that enter the storage room in a hot condition in large, closed packages, may continue to ripen considerably before the fruit cools down, and the ripening will be most pronounced in the center of the package, where the heat is retained longest. The influence of the package, therefore, will be most marked at the time during which the fruit is exposed to the hottest weather and on those fruits that ripen most quickly.

In the experiments of the Department of Agriculture the Bartlett pears were stored in tight and in ventilated barrels, in closed 40-pound boxes, and in slat bushel crates. After three weeks in the storage house the fruit that was stored in barrels soon after picking in a temperature of 32° F. was yellow in the center of the package, while the outside layers were firm and green. Fig. 6 shows the average condition of the fruit in these two positions one week after storing. The upper specimen shows the condition of the fruit in center of a barrel. In this position the fruit cools more slowly than that near the staves or ends and it therefore ripens considerably before the temperature is reduced. The lower specimen shows the condition of the pears at top and bottom and next to the staves of the same barrel. In these positions the fruit cools quickly and the ripening processes are retarded. For quick ripening fruits that are handled in hot weather small packages are preferable. After five weeks in storage the fruit in the center of the barrel was soft and of no commercial value, while the outside layers were still in good condition. The difference was still greater in a temperature of 36° F., and was more marked in both temperatures in fruit that was delayed in reaching the storage house.

FIG. 5—KIEFFER PEARS IN MARCH—REDUCED ONE-FIFTH.

In both the closed 40-pound boxes and the slat crates the fruit was even greener in average condition than the outside layers in the barrels, and it was uniformly firm throughout the entire package.

There was apparently no difference between the fruit in the commercial ventilated pear barrel and the common tight pear barrel.

With the Kieffer, which enters the storage room in a cooler condition and which ripens more slowly, a comparison has been made (in 1902) between the closed 40-pound box and the barrel, and while the difference has been less marked the fruit has kept distinctly better in the smaller package. The fruit in barrels was the property of Mr. M. B. Waite, and was under observation by the Department through his courtesy.

There is a wide difference of opinion concerning the value of ventilated in comparison with tight packages for storage purposes. No dogmatic statements can be made that will not be subject to many exceptions. The chief advantage of a ventilated package for storage appears to lie in the greater rapidity with which the fruit cools, and the quickness with which this result is attained depends upon the temperature of the fruit, its bulk, the temperature of the room, and the openness of the package. The open-slat bushel crate, often used for storing Bartlett pears, with which rapid cooling is of fundamental importance, may be of much less value in storing later fruits that are cooler and which ripen more slowly, and it may be of even less importance in Bartletts in cool seasons.

The ordinary ventilated pear barrel does not appear to have sufficient ventilation to cool the large bulk of fruit quickly.

The open package has several disadvantages. If the fruit is to remain in storage for any length of time its exposure to the air will be followed by wilting, which, in fruits held until late winter or spring, may cause serious commercial injury. The ventilated package, especially if made of slats, needs to be handled with the utmost care to prevent the discoloration of the fruit due to bruising where it comes in contact with the edges of the slats.

FIG. 6—BARTLETT PEARS AFTER ONE WEEK IN STORAGE—REDUCED ONE-FIFTH.

There was little difference in the behavior of the Bartletts in the closed 40-pound boxes and the slat crates at the end of five weeks, and it would appear that a package of this size, even though closed, radiates the heat with sufficient rapidity to quickly check the ripening. Therefore the grower who uses the 40-pound or the bushel pear box for commercial purposes can store the fruit safely in this package, but if the barrel is used as the selling package, and the weather is hot, it is a better plan to store the fruit in smaller packages, from which it may be repacked in barrels at the end of the storage season. While this practice is followed in several storage houses, it is

FIG. 7—KIEFFER PEARS FROM COLD STORAGE ON JANUARY 20, UNWRAPPED.

not to be encouraged, as the rehandling of the fruit is a disadvantage. Rather the use of the pear box should be encouraged as a more desirable package, both for storage and for commercial purposes.

The fruit package question, as it relates to the storage house, may be summed up by stating that fruits like the Bartlett pear and others that ripen quickly and in hot weather may be expected to give best results when stored in small packages. If the storage season does not extend beyond early winter, an open package may be of additional value, though not necessary if the package is small. But fruits like the winter apples and late pears, which ripen in the fall in cool weather

FIG. 8—KIEFFER PEARS FROM COLD STORAGE ON JANUARY 20.

and remain in storage for a long period, should be stored in closed packages to prevent wilting. In such cases the disadvantages of a large package, like a barrel, are not likely to be serious.

INFLUENCE OF A WRAPPER ON KEEPING QUALITY.

The life of a fruit in cold storage is prolonged by the use of a fruit wrapper, and the advantage of the wrapper is more marked as the season progresses. In Figs. 7 and 8 are shown the average quantity of sound specimens of Kieffer pears in unprinted news paper and in parchment wrappers in comparison with the quantity of commercial unwrapped pears in boxes in January, the fruit having been picked October 21 and placed in storage on the following day in a temperature of 32° F. Nearly 50 per cent of the unwrapped fruit (see Fig. 7) had decayed at that time, while that in unprinted newspaper and in parchment wrappers (see Fig. 8) kept in perfect condition. Early in the season the influence of the wrapper is not so important, but if the fruit is to be stored until late spring the wrapper keeps the fruit firmer and brighter. It prevents the spread of fungus spores from one fruit to another and thereby reduces the amount of decay. It checks the accumulation of mold on the stem and calyx in long-term storage fruits, and in light colored fruits it prevents bruising and the discoloration that usually follows.

Careful comparisons were made of the efficiency of tissue, parchment, unprinted news paper, and waxed papers, and but little practical difference was observed, except that a large amount of mold had developed on the parchment wrappers in a temperature of 36° F. A double wrapper proved more efficient for long keeping than a single one, and a satisfactory combination consists of an absorbent, unprinted news paper next to the fruit, with a more impervious paraffin wrapper outside.

The chief advantage of the wrapper for the Bartlett pear, which is usually stored for a short time only, lies in the mechanical protection to the fruit rather than in its efficiency in prolonging its season. Its use for this purpose is advisable if the fruit is of superior grade and designed for a first-class trade. For the late varieties the wrapper presents the same advantages, and has an additional value in increasing the commercial life of the fruit. It is especially efficient, if the package is not tight, in lessening the wilting.

INFLUENCE OF COLD STORAGE ON THE FLAVOR AND AROMA OF THE FRUIT.

There is a general impression that cold storage injures the delicate aroma and characteristic flavors of fruits. In this publication the most general statements only can be made concerning it, as the subject is of a most complicated nature, not well understood, and involving a consideration of the biological and chemical processes within the fruit and of their relation to the changes in or to the development of the aromatic oils, ethers, acids, or other products which give the fruit its individuality of flavor.

It is not true that all cold storage fruits are poor in quality. On the contrary, if the storage house is properly managed the most delicate aromas and flavors of many fruits are developed and retained for a long time. The quality of the late fall and winter apples ripened in the cold storage house is equal to that of the same varieties ripened out of storage, and the late pears usually surpass in quality the same varieties ripened in common storage.

The summer fruits, like the peach, the Bartlett pear, and the early apples, lose their quality very easily, and in an improperly managed storage house may have their flavors wholly destroyed. Even in a room in which the air is kept pure the flavor of the peach seems to be lost after two weeks or more, while the fruit it still firm, much as the violet and some other flowers exhale most of their aromatic properties before they begin to wilt.

It is probable that much of the loss in quality may be attributed to overmaturity, brought about by holding the fruit in storage beyond its maximum time; but it should be remembered that the same change takes place in fruits that are not ripened in cold storage, the aroma and fine flavor often disappearing before the fruit begins to deteriorate materially in texture or appearance.

On the other hand, it is certain that the quality of stored fruits may be injuriously affected by improper handling or by the faulty management of the storage rooms. Respiration goes on rapidly when the fruit is warm. If placed in an improperly ventilated storage room, in which odors are arising from other products stored in the same compartment or in the same cycle of refrigeration, the warm fruit may absorb these gases and become tainted by them, while the same fruit, if cool when it enters the storage room, will breathe much less actively, and there will be less danger of injury to the quality, even though the air is not perfectly sweet. The atmosphere of the rooms, in which citrus fruits or vegetables of various kinds—such as cabbage, onions, and celery—are stored, is often charged with the odors arising from these products, if the ventilation is not thorough. In small houses, in which a single room cannot be used for each product, fruits are often stored together during the summer months, and at this period the storage air is in greater danger of vitiation, since it is more difficult to provide proper ventilation.

The summer fruits, therefore, being generally hot when placed in the storage room, are in condition to absorb the odors which are likely to affect the rooms during the warm season, and as the biological and chemical processes are normally more active in the case of such fruit than in fruits maturing later, the flavors deteriorate more quickly, even in well-ventilated rooms. The fruits that are picked in cool weather and enter the storage rooms in a cooler and less active condition are not in the same danger of contamination.

From the practical standpoint it may be pointed out that summer fruits should be stored in rooms in which the air is sweet and pure. They should not be stored with products which exhale strong aromas, and the danger of contamination is lessened if the fruit can be cooled down in a pure room before it is placed with other products in the permanent compartment provided for it. For the same reason the winter fruits should be stored in rooms in which the air is kept pure, and preferably in compartments assigned to a single fruit.

The experiment furnished no evidence that the quality deteriorates more rapidly as the temperature is lowered. On the contrary, all of the experience so far indicates that the delicate flavors of the pear, apple, and peach are retained longer in a temperature that approaches the freezing point than in any higher temperature.

BEHAVIOR OF THE FRUIT WHEN REMOVED FROM STORAGE.

There is a general impression that cold storage fruit deteriorates quickly after removal from the warehouse. This opinion is based on the experience of the fruit handler and the consumer, and in many cases is well founded, but this rule is not applicable to all fruits in all seasons. The rapidity of deterioration depends principally on the nature of the fruit, on its degree of maturity when it leaves the warehouse, and on the temperature into which it is taken. A Bartlett pear, which normally ripens quickly, will ripen and break down in a few days after removal. If ripe or overmature when removed, it will decay much more quickly, and in either condition its deterioration will be hastened if the weather is unusually hot and humid. In the practical management of this variety it is fundamentally important that it be taken from storage while it is still firm and that it be kept as cool as possible after withdrawal. It is probably true that all fruits from storage that are handled in hot weather will deteriorate quickly, but it appears to be equally true that similar fruits that have not been in storage break down with nearly the same rapidity, if they are equally ripe. The late pears, which ripen more slowly, if withdrawn in cool weather will remain firm for weeks when held in a cool room after withdrawal. If overripe they break down much sooner, and a hot room hastens decay in either case. The same principles hold equally true with apples. The winter varieties, if firm, may be taken to a cool room and will remain in good condition for weeks and often for months and will at the same time retain their most delicate and palatable qualities, but in the spring, when the fruit is more mature and the weather warmer, they naturally break down very much more rapidly.

In commercial practice fruits of all kinds are often left in the storage house until they are overripe. The dealer holds the fruit for a rise in price, but sometimes removes it, not because the price is satisfactory, but because a longer storage would result in serious deterioration. If considerable of the fruit is decayed when withdrawn, the evidence is conclusive that it has been stored too long. Fruit in this condition normally decays in a short time, but the root of the trouble lies not in the storage treatment, but rather in not having offered it for sale while it was still firm. In the purchase of cold storage fruit, if the consumer will exercise good judgment in the selection of sound stock that is neither fully mature nor overripe, he will have little cause to complain of its rapid deterioration.

SUMMARY.

A cold storage warehouse is expected to furnish a uniform temperature in all parts of the storage compartments throughout the season, and to be managed in other respects so that an unusual loss in the quality, color, or texture of the fruit may not reasonably be attributed to improper handling or neglect.

An unusual loss in storage fruit may be caused by improper maturity, by delaying the storage after picking, by storing in an improper temperature, or by the use of an unsuitable package. The keeping quality is influenced by the various conditions in which the fruit is grown.

Pears should be picked before they are mature, either for storage or for other purposes. The fruit should attain nearly full size, and the stem should cleave easily from the tree when picked.

The fruit should be stored at the earliest possible time after picking. A delay in storage may cause the fruit to ripen or to decay in the storage house. The effect of the delay is most serious in hot weather and with varieties that ripen quickly. (See Figs. 1, 2 and 3.)

The fruit should be stored in a temperature of about 32° F., unless the dealer desires to ripen the fruit slowly in storage, when a temperature of 36° or 40° F., or even higher, may be advisable. The fruit keeps longest and retains its color and flavor better in the low temperature. It also stands up longer when removed.* (See Figs. 2, 4 and 5.)

The fruit should be stored in a package from which the heat will be quickly radiated. This is especially necessary in

*30°F. is even better for pears if the fruit is in prime condition when stored, and it is desired to hold it to the extreme limit of its life.—Author.

hot weather and with quick-ripening varieties like the Bartlett pear. For the late pears that are harvested and stored in cool weather it is not so important. Bartletts may ripen in the center of a barrel before the fruit is cooled down. A box holding not more than 50 pounds is a desirable storage package, and it is not necessary to have it ventilated. The chief value of a ventilated package lies in the rapidity with which the contents are cooled, but long exposure to the air of the storage room causes the fruit to wilt. (See Fig. 6.)

Ventilation is essential for large packages, especially if the fruit is hot when stored and ripens quickly.

A wrapper prolongs the life of the fruit. It protects it from bruising, lessens the wilting and decay, and keeps it bright in color. A double wrapper is more efficient than a single one, and a good combination consists of absorptive unprinted news paper next to the fruit, with a more impervious paraffin wrapper outside. (See Figs. 7 and 8.)

The quality of a pear normally deteriorates as it passes maturity, whether the fruit is in storage or not, or it is never fully developed if the fruit is ripened on the tree. The quality of the quick-ripening summer varieties deteriorates more rapidly than that of the later kinds. Much of the loss in quality in the storage of pears may be attributed to their overripeness. The quality is also injured by impure air in the storage rooms, and the warm summer pears will absorb more of the odors than the late winter varieties. The fruit will absorb less if cool when it enters the storage room. The air of the storage room should be kept sweet by proper ventilation.*

The rapidity with which the fruit breaks down after removal depends on the nature of the variety, the degree of maturity when withdrawn, and the temperature into which it is taken. Summer varieties break down normally more quickly than later kinds. The more mature the fruit when withdrawn the quicker deterioration begins, and a high temperature hastens deterioration. If taken from the storage house in a firm condition to a cool temperature, the fruit will stand up as long

*See chapter on Ventilation for suitable means of supplying fresh air and forcing out the accumulating gases arising from pears when in cold storage.

as other pears in a similar degree of maturity that have not been in storage.

It pays to store the best grades of fruit only. Fruit that is imperfect or bruised, or that has been handled badly in any respect, does not keep well.

INFLUENCE OF COLD STORAGE ON THE PEACH INDUSTRY.

Cold storage has not materially influenced the development of the American peach business, and it is not likely to do so to any extent in the future. In the early days of peach growing the industry was localized in sections like the Chesapeake peninsula, New Jersey, and Michigan. The use of the fruit in considerable quantities was then limited to a few nearby markets and to a short time in July, August and September. Now peach growing is rapidly extending to all parts of the country where the climatic conditions and the facilities for transportation are favorable. The refrigerator-car service has brought the peach belts and the distant markets close together, and whenever the crop is general the New York or the Chicago trade may be supplied almost continuously from May till late October with fruit from Florida, Texas, Georgia, the Chesapeake peninsula, New Jersey, the Ozark mountain region, Michigan, New England, California, West Virginia, western Maryland, and other peach-growing sections.

The chief value of cold storage to the peach industry will probably lie in the temporary storage of the fruit during an overstocked market, when, however, there is a reasonable prospect of a better market within two or three weeks. It might be useful also in filling the gaps between the crops of different regions, especially when there are local failures which prevent a continuous supply. It is not now profitable to store the fruit for any length of time, nor under any circumstances unless the conditions of the fruit and the storage conditions are most favorable. The life processes in the peach and the weather conditions in which it is handled make it even more critical as a storage product than the delicate Bartlett pear. In normal ripening it passes from maturity to decay in a few hours in hot, humid weather. The aroma and flavor are most delicate in character and are easily injured or lost, and the influence of any mismanagement of the fruit in the orchard, in transit, or in the storage house is quickly detected by the consumer.

PRACTICAL DIFFICULTIES IN PEACH STORAGE.

Under the most favorable conditions known at present, peach storage is a hazardous business. Before the fruit is taken from the storage house the flesh often turns brown in color, while the skin remains bright and normal. If the flesh is

FIG. 9.—PEACHES IN COLD STORAGE ROOM.

natural in color and texture it frequently discolors within a day or two after removal. There is a rapid deterioration in the quality of stored peaches when the fruit is held for any length of time, the delicate aroma and flavor giving way to an insipid or even bitter taste. Sometimes the flesh dries out, or under other conditions it may become "pasty." Dealers in storage peaches frequently sell them in a bright, firm condition, and shortly afterwards the purchasers complain of the dark and worthless quality of the flesh. It has often been noticed that fruit in the various packages in the same room does not keep equally well, some of it ripening and even softening while the fruit in other packages is still firm. In fact, the difficulties are so numerous that few houses attempt to store the fruit.

It has been the aim in the cold storage investigations of the Department of Agriculture to determine, as far as possible, the cause of the peach-storage troubles and to indicate the conditions under which the business may be more successfully developed.

OUTLINE OF EXPERIMENTS IN PEACH STORAGE.

The investigations were conducted in the cold-storage department of the Reading Terminal Market in Philadelphia, Pa., with Elberta peaches from the Hale Orchard Company, Fort Valley, Ga., and in the warehouse of the Hartford Cold Storage Company, Hartford, Conn., with Elberta and several other varieties grown by J. H. Hale at South Glastonbury, Conn.

In Georgia the fruit was packed in the Georgia peach carriers, left unwrapped, and divided into two lots, one representing fruit that was nearly full grown, well colored, and hard; the other, highly colored fruit, closely approaching but not yet mellow. Three duplicate shipments were forwarded at different times in the two bottom layers of refrigerator cars, and in each shipment part of the fruit was placed in the car within three or four hours after it was picked, and an equal quantity delayed in a packing shed for ten to fifteen hours during the day before it was loaded. Equal quantities of each series were stored in temperatures of 32°, 36°, and 40° F. The transfer from the refrigerator car to the storage house was made by wagon at night, the interval between the car and storage varying from two to five hours.

In Connecticut the fruit represented two degrees of maturity, similar to the Georgia shipments, except that the most mature fruit was mellow when stored. This fruit was grown at an elevation of 450 feet on trees six years old. It was medium in size, firm, highly colored, and of excellent shipping quality. Equal quantities were wrapped in California fruit paper and left unwrapped, and packed in the Connecticut half-bushel basket, in Georgia carriers, and in flat, 20-pound boxes, holding two layers of fruit. The peaches were forwarded by trolley to the storage house, which was reached in two hours after the fruit left the packing shed. Duplicate lots of all the series were stored in temperatures of 32°, 36°, and 40° F.

GENERAL STATEMENT OF RESULTS.

The general outcome of the experiments, both with the Georgia and the Connecticut fruit, is similar and may be summed up as follows:

The fruit that was highly colored and firm when it entered the storage house kept in prime commercial condition for two to three weeks in a temperature of 32° F. The quality was retained and the fruit stood up two or three days after removal from the storage house, the length of its durability depending on the condition of the weather when it was removed. After three weeks in storage the quality of the fruit deteriorated, though the peaches continued firm and bright in appearance for a month and retained the normal color of the flesh two or three days after removal. If the fruit was mellow when it entered the storage house it deteriorated more quickly, both while in storage and after withdrawal. If unripe it shriveled considerably.

In a temperature of 40° F. the ripening processes progressed rapidly, and the flesh began to turn brown in color after a week or ten days in storage. The fruit also deteriorated much more quickly after removal, as it was already nearer the end of its life history. It began to lose in quality at the end of a week.

In a temperature of 36° F. the fruit ripened more rapidly than in 32°, and more slowly than in 40° F. It reached its profitable commercial limit in ten days to two weeks, when the quality began to deteriorate, and after this period the flesh began to discolor.

Fig. 10 shows average condition of Georgia Elberta peaches two weeks in storage after forty-eight hours withdrawal to a warm room. The upper specimen represents the average condition of fruit stored in a temperature of 36° F. The lower specimen represents the average condition of the fruit stored in

32° F. The lower temperature gave better results in every respect.

The fruit kept well in all of the packages in a temperature of 32° F. for about two weeks, after which that in the open bas-

FIG. 10—ELBERTA PEACHES, STORED FOR TWO WEEKS AT 36° F. AND 32° F.

kets and in the Georgia carriers began to show wilting. In the 20-pound boxes, in which the circulation of air is restricted, the fruit remained firm throughout the storage season.

It is necessary that the fruit be packed firmly to prevent bruising, in transit, but if the peaches pressed against each other unduly it was found that the compressed parts of the flesh discolored after a week in storage. A wrapper proved a great protection against this trouble, especially in the baskets of the Georgia peach carrier, and in all of the packages the wrapped fruit retained its firmness and brightness for a longer time than that left without wrappers.

The fruit should be removed from storage while it is still firm and bright. The peach normally deteriorates quickly after it reaches maturity, and the rapidity of deterioration is influenced by the nature of the variety, by the degree of ripeness when removed, and by the temperature into which it is taken. A quick ripening sort, like Champion, is more active biologically and chemically than the Elberta variety, and the warmer the temperature in which either is placed the sooner decomposition is accomplished. It is advisable, therefore, to remove the fruit while firm and keep it in the coolest possible temperature.

The peaches in the top of a refrigerator car that has been several days in transit in hot weather are sometimes overripe and need to be sold as soon as the market is reached, while at the same time the fruit in the bottom layers may still be firm. The rapidity with which the fruit cools down in the car depends on the care with which the car is iced, and on the temperature at which the fruit enters the car. Fruit that is loaded in the middle of a hot day and that has been picked in a heated condition may be 20 or more degrees warmer than fruit picked and loaded in the cool of the morning. Such warm fruit ripens much more rapidly, consumes more ice in cooling down, and takes longer to reach a low temperature. When the temperature in the top of the car is higher than that of the lower part the ripening of the upper layers of fruit will be hastened. If the fruit is destined for cold storage, these upper layers, if more mature, should be piled separately, and sold as soon as their condition warrants it. Under these conditions, if the fruit from this position is mixed in with the rest of the load it may begin to deteriorate before the remainder of the fruit shows mellowing.

The general principles outlined in former pages for the handling of the Bartlett pear apply to the storage of the peach, except that the latter fruit is more delicate and the ripening processes are even more rapid. Every condition, therefore, surrounding the peach in the orchard, in transit, in the storage house, and at withdrawal must be most favorable. The fruit must be well-grown and well-colored but firm when picked. The packing must be done with care to prevent bruising. If the fruit is to be transported in refrigerator cars, it should be loaded soon after picking, and preferably before it loses the cool night temperature. The peaches should be transferred from the cars to the storage house, or from the orchard to the storage house if the latter is near the orchard, in the quickest possible time. The air of the storage room should be kept sweet and pure. The fruit should always be removed to the coolest possible temperature, usually at the end of two weeks, while it is still firm, and it should be placed in the consumer's hands at once.

If the fruit is overripe when picked, or becomes mellow from unfavorable handling before it enters the storage house, it is already in a critical condition and may be expected to deteriorate quickly.

If the conditions outlined are observed in the handling of the peach, it is possible to store it temporarily with favorable results.

CHAPTER XIX.

COLD STORAGE FOR FRUIT GROWERS.*

ADVANTAGES OF LOCAL COLD STORAGE.

The experiments conducted by the United States Department of Agriculture (described in other chapters) to determine the best methods of handling and storing fruits have resulted in securing information of much value. Information before well known to the author and others connected with the industry has been verified by the experiments and put in the form of plain statements of facts. It has been fully demonstrated that better results are secured by the placing of fruit in storage promptly when picked, and that fruit, especially apples, should remain on the trees until well colored and fairly ripened before picking for storage. These facts argue strongly in favor of the fruit grower operating his own cold storage. Prof. G. Harold Powell, who had the experiments in charge, says:

> The local warehouse is ideal for quick storage and for the grower who is competent to handle his own crop. Capital has developed the warehouse business in the large cities, as it is more convenient to distribute the fruit from them and more economical to maintain a plant where a general storage business can be operated. *But as the importance of quick storage at harvest time is more generally appreciated,* it will probably lead to a greater development and concentration of local storage houses and to a greater use and improvement of the refrigerator car service. * * * I believe that one of the developments that will take place in the future is the building of warehouses in the apple producing regions, and the distribution of the product from these warehouses in cooler weather.

The part in italics is used by the author to emphasize the point under consideration, viz.: That best results and greatest profits to the grower can only be secured by placing the fruit in cold storage as soon as removed from the trees. This does not necessarily mean that the grower must have a cold storage

---

*Extracted from a series of articles written for Green's Fruit Grower by the author.

house on his premises, although in many cases this is the best and most practical plan; but the cold storage house should be easily accessible in order to secure the best results. Many fruit growers are at present so situated that their fruit is packed in barrels and shipped by refrigerator car to the nearest storage point, requiring only two or three days in transit. Even this short time causes deterioration of some of the softer varieties of fruit, as the warm fruit going to the car cooled with ice only will not in all probability become cooled below 45° or 50° F. With a local cold storage the fruit requiring quick work may be cooled down rapidly to a temperature of 30° F., thus improving its keeping qualities, and shipped out later in the season when outside temperatures are lower. Many times refrigerator cars are not available and the damage is then much greater.

As an instance of one of the benefits to be derived from home cold storage may be cited the barrel situation during some years. It is often impossible to obtain barrels in sufficient quantity to take care of the crop at harvest time, and it is reasonable to say that many thousands of dollars have been lost to the grower from this reason on account of deterioration of quality of fruit while lying in the orchards waiting for barrels. In many cases total losses have occurred. Apples may be successfully stored without barrels; and boxes and crates are regularly used for this purpose. They may also be stored in bulk, but this is not as good. A grower provided with suitable cold storage facilities does not have to wait for barrels.

Apples to stand shipment long distances before placing in storage must be picked while still somewhat immature. The bothersome apple scald is increased by too early picking, as it has been shown by the experiments and by practice that mature, well colored fruit does not scald to any extent. On this score Professor Powell states: "The experiments indicate that so far as maturity is concerned, the ideal keeping apple is one that is fully grown, highly colored, but still hard and firm when picked. Apples that are to be stored in a local cold storage house to be distributed to the markets in cooler weather may be picked much later than fruits requiring ten days or more in transit. * * * Therefore, to sum up in a general way, the results of the experiments which have been made seem to indicate that the ideal fruit for storage purposes is that which is taken from the tree to the warehouse in the quickest possible time, in order to prevent the fruit from consuming a large proportion of its own life history during the delay that may take place."

COMMERCIAL ASPECT OF THE PROBLEM.

These are some of the benefits of home or local cold storage. Many instances could be cited where large profits have been made by placing fruit in cold storage for a time and selling when the market was comparatively bare, but these seasons are exceptions, and in going into a cold storage proposition, the grower should not expect more than a reasonable profit, amounting to interest and a fair remuneration for the risk assumed. One season with another, a good profit is certain if the business is as well handled as it should be, and none but a careful person of methodical habits will succeed in the operation of a cold storage plant. In the future the grower with modern cold storage facilities will have the advantage over his less progressive neighbor from the fact that his losses will be less and he will be able to place in the hands of the consumer a better preserved and more attractive grade of fruit.

The question may arise as to the probable result of the erection of a much larger number of cold storage houses than are now in use throughout the section of the country where commercial orcharding is largely practiced, and also the probable result of the great increase of acreage of fruit bearing trees. The application of cold storage is still in its infancy. It cannot be said that its use so far has been in any way detrimental to the development of the industry, on the contrary, it has been a great benefit, as fruit growers well know. If the development of cold storage has been beneficial in the past, why should not further development be beneficial? It may be true that the profits will not be as great in the future with more storage houses in use, but the profits will be more certain and regular. The old cry of overproduction has been raised in connection with fruit growing and storing, but with the country only half populated, growing fast, and with developing tastes and rapid improvements in transportation, overproduction is impossible. If there has at times been a temporary overproduction in the past, it has not been due to a surplus, but to lack of facilities in distributing and transportation. Commercial orcharding is rapidly expanding and cold storage will be necessary as an auxiliary. There can be no disastrous glut of the market when cold storage will absorb the surplus at harvest time and distribute it as needed by refrigerated transportation to the markets of the world. With developing civilization and a better understanding of the beneficial results of a fruit diet, stimulated by a rapid increase in the price of meats, the use of fruits as food will surely multiply many times. It is doubtful if the present enormous planting of fruit trees will cause any overproduction. Nearly every one can remember when the cold storage of apples was almost unknown—they were stored in basements, cellars or "fruit houses" without refrigeration. Probably a few are still doing this, but it is safe to say that not more than 20 or 25 per cent of the fruit is so stored for temporary purposes, and storage of this 20 or 25 per cent would save money in improving the quality of the fruit by employing artificial refrigeration. Owing to the considerable investment necessary it is improbable that the construction of cold storage plants will ever be on a scale large enough to cause an oversupply of cold storage space, and the time will shortly arrive when practically all perishable goods will be handled in and sold from cold storage. Those who first provide themselves with cold storage will be the ones to be benefited most largely thereby.

DESIGNS FOR SMALL COLD STORES.

The absolute necessity of cold storage at or near the orchard in order to secure the most perfect results seems unquestionable. What then should a modern cold storage plant consist of? The answer depends largely upon climatic conditions and extent and character of the crop to be handled. We will here consider only the needs of the comparatively small grower who will store say from 200 to 2,000 barrels. The use of natural ice for cold storing of fruit dates back thirty years or more. As has been previously pointed out, and as generally understood among the trade, the natural ice systems with which we are all more or less familiar have not been generally successful for the purpose. The chief objections to these methods were found to be lack of control as to temperature and too much moisture in the air of the rooms. The lowest dependable temperature during warm weather was about 38° to 40° F., oftentimes higher. The moisture in the air was excessive at times, especially during cold weather when the temperature was lowest in the storage room.

FIG. 1—VIEW OF HOUSE FOR FRUIT GROWERS, PLAN No. 1

At the present time a temperature of 30° F. is considered best for apple storage, and any apparatus which cannot produce this temperature cannot be considered for practical purposes as a modern system. Humidity also should be under control. It is for this reason that the ice systems have gone into disuse, and the ammonia or mechanical systems are understood to be the best. The advantages of simplicity and low operating cost when using ice for cooling, combined with the positive control of temperature and moisture obtainable with the ammonia or mechanical system, are all embodied in the Cooper brine system, described in another chapter. This system has none of the disadvantages of complicated machinery, requiring skilled labor, as is necessary with the mechanical or chemical systems.

The buildings here illustrated are planned to meet the needs of those who have a crop large enough to make storing profitable. It is not recommended that a cold storage plant of less capacity than 200 to 300 barrels be built, except under spe-

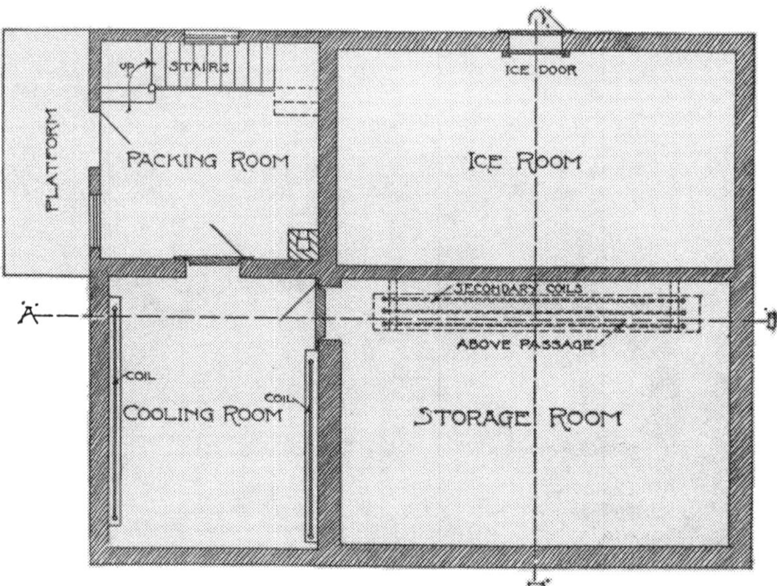

FIG. 2—FLOOR PLAN, HOUSE FOR FRUIT GROWERS, PLAN No. 1.

cial local conditions which might warrant a smaller capacity. The cost of constructing a very small house is greater in proportion as will be seen by the subjoined estimates. The cost of operating is also greater in proportion and the time and care necessary to make a success of a very small plant will operate a much larger one equally well. The relative cost of a plant of 600 barrels capacity and one of 1,500 barrels capacity are here figured with some degree of accuracy for average conditions. The operating cost would be in about the same proportion. The cost of building and operating a house of say 300 barrels would be more than half as much as the house here described for 600 barrels. It will be apparent that an extremely small house is not profitable under average conditions.

Plan No. 1, which is illustrated by perspective, plan and sectional views (see Figs. 1, 2, 3 and 4), is suitable for a capacity of from 200 to 1,000 barrels of apples or other fruit, without change in arrangement of rooms and general plan of building. The cold storage space consists of a large room

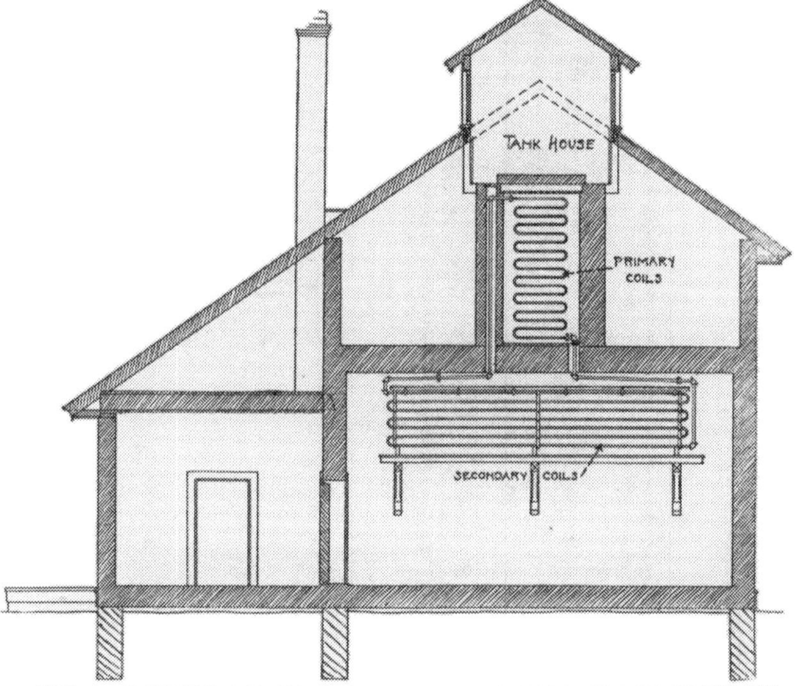

FIG. 3—SECTION A-B OF PLAN No. 1, HOUSE FOR FRUIT GROWERS.

12 feet in height, which may easily be maintained at a temperature of 30° F. during the warmest midsummer weather, and a smaller room, or cooling room, shown in Fig. 2, 8 feet in height, which is used for bringing down the temperature of the fruit partly before placing in the large storage room. Access to the storage room is only had through the cooling room, preventing at all times the inflow of warm air. This cooling room is most

useful during comparatively warm weather, for instance, while storing the summer or winter varieties of fruit, or for cooling and storing Bartlett pears or similar fruit which require quick cooling. By placing the fruit over night in the cooling room a considerable part of the heat may be removed and then, when removed to the storage room, no marked change of temperature in the large room will take place. The cooling room has pipe

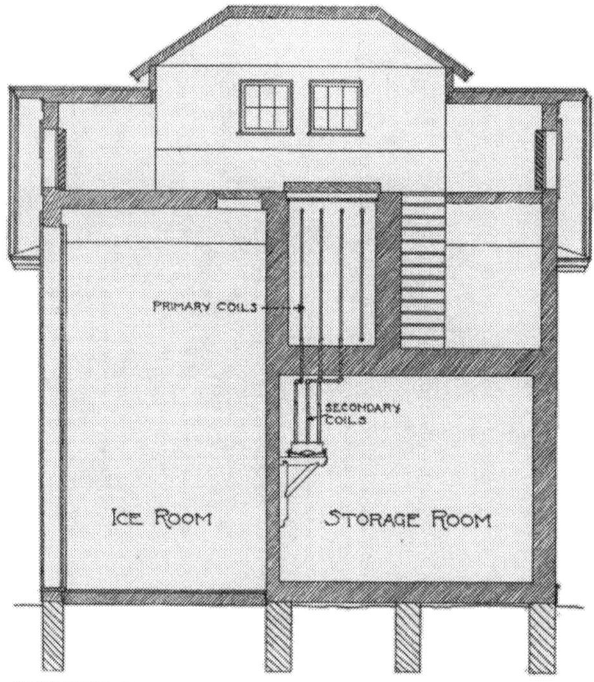

FIG. 4—SECTION C-D OF PLAN No. 1, HOUSE FOR FRUIT GROWERS.

coils of sufficient capacity to carry a uniform temperature of 30° F. during the cold weather of fall and winter and this room may be used for permanent storage of the late winter varieties which are not placed in storage as a rule until cold weather in the fall. The cooling room is entered from a packing or receiving room, as it is generally called. The packing room may be made larger if desired, or it may be omitted if cold storage is to be built adjacent to a fruit packing shed already in use. The packing room is provided with a chimney, so that a fire may be built in extremely cold weather if necessary to prevent low temperature in the storage room and cooling room, or when it is desired to work in packing room in winter. From the packing room, stairs lead up to lofts above. These lofts are useful for the storage of empty packages, etc. The ice room adjoins both the packing and storage rooms. There are no openings from

FIG. 5—VIEW OF HOUSE FOR FRUIT GROWERS, PLAN No. 2.

the ice room to any part of the building except to tank house for the purpose of raising ice to tanks.

Plan No. 2 (illustrated in Figs. 5, 6, 7 and 8) is in most respects like plan No. 1, but is adapted to larger houses. Plan No. 2 may be readily built ranging in capacity from 1,000 to 2,000 barrels. The estimate is based on a capacity of 1,500 barrels. The ice room is placed at one end of the house in this case and the storage room between the ice room on one side and packing and cooling rooms on the other. The storage, cooling and packing rooms bear the same relation to each other and are of the same height and similarly equipped as in plan No. 1.

It should be understood that both these plans include about

as much space in the packing room and lofts as is contained in the storage rooms equipped with the cooling apparatus. In case it is desired to dispense with this storage space for empty packages, etc., as would be the case when the cold storage was built against a barn or fruit house already existing, a consider-

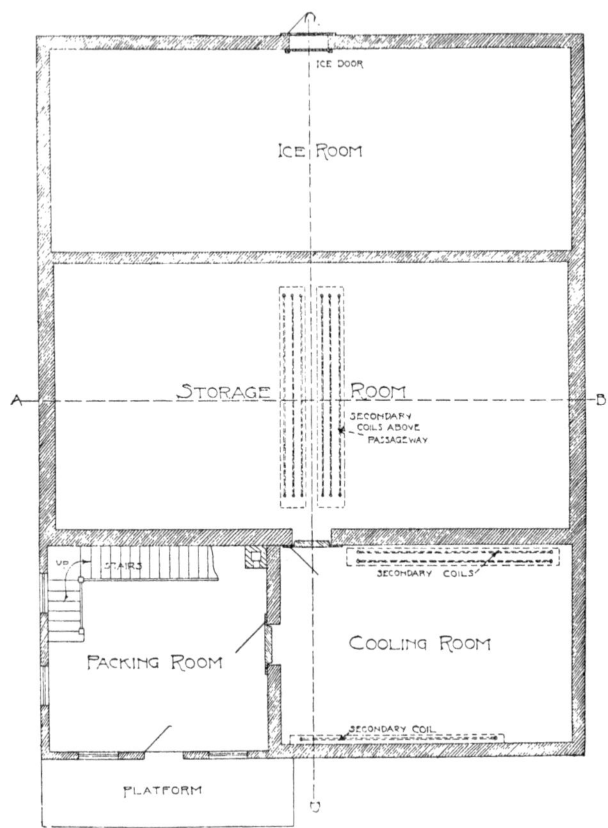

FIG. 6—FLOOR PLAN, HOUSE FOR FRUIT GROWERS, PLAN No. 2.

able saving could be had by some slight changes in plans. If a sidehill location is available, a two-story building is more economical to build, and cheaper to refrigerate, and the access by team to the two different floor levels is a great convenience and saving in the handling of goods for packing and storage. Such a plant which has been built and operated is illustrated further on in this chapter. Old buildings may be remodeled in most cases to good advantage and a handsome saving thereby effected.

The estimates here given are for good, though plain construction, and cold storage houses built in this way will do good services for many years. The estimated cost of constructing and insulating a cold storage house of 600 barrels capacity on plan No. 1 is $1,365. The cost of refrigerating equipment, consisting of piping, galvanized iron work, etc., $650, making a total

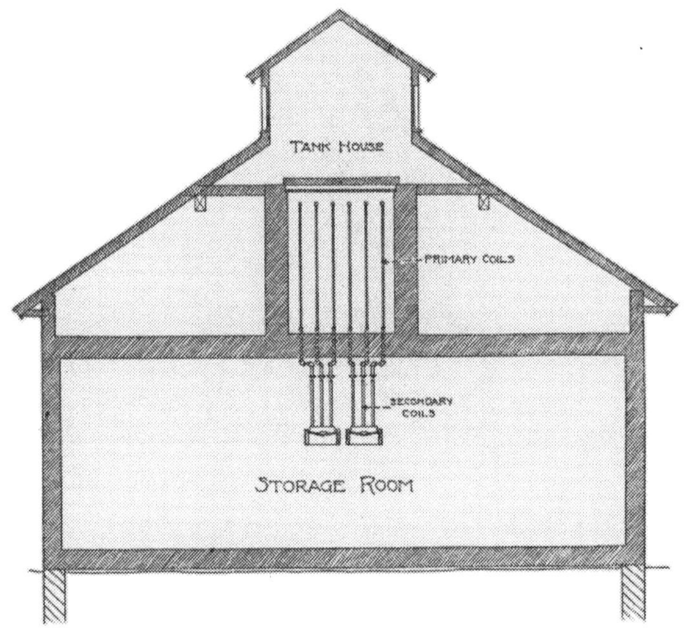

FIG. 7—SECTION A-B OF PLAN No. 2, HOUSE FOR FRUIT GROWERS.

of $2,015. Plan No. 2 is estimated at $2,545 for building and $1,075 for equipment, total $3,620. These figures are based on average costs and conditions, and will of course vary somewhat in different sections. Country locations are usually much cheaper to build in than cities, but this is not always true.

The ice room in this style cold storage house is merely a storage place for ice, and there are no openings from the ice into the storage part of the building. The ice room is to be filled in winter, and will hold sufficient ice not only for the operation of the cold storage plant for an entire season, but for any ordinary farm or family uses as well. No packing material of any sort is used on or around the ice. The floor, sides and ceiling of ice room are well insulated with mill shavings or some similar material. This saves considerable unpleasant labor in taking

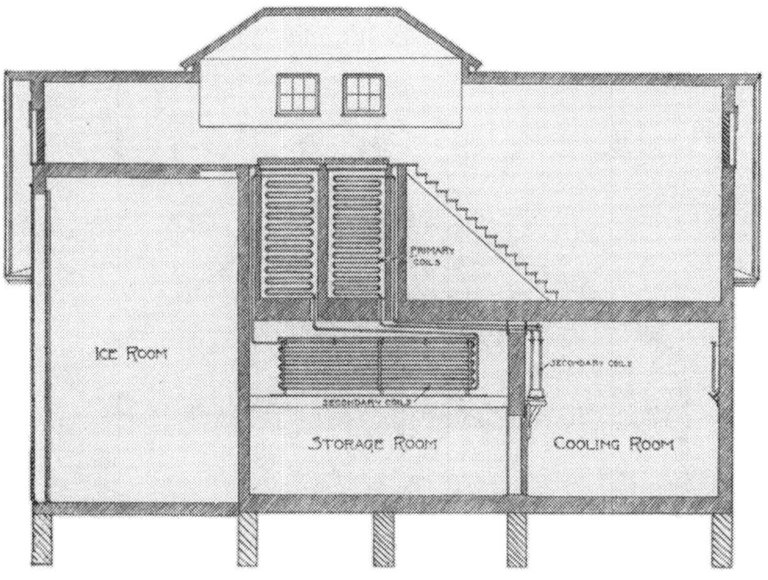

FIG. 8.—SECTION C-D OF PLAN 2, HOUSE FOR FRUIT GROWERS.

out ice, and the ice will keep as well or better than it will in the old style way of covering with sawdust or other material. The ice is also clean and ready for use when taken out. Ice is filled into the ice room through an ice door extending from floor to ceiling, consisting of inner and outer sections which are filled between with shavings or other material after filling the room with ice. Ice for use in the primary tank of the Cooper brine system is first broken or pulverized in the ice room and then raised by a rope through a trap door to the tank house. The operation of this system which cools the room is based on well known natural laws that heat expands and cold contracts. (For description see chapter on "Refrigeration from Ice.")

A MODEL SMALL FRUIT STORE.

The plant of George Smith, South River, N. J., illustrates a desirable type of sidehill construction. The view shown in Fig. 9 is what might be called the front of the building. The photograph was taken while snow partly covered the ground, and before the plant was entirely completed ready for business.

FIG. 9—GEORGE SMITH'S FRUIT STORAGE HOUSE, SOUTH RIVER, N. J.

The door and platform shown give access to the receiving room on the second floor of the building. (See plan and sections for the arrangement of rooms.) The convenience of handling fruit in on the second floor is made possible by the fact that the plant is built on a side hill, and one end of the building is excavated partially into the hill. This arrangement may also be seen in the view showing the rear of the building with ice pond.

The building is 65x40 feet with 20 feet locust posts set on a concrete wall 14 inches in width and three feet in depth. On the front of the building this concrete wall is carried up to the top of the second floor, forming a retaining wall for the earth

FIG. 10—RIVER VIEW—ICE POND IN FOREGROUND.

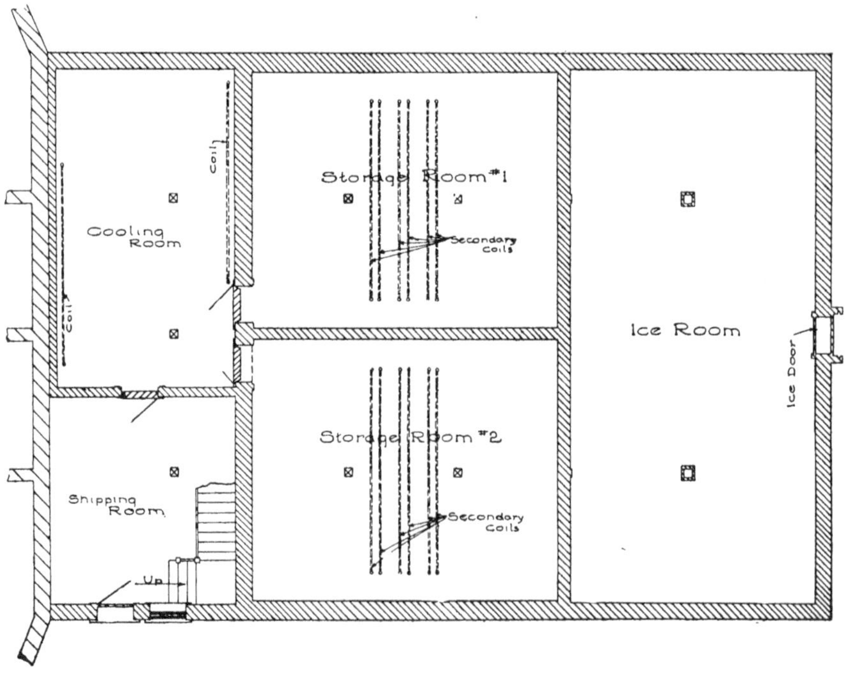

FIG. 11—FIRST FLOOR PLAN.

or grading which gives team access to the receiving room. This retaining wall also results in giving access to the shipping room on the main floor, as may be seen in plans and sections. The ice pond, shown in rear view of building is flooded only in winter, and when the ice is being housed it is floated right to the gig or elevator, where it is hoisted into the building by a horse. After the ice harvest is over the water is let off and the pond affords good pasture during the summer months.

As the fruit comes from the orchards it is taken in on the second floor into the receiving room, where it is sorted and

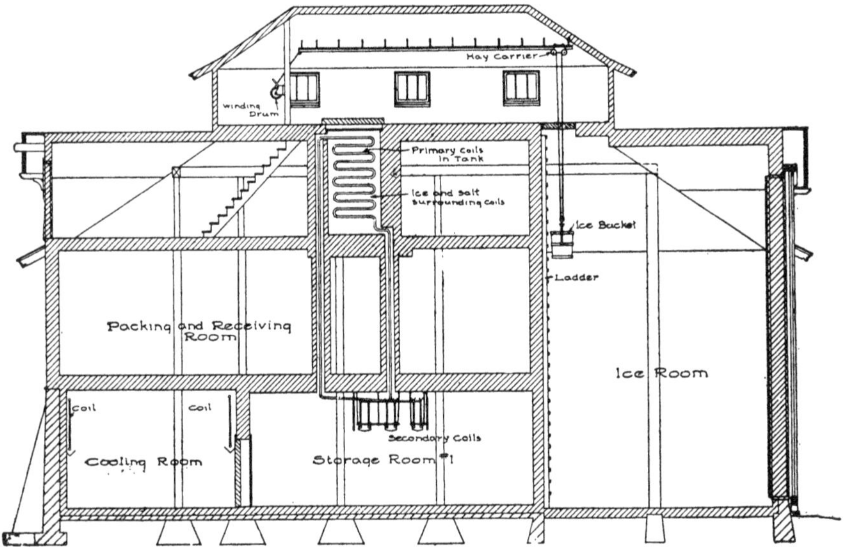

FIG. 12—LONGITUDINAL SECTION.

packed, and then lowered to the shipping room from where it is placed in the cooling room. This room is 26x15 feet, and is useful for partially cooling the fruit before it is placed in the storage rooms proper. The cooling room is chilled to a moderately low temperature by the melted ice and salt. Ice is hoisted from the ice room by means of a hand hoisting drum. A rope from this drum runs through pulleys to a car and ice bucket. After the ice is crushed in the ice room, it is drawn up to the track and is easily pushed along, so as to dump directly into the primary tanks. The ice is broken by hand and shovelled into the bucket.

The ice room has a capacity of about 500 tons, but this quantity is not needed for the two rooms equipped. Five

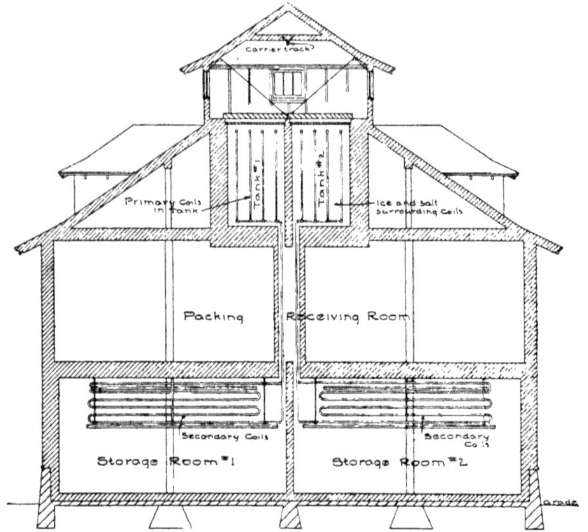

FIG. 13—TRANSVERSE SECTION.

hundred tons would more than operate the complete plant with four rooms equipped for refrigeration for an entire year.

A COUNTRY ESTATE FRUIT STORAGE.

The extensive and beautiful country estate of E. C. Converse of New York, located at Greenwich, Conn., is provided with a modern and practical cold storage plant, designed by the author. No great amount of money has been expended on ornamentation, and the plant, while of fine appearance, is strictly businesslike.

Fig. 14 shows the east or approach view, and gives a very good idea of the exterior appearance. The building is 48x99 feet, and consists of two floors of cold storage and a roomy attic or third floor. The exterior of the building is of rustic stone. The roof is of slate and the building is, therefore, practically fireproof from the exterior. A large loading and sorting platform of concrete occupies the south end and a part of the west side. This platform is equipped with a pipe frame on which a canvas awning may be hung to furnish shade or to protect from storm.

As will be seen by the basement and first floor plan the south end of the building on both floors forms a receiving, packing or shipping room and a power freight elevator and a

FIG. 14—CONYERS FARM, GREENWICH, CONN.—COLD STORAGE AND FRUIT PACKING PLANT.

stairway gives access to both floor and the attic. On the first floor this space is used as a packing room and the same is true of the basement, although as originally planned the basement was intended as a fire engine house on account of the location of the cold storage plant which is near the northeast corner of the property. The attic is used for the storage of empty barrels, boxes, etc., which are used in connection with the business and the rear of the attic over ice room contains a water tank from which water is drawn for mixing spray material, etc.

The cold storage plant consists of five separate rooms, the total capacity, exclusive of ice room mentioned later, being about 5,000 barrels. The two rooms in the basement are called frost-proof rooms for the reason that they are equipped with a comparatively light refrigerating apparatus for maintaining temperature only during cold weather. The three rooms on the first floor are all heavily piped for maintaining temperatures during summer if necessary. The ice room is also equipped

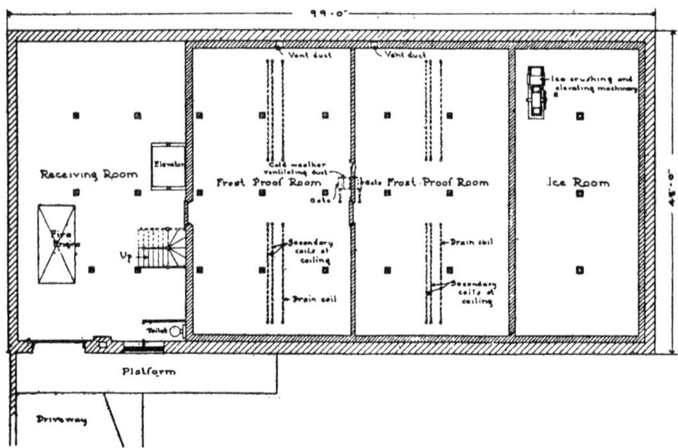

FIG. 15—BASEMENT PLAN.

with refrigerating pipes on the ceiling as it is anticipated that this room will be used for the cold storage of apples as the orchards come more fully into bearing and a new ice room will be constructed at the north end of the building. A very clear idea of the appearance of the inside of the cold rooms may be secured from the two illustrations, one showing barrels of apples in storage and the other boxes of apples in storage.

The longitudinal and transverse sections show the primary tanks of the Cooper brine system, the ventilating fan, power shaft, ice elevator tower, ice spout, etc. The ventilating fan is arranged to draw air from outside during cool or cold weather either for ventilating the rooms and to force out the accumulated gas from the stored product, or for cooling the rooms as required. It is also arranged to force air over a heater for warming the rooms if necessary in an extremely cold time, or when whitewashing the rooms at the beginning of a season's business.

It may be explained in connection with the barrels standing on end that they are not headed up and contain apples direct from the orchards, placed quickly in storage without sorting, and that they will be sorted and repacked as required for shipment. This accounts for their storage on end rather

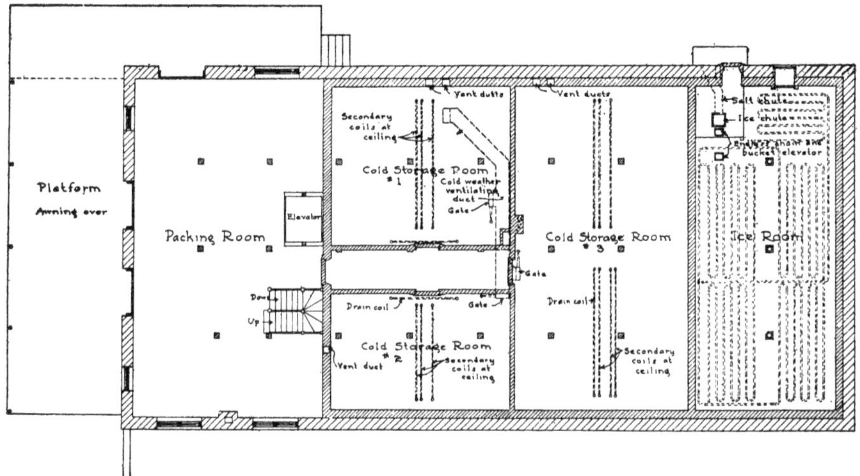

FIG. 16—FIRST FLOOR PLAN.

than being stored on the side as is customary. The apples in boxes are likewise hurried from the orchards to the cold storage room and no attempt made to sort them, except to leave the culls, if any, in the orchard.

The plant was designed for the storage of apples, peaches and grapes and, as reported by Mr. Geo. A. Drew, superintendent, it fills the requirements admirably. A temperature of 32° F. is maintained for apples and grapes, and peaches are stored at about 40° F. Peaches have kept in first class condition for at least three weeks, and there seems to be no reason why with careful handling and a little more experience they cannot be stored for a much longer period. Prompt cooling of the fruit when picked is secured by the location of the cold storage plant within a few minutes drive of the orchard. This question of location cannot be too strongly emphasized in connection with cold storage plants designed principally for the storage of fruit. Conyers Farm is located eight miles from Greenwich, the nearest railroad station, and the advantage of storing fruit on the farm at the orchards is very great. Prompt storage at picking and without the waste of time necessary for sorting will be one of the prime essentials for the successful handling of fruit in future.

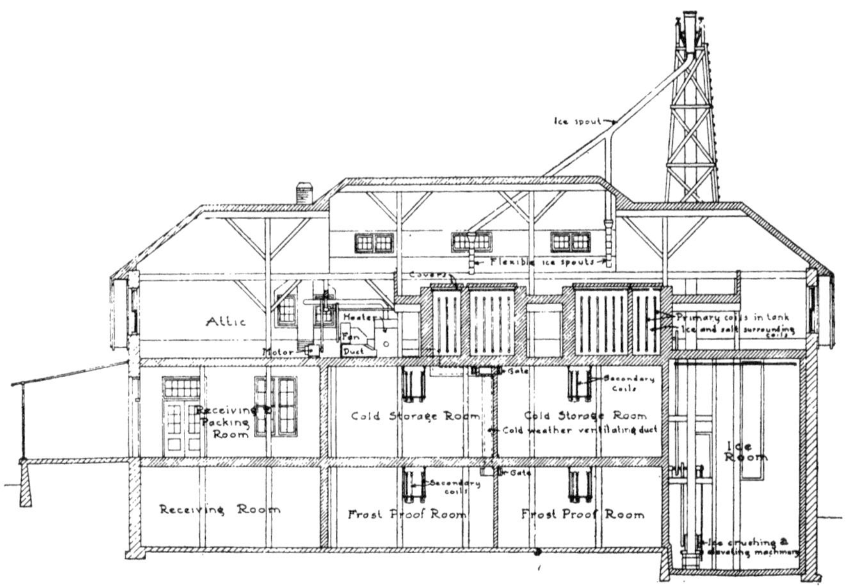

FIG. 17—LONGITUDINAL SECTION.

SUGGESTIONS APPLYING TO LARGER PLANTS.

For cold storage houses of a capacity greater than about 2,000 to 5,000 barrels of fruit, the complete Cooper systems are installed. In addition to the brine system and chloride of calcium process, they include the forced air circulating and ventilating systems, viz., an improved method of circulating the air of the storage over the secondary coils in the storage rooms, and a system for ventilating cold storage rooms by the forcing in of air which has been thoroughly purified, dried, and brought to about the temperature of the storage room. These air circulating and ventilating systems are necessary in larger houses where the arrangement is more complicated and the rooms are larger and the natural circulation of the cooled air is not uniform in all parts of the rooms; thus making advisable the use of a forced air circulation induced by a power driven fan. On account of requiring continuous power, the air circulating system has not been applied to the small houses here described.

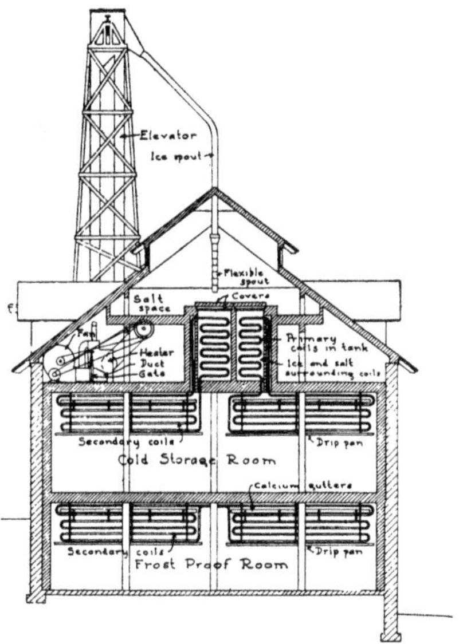

FIG. 18—TRANSVERSE SECTION.

In places where ice may not be had at low cost, and where the capacity is comparatively large, say, 40,000 barrels or more, a machine system of refrigeration is advisable. A skilled engineer should be employed by the year to operate the plant and maintain it in efficient condition.

ADDITIONAL ESTIMATES OF COST.

It is difficult to give accurate figures on the total cost of a cold storage plant as this necessarily involves a great variety of conditions. A plant of one or two cars capacity would mean one which might be used by a small fruit grower, but which would not be adequate for a comparatively large one. A plant of 5,000 cubic feet which would probably store about four carloads of fruit can be built at a cost of about $2,000. This would include plenty of space for packing room, storage of empty packages, ice storage room, etc. A plant of this capacity could be built without allowing much space for these purposes at a cost of probably $1,200 to $1,500. A plant of one to two carloads if all in one room, with perhaps a vestibule refrigerated by drips from primary tank, could be built at a cost of from $700 to $1,000 complete. It is hardly possible that a smaller plant than this could be made profitable. However, an 8x10-foot room, 7 feet high could be built at a cost of $350 to $500 com-

FIG. 19—APPLE STORAGE ROOM SHOWING BOXES USED FOR STORAGE.

plete, and the equipment for the Cooper brine system and the chloride of calcium process could be furnished for $175 f. o. b. cars factory. This would also include complete plans for the construction and insulation of the room. If a room of this capacity were to be built of other dimensions with plans made especially and the equipment built especially the cost would be $25 to $50 more.

REMOVING FRUIT FROM STORAGE.—SUGGESTIONS FOR CORRECT TREATMENT.

When removing perishable goods from cold storage a certain amount of care is essential. The plan of artificially drying air to prevent moisture on the goods is entirely correct theoretically, but practically it can not be made operative. It is hard enough, for instance, to get people to take their goods out of storage two or three days in advance of actual requirement, to say nothing about putting them through a complicated warming up or de-frosting process.

The very best suggestion the author has in connection with this matter is to simply take the goods out of storage; lay down a tarpaulin or canvas or wagon cover on the floor; pile your goods on it; and then cover the goods tightly with the

FIG. 20—APPLE STORAGE ROOM, SHOWING BARRELS ON ENDS NOT HEADED.

tarpaulin or canvas. A tarpaulin is pretty nearly air-tight, and this scheme will protect the goods from direct air contact while they are warming up, and will prevent sweating. This process takes several days depending on the weather, and if a free circulation of air is present they will naturally warm up much quicker.

Not only does the above scheme accomplish the desired result, but it brings the temperature of the fruit or other goods up gradually and thus prevents deterioration which may occur if the scheme of exposing them to artificially dried air and circulating the air over the fruit were adopted. For the welfare of perishable goods it is best not only to cool them rather slowly when placed in cold storage, but to warm them rather slowly when taken out of cold storage, and of course, the condensation or "sweating" must necessarily be prevented if best results are desired. Commercially, even the scheme of covering with tarpaulin is almost out of the question, as people cannot or will not take the necessary time and the exact requirements, or supply needed, is not known from day to day. Goods intended for cold storage should be packed in a package which will protect its contents fairly well from a circulation or direct contact with the outside air and this will to a large extent prevent the trouble experienced from condensation if the package is not opened until the goods have warmed up to the temperature of the surrounding air. The "sweat" is not at all necessary.

CHAPTER XX.

STORING CIDER UNDER REFRIGERATION.

ELIMINATING CHEMICAL PRESERVATIVE.

This is a new subject to most people, and even those who have been in the apple growing and cider making business for some years have had no experience and know little about the keeping of cider by means of cold storage. It is safe to say that the consumption of apple cider could be increased to many times what it is now, if sweet cider without marked fermentation could be had by the consumer for any considerable period of time. As the business is now handled there is a big surplus of cider during the fall, and the only cider available during the winter or spring is that which is bottled or preserved in some artificial manner. The most of the ciders on the market are treated with some preservative chemical which make them more or less unwholesome. Other methods consist of sterilizing by partially boiling, and then bottling. This latter method means an inferior quality of cider so far as palatability is concerned.

With the enormous plantings of apple orchards which are now going forward it would seem that after a few years the production of apples will be so large that growers will find it necessary to secure every possible outlet for the fruit, and for the manufactured products of same, such as cider. Those who are interested in orcharding should bear in mind the possibilities of the development of a fancy cider trade. The preliminary results reported below can doubtless be improved upon, and it would seem that with suitable cold storage facilities at or near the orchards, an almost unlimited market for a well prepared and properly clarified cider could be developed. Apple growers understand fully the difficulties of keeping cider for any length of time without resorting to chemicals. Good cold storage facilities where the temperature can be held under positive control are absolutely essential to the successful storage of cider in its natural state.

The U. S. Department of Agriculture, Bureau of Chemistry, has issued a circular by H. C. Gore on the cold storage of apple cider from which we extract the information which follows. The information given in this circular is thoroughly practical and the experiments, while admittedly preliminary, are doubtless very thorough, and it is probable that further experimentation will bring to light comparatively few additional facts, except possibly the adaptability of different varieties of apples for cider making purposes.

FRUIT USED FOR THE EXPERIMENT.

The apples used by the Government in the experiments referred to were of the grade commercially known as "Seconds." As it was not practical to begin the experiments when the fruit was received, the apples were stored at a temperature of 32° F. when received in Washington. Considerable decay occurred during cold storage in case of the Tolman (Tolman Sweet), Winesap, Yellow Newtown (Newtown Pippin), Ralls (Geniton) and Gilpins. As the apples were not of first grade this was to be expected, but very little decay was found among the Baldwins, Golden Russet, Roxbury Russet and Kentucky Red.

As the fruit was not ground for cider as soon as received, Mr. Gore suggests that the sugar content of the apples was probably higher than it would have been had this been done, and that this is particularly true of the three late winter varieties, Baldwins, Golden Russet and Roxbury Russet, on account of the fact that most fall and winter apples contain starch at picking time which disappears rather rapidly in common storage, and comparatively slowly in cold storage. On the other hand, the acid content would have been much higher had the apples been ground promptly when received, because the acid disappears rapidly during cold storage. These facts have been demonstrated by repeated experiments.

PREPARATION AND HANDLING OF CIDER.

Not less than an entire barrel of apples was used in the experiments, and from that up to six barrels in case of the Baldwins. All rot was removed from each lot of apples before grinding, and Mr. Gore states that the method of preparing the juice closely approximated standard commercial practice. It may be noted in this connection, however, that it is not customary to remove rot from apples before grinding for cider, except in preparation of extremely high grade goods. The fact that all rot was removed in these experiments may doubtless account for the fact that fermentation did not occur quickly. Those who wish to experiment in connection with the cold storage of apple cider should note this fact and be very careful not to grind apples which are found slightly decayed.

For the government experiments the fruit was ground in a rotary apple grater or grinder of the usual type and pressed by powerful hand power presses; racks and press cloths were used, following the usual American method. The racks were 36 inches square, and each cheese or pomace was 32 inches square and about three inches thick, representing the pulp from a barrel of apples. As a cold storage package five gallon kegs were used as containers for the juice from eight varieties, and a 50 gallon barrel for the juice from the Baldwin apples. Containers were sterilized by steaming and then rinsing with cold clear water immediately before using. After the kegs were filled they were placed out of doors over night at a temperature of about 32° F. or in the cold storage room at the same temperature; thus cooling the cider rapidly.

After the casks had been placed in their final position in the cold storage house, a three-eighths inch hole was bored in the head of each, to serve as a vent in case of gas forming, and through which samples could be taken at any time. The holes were temporarily blocked with cotton to keep out foreign matter, and the plugs were removed only when drawing samples, which were taken frequently during the first week of storage and somewhat less often thereafter.

DISCUSSION.

The striking fact brought out in experiments was that cider was kept in cold storage from 36 to 83 days, and an average of 61 days, before beginning to ferment noticeably. We may note in this connection what has already been suggested, that all decay was removed from the apples before grinding, and this probably accounts for the comparatively long period before fermentation commenced. The average for the Tolman, Winesap, Yellow Newtown, Ralls, Gilpin and Baldwin was 50 days. These varieties represent the usual type of American cider apples more fully than do the Russets and Kentucky Reds. From 90 to 125 days were required before the ciders had fermented too far to be called sweet, or an average of 107 days for all varieties, and 99 days for the six varieties just mentioned. No deterioration in flavor of the cider was noticed during cold storage, except in case of the Tolman, which can hardly be considered a cider apple. Although some of the ciders were frozen while in cold storage, yet no perceptible injury in flavor resulted. Not only was the characteristic flavor of the apple varieties maintained, but actual improvement was noted, due to the presence and retention by the low temperature of carbon dioxid or carbonic acid gas. The Baldwin, Golden Russet, Roxbury Russet and Kentucky Red gave the highest grade of ciders. The rate of fermentation increased rapidly in all instances after about fifty days, but the changes in this respect were far slower than those occurring in common storage without refrigeration.

SUMMARY.

A brief summary of the results of the experiments may be stated as follows:

(1) Ciders prepared from apples free from decay chilled rapidly to the freezing point immediately after pressing, and then held in cold storage at 0° C. (32° F.) remained without noticeable fermentation for a period of from 36 to 57 days, an average of 50 days for the Tolman, Winesap, Yellow Newtown, Ralls, Gilpin and Baldwin varieties and of 83 days in

the case of the Golden Russet, Roxbury Russet and Kentucky Red.

(2) These ciders were held for a period of from 90 to 119 days, an average of 99 days for the first six varieties and of 125 days for the last three, before they fermented sufficiently to be considered as becoming "hard" or "sour."

(3) The ciders were found to have suffered no deterioration (with the exception of the Tolman), but rather had become more palatable during storage.

CLARIFYING OF THE JUICE.

While nothing is said in this bulletin about the clarifying of the apple juice preparatory to storing, this subject is treated in Bulletin No. 118 by H. C. Gore, Bureau of Chemistry, U. S. Department of Agriculture, and Mr. Gore has given the details very fully indeed: Fresh apple juice contains a large quantity of solid matter which partially settles on standing. This solid matter or "pomace" as it is known, contains dirt particles or foreign matter and starch grains, as well as fragments of the apples, and albuminous matter. The albuminous matter composes the greater part of the sediment and is very objectionable, as its presence detracts from the appearance of the finished juice, and it contains the elements which quickly cause fermentation.

The removal of the materials which form sediment are, therefore, of the utmost importance in the preparation of a marketable product, and a product which can be successfully cold stored without sterilizing or without treating with chemical preservatives. The sediment may be removed by filtration or by allowing it to settle. It may also be clarified by passing it through a centrifugal separator. Filtration is expensive and slow and out of the question commercially. Allowing the sediment to precipitate by gravity is also slow and imperfect. The best method is by the use of a centrifugal cream separator, and repeated trials have shown that a cream separator will successfully clarify the juice, leaving only traces of sediment in the product. The suspended matter in the juice collects in the bowl of the separator, while the clear juice flows out through the milk and cream openings. After operating the machine for a time the foreign matter and sediment will be found tightly packed in the bowl of the centrifugal machine. A little experience will indicate just how long the separator can be operated before it is necessary to stop and clean out the sediment.

With reasonably clear juice, clarified quickly after pressing, it is doubtless possible to cold store apple cider for several months, if not for periods of very nearly a year. The extreme limit has not been determined, but it is doubtless upward of six months.

CHAPTER XXI.

NURSERY STOCK*

WINTER STORING OF NURSERY STOCK.

It may be stated at the outset that this chapter is written with regard to northern conditions, especially such as would be the average north of the Ohio River. In applying the suggestions and information given to other conditions further south due allowance should be made for variation of winter climate.

It is within recent years that the digging of trees from nursery row in the fall and storing during the winter for spring shipment has come to be an established feature of the nursery business. This subject was brought to the author's attention by a discussion between nurserymen of the advisability of the method. In this discussion the term "cold storage" was used in reference to the cellars or sheds in use for the purpose. Having a great interest in cold storage matters, the author determined to get the best information obtainable from those actually using the storage method. Letters of inquiry were therefore sent out to representative nurserymen. That nurserymen are in the main progressive and liberal minded is evident from the interest shown and the careful replies received. This chapter, therefore, gives no mere theory or opinion by the author, but information carefully gleaned from those actually engaged in the business and put in shape by one who has had a long experience with the cold storage of perishable products.

From the information obtained, it is beyond doubt a fact that a large majority of nurserymen are using retarding houses or frost-proof winter storage facilities of one kind or another.

---

*Originally published in *The National Nurseryman*, by the Author.

A few are using artificial cooling, but as a general proposition, this is not as yet fully appreciated. In time, no doubt, this feature will also come to be permanent, not only for maintaining regular temperatures during winter, but should there be an overstock of certain varieties in the spring, it would result in a great saving to store the surplus over until the next shipping season. Artificial cooling is another step in advance of frost-proof storage in the same sense that fall digging and frost-proof storage is a step in advance of the old method of digging at shipping time in the spring. It is natural that every planter should want his trees immediately as soon as the frost is out of the ground. The result is that they all want their stock at the same time. As a consequence, nurserymen who do any considerable amount of business and have no storage facilities have more than they can attend to in the spring. Even with this almost impossible problem to solve, there are many who are not converted to the storage method, so a few words regarding its advantages and alleged disadvantages will be timely. The advantages may be stated as follows:

1.—*Protection from Loss*:—Every few years thousands of dollars' worth of trees and vines are killed during a severe spell of extremely low temperature during the winter at a time when the ground is nearly bare of snow. It is also believed that nursery stock is in better condition to thrive when dug in the fall and stored in an even temperature approximating the freezing point than if allowed to stand in the nursery subject to wide fluctuations of temperature which will cause injury to a greater or less extent, depending upon severity of the winter and snow protection afforded.

2.—*Prompt Shipment*:—If no storage is provided digging must be done in the spring after frost is out of the ground. Frost is not generally out of the ground in northern regions, until April 1, sometimes later. This means that a large part of the trees are not finally planted until May 1 to June 1, and perhaps not until the leaves have started. Trees set under those conditions do not thrive as well and many die.

3.—*Saving in Labor*:—The shipping season is so short that if trees were all dug and shipped after frost is out of the ground the necessity of having a large and well trained force to get the shipments out promptly would be very expensive. With storage facilities, stock can be graded at convenience, counted and put in bundles ready for packing by cheap help during the winter. Trees may be dug in the fall at a much lower cost than in the spring, owing to more abundant available labor and dryer working conditions. Less hands are required as the labor is more evenly distributed.

4.—*Theoretically Correct*:—Trees dug late in the fall are dormant from natural causes and will stand handling, shipping and planting much better than trees dug after frost is out of the ground in the spring. After frost is out, sap starts and the tree is more liable to be damaged by rough usage and replanting. A dormant tree held at about the freezing point will retain its vitality almost indefinitely.

The disadvantages or bad effects of winter storage as claimed by those who oppose the method, are that trees dry out and mold when stored and that when finally set the percentage of trees which die is greater. It is also claimed that among the stock which survives, the growth is retarded and the trees handicapped by at least a year's growth as compared with freshly dug trees. Plenty of evidence is obtainable from disinterested parties that these effects result in some cases. These bad effects are, however, not from defects in the method, but from careless or unskillful handling, or lack of suitable storage facilities. Farther on we will take up the construction of suitable buildings. It is notable that the advocates of freshly dug trees are almost wholly of the "old line" element who stick to old customs, because some few failures have resulted from the winter storage method. This method, which has barely passed the experimental stage, can not but record some failures on account of improper application.

Nurserymen who practice the selling of freshly dug trees are handicapped in the handling of their business, and the increasing of same to any considerable proportions is practically impossible. In extensive enterprises, where the sales lists reach thousands of people, and where the distribution is made throughout a number of states possessing a variety of soil and climate conditions, the distribution must extend over a very considerable period of time, much greater than is allowed by the normal behavior of the plant; therefore, artificial means must be resorted to in order to hold the nursery stock in suitable condition for shipment, to provide for this wide distribution. From the preponderance of evidence in favor of winter storing, it seems that this will be universal in due time. We have then to consider the most approved methods now in use and suggestions for possible improvements.

COMMON METHODS OF WINTER STORING.

Some of the nurserymen who do not advocate winter storage, admit the need of something better than spring digging by "heeling in" or "trenching" their trees for the winter in a protected place which will drain naturally. They admit that this allows of possible damage to the tops of the trees in severe weather, but it saves time and wet digging in the spring. As an improvement over this it is only another step towards the solution of our problem to put a shed over these heeled-in trees to protect the tops from low temperature during severe weather. This is a common method and was, until quite recently, practiced by some very large nurserymen. A frost-proof cellar or shed is provided in which the trees are heeled-in in the fall, so as to have them ready for spring shipment. The storage shed is kept at the freezing point or somewhat above, so that sorting, grading and packing may go on independent of weather conditions outside, enabling shipments to be made as early as desirable in the spring. Much storage space is needed with this method and under such conditions the trees may dry out or shrivel, but the heeling-in in storage method has the advantage of being more independent of temperature changes than where the stock is piled up with roots exposed. A change of temperature is largely what causes the drying out of trees, owing to the change of humidity with the changing temperature.

Most of the winter storage structures in service are built partly below the surface, but many of the largest are wholly above the ground. Nearly all are insulated by building air spaces into the walls or by a filling of shavings, sawdust or similar non-conducting materials. It is the idea in building partly below ground to secure the protection afforded by the earth. It is a well known fact that at a depth of a few feet below the surface of the earth a nearly stationary temperature of about 55° F. may be obtained winter and summer. This will prevent freezing in winter if the cellar is rightly built, but it will likewise cause a marked rise in temperature whenever a winter thaw occurs and it becomes necessary to close the building tightly. The heat of the earth will then work up into the storage room and a temperature of 40° F. to 50° F. may result. Another disadvantage of the cellar is that when the first trees are stored during the fall, the surface of the earth is quite warm, and it is very difficult to keep the temperature of the cellar low enough. Ventilators, windows and doors are opened on a cold day or night, and in this way the temperature is, after considerable delay, finally reduced to the desired point. A warm spell alternating with cold weather in the fall after storing commences will cause a great deal of damage by causing the temperature of the cellar to vary greatly. A variation of temperature and consequent variation of humidity will not only cause a drying out or shriveling of the trees, but may cause a growth of mold or mildew. A building wholly above ground has many of the disadvantages above mentioned, and also the disadvantage of lack of protection during extremely cold weather. There are, however, advantages in above-ground construction, in that, if the building is built of frame, it will not rot out as quickly, and it may be cooled more readily in the fall, and it is not affected so much by heat from the earth. It is stated by many nurserymen that temperatures are very difficult to maintain in any of the ordinary sheds or cellars in use, especially during the storage season in the fall and during the shipping season in the spring. Winter storage for nursery stock should be so arranged that when natural temperature is suitable, air may be taken from the outside and forced into the room for refrigerating, and when natural temperatures are not suitable, as during a warm spell in fall or spring, or during a winter thaw, artificial refrigeration may be applied. Moisture brought in with stock,—especially if the fall has been a wet and warm one,—might cause mold. A proper cooling and temperature regulating system would prevent this.

DAMAGE FROM TEMPERATURE CHANGES.

From the data at hand, it seems clear that practically all of the damage to nursery stock experienced in winter storing in cellars or sheds as ordinarily practiced, comes from changes of temperature, and a generally too high temperature, which cannot by present methods be avoided. It has been noted that trees dug late in the fall and placed in storage after the temperature of storage room has been reduced to about the freezing point have carried through in better condition than those dug at an earlier date and placed in storage while the temperature of the room was still comparatively high. This may be partly because the wood is more dormant, but is probably largely because it is easier after about November 15 to keep down the temperature of the storage room. A high temperature and frequent changes of temperature will cause stock to dry out and shrivel. This is especially true of vegetation of quick growth, such as peach trees. To prevent this drying out, a spraying with water is often resorted to, but this again leads to mold or mildew if the temperature is high and not very carefully handled. One nurseryman states: "When stock is put in late in October and November, it needs no wetting at all, but stays damp all winter and spring;" another says: "In our own case, we find on account of the ups and downs of temperature, we must sprinkle with water more or less, but we believe that with a fixed temperature that did not vary to any great extent, the water could be omitted." No better argument could be made for low and uniform temperatures. There is no question at all that trees may be dug any time after October 1, or after the tree is dormant from natural causes, placed in a temperature of from 28° to 30° F., held steadily until spring, and come out in better condition for planting than stock allowed to remain in the nursery all winter and dug at the shipping time. Humidity must be attended to, but this is very easy to regulate at the low temperatures mentioned. As to temperatures at which trees should be held there seems to be a wide difference of opinion; no doubt this opinion is largely influenced by the temperatures it is possible for each individual nurseryman to maintain in his storage cellar. Nearly all admit the difficulty of keeping uniform temperatures, and opinions as to correct temperatures vary from 30° to 50° F. No doubt 30° F. will produce better results than any of the higher temperatures. It has been demonstrated in the history of preserving perishable products by refrigeration that the lower the temperature at which any particular product may be carried without damage from such low temperature, the better and longer it may be kept in cold storage. Certainly a temperature of 30° F. cannot injure nursery stock if it is able to withstand severe winter weather with any degree of safety. It seems reasonable, therefore, that this is a suitable temperature to maintain.

HUMIDITY AND TEMPERATURE.

At a temperature of 30° F. the air contains very little moisture, and in fact it cannot hold much, so the possibility of drying out nursery stock is much less when stored in a temperature of 30° F. than at from 40° to 50° F., which many recommend. The capacity of air for moisture is a direct property of its temperature—the higher the temperature, the more moisture air will take up and hold. At 30° F. air will hold less moisture than at any higher temperature. Air which is saturated with all the moisture it will hold at 30° F. contains 1.96 grains per cubic foot. At a temperature of 40° F.; 2.85 grains per cubic foot. This shows the rapid increase in capacity for moisture as the temperature of the air is increased. Suppose we are holding our storage room for nursery stock at 30° F. and a warm spell of weather comes, one which obliges us to close tightly all openings leading to the outside air. After a few days the temperature goes up to 40° F. What is the result? The air, say, was at the 84 per cent relative humidity at 30° F. When the temperature has increased to 40° F., the relative humidity will be 56 per cent. What does this mean? Simply that the air has become comparatively very dry and that moisture-containing products like trees will dry out very quickly. This case is stated to show the operation of this simple natural law in connection with the winter storage of nursery stock. Possibly these exact conditions might not occur in practice, but they would be approximated. The great importance of maintaining uniform temperature and humidity is plainly illustrated, and the cause of the drying out of trees by fluctuating temperatures is readily seen.

To overcome the difficulties of winter storing as above outlined artificial refrigeration should be applied when necessary to maintain sufficiently low temperatures. By the term artificial refrigeration it should not be understood that a complicated ice machine system is necessary. The term is used to express cooling effects other than those produced by outside atmospheric conditions. Such a refrigerating equipment is embodied in the Cooper brine system described in chapter on "Refrigeration from Ice."

IMPROVED BUILDINGS AND APPARATUS.

The accompanying illustrations show a combination winter and summer storage building constructed wholly above ground. The storage space is divided by a partition into two rooms, one small room 30x50 feet, and one larger room 50x80 feet. These rooms are both cooled from one battery of pipe coils, but the air ducts are provided with gates so that the entire refrigerating effort may be applied to the smaller room. The refrigerating equipment is of sufficient capacity to maintain a temperature of 30° F. in the small room during midsummer, and to maintain the same temperature in both rooms during comparatively cold weather, say from November 1 to May 1. Both rooms may be used for winter storage, and during the summer the large room may be shut off and only the small room used. If

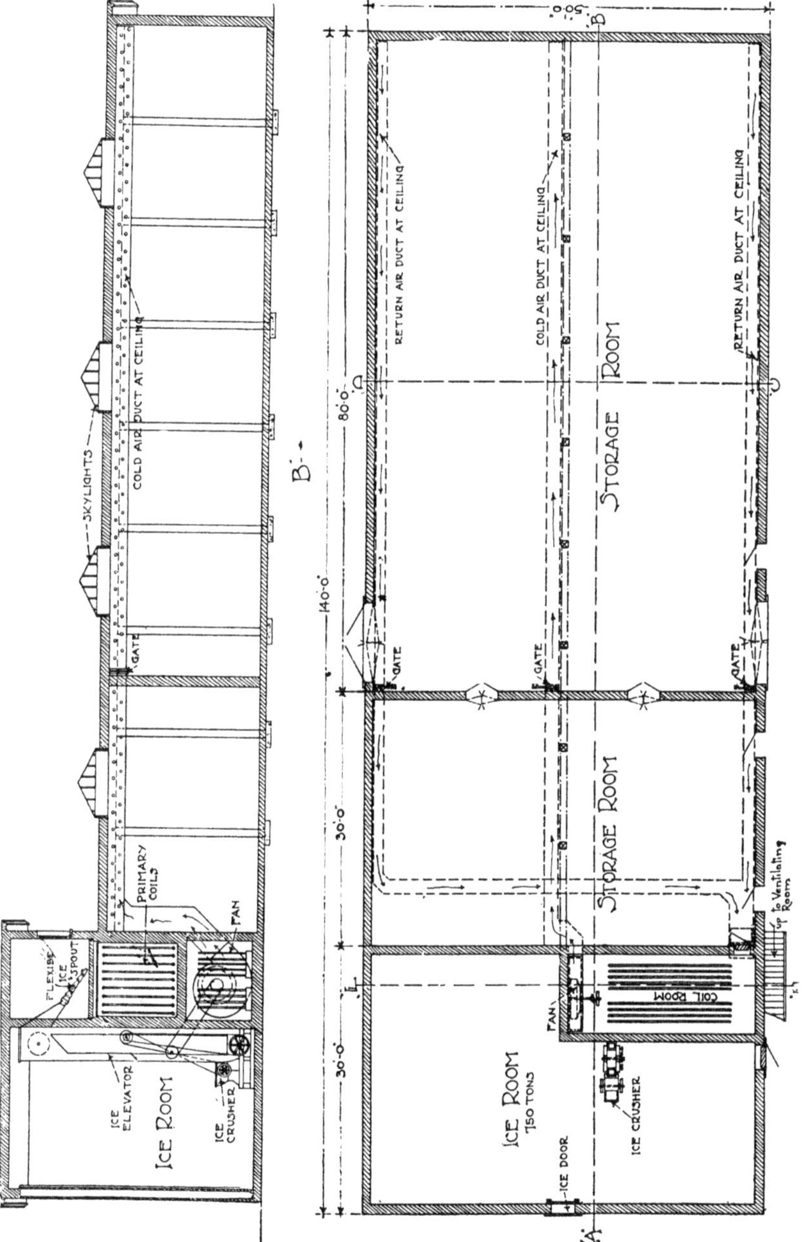

FIGS. 1 AND 2.—LONGITUDINAL SECTION AND FLOOR PLAN, COLD STORAGE HOUSE FOR NURSERYMEN.

it is not desired to store nursery stock during the summer, other goods may be taken for storage if they are to be had, or the plant may be shut down during the summer. No expense whatever is necessary when the plant is not in operation. The main part of the storage building, 50x110 feet, is essentially like many storage cellars or houses now in use, consisting of as plain and as cheap a building as can be built, and roughly insulated. At one end of the storage building is the ice room,

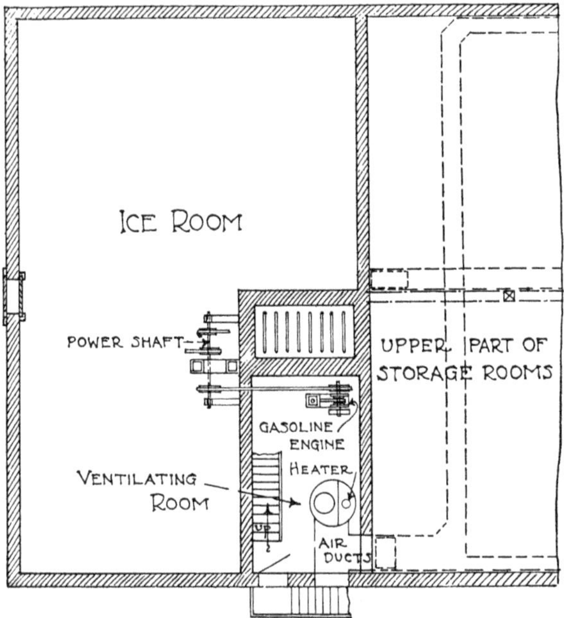

FIG. 3.—PLAN OF VENTILATING ROOM ABOVE COIL ROOM.

which also contains the complete refrigerating and mechanical equipment. The ice room is 50x25 feet on the ground, 30 feet high inside and will hold about 750 tons of ice, which is more than sufficient to maintain the temperature as above stated during the year. The room containing the secondary coils of the Cooper brine system is located on the ground. Above this room is located the tanks containing the primary coils and the ventilating room containing the heater for use during extremely cold weather and at such times as it is necessary to warm or dry the storage rooms. The gasoline engine or other power used for driving the fan for circulating the air through the storage room and for ventilating, is also located in the room above the tank and ventilating room, where access is had to top of tank for filling with ice. On this floor is also provided storage bins for salt. In houses the size of the one here illustrated, or larger, an ice crushing machine and

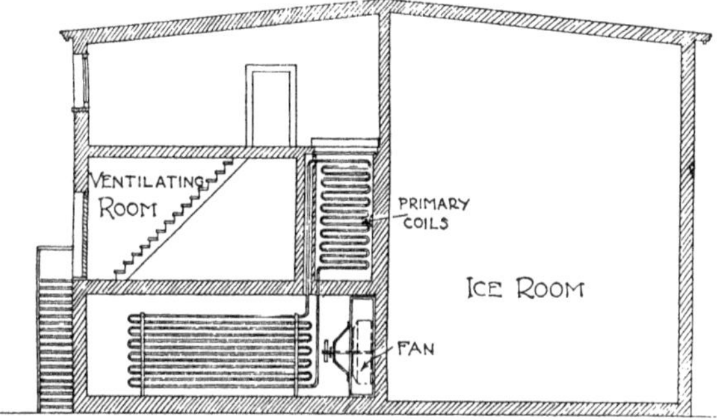

FIG. 4.—CROSS SECTION AT E-F OF FIG. 2.

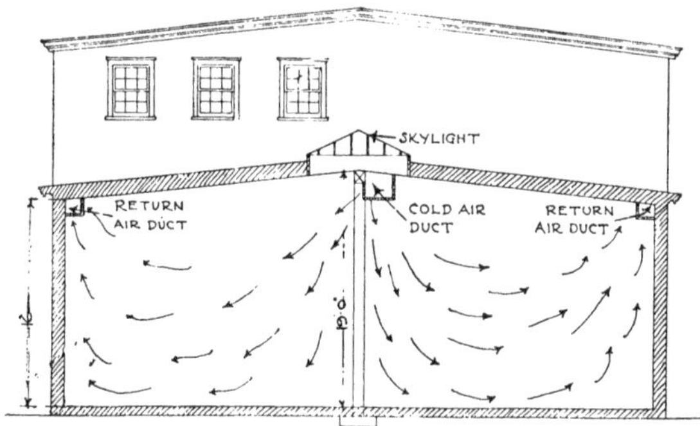

FIG. 5.—CROSS SECTION AT C-D OF FIG. 2.

ice elevator as shown is desirable, especially as the power is at hand for operating the same. In smaller plants this may be dispensed with.

The operation of the plant is as follows: Ice is fed to the ice crusher, which reduces it to about the size of hen's eggs; from the crusher the ice drops into a bucket elevator, which lifts it up above the tank containing the primary coils and drops it into the tank through a flexible spout. It will be noted that very little labor is necessary with this arrangement. As the ice falls into the tank a small amount of salt is sprinkled in. This produces a low temperature in the tank, which cools the chloride of calcium brine in the primary coils and causes a circulation as already described. The actual cooling of the storage rooms is accomplished by drawing the air in through small ducts on the sides of the rooms by means of the fan and causing it to pass over the secondary coils of the Cooper brine system in coil room, where it is cooled; then forcing it from fan into large duct in center, where it is evenly distributed to the rooms. When necessary to heat the storage rooms, the return air to coil room is caused to circulate over the large, jacketed heater in ventilating room, or fresh air for ventilation may be drawn over heater for ventilating and heating at the same time. When weather conditions are right, a large volume of air from the outside may be forced into the storage rooms for the purpose of cooling the rooms. Many times greater cooling results may be secured in this way than by the opening of doors and windows, and the cold air is evenly distributed to the rooms so that no freezing or harm can result, as is possible to goods stored near open windows or doors on frosty nights.

The estimated cost of complete apparatus, aside from the buildings, for a house the size shown, completely erected in place, is from $2,500.00 to $2,800.00.

The plant described will maintain uniformly low temperatures at about the freezing point in the entire building during the cold weather when most of the nurserymen's products are stored, and in one-fourth of the house during the summer. The initial cost of the apparatus is not excessive, the cost of operation almost nominal and the results to be obtained positive. Only a moderate amount of refrigeration is required in storing nursery products, but when required, it is very important, and the cost is so small that it will soon pay for itself in saving of loss and perfection of results possible to obtain. In many cases the nurseryman is a fruit grower as well, and cold storage would be a good auxiliary to add for the purpose of taking care of the softer fruits temporarily and the hardy fruits for a longer term of storage.

This description of a suitable plant for nurserymen is designed for northern locations where the nursery business has had greatest development. In the south or extreme west the mechanical systems of refrigeration would be best adapted; or in a large plant in the north. The other features of the plant would remain the same so far as construction, air circulation, etc., is concerned.

NOTES.

Some suggestions are here added from the *Florists Exchange,* as follows:

Everyone who has handled the Japanese Snowball in Spring knows how quickly its buds swell when Winter ends, and how necessary it is to get it planted the very first thing. It must be classed with the Larch for early planting and shipping, for, unlike some shrubs of early sprouting nature, it does not take kindly to being planted when in leaf, or when near this condition. Because of this, plants of it should be dug in Autumn and placed in cold storage for Spring use. If no building suitable is available—it should be one kept at about the freezing point—the plants should be buried under ground outdoors until Spring opens, selecting a sloping bank for the purpose, or at least a place where water drains away freely.

All early starting trees and shrubs should be treated in this way. The Larch, already mentioned, for one, the Weeping Willow for another, bush Honeysuckles, Spiræa sorbifolia, and several other kinds that will come to mind as well.

Many nurserymen have cold storage houses already in which to store, in Autumn, stock required for localities earlier than where they are, and it works also to the advantage of buyers in colder States as well, for when in the cold storehouse the plants can be kept in good condition almost indefinitely, so that both early and late customers may be supplied.

Florists have found benefit from having shrubs in flower at a later date than usual with the varieties, and retarded plants such as those held back in storage houses are very useful. It is quite common to treat Hydrangea paniculata grandiflora in this way, deferring its planting until late May, or any time desired, which brings it in flower long after the natural crop is over; and this course could be adopted with any and all shrubs useful in the same way as the Hydrangea named. It is in favor of the latter shrub that it is aided by pruning it back when planting it, with the object of making it thrive even if the weather should be hot and dry, a pruning that the Snowball and other shrubs that flower from their last season's growth could not receive.

CHAPTER XXII.

POTATOES.

METHODS OF PRESERVATION.

Indications lead to the conclusion that potatoes will follow the history of cheese and apples in methods of preservation. In the early days of the cheese business, even after cold storage was available, much cheese was stored in cellars or basement storage or in ordinary ice refrigerators of large capacity. Now, practically all cheese made reposes for a time at least in cold storage. Apples were formerly mostly stored in "common storage" or "frost-proof" storage, but now the bulk of the crop not sold promptly for consumption is placed immediately in cold storage. Likewise, potatoes in their early history were, and we might say still are, stored in the cool temperature of a basement or a cellar rather than in artificially cooled space. In recent years, however, quite a large quantity of seed potatoes have been cold stored and this has demonstrated possibilities of successful storage which were not known under the old methods, and will doubtless lead to a general use of cold storage for keeping eating potatoes as well as those used for seed.

SEED POTATOES FOR EARLY CROP PLANTING.

The business of growing early potatoes in the South for Northern market is now an important one and the storage of seed potatoes gives a good income to a number of cold storage houses. In many parts of the South the practice is to raise two crops of potatoes per year on the same ground, the second planting being in June and July, promptly on harvesting of the first crop. It has been found that the first crop of potatoes cannot be used as seed for the second crop, because it is too slow in sprouting, and the second crop tubers have also proved to be unsatisfactory for second planting for reasons not stated.

The best results have been obtained by shipping in Northern grown potatoes and placing them in cold storage until wanted. A temperature of 38° F. to 40° F. has been employed, but it is believed that lower temperature would give still better results, as will be suggested further on.

Around Louisville, Ky., the business is handled somewhat differently than described above. Here second crop potato growing does not mean two crops of potatoes on the same land, but a crop of potatoes follow a crop of cabbage, cauliflowers, etc. Seed from the so-called second crop for the following year's use is put into cold storage, and planted from July 20 to August 15. Late varieties are planted first, and the crop is mostly marketed for eating in February, March, April and May. They are still unsprouted as late as May and in perfect market condition. Early varieties are planted later, and the vines are usually killed by frost about Oct. 1st to 15th, and while somewhat immature, are in good condition for planting after lying dormant for four or five months. Excellent results are reported from this practice. It is not possible to use first crop potatoes from further south for the same year's second crop planting at Louisville for the reason that they have not lain dormant long enough and sprout slowly or rot in the ground. Cold storage seed will come up in a week or less if the ground is moist and warm.

Many Northern growers secure second crop seed from the South, and place it in cold storage for early planting. While the second crop potatoes are somewhat green and immature, yet this does not in any way affect them for quick growing; it seems in fact to assist prompt sprouting and vigorous growing when planted.

It is of the utmost importance that potatoes to be used for seed must remain dormant for a period of four months or more.

"FITTING" FOR STORAGE.

Some of the above remarks apply to the securing of suitable tubers for storage. One of the chief points is that the seed be placed promptly in the cold room. It has been demonstrated that steady temperature, with steady humidity has much to do with the vitality of the tubers for growth when taken out of storage. A potato which has sprouted and dried out to the extent of perhaps one-quarter or one-third its weight, certainly cannot push the new growth like a firm, plump potato which has not started growth.

The "fitting" of the tubers for cold storage is important. Good farmers know that potatoes must not be placed at once in the cellar when dug. There are two reasons for this. The cellar will be dryer and lower in temperature later in the season, and the potatoes must be allowed to dry and ripen up and lose their surplus moisture before storing. It is customary to allow the tubers to lay on the ground only long enough to dry so that the soil will not adhere and then haul them to cover where they are spread out and covered with bags to protect from the light and allowed to dry and ripen. From one to three weeks are commonly allowed for this. If, however, potatoes are not harvested till late autumn and are fairly well matured and allowed to dry off thoroughly in the field before being picked up, they may safely be hauled direct to the cold storage room, where they may be stored in open boxes or crates or barrels without heading up for a few weeks and then placed in the permanent storage package. Potatoes should, for best keeping qualities, be fairly ripe when harvested. Potatoes which "skin slip" will turn black and do not keep well, and have a bad appearance when exposed for sale.

The necessary "fitting" of potatoes for storage depends on conditions of soil when tubers are dug, degree of maturity, and to some extent the character of the soil. Some years during the harvesting season the soil is quite dry, and potatoes dug under these conditions need very little "fitting." Other years with the soil filled with water, the tubers may need two weeks or more of exposure to air currents before placing in cold storage packages.

PACKAGE.

All sorts of packages are used for shipping potatoes, boxes, barrels, crates and bags, and the favorite method of shipping full carloads is to handle in bulk as this is cheaper and a heavier load may be placed in the car, which is an advantage in extremely cold weather. Crates are used largely as a harvesting package as they are convenient, and allow of a circulation of air, and the soil adhering readily drops off. Crates may be used for storage for a few weeks only, but for permanent storage boxes or barrels, or some tight wooden package should be used. Bags or sacks should not be used as a storage package except for short periods, as they lead to too great a bulk being stored together, and besides the tubers may be bruised or even crushed. By building racks or shelves at intervals of about three feet in height potatoes may be stored in bulk for several months at a high relative humidity with fair success. The practice cannot be recommended, but where suitable packages are not available it is permissible. Boxes and barrels are about equally good as a storage package, but preference is given in the storage of potatoes to the old reliable barrel as it is in the storage of apples. Barrels make a strong package which will stand rough handling and which is just about air-tight enough to give the needed ventilation and properly protect its contents. Storing in bulk is permissible for short carry only and cannot be approved for any period of longer than two months.

Whatever packages are used for cold storage purposes, whether boxes or barrels, these may be used over again year after year if local conditions make it advisable. If, for instance, the cold storage house is near the place where the tubers are grown, the storage package may be taken to the fields and filled there. In shipping it is often found more convenient to ship in sacks or in bulk than in barrels or boxes, and for this reason the suggestion to use the wooden package for storage purposes only, is a practical one.

One very important point must be borne in mind if storing in a room where daylight is admitted: Potatoes will become "sun struck" and turn black if exposed to daylight for any considerable length of time. A package, therefore, which will protect the tubers from the light is absolutely essential.

TEMPERATURE.

While the correct temperature cannot be stated with accuracy in the absence of definite data and the result of experiment, 33° F. to 35° F. has been mentioned by those best posted and with the most experience in the business. At the same time the tendency in the storage of many products is downward in temperature, and we doubt not but what potatoes will in time be carried at 32° F. or even 30° F. and possibly lower.

It is said that if subjected to too low a temperature for considerable periods, say a month or more, potatoes gradually become sweet to the taste when cooked. Chemists explain this as being due to an enzymic action, whereby the starch, of which a potato largely consists, is converted into sugar. This action is going on in the tubers even at the low temperature of cold storage, and it seems that the lessened respiration or evaporation of moisture from the tubers more than keeps space with the enzymic action. That is, lowering the temperature reduces the respiration, while the enzymic action is only slightly lessened. At ordinary temperatures the two actions equalize each other to an extent that the sugar is completely oxygenized. The sweetness of flesh caused by low temperatures disappears when the tubers are again subjected to ordinary temperatures. If potatoes were intended for seed purposes this sweetness would be of no consequence.

The lowest temperature which potatoes will withstand without damage has not been determined with any degree of accuracy, but it has been reported that tubers stored in a basement room with an earth floor, during severe winter weather have successfully stood a temperature of 28° F. to 30° F. During a considerable portion of the winter the walls were covered with frost. It would seem, therefore, that the idea that potatoes would freeze and be completely ruined at a temperature of 32° F. is incorrect. It is, however, well known that if potatoes do freeze they are useless for any purpose, and doubtless it is this serious result when frozen that has caused people to believe that they freeze at about 32° F.

When potatoes are first placed in cold storage it may be advisable to carry a temperature of 40° F. or 45° F. and gradually bring it down to the temperature at which it is finally desired to carry the tubers, and a few weeks at the higher temperature might prove beneficial.

HUMIDITY.

No experimental data is available on the subject of humidity for potato storage, but it is well known that the storage room should not be too dry. It is, in fact, generally considered that the more moist a room can be carried within reasonable limits, the better will the tubers keep. It is, of course, possible to have the air so moist that it will cause mould, but in cold storage this is hardly probable. The influence of a greater or smaller quantity of goods in the cold storage room where potatoes are stored would be important, but if the goods were boxed or barreled, this would not have so much effect. A cold storage room, to get the best results, should be filled as full as possible. A room with few goods in it ordinarily means a comparatively dry room, and this does not give best results. A humidity of from 85° to 90° is suggested as correct for a potato storage room with a temperature at from 33° F. to 35° F. The use of chloride of calcium in a potato storage room is essential if a heavy quantity of goods is stored in bulk. If the room should be too damp or the goods carried for long periods, a small amount of chloride of calcium exposed to the air of the room would be beneficial, as it would tend to check a growth of mould and keep the air pure as well as regulate humidity.

MISCELLANEOUS.

E. H. Grubb, of Carbondale, Colo., states that while prospecting in the early days, he met a prospector who directed him to some potatoes which he had stored two years before in an old tunnel at an altitude of 12,000 feet. The potatoes, Mr. Grubb states, were found in perfect condition, being the same as when dug. The tunnel had a circulation of air in it and a temperature of about 40° F. Making due allowance for the well-known preservative effect of Colorado mountain air, it would seem that if the potatoes kept without rotting for as long a period as two years, it was quite remarkable, and as there must have been considerable change of temperature at different seasons of the year, the long keeping possibilities of potatoes were by this circumstance fully demonstrated. This might argue against cold storage rather than for it, but it is probable that the potatoes were not in the perfect condition stated, and while they doubtless looked very good to a hungry prospector, would hardly class as marketable.

It is reported that potatoes shipped in refrigerator cars which are heavily iced, will not be as apt to freeze in very cold weather as they would if the car were not iced. There can be no scientific explanation of such a claim, and the only practical reason is that with the ice bunkers filled with a heavy weight of ice, there is a little more balance on the temperature and a larger body or mass to be cooled before the potatoes in the car will freeze. However, if the ice bunkers are water-tight and water from the melting ice is allowed to stand in them, as some of the cars are arranged, this would be a very good reason why potatoes would not freeze as quickly as they would without any ice or water in the bunkers. The water in the bunkers would freeze before the temperature of the car would be reduced much below the freezing point, and thus the potatoes kept from freezing.

There is another old fashioned idea that potatoes will not freeze in hauling as long as they are kept in motion. It is, of course, a well known fact that any liquid or solid body will not freeze as hard if kept in motion, but it will freeze just as surely in motion as it will at rest and somewhat more quickly. The scientific explanation is that ice crystals form more quickly in liquids or bodies at rest than they will if agitated. The difference in freezing point, however, would be but slight, and this old fashioned idea is, therefore, largely erroneous.

CHAPTER XXIII.

GRAPES.

LONG STORAGE OF GRAPES.

Comparatively little has actually been done in the successful cold storage of grapes for long periods. There is, however, some activity in this direction, and it seems that the large growers who produce grapes on a commercial scale, are experimenting along this line. It has been found practicable to carry some varieties until February, when small fruits are comparatively scarce and prices high. The favorite winter keeping varieties are the Catawbas and the Vergennes. The Concord and other similar varieties do not seem to do as well in storage owing to the fact that they loosen from the stems and decay starts at that point.

For long period storage, grapes are removed from the vines when barely matured, and placed in shallow boxes in the packing house for a few days, until the stems have wilted and the natural evaporation or sweating has taken place. They are then packed in baskets lined with paraffine paper which is carefully folded over the top so as to make a fairly air-tight package. The fruit is carefully selected and carefully handled from the vineyard to the packing house, and carefully handled when placed in the baskets. It is not pressed or crushed into the baskets as is sometimes done where grapes are packed for prompt consumption. The baskets are only jarred gently to settle the grapes firmly into position.

When picked and packed in this way the length of time which grapes can be stored is quite remarkable. The writer has long appreciated the possibilities in this line, but growers have been so conservative in taking hold of the proposition, and unwilling to put any money into a suitable cold storage plant, that the development of the cold storage of grapes has been extremely slow. Of course the amount of business which can be handled in this way is small as compared with the total volume, as only certain varieties and only the best selected stock should be stored; but where the best quality grapes can be grown and where a suitable cold storage house is available, a handsome additional profit should accrue to the grower by handling as above suggested.

EXPERIMENT OF AGRICULTURAL DEPARTMENT.

The Bureau of Plant Industry, U. S. Department of Agriculture, has made some experiments in grape storage and shipping of which the following is a brief summary:

"The importations of fresh grapes from Spain during the present season (1912) amount to nearly 900,000 barrels which have sold at wholesale prices ranging from $2.50 to $7.00 per barrel, or from 5 to 15 cents per pound, the bulk selling at the lower price. Under ordinary conditions, most of the California table grapes must be marketed within a period of a little over two months and the early attempts to hold them in storage for the holiday markets did not prove entirely successful.

"The Bureau investigations have shown the importance of handling grapes with care to insure their being packed in sound condition. It has also been found that it is impossible to hold the varieties of grapes that are commercially grown in California any appreciable length of time without a filler of some kind. The Spanish grapes are packed with a filler of ground cork. As this material is both scarce and expensive in California, special efforts were made to obtain a satisfactory substitute. Many different materials were tested but only one has thus far proved wholly satisfactory. This is redwood sawdust, which is a waste product of the California sawmills. Much to the surprise and gratification of the Department investigators this material has proven even superior in many ways to the ground cork. It is found that the grapes hold longer and in better condition when packed with the redwood sawdust. Great pains have been taken to corroborate the results and the data have been consistent throughout. It was necessary to learn how to prepare the sawdust in order to have the grapes remain in attractive and salable condition. The sawdust must be perfectly dry and the finer particles must be removed.

"A number of varieties have been under investigation, and naturally their behavior under storage conditions has been different. Of the varieties grown in commercial quantities, Red Emperor, Malaga and Flame Tokay have been found to hold best in storage. The lengths of time which these varieties may be held vary from sixty to seventy days for the Flame Tokay and Malaga, and from ninety to one hundred and ten days for the Emperor.

"In the commercial test of the application of this work during the past storage season the grapes were packed in drums holding about twenty-seven pounds, and the work of packing and shipping was done largely under the supervision of one of the bureau representatives. The drums were forwarded from California to Chicago and New York under refrigeration where they were held at a temperature of 32 degrees in cold storage. The Emperors proved to be the best for storage purposes and formed the bulk of the grapes sold for the Christmas trade. The best grapes of Flame Tokay may be held until Christmas, but the ordinary run of this variety will not hold in first-class condition beyond December 1. The Malaga varies considerably in its behavior in storage, depending upon the conditions under which it is produced. Some lots of this variety have been held in first-class condition until January 1 in past years, while others are not safe beyond December 1.

"The value of this work to the grape industry of California is apparent when the full significance of the extension of the marketing season is appreciated. The production of table grapes in California is increasing and unless some way can be found either to broaden the area over which the fruit may be distributed, or to lengthen the marketing season, the industry will be face to face with a serious problem of over-production. When it is considered that this country uses large quantities of imported grapes, the demonstration of the possibility of replacing the foreign product by one home grown, is worthy of the most strenuous effort.

"The possibilities of packing California grapes with the redwood sawdust filler for export are also recognized and efforts are being made to extend the marketing area by this means. A small test shipment of California Tokay Grapes shipped to England was made during the past season and the fruit arrived in excellent condition. The sawdust pack in drums is well adapted to ocean transportation, because the necessarily rather rough handling in loading and handling aboard does not affect the grapes when packed in this way, while the ordinary open crates are too weak to withstand rough handling, and in addition the grapes deteriorate during a long trip unless a filler is used."

More recent reports are that grapes have been stored in October and carried as late as January 17th, or a period of three months. It is presumed that this is about the commercial limit of the cold storage of grapes, as they are rather delicate and not naturally substantial enough to handle or cold store for any great length of time. It is to be regretted that in reporting these experiments more information has not been given as to temperature at which they were carried and some discussion given on the subject of humidity, conditions of storage, etc.

Some recent experiments conducted by the U. S. Department of Agriculture in California in the shipment and storage of grapes from the Fresno District indicate that the Emperor and Almeria varieties of grapes may be packed in redwood sawdust, stored at a temperature of 32° F., shipped to Eastern markets and stored for a period of five to six months in sound and useful condition. Muscatel grapes have been cold stored for four months successfully as reported by W. E. Alexander of Escondido, Calif. It seems that this method of packing and storage is likely to develop into an important industry.

CHAPTER XXIV.

THE PRE-COOLING OF FRUIT.

OUTLINE.

The term "pre-cooling" as applied to fruit means the process of rapidly and permanently reducing the temperature of the fruit immediately after picking and before it is put into cold storage proper or loaded into cars for transportation. Fruit when removed from the tree immediately becomes a substance without life, and the processes of decay are at once started and continue more or less rapidly during the natural history of the fruit, depending upon the temperature. With the softer fruits, especially as represented by peaches, which form an important article of commerce, it is of the utmost importance that the heat be taken out of the fruit promptly when picked in order to transport it for several days to market in prime condition. Cooling is accomplished in various ways, which will be described further on in this chapter either by mechanical refrigerating methods or by the use of ice or ice and salt. Pre-cooling is necessary mostly because the so-called refrigerator cars which are largely in use are not equipped with sufficient cooling capacity to take the heat out of the goods after loading into the cars.

ADVANTAGES OF PRE-COOLING.

It has been demonstrated by actual tests by the representatives of the U. S. Department of Agriculture that warm fruit loaded into a refrigerator car required from three to four days to reach a temperature of 45° F., and even then the fruit in the car was not uniformly cooled, that in the top of the car being from 10° to 25° higher in temperature than that located at the floor and near the ice bunkers. During this period of three or four days the high temperature of the fruit and the moisture and gases in the car result in conditions which promote decay and rapid deterioration. Fruit slightly blemished or injured in picking under these conditions will on its arrival in the market show up in poor condition and be unsalable at the highest market price. It is claimed that a loss of two million dollars has resulted to the citrus fruit growers of the Pacific Coast in one season through failure to properly cool and transport their product to the ultimate consuming market. This is the shipper's viewpoint. From the viewpoint of the transportation companies they are interested in pre-cooling from the fact that an ordinary refrigerator car as generally loaded will transport only about 400 boxes of oranges, whereas the same car loaded with fruit properly pre-cooled may hold from 550 to 600 boxes. More will be said on this score in connection with suggestions for shipping oranges which have been pre-cooled before shipment.

Prior to the year 1905 the peach growers of Georgia and California were unable to ship ripe peaches in refrigerator cars to distant markets without incurring heavy losses because of over ripening and decay during transit. The same trouble was experienced more or less with California oranges and grapes, and other products of lesser importance have come in for their share of losses and unsatisfactory results. Pre-cooling where properly applied will result in saving a greater part of the losses which have heretofore been encountered, but necssarily pre-cooling will not take the place of care and common sense and experience and skill in the handling of any perishable product. Pre-cooling is only one of the steps necessary to attain an approximately perfect result in giving the ultimate consumer a first class quality of fruit. What is said above should not be construed as meaning that pre-cooling facilities are now available to fruit shippers where needed. Such is far from the case, and as a matter of fact pre-cooling is only just beginning, and probably not one-tenth of the fruit shipped to market is subject to adequate pre-cooling conditions.

METHODS OF PRE-COOLING.

There are two general methods of pre-cooling in use:

First—The car pre-cooling method, and

Second—The warehouse pre-cooling method.

Pre-cooling fruit in cars after loading has been advocated and put forward by the transportation companies to a considerable extent, and they have erected some very large and expensive pre-cooling plants for this purpose. No doubt they have been influenced in this by the control which this would give them of the business as well as the revenue which would result. It is the positive opinion of the author, however, that this method is not theoretically correct, nor is it practically efficient, nor is it likely to come into general use. The disadvantages of this method are so great as compared with the warehouse method of cooling that it will doubtless fall into disuse, and those plants which have already been erected for this purpose will probably be put to other uses.

CAR PRE-COOLING.

Car pre-cooling is accomplished by setting the cars loaded with fruit on a side-track adjacent to the cooling plant. Cold air ducts or chutes are attached to the trap doors through which ice is loaded into the ice bunker, and in some cases the cold air ducts are attached to the doors of the car. Suitable cold air and return warm ducts are provided. A fan circulates the cold air at a low temperature rapidly through the ducts which results in circulation of the air through the car removing the heat to a coil room or bunker room containing refrigerating pipe coils where the heat is absorbed, and the air after being cooled is again sent on its mission of cooling. This results in a very constant cooling or chilling, but it may be said that the fruit is not uniformly cooled throughout the car by this method, as it is not possible with ordinary waste of loading to secure a uniform flow of air throughout the body of goods in the car. Some very remarkable quick cooling results have been obtained by this method of cooling, the temperature being reduced from that of the ordinary outside air, say 80° to 90° F. down to 35° or 40° F. in from one to three hours. Very rapid cooling is absolutely necessary with this method, for the reason that it is not permissible for obvious practical reasons to have loaded cars standing on sidetrack for any considerable length of time. A rapid cooling or chilling of fruit is positively detrimental to its quality, and it is one of the reasons why car cooling cannot produce as perfect results as may be had with warehouse cooling. Added to this the lack of uniformity in cooling as above suggested, and added still again to this the fact that the cars may stand on the siding for some hours waiting their turn at the pre-cooling plant, it may be seen that the car cooling, practically, is a very difficult thing to work out, to say nothing of the inferior cooling results secured.

A car pre-cooling plant must be equipped relatively with a very large refrigerating capacity in order to accomplish the very rapid cooling required on account of the limited time available. This means a very high first cost for a plant of this type, and the construction of numerous plants at points where but few cars are to be cooled is impracticable for this reason. The relative high cost of the machinery required and the short time each day that this machinery can be used from a commercial standpoint makes the car pre-cooling plant a difficult problem. If, as has already been established at several places, it is attempted to concentrate the car cooling plants at chief shipping points, the delay and additional cost of switching cars to such plants, is a further disadvantage. There are other practical objections to car cooling also, and the difficulty of attaching the removable ducts to the cars without much loss of refrigeration due to leakage of cold air, has not been overcome.

WAREHOUSE PRE-COOLING.

Pre-cooling rooms arranged for this purpose need not be essentially different than regular cold storage rooms except that it is desirable to take the heat out of the fruit as rapidly as practical, and therefore, a larger refrigerating capacity is necessary and a fan system of air circulation desirable although not absolutely necessary. In warehouse pre-cooling the time element is not so important, and the fruit may be advantageously kept in storage under refrigeration for several days if necessary. Three days or 72 hours is considered correct by this method for citrus fruits and the longer keeping varieties and 24 to 48 hours for peaches and softer fruits. A reasonable length of time should be taken for pre-cooling because a sudden cooling or chilling of warm fruit means a rapid change in its cell structure and this hastens deterioration. Therefore, the longer period practically that is consumed in fruit pre-cooling, the better the results to be expected. In warehouse cooling the air need not be as cold nor circulated as rapidly as in the car method. The room may be better insulated and better constructed, and there will be very much less loss of refrigeration, and the cooling is accomplished at much lower cost. The warehouse type of plant is the only one practicable for the shipper who desires to pre-cool his own fruit, or for fruit associations who desire to put up their own plant for this purpose.

Another advantage of the warehouse method of pre-cooling is the possibility of loading a much greater quantity of fruit into a car of a certain given capacity. With the car cooling method the fruit must be piled open so as to allow a free circulation of air throughout, whereas with the warehouse method wherein the fruit is cooled before loading, it is not necessary to leave any space whatever between the packages of fruit in the car, and this results in increasing the capacity of each car by 25 per cent to 50 per cent as has already been suggested.

POSSIBILITIES OF FRUIT PRE-COOLING.

Aside from the advantage of pre-cooling fruit to prevent deterioration and to have it arrive on the market in the best possible condition there are some practical advantages which are not yet generally understood.

It has apparently been assumed that pre-cooled fruit after loading into refrigerator cars, must necessarily be iced in transit to maintain temperatures in the cars and deliver the fruit in prime condition at destination. This assumption is based on the use of the average refrigerator cars now in service. If suitably insulated cars were provided, no ice bunkers or other means of refrigerating would be necessary for a ten days' trip in the warmest weather. By suitable insulation is meant insulation equivalent to what is now used in our best cold storage houses.

Assuming a suitably insulated car, and that it will be loaded full of fruit, the more the better for our purpose. If the fruit is cooled before shipping, to 32° F. and the doors and ventilators tightly closed, this car will arrive at its destination during average summer weather at a temperature of from 45° F. to 50° F. This would be a very suitable temperature for unloading the fruit into the average temperature to be encountered during the spring or summer season. Of course, in shipping during the winter when low temperatures are likely to be encountered in transit or when unloading, it would be better to pre-cool the fruit to only 35° F. or 38° F. These statements are approximate, and are based on theoretically correct figures and can be depended upon in practice.

To insure best results under the above suggested method, not only should the fruit be loaded tightly into the car without air spaces between or around the fruit, but it should be protected on top if much space is left between the fruit and the top of the car, by means of thick, heavy paper spread across the car and secured by battens. If the car is loaded nearly to the ceiling this need not be done. Of course, it is unnecessary to state that if the car is loaded during warm weather, it should be blown out for an hour or so with cold air before loading fruit into it, and a suitable loading vestibule reaching to the sides of the car to not only protect the fruit while loading, but to keep the car cold, would be desirable.

It must be understood that this applies to fruit which is pre-cooled before loading and which is loaded into an insulated car, which is suitably insulated equal to the best modern cold storage insulation. Under this method of handling the ventilator of the car should not be opened at any time while en route to destination. It is assumed that at least 600 boxes of oranges would be loaded into one car.

It will be readily appreciated that the above suggested scheme opens up some possibilities in fruit shipping which were not dreamed of up to very recent date. It would seem that icing by the railroads with their arbitrary and outrageous charges therefor would be entirely eliminated from the shipping of fruit for long distances in future. The plan proposed of pre-cooling and carrying through to destination without icing or without supplying refrigeration will not only be a great saving to the shipper in loss of fruit by damage while in transit, but it will also be a big saving to the railroads on account of not being obliged to haul ice or other means of cooling, and lose time by the stopping of trains at intervals for re-icing. The entire proposition looks so simple and sensible that there can be no real argument against it except preconceived ideas and former practice. It will be interesting to see who will be first to adopt and work out this suggestion.

Pre-cooling has already become a very important feature of fruit production and marketing, but the developments are unimportant as compared with what they will be in future years. The general process has been thoroughly demonstrated and is by the best informed and best qualified judges of the situation, admitted to be a necessity for proper marketing. Adequate facilities will doubtless be provided after a time, but this will be when those who are most vitally interested, the growers and shippers, are thoroughly convinced that it will be profitable, and when they are better able financially to provide their own pre-cooling facilities. It will not do at all to depend on the transportation companies for means for doing this work. They are chiefly interested in getting their pay for transportation, and as long as they can get this with present facilities and equipment, it is not at all likely that they will see fit to improve them materially. What is said here applies particularly to refrigerator cars and their construction which we will now consider separately.

REFRIGERATOR CARS.

The refrigerator car service of this country is deficient and unsatisfactory, and this fact will doubtless be admitted by those who really know and at the same time are disinterested. While the average refrigerator car is certainly better than a box car for shipping perishable goods, it is far from satisfactory and far from being as perfect as it might be, and the cars now in use fall far short of what they should be considering the length of time through which they have been developing.

If reasons are sought for the present inferior refrigerator car service, the answer by those in control of the business is that they are handicapped by present types of construction and former practice and by the necessity of keeping down the cost, as well as by the contracted area or space available. These reasons are plausible and to some extent reasonable, and they are based on facts, but the reasons are nothing more than reasons, and they may all be overcome. It is, of course, necessary to keep within the natural practical limits of car dimensions, but these are not all arbitrary within reason. Present construction and former practice except as it applies to methods of icing, loading, etc., need not stand in the way of radical changes if they are found to be improvements. Keeping the cost down is all right enough from a business standpoint, but if great benefits are to be derived from improved construction with increased cost, the first cost of a refrigerator car need not stand in the way of improvements.

There are, as the author sees it, two chief reasons which are not mentioned above, and which probably have more bearing on the present inferior refrigerator car construction and service, practically considered, than the ones given:

First: The refrigerator car service is mostly in the hands of two or three strong companies. Improved devices suggested by outsiders must first be approved by those in authority; and this means the approval by persons generally with limited experience in refrigeration and with no scientific training. Add to this the possibility that railroad officials and their friends are interested in the building of refrigerator cars now in use, and a good reason for lack of progress is plain.

Second: The refrigerator car lines are in business to make money. If they can get as much for the use of a poorly insulated car of inferior construction as for a first class car, why should they provide something better? It would be an awful thing if they should spend $200 or $400 more in building and insulating a more perfect car. The shippers pay the bill for icing, and why should the cost of cooling interest them? And if the fruit partly spoils or deteriorates as it frequently does, it is the shippers' loss, not the car lines'.

There are, of course, other things to be considered like lack of proper skill and knowledge of what constitutes suitable insulation for a given work, but the above may be considered as chief. Considering the fact that shippers the country over have been paying a price high enough for the best of equipment and service, the first cost of a refrigerator car should be of secondary importance. Surely there is no excuse for economy to the extent of a few hundred dollars on each car, when a positive saving in operating cost to much more than pay big interest on the investment, may be shown, to say nothing of the saving in damage to goods in transit.

We now get back to pre-cooling. As has been already suggested above, pre-cooled oranges (*pre-cooled during a period of two or three days in the packing house and not for as many hours at the railroad car-cooling plant*) are now being shipped without icing during the cool weather of March and April in the inferior refrigerator cars now provided. If suitably insulated cars were obtainable, *oranges properly pre-cooled could be shipped from California to New York in any weather without icing,* and would arrive in really better condition than when shipped with ice in the regular way, and there would not be any necessity for icing *anywhere* if the car insulation were as good as the average cold storage warehouse. The insulation of the present refrigerator car is only an excuse for insulation, as it is not more than one-quarter to one-third the insulation considered correct for a modern cold storage plant of small capacity.

It will be claimed that additional insulation means the loss of so much space, as well as increasing the weight to be hauled. This will be such a small fraction of the possible saving resulting from loading 25% to 50% more fruit into a car *when the fruit is properly pre-cooled,* that it needs no answer.

The National League of Commission Merchants assembled in convention early in the year 1913 offered resolutions condemning the present equipment of the private refrigerator car lines as inadequate and obsolete, and suggesting that this feature of the transportation business be put under the jurisdiction of the Interstate Commerce Commission. They also passed resolutions favoring the building of pre-cooling plants at shipping centers. This fact is mentioned to show that refrigerator cars and pre-cooling are closely allied, and a representative body of produce dealers like the National League of Commission Merchants may be considered qualified to pass on the results which have been secured in fruit transportation and shipping.

Criticism is perhaps unfair without suggestions for improvement, and the author, therefore, offers the following suggestions for improving the insulation of refrigerator cars. It is not practicable to offer suggestions for improved refrigeration as this involves complicated mechanical details, but improved insulation is such a crying need that it is very easy indeed to suggest improvements along this line.

It may be interesting to note that the present average insulation of a refrigerator car consists of not more than one to two inches in thickness of insulating material, and not that much where the framework of the car interferes. There are, of course, some air spaces, but air spaces are obsolete as insulation, as is now well known by competent refrigerating engineers. The suggestion, therefore, is made that the insulation be increased to at least six inches of some of the better insulating materials like hair felt, or sheet cork, and that the frame of the car itself be filled with some material like mill shavings or granulated cork. This material must, of course, be properly protected from access and penetration of air and moisture, and this may be easily accomplished by using the best grades of insulating paper.

The question at once comes up as to what additional sum of money this would mean in increased cost of a car, and roughly speaking, assuming the exposed surface of an ordinary refrigerator car at 1500 square feet, the extra expense of suitable insulation over what is now being used would be in the neighborhood of $400 per car. Ample insulation in a refrigerator car is especially necessary on account of the large outside exposure compared with the cubic capacity, also because the car is exposed to the direct rays of the sun.

If the refrigerator car service were handled as most other businesses are necessarily handled, and those who own and operate the refrigerator cars were obliged to pay for the icing of same out of their own pockets, the cost of increased insulation would be paid in a year or two in the actual saving in ice consumption to say nothing about the great saving from deterioration of the perishable goods which are transported in the car. Those who are responsible for the building and operating of refrigerator cars must be induced by peaceable means, if possible, or by compulsion if more reasonable methods fail, to provide adequately insulated cars for both summer and winter service.

What constitutes adequate insulation is necessarily subject to a difference of opinion, but this can all be figured out in dollars and cents when it is reduced to the actual loss from ice melting, and if something may be allowed for unnecessary loss from damage to perishable goods shipped, the showing is in favor of a heavy insulation on refrigerator cars instead of the present flimsy and inadequate quantity which is provided as a mere excuse and called suitable insulation. The refrigerator car service up to the present time is almost as bad an abuse and an imposition on the shippers of perishable goods as the express abuse has been on the shippers heretofore.

SUGGESTIONS ON SHIPPING ORANGES WHEN THOROUGHLY PRE-COOLED IN PACKING HOUSE.

During March pre-cool to 36 or 38 degrees only.

After March, pre-cool to 32 degrees. Do not use ice in the ice bunkers of refrigerator cars.

Load tightly in the car and load as heavy as the rules of the railroad, or circumstances, will permit. It should not be necessary to leave more than a foot of space above boxes at the top of the car.

Put paper on top tier of fruit if not loading up to capacity. Run the paper crosswise, lapping about 6″ to 12″ and fasten with lath on both sides. Use as wide paper as obtainable to avoid multiplicity of joints.

Paper as tightly as convenient the openings from the body of the car to the ice bunkers. This is to be done, of course, only when the shipments are not iced either before shipment or in transit. Use a heavy and rather porous paper rather than the glazed insulating paper.

Protect fruit fully from contact with air while moving from cooling rooms to the car. Extreme care is especially needed at the car door. If a suitable canvas vestibule is provided, the car may be too dark to work in and an extension electric light may be used.

It may be possible to cool the car partly in warm weather before loading fruit into it, by attaching suitable cold air spouts for an hour or two, much as is done at the railroad precooling plants.

The idea in not cooling to 32 degrees in March is to provide some resistance to frost, if below freezing temperatures are encountered in transit. Also from the fact that higher temperatures will answer as well during cool weather.

It seems that we are not near the car capacity with even as many as 600 boxes. If so, we are limited only by the number of boxes possible to load into the car. There may, however, be some business reason for not loading a larger number of boxes than at present.

By papering the openings from the car to the ice bunkers and loading heavily, a mass of cold fruit is obtained which will aid much in carrying the low temperature for several days. Note that these suggestions apply only to pre-cooled fruit, which is not to be iced in the car either at time of shipping or in transit.

CAR COOLING METHOD ILLUSTRATED.

The plan and section shown herewith represent on a small scale the principle of the car method of pre-cooling. The essential parts of such a plant consist of the primary means of cooling located in coil room or bunker room; a fan for circulating the air over the primary means of cooling, and suitable discharge and return ducts for conveying the air to and from the car to be cooled. In the plant here illustrated flexible detachable ducts are arranged so as to be attached to the trap doors of the ice bunker of the ordinary refrigerator car. This cooling plant was designed by the author and has been subjected to a satisfactory test, and the following description gives the result of same:

"The car is so connected up to the cooling room that the air goes in at one end, through the ice bunker door and out at the opposite end, then back through the room to be re-cooled. In this method the fruit is given air at a temperature of ten or fifteen degrees colder than the temperature of the car and within an hour the temperature of the air going into the car is reduced to about 35° F. During the next hour the temperature of the air is reduced to 32° F. The cooler and fan are of sufficient capacity to bring the temperature of the car down to 32° in from three to six hours, according to the pack of the fruit and the temperature in the beginning.

"The efficiency of this system of pre-cooling was tested by the Northern Pacific Railway Co., who had one of their

FIG. 1—FRUIT PACKING HOUSE AND PRE-COOLING PLANT, UPLAND HEIGHTS ORANGE ASSO., UPLAND, CAL.

most modern insulated refrigerator cars loaded with pears and placed on track in the state of Washington for pre-cooling. A cold blast was forced through this car for twelve hours until the temperature of the air in and out showed the same, 34° F.; then the ice bunkers were thoroughly filled, all doors and openings to the car were locked and the keys turned over to an attendant who accompanied the car to Minneapolis to see that no changes were made enroute. At Minneapolis the ice bunkers were well filled, temperature at the top and bottom of the fruit at the middle of the car still remained 34°, and the fruit was in prime condition. The car was forwarded to Chicago, where the same conditions were observed as to ice, temperature and condition of fruit. From here it was continued to New

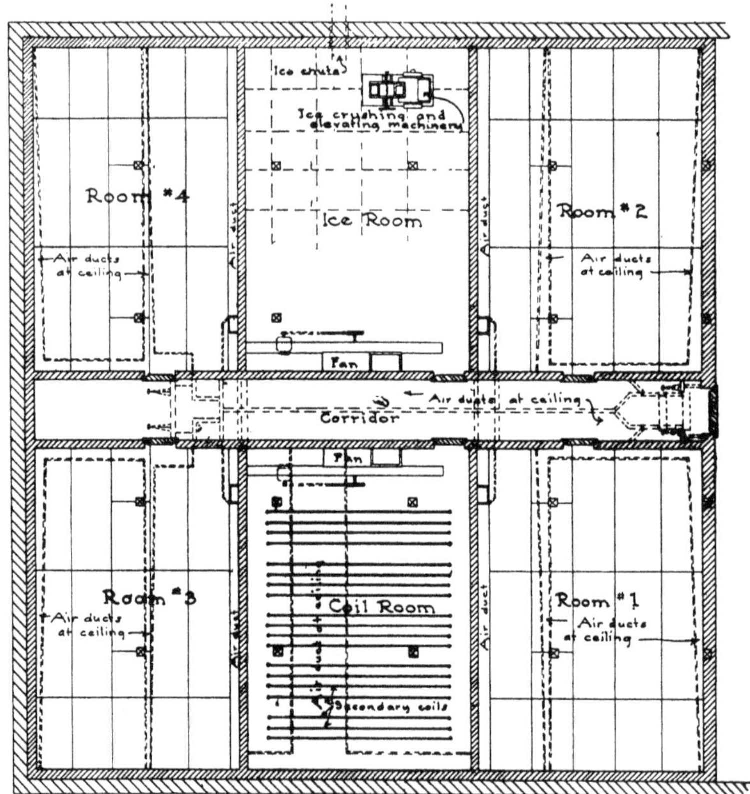

FIG. 2—PLAN OF BASEMENT.

York city, where an inspection showed one ice bunker half full of ice, one two-thirds full and the temperature of fruit at the top warmed up to 36°, while the temperature on the floor still remained 34° and the fruit in prime condition. By this pre-cooling process peaches have reached Texas and Philadelphia in first class condition."

WAREHOUSE METHOD OF PRE-COOLING.

The pre-cooling plant or apparatus of the Upland Heights Orange Association shown in the accompanying illustrations is located in the basement of the building entirely below the platform level, and a very good idea of the arrangement of the pre-cooling rooms is shown in the basement plan. The transverse and longitudinal sections also show the relation of the pre-cooling plant to the packing room on the main floor of the building.

To those who have never visited a fruit packing plant, the complicated machinery used for handling, grading, sorting, packing, storage and loading of fruit into cars is something of a novelty, not to say a revelation of the possibilities of automatic machinery. A bird's eye view of the graders, as they are called, is shown in the view of the interior. These consist of belts which carry the fruit along until it drops into the receptacle arranged for its particular size. Before passing to the graders, the fruit is delivered onto traveling belts where it is sorted by experts. From the graders where the fruit is

FIG. 3—VIEW IN PACKING ROOM, UPLAND PLANT.

graded into sizes, it is taken by the packers, wrapped with paper and packed into boxes. After having covers nailed on, the fruit is placed on a conveyor which delivers it down to the pre-cooling rooms in the basement. The fruit enters through the corridor, and may be delivered to any one of the four pre-cooling rooms. Each room has a capacity of three car-loads, and this represents the daily pre-cooling capacity of the plant. For instance: Room No. 1 may be loaded with fruit today; room No. 2 tomorrow; and room No. 3 the third day. While room No. 1 is being discharged on the fourth day, room No. 4 is being loaded; so that allowing three days

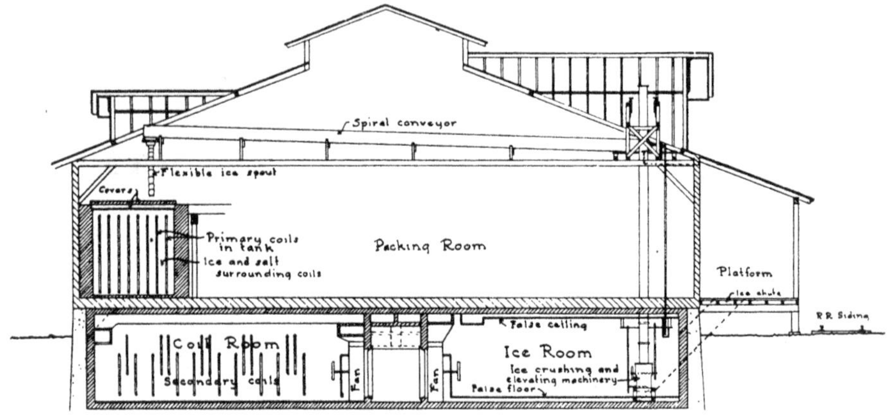

FIG. 4—TRANSVERSE SECTION.

for pre-cooling and with four rooms, one extra room is given for the loading and unloading of fruit. After being pre-cooled, the fruit can pass out through the corridor, and by means of endless chain conveyors and elevators is delivered directly up to the car platform shown in the exterior view, where it is loaded directly into a refrigerator car without exposure to warm air.

The pre-cooling apparatus and system consists of: first, the ice room; and second, the coil room, both shown in the basement plan. The ice room is to be kept filled with ice which is delivered in cars as required. Air from the ice room is, by means of a fan and suitable air ducts and gates delivered to any one or more of the four pre-cooling rooms and for any period required. The air from the ice room is used for cooling for a period of from one to two days and the temperature of the fruit brought down to about 40° to 45° F. After bringing the fruit down to this temperature by means of air from the ice room, again by a fan and suitable arrangement of air ducts and gates, air is used from the coil room until the temperature is brought down to the desired point for shipment, about 32° to 35° F. The length of time required depends necessarily on temperature of fruit when stored and quantity of fruit under pre-cooling at one time. When cooling three cars per day it has been found that

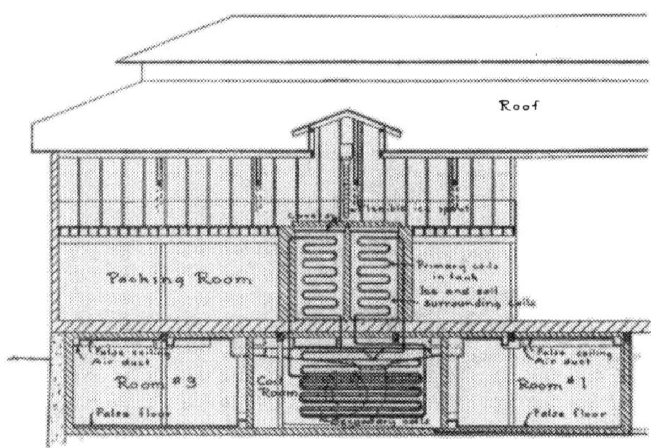

FIG. 5—LONGITUDINAL SECTION.

in about forty-eight hours air from the ice room will reduce the temperature of the fruit to 45° F., and that in about twenty-four hours air from the coil room will still further reduce the temperature of the fruit to 35° F.

When ice is delivered from the cars it is loaded into the ice room by means of a lowering rig. Located in the ice room is the ice crusher. This is arranged to deliver ice to an endless chain bucket elevator, which in turn delivers it to a spiral conveyor, which is shown in the transverse section. From the spiral conveyor by means of a flexible spout the ice is delivered directly to the primary tanks of the Cooper brine system.

The pre-cooling plant is equipped with the complete Cooper Systems with the exception of the ventilating system. Each one of the four rooms is equipped with the false floor and false ceiling system of air circulation. The coil room contains the secondary coils and these coils are equipped with the Cooper calcium process for preventing frost on the cooling pipes and purifying and drying the air of the room.

PLANT OF POMONA VALLEY ICE CO.

In and near Pomona, which is one of the great centers of the orange industry, a number of the Orange Growers' Associa-

FIG. 6—PRE-COOLING AND COLD STORAGE PLANT, POMONA VALLEY ICE CO., POMONA, CAL. WAGON APPROACH AND RECEIVING PLATFORM.
Mountains a mile high may be seen dimly to the left.

tions saw the advantage of pre-cooling their fruit before loading it into the cars, and the further advantage of being able to hold in cold storage for a time, some of their fruit instead of shipping it all as fast as packed. In this way a considerable business was offered to the Pomona Valley Ice Company, which has a well equipped ice factory at this point. For a time the oranges sent to this concern for pre-cooling or storage were handled in one section of their ice storage rooms, but as that space was needed for the storage of ice, the company decided about a year ago to erect a pre-cooling and cold storage house especially designed for the handling of oranges.

Views of this plant are shown herewith. The building is approximately 60x90 feet, with a basement and two floors above ground. Commodious hallways cross through the building from the wagon platform to the railroad side of the build-

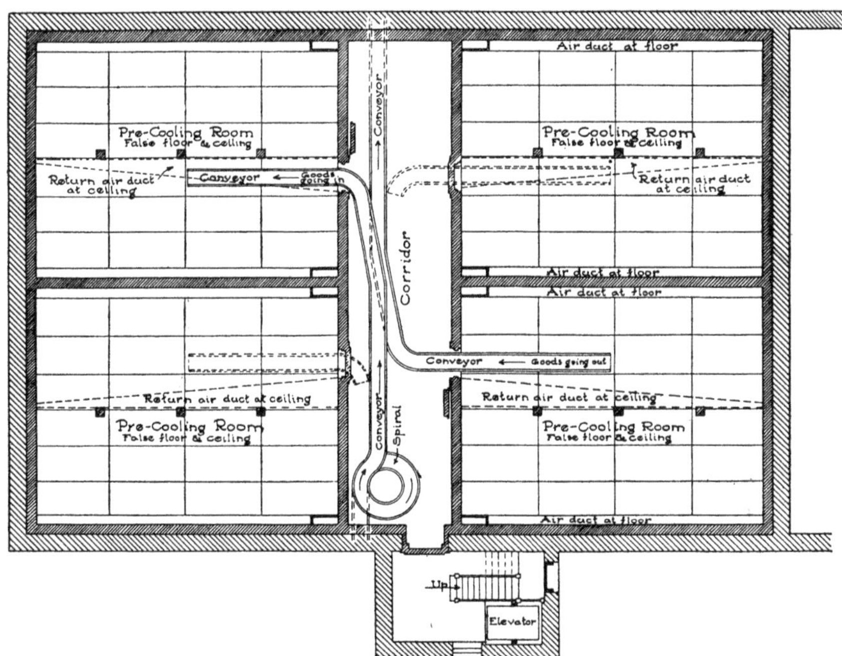

FIG. 7—BASEMENT PLAN.

ing, and on either side of this hallway, in the basement, are two storage rooms, each of a capacity of from five to seven carloads, according to the height to which the boxes are piled. The same arrangement of rooms occurs on the first floor. On one side of the hallway on the second floor are two more storage rooms, and on the other side next to the factory are four coil rooms. The general arrangement of plant is shown by the plans and sections.

As will be seen in the accompanying view, each coil is made of a continuous bent pipe. The Cooper forced air circulating system is used, air being driven by a special fan located in each coil room through air-ducts leading to the various storage rooms. In one of the accompanying views these air-ducts can be seen, also the perforations in the false floor through which the air is admitted to the storage rooms. One can

FIG. 8—POMONA VALLEY ICE CO. PRE-COOLING PLANT.
Railroad side showing platform on which ice is brought from ice storage to the left and loaded into the bunkers of refrigerator cars.

also see in this same picture, the corresponding openings in the false ceiling through which the air is returned to the coil rooms to be again chilled after it has been somewhat warmed by doing its work of refrigeration. No difficulty is found in maintaining any desired temperature. There is a surprisingly small difference in temperature between the coil rooms and the storage rooms, and even when it is found desirable to not operate the refrigerating machinery for some hours, practically no rise of temperature occurs in the storage rooms containing fruit which has been thoroughly cooled. This arrangement of combined ice making and pre-cooling seems to be a mutually advantageous scheme for an ice plant located in a fruit growing district, and for growers located near such a plant. It makes business for the one and furnishes very desirable facilities for the other, and at storage rates which the experienced manager of a refrigerating plant can readily understand are less than could be obtained by a fruit shipping concern which operates its plant only for pre-cooling and is therefore able to use its full capacity during only a comparatively small portion of each year.

FIG. 9—VIEW IN ORANGE PRE-COOLING AND STORAGE ROOM. Equipped with the Cooper False Floor and False Ceiling System of Air Circulation.

SOME FIGURES ON FRUIT PRE-COOLING.

Facts and figures are always interesting even though they may be fragmentary and somewhat incomplete, and as a general thing facts and figures cannot be made to cover all cases. The following, however, will be of some general assistance:

SHIPPING ORANGES WITHOUT ICING.

As applied to the California orange shipping service it has been already suggested in this chapter that oranges might be shipped after being properly pre-cooled, in a suitably insulated car for several days without the necessity of re-icing, and a few figures covering the possibilities of this subject will, therefore, prove useful. Assuming a car of 600 boxes of oranges weighing 70 pounds each, to be loaded into a refrigerator car with insulation equivalent to what might be called first class cold storage insulation, and such as the author has commonly designed for cold storage work, the cooling results could be ex-

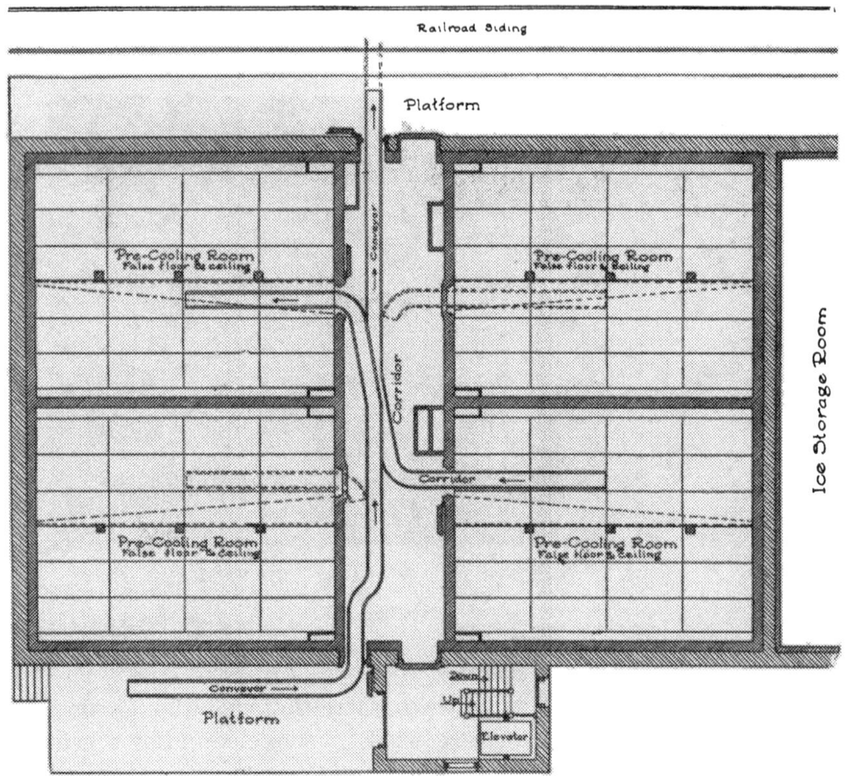

FIG. 10—FIRST FLOOR PLAN.

pected if the fruit were cooled to 32° F. before loading it into the car, and the car chilled by blowing it out with cold air before the fruit was loaded, and assuming that the car in transit would be exposed during a period of eight days to an average temperature of between 80° F. and 85° F.; the oranges under these conditions would arrive at their destination at a temperature of about

50° F. In other words the stored up refrigeration in the 42,000 pounds of oranges cooled to 32° F. would be sufficient to carry the refrigeration for eight days with a final temperature on the fruit of 50° F., if the car were exposed to an average temperature of 80° to 85° F. while in transit. It would be better for the oranges to arrive on the market at a temperature of 50° F. than 32° F., and they would be in better condition for unloading and exposure to a comparatively high temperature. These figures are approximate, but they are sufficiently accurate to show the possibilities of fruit pre-cooling and transportation if suitably insulated cars are provided. In other words re-

FIG. 11—VIEW IN COIL ROOM SHOWING FROSTED AMMONIA COILS. SECOND FLOOR PLAN.

frigerator cars would not be necessary, but only insulated cars providing the insulation were of sufficient value and equal to the author's standard cold storage insulation.

QUANTITY OF ICE REQUIRED FOR PRE-COOLING.

The cost of pre-cooling may be figured on a basis of ice costs, and the following figures will prove useful in this connection:

Taking gross weight of the average orange box, (85 pounds) and 448 boxes to the car, this would give us 38,080 pounds. The specific heat of oranges is .92, and the tons of refrigeration required for cooling a car would be computed as follows:

$$\frac{38080 \times .92 \times 36^\circ}{284{,}000} = 4.44 \text{ tons}$$

In the above 36° represents the range of temperature, or say from 70° F. down to 34° F. The product of the number of pounds multiplied by the specific heat, multiplied by temperature range, and divided by 284,000 (the number of heat units representing a ton of ice melting) gives the number of

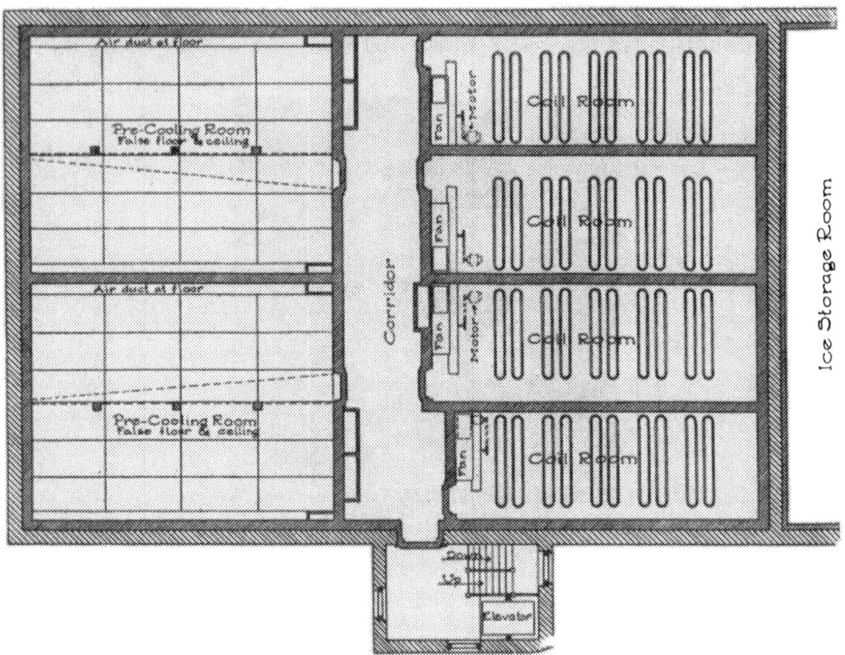

FIG. 12—SECOND FLOOR PLAN.

tons of refrigeration required to do the work of cooling; or 4.44 tons. This is doubtless not exact, as the estimated gross weight of a box of oranges is taken, and the specific heat of the wood in the box is not exactly the same as the fruit itself. The figures, however, are near enough for any practical purpose. It must be understood, however, that this does not represent accurately the amount of ice required to do the entire cooling, for the reason that no heat leakage through the insulated walls of the cooling room is represented in the above calculation, and as this would be a variable quantity depending on outside temperature, insulation, etc., it may be neglected.

PRE-COOLING GRAPES.

Assuming a carload of 24,000 pounds of grapes is to be cooled from 85° F. to 35° F. or through a range of 50° F. This will mean approximately that a pound of ice will cool nearly three pounds of fruit, and it will, therefore, take about four tons of ice to cool a car of grapes. If the ice costs $4.00

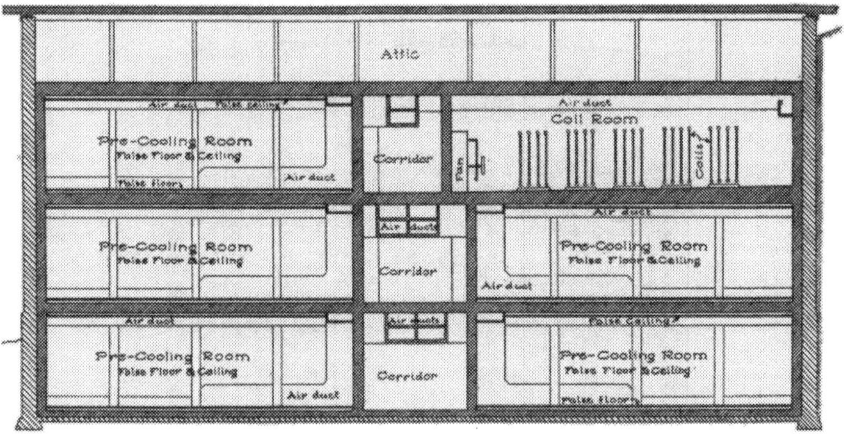

FIG. 13—LONGITUDINAL SECTION.

per ton it would require $16 worth of ice to do the actual cooling of grapes. To this would need to be added about 25% for heat leakage and other losses, and on this basis about $20 would be the cost of pre-cooling a carload of grapes through the range of temperature indicated. In many places ice may be had at a much lower cost than this, especially in natural ice territories, and in other places the ice costs may be higher.

SOME ADDITIONAL COOLING COSTS.

At Pomona, California the Fruit Growers' Exchange formerly secured their refrigeration from the local ice plant and paid for same on a basis of four cents per box. They also figured that it cost them $2.00 per car to place the fruit in the rooms and remove it to the car on gravity carriers. They paid from $3.00 to $3.25 per ton for ice in the bunkers of the car, and their total pre-cooling and pre-icing charge was figured at about $32.50 per car, allowing for interest, depreciation and taxes. This plant handled between 400 and 500 carloads per year, and their rooms had a total capacity of about 42 cars of fruit. These rooms were cooled by a fan system of air circulation from a bunker room with direct expansion ammonia piping, and the total cost of cooling rooms and plant represented an investment of between $25,000 and $30,000.

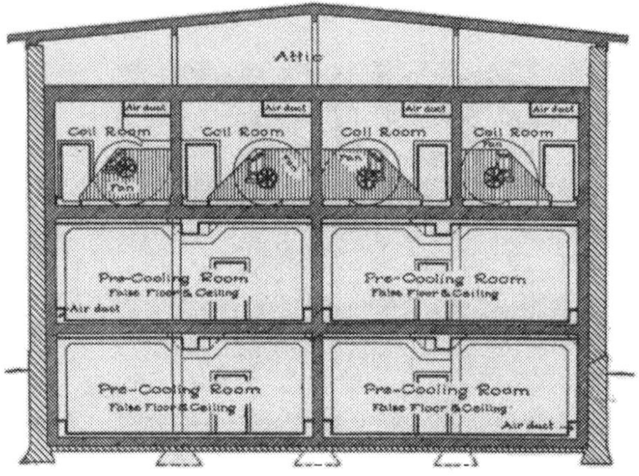

FIG. 14—TRANSVERSE SECTION.

At East Highlands, California, the Fruit Association has a cold storage and ice making plant complete. There are six rooms with a combined storage capacity of about 24 carloads, with five carloads per day capacity for pre-cooling. About 48 hours is consumed in reducing the temperature of the fruit to 33° F. The method of cooling is gravity air circulation, with the refrigerating pipes located above the fruit room. The total cost of this plant was about $50,000.

At the plant of the Upland Heights Orange Association, described in preceding pages about 150 carloads of fruit are handled per year, and there are four cold storage rooms with a combined capacity of about 12 carloads, and the plant has a cooling capacity of about three cars per day. The ice required for pre-cooling and pre-icing costs about $3.75 per ton freight paid to Upland. The total cost of this plant including equipment and insulation of the rooms, but not including any part of the building cost was about $11,000. G. Harold Powell,

FIG. 15—VIEW IN CORRIDOR IN BASEMENT.
In the background, spiral of incoming gravity conveyor is shown and sections of roller conveyor leading to outgoing elevator.

the well known fruit expert, has the following comments on this system: "The advantage of this system lies in the fact that the shipper is relieved of the management and maintenance of a complicated refrigerating plant. It is easily operated. The depreciation is comparatively small. The practicability of the system depends on the prices at which the shipper can purchase the ice. It may cost more per ton for the ice for this type of plant than the cost of ice manufactured at the shippers' mechanical ice plant. The operating expense and the maintenance of the gravity brine system will be less and he will have to balance one against the other in considering the comparative merits of the two systems."

THE RAPIDITY OF COOLING FRUIT.

To those interested in pre-cooling of fruit we commend a careful perusal of what G. Harold Powell has to say on the above subject. Mr. Powell has, doubtless, had more experience in fruit pre-cooling and shipping than any other living person, and what he says may be considered authoritative.

"The rapidity of cooling the fruit depends primarily on the difference between the temperature of the cold air and the temperature of the fruit; secondly, it depends on the method of circulating the air over the fruit. Low temperature and rapid circulation means quick cooling, but there is a limit in both temperature and rapidity of circulation beyond which it is not safe or economical to go. This is still in the experimental stage. As a general principle, the air should be as cold as the fruit will stand without injury. When the fruit is warm the temperature of the air may be below the freezing point without danger to the fruit. How low it is safe to run the temperature under these circumstances is still in the experimental stage. When low temperatures are used, it is necessary to provide heavier refrigerating machinery and better insulation to insure against an excessive loss of refrigeration through the walls of the building. To cool the fruit uniformly in all parts of a package requires 24 hours or more of time. Quicker refrigeration by extremely low temperature subjects the exposed fruit to the danger of freezing. It results in the uneven cooling of the fruit in the packages and in the different parts of the room and is expensive on account of the comparatively large loss of refrigeration through the walls of the chill room during the cooling of the fruit."

AMOUNT OF REFRIGERATION REQUIRED TO PRE-COOL ORANGES.

Mr. Powell is again quoted as follows:

"The amount of refrigeration required to cool a carload of fruit to a desired temperature depends on the initial temperature of the fruit; the temperature to which it is to be reduced, and the insulation of the plant. To reduce the temperature of the fruit and the package alone will require an amount of refrigeration as set forth in the following table:

Range of Cooling	Tons of Refrigeration Required to Cool Oranges. 448 Boxes
100 to 35 degrees	5.94 tons
90 to 35 degrees	5.03 tons
80 to 35 degrees	4.11 tons
80 to 35 degrees	4.11 tons
70 to 35 degrees	3.20 tons

The figures above represent the actual amount of melting ice required to reduce the temperature of the fruit and packages. To these figures should be added at least one-half more refrigeration, depending on the insulation and construction of the plant. It is probably safe to estimate that it would require six tons of ice or refrigeration to reduce a carload of 448 boxes from 80° F. to 35° F. To this must be added about 6½ tons for the initial icing of the car, making thereby a total of about 13 tons of ice required to cool a carload of fruit over the range of temperature specified and to fill the bunkers of the car."

CHAPTER XXV.

SHIPPING PERISHABLE PRODUCTS.*

PROTECTION FROM INJURIOUS TEMPERATURES.

The information following is largely a compilation of the opinions of farmers, merchants and shippers in all parts of the country, which were received in reply to a circular letter sent out by the United States Weather Bureau. The principal kinds of goods which are considered perishable, and for which protection from excessive heat or cold is necessary are: All fruits and vegetables, milk, dairy products, fresh meats, poultry, game, fish, oysters, clams, canned fruits and vegetables, and most bottled goods. In the transportation of perishable freight there are three primal objects to be attained:

1.—The protection of the shipment from frost or excessive cold.

2.—The protection of the same from excessive heat.

3.—The circulation of air through the car, so as to carry off the gases generated by some classes of this freight.†

The degree of cold to which perishable goods may be subjected without injury varies greatly with different commodities, and depends somewhat on the time the shipment will be on the road, its condition when shipped, whether it is kept continually in motion, and also on whether it is unloaded immediately upon arrival at its destination, or allowed to stand some time. The direction of shipment, whether toward a cold area or away from it, should also be considered.

---

*Abstracted from Farmers' Bulletin No. 125, United States Department of Agriculture.

†What is meant, doubtless, is ventilation or the allowing of outside air to circulate through the car by opening vents or ice bunker covers. This is desirable where the fruit is loaded in a heated condition, but not at all necessary when fruit is properly pre-cooled as it should be.—Author.

CARS, APPLIANCES AND METHODS OF SHIPPING.

Precautions taken in shipping to protect from cold are packing in paper, straw or sawdust, boxing, barreling with paper lining, shipping in paper lined cars, refrigerator cars, and cars heated by steam, stoves and salamanders.

Shippers and agents concur in the statement that danger in transportation by freezing can be practically eliminated by the shipment of produce by modern methods; the lined car suffices in spring and autumn, and usually during winter, while in extremely cold weather specially built cars are used.

In ordinary freight cars perishable goods can be shipped with safety with the outside temperature at 20° F., and in refrigerator cars at 10°. In the latter these goods may be safely shipped with an outside temperature of from zero to 10° below, if the car is first heated, and at the end of the journey the goods are immediately taken into a warm place without being carted any great distance.*

To protect goods shipped in an ordinary car, the sides of the car should be protected by heavy paper tacked to the wall, and by the addition of an inner board wall, a few inches distant from the outer one. A car thus equipped and packed with produce, surrounded by straw, will retain sufficient heat to prevent injury for twenty-four hours, the average air temperature inside the car being at least twelve degrees higher than the outside air. Cars are sometimes warmed by steam from the locomotive when in motion, and by stoves when steam is not available. Cars, after being loaded, are carefully inspected as to temperature within; their destination is considered; and, if the weather is exceedingly cold, or is liable to be, the car is often accompanied by an attendant; otherwise it is inspected

* Any statement cannot be as positive as this and be accurate when applied to so varying a subject as shipping of perishable goods. The protection of food products in shipment during extremely cold weather depends on several things with a great variation of conditions. Fully as much depends on the temperature of the goods themselves as on the temperature of the car and the use of insulating substances for packing the goods or the use of an insulated car. Take as an example the shipping of eggs: If loaded into a good refrigerator car at a temperature of 30° F. (as when loading from the cold storage room) no amount of protection or the use of an extra well insulated car will prevent freezing if on the road for several days with an outside temperature below zero. On the other hand, if started at a temperature of from 45° to 50° F. a moderate protection will suffice, and the regular refrigerator car will take them through safely.—Author.

from time to time on the road. Lined cars—that is, cars lined with tongued and grooved boards on the sides and ends--are considered the best for shipping potatoes, as they can be heated by an ordinary stove and will stand a temperature, outside, of 20° below zero, when a man is in charge to keep up the fires.*

REFRIGERATOR CARS.

The better class of refrigerator cars will carry all perishable goods safely through temperature as low as 20° below zero, provided they are not subjected to such temperature longer than three or four days at a time; but with the ordinary refrigerator cars a temperature of zero is considered dangerous, especially if the goods they contain be of the most perishable kind.

In winter time refrigerator cars are used without ice in forwarding goods from the Pacific coast; in passing through cold belts or stretches of the country the hatches are closed, and the cars being lined and with padded doors, the shipment is protected against the outside temperature; in passing through warmer climates the ventilators are opened in order to preventing the perishable goods from heating and decaying.

It is stated, however, that for the shipment of fruit the ordinary refrigerator car is not entirely satisfactory, and that there is a strong demand for a better refrigerator car than can now be obtained.†

A car is wanted that will carry oranges, bananas, etc., without danger of chill through the coldest climates of the country, as the delays in housing are injurious to the keeping

*The most approved arrangement in a potato shipping car is a false floor and a partial false ceiling to allow of a circulation of air. The stove is placed in the center and the warm air ascends to the ceiling where it passes along to the ends of the car, descending and returning under the false floor to the stove in center of car.—Author.

†The author knows this to be a fact. The refrigerator cars now in use have been designed for the most part by men of no modern scientific or mechanical knowledge; they are inferior in many ways and present great opportunity for improvement. They have, for the most part, less than half the insulation needed, and even less than half the insulation which would be used in a stationary cold storage room of similar size. Owing to the nature of the companies controlling the refrigerator car business, the practical engineer has little opportunity of introducing improved methods in the construction of refrigerator cars.—Author.

qualities of the fruit, and the dealer is also kept out of the use of his goods.

The following is a descriptin of a much used patent refrigerator car:

"The car is double lined and has at each end of the interior four galvanized iron cylinders, reaching from the floor to near the top. Ice is broken to pieces about the size of the fist, and the cylinders filled with this ice and salt, the whole being tamped down hard. It is claimed that cars iced in this manner do not need re-icing in crossing the continent, as other styles of cars do. The car is iced in winter in the same manner as in summer, as such icing prevents freezing."*

The car that has the most floor space and will hold the greatest quantity of ice is preferred by most shippers.

Mistakes are often made in building fires in roundhouses where cars of produce are stored, unnecessarily heating it, a uniform temperature, just above the danger point, being the most favorable.

VENTILATED CARS.

In 1895 an experiment for testing the advantages of different modes of ventilation during the shipment of fruit was made under the direction of the Riverside Fruit Exchange, of Riverside, Cal. Five cars loaded with oranges were shipped a distance requiring a seven days' run. Four refrigerator cars and one ventilated or fruit car were used. Two of the refrigerator cars had the ventilators closed from 4 a. m. till 8 p. m. each day, and open the remainder of the time. The other two and the fruit car had ventilators open during the entire trip. Observations were made of the outside and inside temperatures at 4 and 9 a. m. and 3 and 8 p. m. In the first two cars the inside temperature ranged from 46° and 42° F. minimum to 56° and 58° F. maximum, respectively; in the second two, from 48° and 44° F. minimum, to 58° and 62° F. maximum, respectively; and in the fruit car from 42° minimum to 68° maximum. The outside temperatures ranged from eight

*An absurd statement. Icing and salting will not prevent freezing, and there is no use in icing during cold weather. If tanks could be filled with water, freezing of goods in the car would be, in some cases, prevented.—Author.

degrees lower to nineteen degrees higher than the inside. It was found that the temperature varied less in the refrigerator cars than in the fruit cars, owing to the fact that they were better insulated. It was also found that the fruit in the cars which had the ventilators closed during the day arrived in much better condition than that in the cars which had the ventilators open.

OUTSIDE AND INSIDE TEMPERATURES.

The relation between the outside air temperature and the temperature within the car varies largely, depending on the kind of car, whether an ordinary freight or refrigerator car, whether lined or not, whether standing still or in motion; and also on the weather, whether windy or calm, warm or cold. In an ordinary freight car the difference ranges from two to fifteen degrees, and in a refrigerator car from fifteen to thirty degrees. If the latter be provided with heating apparatus, the temperature in winter can be kept at any required degree.

From six observations taken at intervals of ten minutes, it was found that on a warm day, when the mean of the six readings outside was 68°, it was 66° F. on the inside of an ordinary freight car, and 63° F. inside of an uniced refrigerator car. On a cold day the mean of six observations was 38° F. outside and 35° F. inside of an ordinary car, and 36° F. inside of a refrigerator car; the cars were stationary.

Freight from the Pacific coast to the Mississippi valley, or to the Atlantic coast, has to pass through several varieties of climate at any time of the year, so that at one time the temperature inside the car will be materially above the outside temperature, while perhaps a few hours later it will be below.

Products sent loose in a car are packed in straw on all sides, particular attention being paid to the packing around doors, and to see that the car is full. Manure is largely used to protect perishable goods, the bottom of the car being thickly covered with it, and in some cases it is put on top of the goods.*

*No sane man would use manure in a car with perishable food products unless they were in some sealed package like cans or bottles. In any case straw, or better still, mill shavings are better than manure for any purpose of this kind.—Author.

The temperature of the produce when put into the car is quite a factor to be observed. If it has been exposed to a low temperature for a considerable time before, it is in a poor condition to withstand cold, and the length of time so exposed should be taken into account. It is also claimed that a carload of produce, like potatoes will stand a lower temperature when the car is in motion than when at rest.*

Goods at a temperature of 50° to 60° F., packed in a refrigerator car, closed, have been exposed to temperatures 10° to 20° below zero for four and five days without injury.

FRESH MEATS.

In shipping fresh meats the almost universal practice is to ship in refrigerator cars where the temperature can be maintained at any desired degree, a temperature from 36° to 40° being considered the best.

Beef.—Fresh beef for shipping should be chilled to a temperature of 36° F., although under favorable conditions it will arrive in a good state if chilled to only 40° F. The cars should be at the same temperature as the chill room, and it is considered very important to have an even temperature from the time the beef is taken from the chill room until its arrival at its destination.

In shipping long distances in summer, it is necessary to re-ice the cars, the frequency depending on the prevailing temperature, so that no fixed rule can be given. In winter the temperature is kept up to 36° F. by means of stoves or oil lamps.

If refrigerator cars are not used, the meat should be wrapped in burlaps, and the carcasses hung so as not to touch each other. With an outside air temperature of 50° F., or below, in dry weather, meat that has been thoroughly cooled will keep a week if shipped in an ordinary box car.

Pork.—Pork is injured more quickly by high temperature than other meats, and greater care should be taken with it in storing and shipping. Sudden changes in temperature of

*One of the old popular ideas without material foundation. Men and animals will withstand low temperature best when in motion, but this does not apply to perishable goods.—Author.

from 10° to 20° F. are very injurious to fresh meats, and should be provided against when possible.

Poultry.—Poultry, if shipped at a temperature of 50° F. or higher, should be packed in ice and burlaps; and if under 50° F., in dry weather, no extra precautions are needed. In shipping live poultry the coops are frequently overcrowded, resulting in the death or great deterioration of many of the fowls.

DAIRY PRODUCTS AND EGGS.

Milk.—Milk for shipping requires great care to prevent souring; it should be reduced after drawing to a temperature of 40° F., which extracts the animal heat. It should never be frozen, as it becomes watery and inferior in quality when thawed out.

Eggs.—Eggs are packed in crates with separate pasteboard divisions, with a layer of excelsior top and bottom. Pickled eggs are injured by cold sooner than fresh ones.

A prominent wholesale dealer in butter, eggs, and cheese at Chicago, says:

> Eggs in storage and transportation cannot stand a lower temperature than 28° F.; if packed well in cases and loaded in a refrigerator car they usually come through in good condition at from 5° to 10° below zero, and at 10° above zero in common cars, if not exposed more than forty-eight hours.

Butter and Cheese.—A wholesale butter and cheese firm of Chicago writes as follows:

> Butter is probably unaffected by extreme cold. We have never experienced any damage by butter being too cold; in fact, in carrying it in cold storage, it is carried at from zero to 10° above; but extremely warm weather is very injurious and damages the article to a considerable extent. To preserve butter it should be kept as cold as possible, as we state above, all the way from 32° above down to zero. It all depends upon what the facilities are for carrying the same. Of course, when we place it in cold storage the temperature we would require would be zero to 10° above, and, of course, that temperature we cannot have in handling it when we come to sell it out in our store, but we take great care not to take out of storage any more than can be readily sold. In regard to cheese, extreme cold and extreme heat are both injurious to same. For instance, extreme heat will cause cheese to swell and ferment [Not if the cheese is well made. Extreme heat injures cheese by starting the butter fat, which causes the cheese to become dry and crumbly.], while extreme cold will freeze it; the curd becomes dry and like sawdust, and it will never again be firm and stick together, but will crumble. It takes quite a temperature to freeze cheese, say 10° above for one or two days out on the road would freeze it. It is very slow in freezing and very slow in thawing out. A skim-milk cheese will freeze quicker than a full cream cheese.

FISH AND OYSTERS.

Fish.—Fish are shipped by express and also by freight. When shipped by express they are packed in barrels with ice. When shipped by freight they are packed in casks holding 600 pounds each, or in boxes on wheels, holding about 1,000 pounds each. When shipped in carload lots they are packed in bins built in the car and thoroughly iced. The amount of ice supplied should equal one-half the weight of the fish. Fish keep best when the temperature of the box in which they are stored is about that of melting ice. Under favorable conditions fish remain sound and marketable for thirty days after being caught and packed in ice. The entrails of fish should be removed before shipping, as they are the parts that most readily decay, and taint the flesh of the fish. This is especially necessary in shipping long distances.

Oysters.—Shucked oysters, shipped in their own liquor in tight barrels, will not spoil if frozen while in transit. Thick or fat clams or oysters will not freeze as readily as lean ones, as the latter contain much more water. Oysters will not freeze as readily as clams. It is safer when oysters or clams in the shell are frozen to thaw them out gradually in the original package in a cool place.

In freezing weather oysters and clams, in the shell, are shipped in tight barrels lined with paper.

FRUITS.

It is important to note that in shipping fruits, etc., many of the precautions taken in packing to keep out the cold will also keep in the heat, and there is really more danger in some instances from heating by process of decomposition than from cold. All fresh fruit tends to generate heat by this process. A carload of fresh fruit approaching ripeness, closed up tight in an uniced refrigerator car, with a temperature above 50° F., will in twenty-four hours generate heat enough to injure it, and in two or three days to as thoroughly cook it as if it had been subjected to steam heat.*

*This heating action is of small moment if the fruit is cooled before placing in the car to a temperature of 40° F. or lower.—Author.

Suitable refrigerator transportation must, therefore, provide for the heat generated within, as well as the outside heat. The perfection of refrigeration for fruit is not necessarily a low, but a uniform temperature; a temperature from 40° to 50° F. will keep fruit for twenty or thirty days, if carefully handled. Strawberries have been transported from Florida to Chicago, tranferred to cold storage rooms, and remained in perfect condition for four weeks after being picked.*

Fruit intended for immediate loading in cars should be gathered in the coolest hours of the day, and that which has been subjected to a high temperature before being shipped should be cooled immediately after being loaded. Ordinary refrigeration will not cool a load of hot fruit within twenty-four hours, and during that time it will deteriorate in quality very much. It should be cooled in four or five hours in order to prevent fermentation. It is stated that the more intelligent of the large shippers of fruit in the south have about concluded that it is impracticable with any car now in use to load fruit, especially peaches and cantaloupes, direct from the orchard into the car with assurance of safety. In deference to this opinion one southern railroad has announced its intention of establishing at the largest shipping points along its lines, cooling rooms for the purpose of putting the fruit in satisfactory condition for transportation before being loaded.

Shipments of tropical fruits in ordinary freight cars cannot be safely made when the temperature is below 30° F., except in cases where the distance is so short as not to expose them for a longer period than twelve hours, and even then they must be carefully packed in straw or hay. The hardier Northern fruits and vegetables can be safely shipped in a temperature of about 25° F., but the same protective measures must be employed as in the case of tropical fruits when lower temperatures prevail. Long exposure to temperature of 20° F. is considered dangerous to their safety. Foods preserved in cans or glass should not be shipped any distance when the temperature is below the freezing point.

*An uncommon or trial shipment. These results cannot be duplicated on a commercial scale.—Author.

Oranges and Lemons.—Oranges shipped from Florida to points as far north as Minnesota are started in ventilator cars, which are changed at Nashville to air-tight refrigerator cars, the ventilators of which are kept open, provided the temperature remains above 32° F., until arrival at St. Louis, from which point the ventilators are closed and the cars made air tight. Lemons and oranges are packed in crates. Each layer of crates in the car is covered by and rests upon straw, usually bulkheaded back from the door and car full. Oranges loaded in ventilated or common cars should be transferred to refrigerator cars when the temperature reaches 10° above zero; in transit, with a falling temperature, the ventilators should be closed when the thermometer reaches 20° F., and with a rising temperature the ventilators should be opened when it reaches 28° F. For lemons, the minimum is 35° F. for opening and closing the ventilators, and for bananas 45° F. for opening or closing. Some shippers say that ventilators on cars containing bananas, lemons and other delicate fruits should be closed at a temperature of 40° F.

Bananas.—In shipping carloads of bananas a man is usually sent in charge to open and close the ventilators. Bananas should be put in a paper bag and a heavy canvas bag, and then covered with salt hay, unless put in automatic heaters. when the fruit is packed only in salt hay. Bananas are particularly susceptible to injury by cold, and require great care. If exposed to temperatures as low as 55° F. they almost invariably chill, turn black and fail to ripen. Cars containing them are sometimes, in extremely cold weather, protected by throwing a stream of water on them, which, freezing, forms a complete coating of ice. The method adopted by some firms, of shipping this fruit in winter, is to heat refrigerator cars to about 90° F. by oil stoves, remove the stoves and load the fruit quickly, put the stoves back and heat up to 85° or 90° F., then remove the stoves again, close the car tight, and start it on its way. Bananas shipped in this manner are held to be safe for forty-eight to sixty hours, even though the temperature goes to zero.

Quinces, apples and pears are packed in barrels, each layer of barrels covered with and resting on straw.*

VEGETABLES.

Potatoes are packed in straw, bulkheaded back, the center of the car left empty, and the car filled as high as the double lining. When the temperature is 12° F. or more below freezing, the rule is to line the barrels with thick paper, and at extremely low temperatures, as a matter of extra precaution, the barrels are covered over the outside with the same kind of paper.

In shipping early vegetables to a northern market from the South, for distances requiring more than forty-eight hours to cover, openwork baskets, slatted boxes, or barrels with openings cut in them should be used to allow a circulation of air.

As a rule, truckers will not haul vegetables to the cars for shipment when the temperature reaches 20° F. or lower, and in no case when it is near 32° F. if raining or snowing.†

USE OF WEATHER REPORTS.

In connection with the shipment of food products liable to injury by heat or cold, much benefit may be derived from an intelligent use of the information contained in the daily weather reports and forecasts published by the Weather Bureau, which show the temperature conditions prevailing over the whole country at the time of the observations, the highest and lowest temperatures that have occurred during the past twenty-four hours, and the probable conditions that will prevail during the next twenty-four or thirty-six hours. These re-

*Straw is really only necessary on the bottom, top, sides and ends of car; no useful result is obtained by packing straw between barrels.—Author.

†A point in connection with the transportation of perishable goods not touched on is the importance of not overloading refrigerator cars with fruit or other goods of like nature unless pre-cooled before loading (see chapter on Pre-Cooling). The warm air from goods will accumulate in the upper part of the car, and no refrigerator car now in service so far as known to the author has a circulation of air sufficiently perfect to give even approximately uniform temperatures. It is generally necessary to leave at least a foot or eighteen inches space at top and space between packages for air circulation. California fruit shippers fully appreciate this and always tack strips of wood between packages, which holds the packages in place and allows of good air circulation.—Author.

ports and forecasts are received at nearly every Weather Bureau office, of which there is one or more in nearly every State and Territory, and published on maps and bulletins, which are posted in conspicuous places in the city where the office is located, and mailed to surrounding towns. The reports, or a synopsis of them, are also generally published in the daily papers.

Fuller information than is obtainable from either of these sources may be had at the Weather Bureau office itself, from the observer in charge, or, where none of these means is available, arrangements may be made with the observer to supply special information by mail, telephone or telegraph. In the large cities of the country, dealers in perishable goods are guided in their transactions very largely by the information thus obtained. The temperature of the region to which shipments are to be made is carefully watched, and the shipments expedited or delayed, according as the conditions are favorable or unfavorable. Shipments on the road are protected from injury by telegraphic instructions as to the necessary precautions to be taken. As shipments in ordinary box cars, or as freight, are less expensive than in refrigerator cars or by express, advantage is taken of a favorable spell of weather to use the former methods.

Information as to the altitude of the regions traversed by the shipping routes, such as may be obtained from the contour maps published by the United States Geological Survey, the location and capacity of the roundhouses along the routes, and the points on the railroads where transportation is liable to blockage by snowdrifts, in connection with that given by the daily weather maps, will prove of value to the shipper in the supervision of his consignment.

In shipping early vegetables North from Southern ports the weather reports are utilized to determine whether to use water or railroad transportation, the former being the cheaper. Dealers in certain kinds of produce, by careful attention to the daily weather reports and the weekly crop bulletins, keep themselves informed as to the sections where conditions most favorable for large crops have prevailed, and are thus enabled to judge of the probable supply and to know where to purchase to advantage.

As illustrations of the manner in which advantageous use may be made of the weather reports, suppose a merchant in Ohio has an order in January for a load of apples or potatoes to be shipped to St. Paul; when his shipment is ready he may ascertain by personal inquiry at the Weather Bureau office, or by a study of the published reports and forecasts, the probable temperature conditions between Ohio and Minnesota for the period that the shipment is likely to be on the road, and regulate the same accordingly. If neither of these means of information is accessible to him, he may telegraph the observer at the nearest Weather Bureau office, Cincinnati, Columbus, Cleveland, Sandusky, or Toledo, as the case may be, requesting the information, or he may arrange beforehand with the observer to be informed by telegraph when the conditions are favorable for making the shipment, the cost of all telegrams, of course, to be borne by himself. While the consignment is on the road he should still keep himself informed as to the temperature conditions of the region through which it passes, and if injuriously low temperatures are likely to occur, may telegraph to have it housed or otherwise protected until the conditions are again favorable. By the use of similar means, a packer having a large number of hogs to slaughter may ascertain in advance when temperatures favorable for that purpose are likely to prevail in his locality; or a Southern merchant having a consignment of tropical fruit on the road to the North may insure its protection from injuriously high or low temperatures by telegraphic instructions as to the opening or closing of ventilators, or the use of ice or artificial heat.

During the season when cold waves are liable to occur, a careful watch of the reports and forecasts will often enable dealers and others to protect from injury large quantities of produce in storage. Instances are numerous where the use of such information has resulted in large pecuniary benefit.

During the severe cold wave of January 1 to 5, 1896, which overspread nearly the entire United States east of the worth of property was saved from destruction by the warnings of the Weather Bureau, which were sent out in advance of the wave.

TEMPERATURE TABLE.

In the following table are given the highest and lowest temperatures which perishable goods of various kinds will stand without injury, whether packed in ordinary packages, stored in freight cars or placed in regular refrigerator cars.*

LOWEST AND HIGHEST TEMPERATURES TO WHICH PERISHABLE GOODS MAY BE SUBJECTED WITHOUT INJURY.
(The — sign denotes temperature below zero Fahrenheit.)

Perishable Goods.	Lowest Outside Temperature.			Temperatures Above Which Injury Occurs.	How Packed.
	Articles in Ordinary Packages.	In Ordinary Freight Cars.	In Refrigerator Or Specially Prepared Cars.		
	°	°	°	°	
Ale, ginger	30	20	—10	....	
Apples, in barrels	20	10	—10	75	Covered with straw.
Apples, loose	28	15	—10	75	Packed in straw.
Apricots, baskets	35	24	10	70	
Aqua ammonia, barrels	30	20	—10	....	
Asparagus	28	22	....	70	In boxes covered with moss.
Bananas	50	32	....	90	In boxes with straw.
Beans, snap	32	26	....	65	In barrels or crates.
Bear	Zero	—20	....	65	Shipped loose.
Beef extract	25	15	—10	....	
Beer or ale, kegs	32	20	Zero	75	In manure and shavings.
Beets	26	20	....	70	In crates.
Bluing	30	20	—10	....	
Cabbage, early or late	25	20	Zero	75	Barrels or crates.
Cantaloupes	32	25	10	80	
Carrots	30	25	20	....	
Catsup	25	15	—10	....	
Cauliflower	22	15	....	70	In barrels with straw.
Celery	10	Zero	....	65	Packed in crates.

*The temperatures given seem to the author to be too arbitrary and in some cases incorrect but are useful as a guide. There are many things to be considered in fixing the lowest and highest safe temperatures for perishable goods, chief of which are: First.—Initial temperature of goods when loaded into car. Second.—Temperature to which exposed en route. Third.—Time on the road. Other conditions, like ripeness of fruit and variety, have much to do with the temperature it will withstand without injury.—Author.

LOWEST AND HIGHEST TEMPERATURES TO WHICH PERISHABLE GOODS MAY BE SUBJECTED WITHOUT INJURY—CONTINUED.
(The — sign denotes temperature below zero Fahrenheit.)

Perishable Goods.	Lowest Outside Temperature. Articles in Ordinary Packages.	In Ordinary Freight Cars.	In Refrigerator or Specially Prepared Cars.	Temperatures Above Which Injury Occurs.	How Packed.
	°	°	°	°	
Cheese	30	25	10	75	
Cider	22	18	—10	70	
Clam broth and juice	30	20	—10	80	
Clams in shell	20	10	—10	65	In barrels.
Cocoanuts	30	20	Zero	90	In barrels or crates.
Crabs	10	Zero	.....	65	In baskets and barrels.
Cranberries	28	20	Zero	.....	
Cucumbers	32	20	.....	65	In boxes with moss.
Cymlings, or squashes	32	22	.....	75	In crates.
Deer	Zero	—20	.....	65	Shipped loose.
Drugs (non-alcoholic)	32	28	Zero	.....	
Eggs, bar'led or crated	30	20	Zero	80	
Endive	10	Zero	.....	70	In boxes or crates.
Extracts (flavoring)	20	15	Zero	.....	
Fish	10	Zero	.....	65	In barrels always iced.
Fish, canned	18	15	—10	.....	
Flowers	35	20	—10	.....	Packed in moss.
Grapes	34	20	Zero	.....	Packed in cork.
Grapefruit	32	20	Zero	.....	
Groceries, liquid	32	20	Zero	.....	
Ink	20	15	—10	.....	
Kale	15	Zero	.....	65	In boxes or crates.
Leek	28	20	.....	65	In boxes.
Lemons	32	20	10	75	In boxes or crates.
Lettuce	26	15	.....	70	Do.
Lobsters	25	20	Zero	.....	
Mandarins	32	20	Zero	75	In boxes.
Medicines, patent	32	28	Zero	.....	In sawdust.
Milk	32	28	Zero	75	
Mucilage	25	15	Zero	.....	
Mustard, French	26	20	—10	.....	
Okra	25	20	.....	75	In baskets or boxes.
Olives, in bulk	28	25	Zero	.....	In barrels.
Olives, in glass	25	20	Zero	.....	
Onions	20	10	Zero	80	In barrels or crates.
Oranges	28	20	Zero	80	In baskets, bar'ls or crates
Oysters, in shell	20	10	—10	65	In barrels.
Oysters, shucked	30	20	Zero	70	Do.
Parsley	32	20	.....	75	In baskets.
Parsnips	32	20	.....	70	In baskets or barrels.
Partridges	10	Zero	.....	65	In bunches in boxes.
Paste	32	25	10	.....	In barrels.
Pears	32	20	10	80	

LOWEST AND HIGHEST TEMPERATURES TO WHICH PERISHABLE GOODS MAY BE SUBJECTED WITHOUT INJURY—CONCLUDED.
(The — sign denotes temperature below zero Fahrenheit.)

Perishable Goods.	Lowest Outside Temperature.			Temperatures Above Which Injury Occurs.	How Packed.
	Articles in Ordinary Packages.	In Ordinary Freight Cars.	In Refrigerator or Specially Prepared Cars.		
	°	°	°	°	
Peaches, fresh, baskets	32	20	10	80	
Peaches, canned	20	15	Zero	....	
Peas	32	20	.....	80	In baskets or barrels.
Pickles, in bulk........	22	18	—10	.....	In barrels.
Pickles, in glass.......	20	16	—10	.....	
Pineapples	32	25	Zero	75	In barrels or crates.
Plums	35	32	Zero	75	In boxes with paper.
Potatoes, Irish	33	25	10	80	In barrels or baskets.
Potatoes, sweet	35	28	10	80	Do.
Preserves	20	10	—10	.....	
Radishes	20	15	.....	65	In baskets.
Rice	20	10	.....	90	In barrels and sacks.
Shrubs, roses or trees.	35	10	—10	.....	In canvas or sacking.
Spinach	15	15	.....	75	In barrels or crates.
Strawberries	33	25	—10	65	
Tangerines	25	15	Zero	70	In boxes.
Tea plants............	28	20	.....	95	In boxes.
Thyme	20	10	.....	90	In small baskets.
Tomatoes, fresh	33	28	10	90	
Tomatoes, canned	28	25	—5	.....	In boxes.
Turnips, late	15	Zero	.....	75	In barrels.
Vinegar, barrels.......	22	18	—10	.....	
Watermelons	20	10	.....	85	In barrels or in bulk.
Waters, mineral.......	28	25	Zero	.....	
Wines, light..........	22	15	Zero	.....	
Wild boar	Zero	—20	.....	65	Shipped loose.
Wild turkey	Zero	—20	.....	65	Do.
Yeast	28	25	Zero	65	

SHIPPING BEEF AND MUTTON.

The following description regarding the practice of handling and shipping beef and mutton from Argentina to England offers some useful and practical suggestions:

The cattle and sheep are killed as near their own pastures as possible and the carcasses (unless the atmospheric conditions are favorable for natural precooling) pass at once to the refrigerating chambers where they are frozen at a temperature between 5 and 10 degrees F. The hard frozen carcasses are held at a temperature of about 15 degrees from the time they are frozen in the various works in the antipodes, and in the Argentine, until they are thawed out for the consumer in Great Britain.

Chilled beef, large quantities of which are sent to Great Britain, principally from the United States and the Argentine, needs to be handled in a totally different manner. In the case of hard frozen meat, the proposition from a refrigerating point of view, after once the meat is thoroughly frozen, is comparatively simple. Within certain limits, any temperature well below the freezing point of meat, will maintain it in perfect condition for lengthly periods; and although 15 degrees F. has been mentioned as a suitable temperature, a range of temperature, say between 10 degrees F. and 18 or 20 degrees F., makes but little difference to its keeping qualities. With chilled meat, however, it is quite another story, and for long storage the temperature has to be maintained with as little range as possible, a variation of even 1 degree having some influence upon its keeping qualities. The freezing point of beef, that is the temperature at which the liquids are completely changed to solids, is between 28 and 29 degrees F. It has been found that a temperature slightly above this, say 30 degrees F. or thereabouts, is the best and most satisfactory for the storage and carriage of chilled beef. This temperature must be maintained as steadily and with as little variation as possible, and even with such precautions the length of time during which chilled beef can be held in perfect condition is only about six weeks.

ON ICING POULTRY.

Relative to the poor condition in which iced poultry sometimes reaches the market, the *New York Produce Review* publishes the following:

> At this season of year dressed poultry dealers experience much trouble, owing to the fact that stock arrives out of condition and large amounts of money are lost because of the low and unprofitable prices realized for this poor conditioned poultry.
>
> During cold weather shippers send their dressed poultry dry-packed, but as soon as the weather becomes warm the poultry is iced and there seems to be great difficulty at times in getting the iced poultry through in fine condition. The stock will arrive with ice almost melted off, and often entirely gone, and the poultry more or less out of condition. There seems to be more poultry spoiled in transit during this season of year than at any other time, the quantity

even exceeding the amount damaged during the very warm weather, and the natural inference is that the fault lies with the shipper and is largely due to carelessness.

It is always difficult to get the animal heat entirely out of the poultry, as well as other meat, and it is thought that much of the stock which arrives out of condition has not been thoroughly cooled before icing, the comparatively cool weather doubtless causing packers to give this important matter less attention than they should. But the main trouble is the lack of ice used by shippers.

During really hot weather the shipper ices the stock thoroughly, and it usually comes through all right, but while weather is cool, as in the Fall, shippers use less ice to carry the stock, and while it reaches here in good shape if weather keeps cool, every warm spell, or, in fact, every warm day that appears rapidly melts the ice, and the poultry is ruined before it reaches the market, so far as top market prices are concerned, as it has to be forced off to cheap trade for what it will bring.

It is certainly penny wise and pound foolish policy for shippers to try and save a little on their ice accounts at the expense of their poultry. The loss incurred every few shipments by having their stock arrive out of condition is much greater than the cost of a little more ice with each shipment, and it is hoped that some effort will be made by shippers to remedy this long-time evil, which is a drain on the larger and regular shippers as much, if not more, than on the smaller shippers. With little care this loss could be avoided by operators and it would be a great saving to the shipper and receiver of both annoyance and money.

DRESSED POULTRY IN TRANSIT.

The following interesting suggestions by George B. Horr is taken from *The Butchers' Advocate*:

The development of the system of refrigeration has demonstrated that proper packing of poultry for shipment is very essential to insure its arrival at destination in good condition. To accomplish this sufficient ice must be used in packing to last while the poultry is in transit. Most shippers pack their poultry in alternate layers of crushed ice and poultry, placing a large cake of ice on top and covering all with burlap. Usually from 175 to 200 pounds of poultry are packed in an ordinary sugar barrel, using about the same quantity of crushed ice. Some shippers take the precaution of lining the barrels with brown or parchment paper covering the top cake of ice in the same manner before putting on the burlap. The quantity of ice required in proportion to the quantity of poultry in each barrel depends upon local conditions—that is, it would not be necessary for a shipper located in Illinois and desiring to ship to New York to use as great a quantity of ice in packing as a shipper in Kansas. With the ordinary treatment which is furnished by the various refrigerator transportation companies the temperature of a refrigerator car ranges from 35 to 45 degrees when crushed ice and salt is used, the temperature depending on the construction and condition of the car. It follows that there must be some melting of ice in the barrels of poultry, making it necessary to use a greater quantity of ice in packing, according to the distance poultry is to be shipped, and for this reason, also, the maintenance of time schedules is of great importance.

A considerable quantity of dressed poultry is forwarded in what are termed "pick-up" cars and poultry packed for shipment in such cars requires a greater quantity of ice than when shipped in through cars. A pick-up car is one scheduled to pick up small shipments of butter, eggs and poultry at designated local stations between terminal points.

The method of shipping dry packed has come into use largely during the past few years until now more poultry is shipped dry-packed than scalded, but the prompt removal of the animal heat from the dry-packed poultry requires a plant specially equipped for the purpose and calls for a larger investment of capital than the other method. The removal of the animal heat from dry-packed poultry is accomplished by placing the poultry on racks in a cooling room, whose temperature is held at thirty-two to thirty-five degrees. The poultry remains in this room from twenty-four to forty-eight hours. Some shippers reduce to a minimum the chance of forwarding poultry not thoroughly cooled by using a thermometer, as previously described, except that the temperature of dry-packed poultry must be reduced to forty degrees. After the animal heat is removed the poultry is wrapped in parchment paper, either by wrapping each bird separately or by lining the boxes and placing paper between the layers of birds. When the packing is so completed the lot of poultry is held in a cold room having a temperature of thirty-two to thirty-five degrees until car is ready for loading.

In the case of pick-up cars which are sometimes iced and started from small stations it is not always practicable to use crushed ice and salt. In such cases cake ice without salt is used, the car being iced a longer time before loading and at the first re-icing station the remaining ice is broken up and crushed ice and salt added. Most of the large railway systems between the Mississippi Valley and the Atlantic seaboard are well equipped for re-icing cars, this being done practically every twenty-four hours while cars are in transit. The method of re-icing is first thoroughly to tamp down the ice remaining in the tanks and then fill the tanks with crushed ice and salt. Drip pipes and traps are also examined and cleared of any refuse. The way bill or card on the car indicates contents and also stations where it is to be re-iced. The system is so well safeguarded that it is almost impossible for a car to pass a re-icing station without receiving proper attention. These stations are so constructed that a train-load of refrigerator cars can be re-iced in from thirty to sixty minutes.

Since the perfection of the system of cold storage and the construction of cold storage houses at large centers, a much greater quantity of frozen poultry has been transported. The greater part of this is handled in refrigerator cars iced in the same manner as for dry-packed poultry. A small portion is transported in un-iced cars, plenty of straw being used around doors and other openings, the theory being that as the poultry is frozen it will remain in that condition if so packed that the outside air cannot reach it.

Regarding the packing of poultry for shipment Dr. Pennington says:

For long hauls, that is, for 5 days or over, the bird should be packed at a temperature not to exceed 32 degrees F. How much lower the temperature can be depends entirely on the sort of refrigerator car that is to be used. The great majority of the refrigerator cars in service do not maintain a temperature of less than 40 degrees F. in the middle of the car at the top of a three to four foot load. The temperature at the bunker ends of this car, refrigerated with a mixture of ice and salt, may go to 10 degrees F. and hard freeze the poultry at the bunker end on the floor of the car.

REGARDING EGGS.

The effect of good handling and refrigeration on the output of southern eggs has been even more marked than the effect on poultry. Tennessee and Kentucky ship eggs north during the winter months when the supply from other sections has almost ceased. When warm weather comes the eggs in the past have gone still further south, where standards in eggs are not so high, or into the fertilizer pile. Last summer a few shippers, provided with artificial chilling facilities, shipped eggs north for a long part of the summer, and found it profitable. They combined a campaign for careful, quick handling and maintained low temperatures as soon as possible after laying. They found that once thoroughly heated, so that the processes which make for incubation had begun, or in the infertile egg, the deteriorative course induced by heat, refrigeration cannot check nor even greatly slow such changes. Hence, eggs which had been subjected to unfavorable conditions would change en route, even though refrigerated, to such an extent that the packer would not recognize them when they reached their market. If, however, they were well chilled when fresh, deterioration during an average haul under refrigeration was almost a negligible quantity, commercially speaking.

All of our experimental shipments of eggs have confirmed and emphasized our observations on the results obtained by the industry. We find that such factors as dirty shells, wet nest, damp cellars, etc., etc., cannot be overcome by refrigeration, and that the egg must go to the cooler in good condition whether it be for prompt marketing or for long storage, if the maximum benefit of the low temperature is to be secured.—Dr. Mary E. Pennington.

LOADING CARS.

Concerning this subject M. C. Spatz, Linfield, Pa., in the *Egg Reporter* offered the following pertinent suggestions:

Begin in one corner of the car.

Set case lengthwise, and tightly against end and side of the car. End case, set on floor, tightly against first case and against end of car.

Continue this layer entirely across car, seven or eight cases as the space may allow.

Now follow with second layer and set cases exactly same way as first layer, so that one case sets squarely upon the other.

Continue these layers until high enough to accommodate the number of cases to be loaded evenly over entire car.

This will nearly always leave some space open on opposite side of car from which we started.

Now the second row: Begin on opposite side of car from the one we started with first row.

Pile same way as first row, not forgetting to load tightly. This will leave an equal space open on opposite side of car from which we found such space in first row.

Third row: Begin on same side of car as we did with first row, so that the space left open will be found on same side again as of first row.

Continue this method until within 3 or 4 feet of middle of car. Now measure carefully with some cases the space not occupied and find how to arrange the balance of cases, so as to fill out this centre of car tightly. Sometimes it is necessary to put three cases crosswise in car, but avoid putting cases crosswise if possible.

One good way is to start all the rows for which space is yet left at one time, on one side of the car, and thus finish a space only one case wide at a time, being particular to push all cases of all rows tightly towards one end of the car.

Now, there may be a few inches of space left between the last started row, and the one already piled all the way across.

Therefore push the second width of cases in the newly started rows all tightly towards the opposite of the car from which you pushed the first width.

Place the third width of cases same as first width, the fourth same as second, etc.

My experience in loading and unloading during the past nine years is that not once has a car of eggs loaded in the above described manner been found in bad condition at destination.

It is, however, very seldom that a car from the west comes loaded in this manner.

It is a mistake to leave an open space between every case of the floor layer, so as to let the cold from ice chambers, pass under goods.

These floor layer cases will generally be squeezed apart, thus damaging both cases and eggs, making unnecessary expenses and much trouble to all concerned.

MIXED CARS.

Cars containing both butter and eggs should be loaded with the butter in the ends, for the following reasons:

1st. Butter tubs do not pack tightly and thus leave space for the cold air from ice chambers to pass through to the eggs.

2nd. Many cars have improperly constructed ice chambers and thus water is splashed against the goods. This will not injure butter as it would eggs.

If both butter and eggs are properly loaded, I do not see why there is any more danger of damage to goods from bumping of cars than if butter is placed in middle.

Before a car of eggs is started to be loaded, the ice chambers should be carefully examined. Dirt in drip pan should be removed, and drip pipes cleaned.

This may often avoid much annoyance and expense to shippers, receivers and the railroad companies.

When using ice in cars, eggs should be placed on flat solid floor racks, that are about 2 or 3 inches high.

The round or oval strips nailed to the floor in some cars are no good and permit injury to the bottom layer of cases. They are not a preventative of water getting into the eggs.

CHAPTER XXVI.

FURS AND FABRICS.

A DEVELOPING BRANCH OF THE COLD STORAGE INDUSTRY.

The use of refrigeration for the protection of furs, fur and woolen garments, rugs, carpets, trophies, fine furniture, etc., against the ravages of moths or carpet beetles is comparatively recent, and prior to the year 1895 no business of consequence was done in this line. Now many of the larger household goods warehouses, and most of the regular cold storage houses, have rooms devoted to this purpose, and numerous large concerns, both in America and Europe, are operating refrigerating equipments exclusively for the preservation of furs and fabrics. The large department stores are rapidly being equipped with cold storage facilities to care for their furs, woolens, etc., during the heated term.

The use of cold storage for this line of goods is not as yet fully developed. The prejudice of furriers has been largely responsible, and when the cold storage manager first endeavors to obtain business in this line, he usually has a struggle with the furrier. The time honored method of caring for furs, etc., during the heated term, has been to periodically beat, brush, comb or treat them with various chemicals or liquids for the purpose of destroying or preventing the hatching of the egg which produces the larvæ of the destructive miller and beetle. These pests are very generally known as moths. The care of furs during the hot weather of summer has been one of the sources of the furrier's income during his dull season. Naturally, therefore, he looks upon any new method of protecting furs with suspicion and in an unfriendly light. In nearly every instance where the author has obtained the experience of warehousemen on this subject, the same conditions prevail.

In some instances where cold storage is largely in use for fur storage, it has been introduced by the cold storage warehouseman interesting a prominent local furrier, and making concessions which would attract his business. This furnishes

FIG. 1.—COLD STORAGE ROOM FOR RUGS AND CARPETS.

a good reference. After acquiring such a customer, business may be solicited by distributing attractive descriptive advertising matter from house to house. A number of warehousemen known to the author have secured their business almost wholly by advertising directly and without the help of local furriers. It is only a question of time when the prejudice of furriers will be overcome, and they will become the heaviest customers of the cold storage house; but, for the present, their preconceived ideas and fancied financial interests make them the competitors, in some cases, of the cold storage house.

FIG. 2.—COLD STORAGE ROOM FOR FABRICS AND TROPHIES, SHOWING COOLING PIPES.

FUR STORAGE PROFITABLE.

The storage of furs and fabrics pays better per cubic foot than any other class of goods, and cold storage houses located in or near the residence portion of cities, in latitudes where furs are worn, should make an effort to obtain this business. The detail of looking after it is considerable, but it works in nicely with other business. So far the business has been largely developed by the household goods warehousemen, and at present the largest and most successful businesses in this line are conducted by such houses, chiefly because these already have a clientage from whom to draw business, and are equipped with facilities for collecting and delivering goods. To the warehouseman who handles both household goods in dry storage and perishable goods in cold storage, the setting aside of a room for the purpose is a comparatively inexpensive experiment, and it is likely to result in a good business. The largest and most successful houses handling these goods have fire-

FIG. 3.—COLD STORAGE ROOM FOR RUGS.

proof buildings. The large value stored in a small space makes the fireproof building especially desirable for this class of goods.

TEMPERATURE.

The correct temperature for a fur and fabric room has not been accurately determined as yet. Rooms are in operation ranging in temperature from 15° to 40° F. It has been demonstrated that a temperature of 40° F. will prevent the operation of damaging larvæ, but does not destroy them as shown by Dr. Read's experiments described at the end of this chapter.

A safe working temperature for the cold room would be anywhere between 25° and 35° F., and it is believed that the latter temperature is amply low, if continuously maintained.

Raw silk has been placed in cold storage for other reasons than to prevent the working of damaging moth. When stored at ordinary temperatures a loss of weight and lustre results, caused by the evaporation of the natural moisture and volatile matter contained in the silk. A temperature below 30° F. prevents the evaporation and maintains the lustre. Inferior grades are especially liable to damage when exposed on the

FIG. 4.—COLD STORAGE ROOM FOR RUGS AND CARPETS.

shelves for a time and cold storage is necessary to a successful holding.

HUMIDITY.

Furs and fabrics should not be stored in a room with goods giving off moisture, as at times the moisture in such a room may be excessive and harm result. A room containing nothing but furs will be comparatively dry, because furs do not give off moisture, and the only source from which moisture may be added to the air of the room is by air leakage, opening doors, and the exhalation from persons working in the room. A well insulated fur room, protected by a properly designed air lock or corridor, is so dry that the pipes rarely show white, the coating of frost is so very light. It has been advanced as a theory that a very low temperature, like say zero or 10° above, would be detrimental to the skins or leather of furs, causing them to dry out. Evaporation is caused by a low relative humidity, entirely independent of temperature, so this theory is not tenable. (See chapter on "Humidity.") The average

FIG. 5.—COLD STORAGE ROOM FOR GARMENTS.

humidity during winter, when furs are in use, is much lower in most localities where furs are worn than that of a cold storage room under ordinary conditions. No very accurate data are at hand regarding the humidity at which fur rooms should be carried, but it is no doubt lower than for goods which throw off moisture; that is, the room should be dryer. It may happen that furs removed from a refrigerated room and taken into a comparatively warm atmosphere will show dampness on their outer surfaces. This is not from any fault of the storage room, but because the moisture is condensed from the warm air upon the cold surface of the goods. This may be avoided by packing the goods inside the cold storage room in tight paper boxes or bags before delivery, so that the goods will be warmed slowly and condensation prevented. If furs and fabrics are kept in a room by themselves, no harm will result from the moisture, unless conditions are radically wrong. If, when removed from storage, goods show a condensation of moisture, they should be thoroughly aired until dry before delivering, by placing where a gentle current of air will flow over them, as customers may think the moisture was caused by some defect in the system of cold storage.

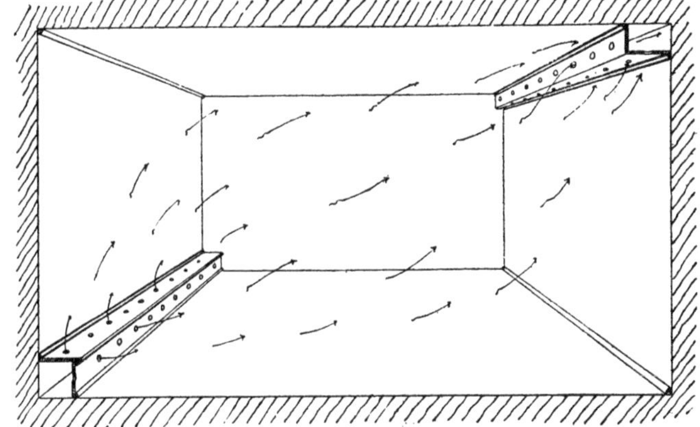

FIG. 6.—AUTHOR'S DIAGRAM, SHOWING DUCTS FOR AIR CIRCULATION IN FUR COLD STORAGE ROOMS.

AIR CIRCULATION.

The forced air circulation system is particularly applicable to the storage of furs and fabrics, and it is recommended, not especially as a matter of purifying the rooms or producing greatly improved conditions, but as a means of avoiding the use of cooling pipes, placed directly in the rooms. Pipe coils on the walls or ceiling of a room may drip at times and cause a spattering of water, which will damage the goods. Space will also be saved, which is an important item, especially in expensive fireproof warehouses. The accompanying illustrations of rooms used for fur storage show clearly the large space occupied by piping. It is not only the loss of space actually occupied by the pipes, but also that the goods must be stored at a safe distance from them. Thoroughly distributed circulation of air is not essential when using the forced circulation system for furs; all that is necessary is a distribution of air which will produce uniform temperatures. A cross-section of the ducts arranged in a fur room designed by the author is shown in diagram. The perforations in these ducts are on the sides of the flow and return ducts. No marked difference in temperature can be noted in different parts of the room when the fan is kept in continuous operation. This arrangement of

FIG. 7.—FUR STORAGE ROOM—COOLED BY FAN SYSTEM OF AIR CIRCULATION—NO PIPES IN ROOM.—FRANKLIN REFRIGERATING COMPANY, SARANAC LAKE, N. Y.

air distribution is not recommended for any goods which throw off moisture, but is sufficient for furs and fabrics. The first rooms to be used exclusively for the storage of furs and fabrics were equipped with brine piping directly in the room, and such an arrangement is still largely in use, but the forced circulation or indirect system outlined above is rapidly coming into use.

In connection with the fan or forced circulation of air for fur storage rooms, fireproof shutters or dampers held open by fusible links, should be povided in the main air ducts. A very disastrous fire occurred in New York, where it was thought that the damage was much augmented by the fact that the fan system was in use. This particular house, after the fire, substituted direct brine piping, and also subdivided their space. In a comparatively large plant several smaller rooms are in any case much more desirable than one or two large rooms.

VENTILATION.

The ventilation of fur rooms may be easily accomplished; and while not absolutely necessary to the welfare of the goods,

FIG. 8.—COLD STORAGE ROOM FOR FURS, SHOWING PIPING ON SIDE WALLS.

it is much better to have a nice sweet smelling room to show prospective customers than one which has the lifeless and impure atmosphere encountered in some fur rooms. The warm weather ventilating system invented by the author for use in summer is desirable at frequent intervals. (See chapter on "Ventilation.") At one of the cold storage plants designed by the author a quantity of clothing containing moth balls was received, and the fact was not discovered until the room was well scented. A few hours' operation of the warm weather ventilating system was sufficient to sweeten the air of the room perfectly. The rooms may be blown out and thoroughly ventilated by forcing in fresh cold air from the outside by using the cold weather ventilator in winter.

EDUCATION OF CUSTOMERS.

One of the first difficulties of the cold storage manager is to educate his customers to do away entirely with the use of

FIG. 9.—COLD STORAGE ROOM FOR GARMENTS, SHOWING PIPING.

moth balls, camphor balls, tar camphor, carbolic camphor, powders, tar paper or any of the ill smelling trash of various kinds which has for years been used to keep out the damaging moths. Some warehousemen have also been troubled by the stable odor from robes and coachmen's garments. Goods received containing these objectionable odors should be carefully aired for some days before placing in the cold storage room.

If the odors cannot be eradicated, the goods must be isolated in a room by themselves, or rejected for storage and returned to the owners. It should be the warehouseman's study to return goods in as good or better condition than when received. To this end, all objectionable goods must be excluded from the storage rooms.

For his own protection the warehouseman will note condition of all goods when received for storage. The unreasonable or dishonest customer is always with us, and he may expect his furs back in prime condition when they were really damaged at the time they were delivered to the storage house. The services of an expert furrier are provided in some cases, and where the volume of business is sufficient, one may be regularly employed. Any bright young man may be trained to inspect furs on arrival at the storage house. Any blemishes or imperfections should be noted on the receipt given to the customer. All furs should be carefully beaten, dusted and

FIG. 10.—COLD STORAGE ROOM FOR CARPETS, SHOWING PIPING.

aired before placing in the refrigerated rooms, and properly placed or stored to keep them in the best possible condition.

HANDLING AND STORAGE.

Trophies like stuffed animals, heads, skins, etc., are best hung or laid on racks. The best method of storing coats, cloaks, etc., is to hang them on forms or shoulder stretchers to preserve the shape and hang of the garment. If any metal hooks with shoulders are used they should be wrapped with tissue paper to prevent discoloring light colored furs or garments. The forms are suspended from racks, and the whole covered by a piece of heavy unbleached sheeting. This arrangement is plainly shown in the accompanying view of a cold storage room (Fig. 8). The illustrations (Figs. 1 to 10) also show the method of storing fur rugs, stuffed heads, carpets, trunks, etc. In some cases each individual garment is encased in a separate cloth cover. Separate closets are sometimes provided for the use of single individuals, furriers and large customers, or for the storage of especially valuable garments, the keys to which may be carried by the customer. Such closets are usually made of slat or open wire work, to allow of a free circulation of air.

SUGGESTIONS FOR ADVERTISING MATTER.

A few of the advantages of refrigeration for the protection of furs and fabrics are here concisely stated for the benefit of those preparing printed matter for distribution:

FOLIO —20—

Cold is instrumental in the production of furs, and it is as necessary to their preservation.

Cold develops and enriches the fur when on the animal's back and preserves its color and gloss when manufactured into useful coverings. Cold storing is like putting the fur back into its native element.

Cold prolongs the life of the fur by retaining the natural oils, which are evaporated by the hot, dry air of summer.

Not only is the appearance of the fur improved, but the flexibility and softness of the leather which supports it are retained.

Carpets, rugs and other woolens lose color and life in the hot summer air. A cold atmosphere revives the colors and rejuvenates the fiber.

The wear and tear on furs, carpets and rugs by excessive heating is entirely eliminated.

Garments stored on forms in refrigerated rooms are ready for immediate use; in fact, can be removed from cold storage, worn for a single night, and returned.

Curtains or draperies may be suspended from racks, avoiding damage from folds.

Furriers who have used the system heartily indorse it.

Obnoxious odors from use of moth preventives are avoided.

Cold storage rooms are dust proof.

Cold storage gives absolute security against moth.

RATES FOR STORAGE.

The usual storage rates for fur and fabric cold storage are given below. The prices are in some cases less, in fact, sometimes only one-half those given. Each warehouseman must be governed by local conditions and competition, and in most cases the charges made by the furrier under the old method must be approximately met. A furrier's charges nearly always include a guarantee against fire and moth.

SEASON RATES ON FURS.

Muffs	$0.75 to $1.00
Boas, caps or gloves	.75 to 1.00
Collarettes	1.00 to 1.50
Capes not exceeding 20 in. in length	1.00 to 1.50
Capes or sacques not exceeding 24 in. in length	1.00 to 1.50
Capes or sacques not exceeding 28 in. in length	1.50 to 2.00
Capes or sacques not exceeding 36 in. in length	2.00 to 2.50
Garments, such as dolmans, long sacques, etc	1.50 to 2.50
Overcoats, etc., not exceeding 40 in. in length	1.50 to 2.50
Garments exceeding 40 in. in length	2.00 to 3.00
Lap robes	1.50 to 2.50
Rugs, according to size	1.00 and up
Stuffed animals, birds, mounted heads, etc	1.00 and up

Monthly rate, one-third of season rates.

Season, nine months.

SEASON RATES ON WOOLENS, ETC.

Woolen garments stored in same manner as fur garments, two-thirds of fur rates.

Blankets, clothing or other garments stored in trunks or boxes, rate of seven cents per cubic foot per month, or fifty cents per cubic foot per season of nine months.

Carpets and rugs not in boxes or trunks, four cents per cubic foot per month.

Suits and dress suits, $1.50 per season.

Furniture, forty cents to sixty cents per cubic foot per season.

Monthly rate, one-third of season rate. Season, nine months.

Some warehouses make the season only six months, but the usual season is nine months, and in most cases is figured to end January 1. Goods carried beyond January 1 are usually charged for at a short season (*i. e.,* January 1 to April 1) rate, at one-third the long season rate. It is customary for warehousemen to make a rate which insures against fire, moth and theft, although this is by no means a universal rule. When so done the usual rate is 1 per cent per season on valuation, insurance rate to govern to some extent.

WAREHOUSE RECEIPTS.

The form of warehouse receipt here shown has been found in practice to answer the purpose for furs, etc., very well, but is subject to many limitations and modifications. The words "Not negotiable" should be printed or stamped across the face of the receipt. It need not necessarily take this form, but should include the items mentioned and should read somewhat as follows:

TWENTY-FOUR HOURS' NOTICE REQUIRED FOR WITHDRAWALS.

NEW YORK STORAGE COMPANY.

COLD STORAGE DEPARTMENT.

New York

RECEIVED

...............191... as per Schedule below, contents of packages unknown. To be stored in Cold Storage.

Lot No......... For the account of.......................

....

for which the sum of $....................per month $............ is to be paid, from the date hereof until January First, 191.....

The responsibility of this Company for any piece or package or the contents thereof, stored in this department, is limited to the sum of one hundred dollars, unless the value thereof is made known at the time of storing and receipted for in the schedule, and an additional charge be paid for a higher valuation.

On consideration of the above sum, the said New York Storage Company agrees to protect the said articles from loss or damage by moths, fire and theft to the extent of such valuation.

Should these goods be withdrawn before the expiration of the above term of storage, no portion of the charge shall be remitted, and if continued longer, it shall be deemed a renewal under the same conditions, for which a like rate shall be chargeable.

The said goods are hereby valued for the purpose of insurance, at the sum specified in the schedule.

When the property covered by this receipt is withdrawn, this receipt should be surrendered to the Company.

A written order should be given when others are to have access or when goods are to be removed.

No person is authorized to make any other agreement or condition on behalf of this Company.

..................Supt.

DR. READ'S EXPERIMENTS.

This chapter would not be complete without the addition of an extract from the article on the "Cold Storage for Fabrics," by Dr. Albert M. Read, of the American Security and Trust Co., Washington, D. C., published in full in the June, 1897, issue of *Ice and Refrigeration.* Dr. Read has taken up the subject and handled it in a masterly and exhaustive way. Credit is also due to Walter C. Reid, of the Lincoln Safe Deposit Co., New York, and Albert S. Brinkerhoff, of the Utica Cold Storage and Warehouse Co., Utica, N. Y., for assistance in securing much of the information contained in the foregoing. Following is a portion of Dr. Read's article, referred to:

COLD STORAGE FOR FABRICS.

In order to conduct our business intelligently, it became necessary to ascertain the effect of low temperatures upon the moth and beetle in the various forms of egg, larvæ and perfected insect. We, therefore, had a small room fitted with brine pipes, divided into several sections by stop-cocks, so that the temperature could be controlled to within about 5°, and began operation at from 20° to 25° F. When we thought we had exhausted the subject at these temperatures, we took the next in order, from 25° to 30° F, and progressed in this manner upward until the temperatures of from 50° to 55° F. were reached. Each of these tests necessarily consumed considerable time, the series having occupied the full period of two years. Early in the first year, however, we learned that the line of safety lay somewhat higher than the freezing point (32°), and our plant was run for the balance of the season at that temperature.

The Egg.—It is probable that the egg of the moth and beetle requires a temperature somewhat higher than 55° F. for hatching. I say probably, because, owing to the difficulty of obtaining the eggs, the experiments on them have not been sufficiently numerous to allow of positive conclusions, although those that have been made point strongly to the possibility stated.

The Larvæ.—The larval condition is the one in which all the damage to fabric is done by the insects in question. In passing from the egg through this condition to the perfected insect, the fiber of the wool, fur, etc., is eaten by the larvæ of both the moth and the beetle for the grease and animal juices in it, these constituting the principal source of food, and the larva of the moth for material out of which to spin the web that constitutes a large proportion of the cocoon used for its protection. In the larval state it was found that any temperature lower than 45° F. was sufficient to keep the insect from doing damage to fabric, although all that temperature, and at a temperature as low as 42° F., there was slow and sluggish movement of the animal. At temperatures below 40° F. movement was suspended, and the larva became dormant. At temperatures above 45° F. the movements of the larva became active, and it began to work upon the fabric, the amount of this work and the quickness of movement increasing with each degree of temperature up to 55° F., when the normal condition of activity appeared to be reached.

The Perfected Insect.—The miller and beetle, when subjected to temperatures below 32° F., were soon killed, as they were also after a longer time at all temperatures between that and 40° F. At temperatures between 32° and 40° F., however, when the insects were placed in the center of a roll of heavy woolen rugs, they appeared to enjoy an immunity from death for several weeks, although during this period they were entirely dormant.

It will be seen from the above that the investigations made have quite conclusively proven that cold storage rooms for the preservation of furs and fabrics from the ravages of moth and beetle may be kept at a temperature as high as 40° F. with perfect safety, so far as these insects are concerned. There may, in some plants, however, be trouble from the drip from the cooling pipes of the storage room at this temperature, which will, of course, be very objectionable, and should be obviated by a slight lowering of the temperature of the room. We have run our rooms at temperatures varying from 35° to 40° F. without trouble in this regard.

In the course of the investigation some matters of interest in connection with the effect of cold upon these insects came to my notice. As these may prove of value in the future, I will state them in a few words. It was found that the larvæ of both the moth and the beetle had the power of resisting temperatures as low as 18° F. for a long period without apparent harm, and that they came out of the dormant condition superinduced by the low temperature in the same physical condition as when they entered it, and apparently took up their natural avocation at the precise point where it was interrupted. When these larvæ were alternately exposed to low and higher degrees of temperature, so that they passed from the dormant to the active condition and back again several times in succession, their power of resistance was considerably lessened, and they died much sooner than when kept dormant in a low temperature continuously. This would indicate that a winter, during which short periods of cold are followed by similar periods of warm weather, would be followed by a summer of decreased insect life.

CHAPTER XXVII.

WILD FERNS.

The storage of wild ferns has developed into an important feature of cold storage industry in parts of New England, and doubtless there will be some demand elsewhere for space for their proper handling. Comparatively little is accurately known about suitable packages, handling and storage.

USES AND MARKET.

Ferns are in great demand by florists during winter, and the holiday season in particular, as a background or "backing" for cut flowers. The greater part of ferns so used are picked in the mountainous parts of New England, the Berkshire Hills of Western Massachusetts being especially productive, with Hinsdale the chief center. They are stored during the entire year and shipped out on orders to practically all parts of the country. New York city is a great consumer, but other Eastern cities also use large quantities, and they are shipped in large lots to Chicago and even the far west.

STORAGE METHODS OLD AND NEW.

In the early history of the fern shipping business it was customary to store them in barns, in beds about 10x4x1 ft., covering them with moss. Others were stored in cellars, sprinkling with water from time to time to prevent drying out, but owing to uncertain temperature, and various other conditions not under control, considerable loss resulted from wilting and rotting. It was found that as business increased picking must be commenced earlier in the fall, when the weather is too warm to keep the ferns properly, and cold storage was then resorted to. Cold storage has proved to be the salvation of the business on a large scale. At first undertaken in a small and experimental way, the results were so much of an improvement that now the business is wholly handled through cold storage. Under the right handling, packing and temperature, the ferns come out of storage with the fresh, *green* appearance which makes them so attractive when used by the florists in combination with cut flowers.

PICKING TIME AND DETAILS.

Exact data are not now obtainable as to the best time of picking, but it is generally understood among the pickers that the work should not begin until the first frost, and the picking, therefore, takes place mostly during September and October. Picking begins about June 20th, but not for storage. It seems that the ferns are toughened by the cool nights of fall, which probably act to lower the moisture content or dry them to some extent. Anyway, it is well understood that the picking for storage should not begin until the first frost, and experience has demonstrated that ferns picked before are not properly matured for storage and shipping purposes. The picking is done mostly by women, children and old men, and as the work is done during a time of the year when rural occupations are least pressing, the work is very acceptable and forms an important industry and source of income in some sections. One and one-quarter to two cents per bunch of twenty-five is paid according to variety, care in picking and quality, and even at this seemingly low rate some of the most expert pickers earn as high as $7 per day.

PACKING, SORTING, ETC.

The wild ferns usually collected are commonly divided into two grades or varieties, "dagger" and "fancy" ferns. The "dagger" variety is the more hardy and easier to pick, and less loss is usual in storing. The leaves are of waxy appearance and coarser than the "fancy" ferns. As the name indicates, "fancy" ferns are very delicate of leaf and finer in every way and easily damaged in picking and handling. The greatest care and skill is necessary in preparing the bunches or bundles. Not all the ferns are marketable, by any means, and the careless picker is penalized. Buyers located at the railroad station receive the ferns from the pickers, and if necessary carefully sort them. Bunches of 25 are standard, and the ferns must be carefully arranged so as to lay flat, to avoid crushing or bruising. None broken or decayed or badly discolored are packed for shipment. As a storage package a wooden box is used (mostly second-hand shoe boxes), and the ferns are packed in layers with moss on top, bottom and sides, about 5,000 to 10,000 to the box. The object is to pack in such a way as to retain the moisture and exclude the air.

TEMPERATURE AND COLD STORAGE TREATMENT.

A close determination of the most suitable temperature has not been made, but it is more than probable that what is wanted is a temperature which will freeze the moisture in the packing material and still leave the ferns unfrozen. Satisfactory results have been obtained at 30° F., but a temperature of 25° F. to 28° F. is suggested as more suitable, and experimental work along this line is recommended. Ferns picked in August and early September should not be stored at as low a temperature as those picked later. If the ordinary frosts up to say October 15th in the Berkshire Hills will not damage the ferns, it would seem that 25° F. in cold storage should not, but still in the presence of moisture soaked packing material the effect may be different. The moss used in packing is pretty well soaked, and 28° F. to 30° F. will, of course, freeze the packing material and leave the ferns unfrozen.

It is absolutely essential to best results that ferns after picking should be promptly sorted and packed and placed in cold storage. If they are shipped and on the road for several days, heating is likely to result and ruin the ferns. If placed in cold storage the same day, so much the better, but in cool weather a day or two may elapse without damage.

If shipment by rail to a cold storage house is necessary, by all means use refrigerator cars. Far better results may be had, however, by cold storing where picked, or sufficiently near, that the ferns packed one day may be in storage the next. A cool or cold room for sorting and packing is almost a necessity, and this may best be obtained in connection with a cold storage plant.

In piling the boxes in cold storage it is advisable to provide a two-inch air circulating space on top, bottom and sides of the box, so as to allow a quick cooling and freezing. After a week in storage they may be piled more tightly, keeping them from side walls and floor at least two inches.

Results from cold storing have been reported as very irregular, and, as stated at the beginning of this chapter, the accurate information available is small. It would seem that the irregular results must be due to the condition of the ferns when picked, or the exposure to varying conditions before storing, as it is comparatively easy now to hold uniform temperatures in cold storage. The lack of uniformly successful results cannot be blamed to cold storage, but rather to lack of uniformity in the product stored, probably too early picking or too late picking, or too careless or unintelligent handling before storing.

CHAPTER XXVIII.

LILY OF THE VALLEY BULBS.

COLD STORAGE BULBS PRODUCE BEST RESULTS.

Lily of the Valley is one of the most important objects of culture among florists, and cold storage is now recognized as an absolute necessity in connection therewith. During the winter season the announcements of wholesalers may be seen in the florists' papers, advertising "Cold Storage Valley." Thus do the florists acknowledge their dependence on cold storage for bulbs, or "pips," as they are called, of high quality and with good vitality for developing the desired bloom. The advantage of stock from cold storage is that it may be brought into bloom in three to four weeks without the application of a high degree of heat or "forcing," as it is called. This has two advantages: First, less bench room or space required in the greenhouse, because of less time required for developing the mature bloom, and second, the leaves and blossoms of the "Cold Storage Valley" are stronger and fresher, and as they are developed or grown at a lower temperature, they endure better and with less wilting.

WHAT "COLD STORAGE VALLEY" REALLY IS.

The bulbs do not become cold storage bulbs until they are about a year old, or at least, as generally handled they do not. When received from Europe in October or November, the bulbs in bunches of twenty-six or twenty-eight are generally repacked in convenient sized boxes with sand or soil around and between the bunches. The boxes are then placed in outdoor or cold frames, where they may be got at as needed for successive forcing. (Please note that the bulbs are fresh and new, and at this age they really require high temperature and real forcing.) After about February 15th to March 1st, it is not safe to leave the bulbs in the outdoor frames longer, and they are then repacked, this time for cold storage. The bunches, after having been immersed in water, are stood upright in a closely packed tier with moss all around them, under, between and over the tips of the bulbs. The boxes are usually nailed up with lath and then piled in the cold storage room one above the other, but a better way would be to use a tightly nailed box, as a better protection against changes in temperature would result.

It is, of course, possible, (and in fact the larger importers do this) to repack the bulbs promptly when received from Europe in the fall, and at once place in cold storage where they remain during the winter and following summer or until wanted for bringing into bloom. Better and more certain results are obtainable in this way.

TEMPERATURE.

It is understood that the correct temperature at which to store the pips depends on the kind and on the time to be held in cold storage. Those adapted for short storage are carried at 26° F. to 28° F., while those to be carried longer are best carried at 23° F. to 25° F. A very uniform temperature is recommended, but here again the question of a suitable tight box might be of much help in overcoming some of the bad effects of change in temperature.

SELECTION OF BULBS FOR COLD STORAGE.

Those grown on sandy soil are reputed to be good for early forcing but not good for cold storage. The so-called Berlin pips are of this type. They have a bunch of long fibrous roots and tapering crowns of pinkish hue, and are generally more fully ripened and may be forced earliest of all, while for long keeping in cold storage they are not the best by any means. An experienced florist thus describes the crowns best suited for cold storage: "The individual pip should be thicker and more stubby, and generally the crown should be characterized by a ranker and more vigorous and coarser appearance than the

Berlin pips used for Christmas forcing. They sometimes have a purple or violet tint. By these characteristics they plainly show that they have been grown on a very heavy rich soil, and that in point of maturity they are far behind those grown on sandy soil."

SUGGESTIONS.

If a suitable selection of high quality bulbs is secured it would seem that the chief requirements of successful cold storage would be proper packing and a low and uniform temperature. The bulbs should be wet thoroughly and packed in carefully dampened moss. Then they should be placed in a room at not above 25° F. if for long storage, where they should be piled loosely, with a couple of inches of space all around, for a week; then they may be piled up in a solid pile, but kept from walls and floor at least two inches. It would surely seem that a tight box, possibly lined with paper, would be preferable to the slatted covers, as it is necessary to keep from the air and prevent drying out, which means loss of vitality.

It is also suggested that the shorter the time the bulbs remain under refrigeration the higher temperature and longer time required for blooming; and the reverse is also true that those bulbs long in cold storage bloom quickly, and therefore, should not be exposed to the high temperatures in use for forcing, as it is destructive to the strength and durability of the leaves and flowers. A temperature of 60° F. is high enough.

CHAPTER XXIX.

WILD RICE SEED.

One of the products which cannot be handled without cold storage is wild rice to be used for seed. This grain has great food value and is one of the chief foods of wild ducks and other game birds in some localities. It may be grown over large areas and even in water which is slightly brackish. It does best in the fresh water lakes and river sloughs. The water should not be stagnant, and yet not with too much current, and it does best where the bottom of the stream or lake is soft and muddy. It is important that there be a slight change in the water level from one season to another, but this variation should not be greater than eighteen inches to two feet, and the water should be constantly freshened by slight movement, with its resulting aeration. The plant is an annual, and is propagated from year to year by the seed falling directly into the water, where it lies until the following spring when it will grow if conditions are favorable.

Many failures have resulted from the planting or sowing of wild rice seed. This has been due to the seed becoming dry while stored. The fact that this destroys its vitality was not understood for some time and a large amount of seed was distributed from which practically no results were obtained. It is natural for the seed to fall into the water as soon as matured, therefore the natural method in preparing for storage is by immersing in water.

The Department of Agriculture has done some work in ascertaining the facts in connection with the saving of seed and growing of wild rice, from the report of which are extracted the following facts: The seed should be gathered as soon as matured and placed loosely, with the hulls still on, in burlap sacks and sent at once to the cold storage house. It is perhaps needless to say that seed which has been threshed, cleaned or parched for food will not germinate. If the wild rice fields are some distance from the cold storage the sacks should be sent by express unless time of transit can be guaranteed by freight. It is very important that the period from time of harvesting and the placing in cold storage be as short as possible. If this time is long, fermentation results, or the seed on the outside of the bag will become dry. In either case its vitality is deteriorated. It is not practicable to give any definite length of time which may elapse between harvesting and storing, as conditions of temperatures, humidity, etc., must be taken into consideration. It is, however, important that the seed remain moist and also that it does not ferment. Here is where cold storage comes in.

To prepare the seed for cold storage it should be taken while still fresh and moist, and before any fermentation has commenced, and put into any clean water-tight receptacle without tight covering. Tin packages or barrels or vats may be used. If there is any quantity of light immature seed or sticks or other refuse mixed with the good seed it will float to the top and can be removed. Water-tight barrels are mostly used, standing on end without heading up and placed one tier above another. The mature seed from the harvest fields is first dumped into the barrels and they are then filled with water. The temperature of the storage room for wild rice seed when prepared in this way should be just about the freezing point, 33° to 34° F.

When removing from storage the seed should not be allowed to dry out before planting, as a few days in a warm, dry air will destroy every germ. Seed properly stored has been held for more than a year and more than 80% germinated. It must be remembered that the vitality of the cold stored seed is more quickly destroyed than that of fresh seed, owing to the fact that it has been soaked in water for so long a period. It is, however, for this reason in better shape to start growth when planted.

For shipping after storage the seed should be carefully packed in dampened sphagnum moss, or fine excelsior, or planing mill shavings in a loosely covered box or crate. No special precaution needs to be taken in connection with temperature if it is not on the road more than five or six days. Should some of the seed germinate or sprout during transportation, it will grow just as well, providing it is quickly sown after receiving. If the time of transportation is long it is recommended that some provision for reducing temperature, like shipping in refrigerator cars, or icing the packages, should be provided. Fair results, however, may be obtained by handling and shipping as above described.

The seed may be sown either in the autumn as soon as harvested or in the spring. Spring sowing is preferable, as the seed is more likely to stay where it is desired to have it grow. It should not be sown in water more than four feet in depth and preferably a depth of one foot to three feet.

The possibilities of development in the growing of wild rice as a food product and as an attraction for wild waterfowl, make the facts which have so far been ascertained of considerable value, and they have, therefore, been given here in some detail.

CHAPTER XXX.

POULTRY.

DRAWN VS. UNDRAWN METHODS OF HANDLING.

It was long a mooted point as to whether poultry should be handled, shipped, frozen and stored in a drawn or undrawn condition. Drawn poultry means that from which the entrails or viscera have been removed, and it was thought by many to be the only correct way of handling, and some of the larger cities even went so far as to pass laws forbidding the sale of poultry which had not been drawn. This brought such a storm of protest from the poultry handlers that laws of this kind did not live long, but it was not until the Department of Agriculture through its Bureau of Chemistry, as represented especially by Dr. Mary E. Pennington, carried on during the season of 1909 and 1910, a series of studies to determine the relative rate of decomposition and deterioration in undrawn poultry as compared with that from which the viscera had been either completely or partly removed, that guesswork was set aside, and some actual facts determined. The tests began at the packing house where the poultry was killed and did not end until it was sold through the retailer direct to the consumer. Actual observations and records were kept at every stage in the marketing. The aim was to compare the relative keeping qualities of the drawn and undrawn poultry under actual market conditions, and to place each method of dressing strictly on its own merits.

In these tests and experiments, temperature conditions were one of the most important points of observation, and the temperature records were made by thermographs which followed the shipments of poultry from start to finish. The experiments extended over a period of six months, from midwinter to midsummer.

The dressing of the carcasses was done according to three methods known as "full drawn," "wire drawn" and "Boston drawn;" the "wire drawn" and "Boston drawn" being a sort of partial step toward "full drawn." The undrawn fowls were shipped with the heads and feet on. The birds were cooled at an average temperature of 34°F.; wrapped in parchment paper; boxed and shipped in a refrigerator car which had been iced and salted, and which was on the road averaging 7½ days. From the refrigerator car the goods were handled through a chillroom at between 32° F. and 33° F. At the retailers the average temperature of the exhibition window was 48° F.

An elaborate set of charts was prepared by Dr. Pennington showing the history of drawn poultry and undrawn poultry from the beginning, and the comparative keeping qualities of each. The conclusions reached were that undrawn poultry decomposed more slowly than either the wholly or partially drawn, and that the full drawn as completely eviscerated poultry decomposed most rapidly, and that the "Boston drawn" and "wire drawn" stood midway between the undrawn and "full drawn" in the rapidity of decomposition. These deductions are based on a number of shipments of dry packed, unwashed fowls and were studied at every stage of marketing from the shipper to the consumer, and the fowls used in the experiments were handled promptly, as ordinarily understood. It is, of course, understood that for best results poultry for slaughter should not be fed for 12 hours prior to killing. There is then little food in the crop and entrails to ferment and sour.

As it is, therefore, now fully understood and determined that undrawn poultry is the only correct way of handling for commercial purposes, we will consider that the application of refrigeration as explained more fully in what follows applies to undrawn poultry. The proper feeding, killing, bleeding, picking, etc., of poultry is a separate subject and should be studied separately. What is said here applies to the proper cooling and freezing mostly, assuming that the poultry has been properly handled according to modern practice.

COOLING AFTER PICKING.

The tendency toward "dry picking" and "dry packing" of poultry for shipment and marketing, is so marked as to indicate that the scalding of poultry to facilitate picking, and the packing of poultry in ice for shipment to prevent deterioration, will be abandoned in favor of the more cleanly and sanitary "dry" methods.

The fact that a picker will pick from four to ten times as many fowls when scalded, is certainly a strong temptation to handle in that way, but the big and increasing demand for the "dry" poultry, and the decidedly higher price which it brings, more than pays for the increased cost. In dry picking and dry packing prompt cooling after killing is a prime essential, and poultry shippers are beginning to appreciate the fact that they cannot get along without artificial refrigeration of some kind to take the heat out of the birds quickly after killing. Many shippers have tried to get along with a makeshift outfit, using an ordinary ice refrigerator or direct ice cooling arranged in some way, while others have provided a real cold storage outfit of one kind or another. Large capacity for cooling is necessary, and the space used for cooling should be divided into at least two rooms, and three rooms are even better than two. The reason is that it will not do at all to put warm, freshly killed poultry into a room with poultry which has been cooled down to correct packing and shipping temperature—30° F. to 35° F.—even if it is partially cooled down. If this is done the result is that moisture or steam from the freshly cooled poultry will condense on the colder poultry in the room, making it flabby and damp, and resulting in poor color, liability to deteriorate in transit or after arrival on the market, and thus losing one of the great advantages of the "dry" method.

It has been thought by some that by providing a large room in proportion to the work to be done that the temperature could be held without serious danger, and that the cooling and packing could be handled in one room. This theory has proved to be incorrect, and it is generally understood that at least two rooms must be provided in order to produce the best possible results in cooling, packing and shipping. Where two rooms are in service one of the rooms can be used today and the second room tomorrow, or after the poultry is partly cooled in the first room it can be removed to the second room, which may be used as a packing room and from which the poultry may be transferred to refrigerator cars for shipment. If it is desired to freeze poultry on the premises three rooms should be provided, two of them to be used as suggested, and the third room or freezer to receive the poultry after cooling and packing in the cooling rooms, and freezing it to the required temperature.

TEMPERATURES.

It is not necessary that the room in which the poultry is first placed should be maintained at a very low temperature, but a large refrigerating capacity should be available so that the moisture and heat may be taken up quickly. The second room for receiving poultry after being partially cooled down, say to 40° F. to 45° F. or 50° F. should be maintained at a temperature of 30° F. to 32° F., as near as practicable, and the poultry should be reduced to 30° F. to 35° F. before packing into boxes ready for freezing.

The correct temperature for freezing and storing poultry has not been determined with sufficient accuracy so that it is possible to name any particular temperature, but where poultry is killed, cooled, packed and frozen at one place and where the various stages are handled properly, promptly and carefully, a temperature of from 10° F. to 15° F. for freezing will produce completely satisfactory results if poultry is packed in small boxes, not more than 10 inches in thickness for the heavier poultry, and proportionately less for the lighter birds. The large city cold storage houses maintain their freezers at zero and even below, but these temperatures are not necessary for freezing poultry in the country or at the point where killed. These low temperatures are useful and necessary in big cities as the poultry is very often greatly deteriorated before placing in the freezer, and it often comes to the city freezers in large boxes and barrels which are slow in giving up their heat, and the poultry, therefore, takes a long time to freeze to the center of the package.

After being properly cooled and packed at a temperature of 30° F. to 35° F. poultry may be frozen, as above stated, in a temperature of 10° F. to 15° F. In placing the boxes in the freezer they must not be piled tightly, and a circulation of air of at least 2 inches between all packages and under the packages on the floor should be allowed. A good method is to set the boxes on end on 2x4's on the floor, allowing a space of 3 to 4 inches between the boxes, and piling them but one tier in depth. The boxes should be set edgewise to the circulation of air or toward the coils if the room is equipped with direct piping. If the room is equipped with the Cooper false floor and false ceiling system of air circulation, it is not necessary to leave more than 2 inches of space between the packages and no attention need be paid to just how the packages are placed in the room except to keep them 2 inches apart all around, and an inch off the floor.

After the poultry has been thoroughly frozen to the center of the box it may be stacked up solidly in the room, but a space of 1 inch to 2 inches must be left on the floor and a similar space around the sides of the room and at the ceiling. Much damage has been caused on other frozen products as well as poultry, by piling them tightly against the outside or warm wall, or failing to leave space on the floor for the cold air to circulate and thus preventing the heat from entering. A thawing or high temperature will result from improper piling, and mouldy, decayed and spoiled poultry results.

METHODS OF COOLING.

Those who are to continue in the poultry business must, in future, provide artificial means of cooling, or they cannot possibly compete with their better equipped competitors. Artificial means of cooling does not necessarily mean a complicated refrigerating machine system. The same results can be had with the Cooper brine system, using ice and salt for cooling. Any method of cooling which will give a temperature of about 30° F. in the cooler is all that is required, but this cannot possibly be had with any method of cooling with air directly off from melting ice.

Satisfactory results can be had in the cooling of freshly killed poultry by locating the cooling pipes directly in the room, providing the room is of fair height and the pipes are hung from the ceiling so that a good circulation of air will be present. If the room is a low one and rather large and the pipes located around the walls, the circulation will be defective and the results from cooling inferior. If the pipes are to be located in the room, they should be hung edgewise from the ceiling and located not more than a pair together, so that drip pans may be arranged under them, and so that the cold air will drop off the coils very quickly, and thus result in a good circulation and uniform temperatures in the room. Then the warm and moisture laden air will rise to the top of the coil and the moisture be condensed thereon. The Cooper chloride of calcium process is especially desirable in a poultry cooling room on account of the excess of moisture which must be disposed of.

Better than any direct piped room, however, is the fan system of air circulation. This does not mean to cool a room by means of piping in the room and then simply to put in one of the small electric fans. This is not air circulation by any means and only a poor makeshift of doubtful value. The very best method of circulating air for cooling purposes, as already mentioned, is the Cooper improved false floor and false ceiling system of air circulation, wherein cold air is distributed through a perforated false floor and the comparatively warm and moisture laden air drawn out of the room through a perforated false ceiling. This gives a perfectly uniform circulation of air in all parts of the room, resulting in a quick cooling and positive prevention of condensation of moisture on the poultry. This system will quickly dry the fowls as well as cool them. The air from the false ceiling is taken to the coil room or bunker room in which cooling pipes are located, and these pipes may be cooled by a refrigerating machine, or they may be the secondary coil pipes of the Cooper brine system, using ice and salt for cooling. On account of the temporary or transient nature of the work, the Cooper brine system, which needs no skilled engineer nor slow and expensive overhauling of plant prior to starting up, is preferable to an ice machine.

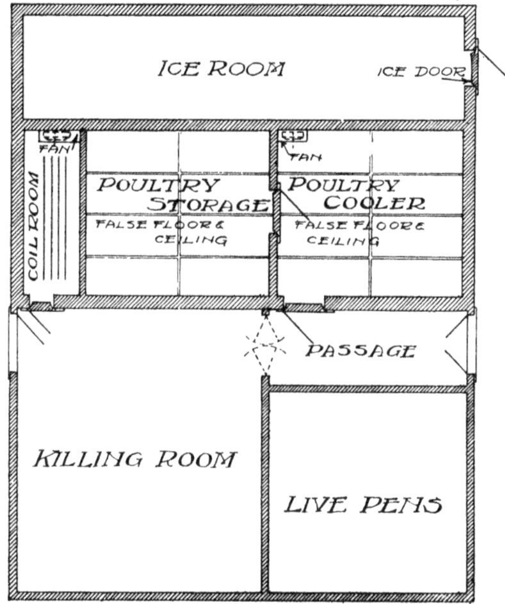

FIG. 1.—MAIN FLOOR PLAN.

The length of time required for cooling poultry depends on its temperature when placed in the room, capacity of refrigerating system, etc. It is really not desirable to turn a blast of frosty air on warm goods. Poultry should be reduced in temperature rather slowly, and from 24 to 36 hours would be the natural period through which to cool it from a temperature of 80° F. or 85° F. down to 30° F., the temperature suitable for packing.

If killing a large amount of poultry and necessary to crowd the apparatus to the utmost, 24 hours or even 18 hours in the chill room is permissible, but otherwise 36 hours is advocated as more desirable for the best results. The very best of results are only obtainable by using two separate methods of cooling and two separate rooms, so that when necessary to put extremely warm poultry in the cooling room during warm and humid days, the poultry is first exposed to only a moderately cold air, say at a temperature of 40° F. to 50° F. After taking the high temperature out of the poultry a colder circulation of air may be employed and the poultry reduced to the correct packing temperature, 30° F.

Several different arrangements may be made to produce these results. An example of a correct plant is illustrated in Fig. 1. This, as will be noted, consists of two poultry rooms, one a poultry cooler and the other a poultry storage room. An ice room extends along one side of these rooms. Both rooms

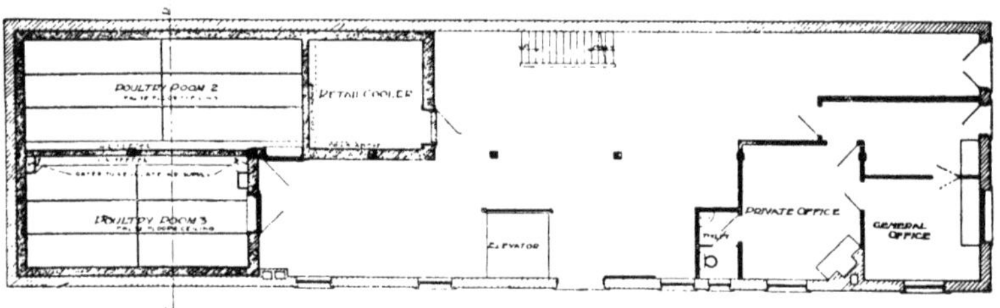

FIG. 2.—FIRST FLOOR PLAN.

are equipped with the Cooper false floor and false ceiling system of air circulation and the air handled by means of the fan system. When first killed the poultry is introduced into room marked "Poultry Cooler," where it takes its initial cooling, and after being reduced to 45° F. or 50° F. it is run into the poultry storage room where it is still further cooled and packed at a temperature of 30° F. The poultry cooler is cooled by air directly from the ice room, while the poultry storage and packing room is cooled by air from the coil room of the Cooper brine system, using ice and salt for cooling. The room marked, "Poultry Storage" is used for accumulating a carload and shipping it out in the best possible condition. This room may be used for egg storage during the egg season and when not in use in connection with the poultry business.

Figs. 2 and 3 show an entirely different arrangement, although operating on the same general principle. There are two poultry rooms which are used for the cooling of poultry for shipping, but the poultry is not moved from one room to the other as in Fig. 1. The same result is obtained by an arrangement of gates. For instance: Freshly killed poultry is run into room No. 3 and has the direct air from ice room on second floor turned on for a period of from 6 to 12 hours. After cooling the poultry to a temperature of 40° to 50° F. the direct air is turned off and air from the coil room containing

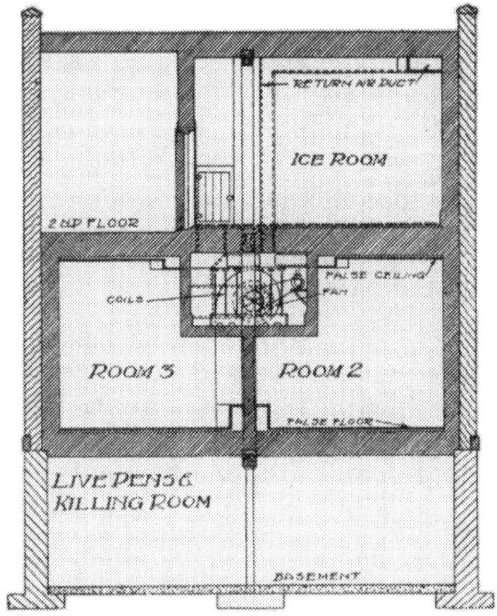

FIG. 3.—TRANSVERSE SECTION "C-D."

secondary coils of the Cooper brine system is turned on and the poultry reduced to a temperature of 30° F. to 32° F. It is then packed for shipping without removing from room and held at a correct temperature until shipped. By using the rooms alternately, with direct air for a portion of the day, and the Cooper brine system for another portion, poultry may be cooled continuously day after day. Should it be desired to hold the poultry for some days before shipping it can be removed from one room to the other essentially as in Fig. 1.

Plan shown in Figs. 2 and 3 has a retail cooler as this plant is located in quite a large town where the local business is considerable. This room is cooled by the Cooper brine system and maintained at a temperature of 32° F. or lower if desired.

With both plans the fans are so arranged as to take the air directly from outside when the temperature is sufficiently low to make this practice desirable. This makes a very rapid method of cooling, much more so than opening windows and doors, as the distribution of air is more thorough.

The false floor and ceiling system is the most perfect theoretically of any in service as the cold air is distributed over the entire surface of the floor and the comparatively warm and moisture-laden air is drawn off from the entire surface of the ceiling, making a very free and penetrating circulation. This results in a very thorough and quick cooling; yet the circulation is not so rapid nor strong as to be objectionable.

RACK FOR HANGING POULTRY FOR COOLING.

For cooling poultry quickly and for handling it economically, it is best to use metal racks which are mounted on casters or wheels, and on which the poultry may be hung in the killing room, then run directly to the cooling and packing rooms.

Such a rig has been described by Dr. Mary E. Pennington in connection with her poultry work for the Bureau of Chemistry, U. S. Department of Agriculture, and illustrated herewith. These rigs are now being made as a regular article of manufacture and they may be had at reasonable cost all ready for business.

The specifications for the rig are as follows: Height over all, 69 inches; width over all, 38 inches; width of base, 38 inches; length of base, 61 inches; width of top of frame, 33 inches; height to top of frame, 68 inches; end supports, four inches apart at base; bend in end supports, 19 inches from floor; first cross bar, 29 inches from floor; cross bars eight inches apart; two bottom cross bars, nine inches apart; end cross brace, 26 inches long and 57 inches from floor; center brace rods, 76 inches long; top of base, 8 inches from floor; corner brace plates, 10 inches on square edges; end brace plate, 10 inches wide, 9 inches high (upper corners beveled); cast-

FIG. 4—POULTRY COOLING RACK.

ers, 6 inches in diameter; base frame, 2x¼-inch angle iron; end supports, 1½ by ⅛ inch angle iron; cross bars, 1¼ by ⅛ inch angle iron; end cross bars, 1½ by ¼ inch flat iron; center brace rods, ⅜ inch round iron, threaded at both ends, one end passing through center of end cross bars, the other end replacing one of the bolts fastening end supports to end brace plate; corner and end brace plates, ¼ inch flat iron; fingers for holding feet of birds made of No. 10 tinned steel wire. The fingers are formed by bending the wire continuously around three pins placed in a triangular position. The fingers are 3½ inches long and approximately 1⅞ inches apart, center to center, being spaced so as to allow 30 openings to hold 15 chickens within a distance of 58¼ inches, which is the clear distance along the cross bars between the upright supports on either end. The slots between the fingers are ⅜ inch wide at the bottom and then widen to a round point at the top as shown.

The finger wire is fastened on with ½ inch and ¼ inch oval-headed bolts, galvanized, with washers. It requires 32 bolts to each cross bar. All other bolts or rivets are ¾ by ⅜ inch, galvanized, except the caster bolts which are 1 by ½ inch, galvanized.

The upright supports are fastened to end brace plates with two bolts each. The corner base plates are riveted to base.

The caster spindles are ⅝-inch in diameter, turning in an extra strong socket, as considerable strain comes at this place.

All metal is galvanized except the finger wire, which may be either tinned or galvanized.

These dimensions are adapted for a truck that will pass readily a refrigerator door four feet wide and six high.

One inch of space in width can be saved by moving the end supports together at the base, making them three inches apart from outside instead of four inches as stated.

In this truck, all four wheels are casters. This makes the truck harder to handle by one man. Wherever plenty of refrigerator and floor space is available, two of the casters may be replaced by fixed wheels, which makes the truck steer much more easily.

For smaller doors than 4x6 feet, the base could be narrowed and the top narrowed and the height decreased to use only five cross bars instead of six. This, however, also cuts down the capacity of the truck.

As it now stands, these trucks will hold 180 broilers, fowls or rabbits, and 48 turkeys each.

The advantages of hanging poultry by the legs while cooling are many, such as more rapid cooling, better circulation of air around the bird, the skin of the bird is not torn by com-

FIG. 5—COOLING POULTRY IN A ROOM EQUIPPEED WITH COOPER FALSE FLOOR AND FALSE CEILING SYSTEM OF AIR CIRCULATION.

ing in contact with shelves, the heads of the fowls are easily wrapped, and the grader can see the stock more easily. Also, the method is more cleanly.

MISCELLANEOUS.

A practical poultry operator in Chicago offers some suggestions with reference to preparing poultry for freezing, which are worth considering. He says that much of the trouble experienced in freezing poultry is due to faulty handling by the commission men; that they expose it on sale all day in warm weather, pack it up toward evening, and put it into the freezer. The result of this is that it may look all right when taken out of the freezer, but as soon as the frost comes out of it, it becomes sticky and slimy and not fit for food. This practical poultry man states that his experience in putting away matured and well fed turkeys, capons and roasting chickens is that they should be placed in the cooler promptly when killed, and the animal heat taken out of them within twenty-four hours, and *then* frozen. He says that such poultry will be far better when removed from the freezer in the spring than freshly killed poultry after running all winter. He says further that spring chickens put away before they are matured may not hold their flavor like properly matured birds, and that while they appear all right, the delicate flavor of freshly dressed stock is gone.

Some actual records made by Dr. Mary E. Pennington of the Bureau of Chemistry, U. S. Department of Agriculture, will doubtless be interesting. They may be summarized as follows:

The tests covered various temperatures and various periods of storage, from carrying poultry in a temperature of 65° F. to 75° F. for three days, to sixteen months in a freezer at a temperature of 10° F. It is reported that chickens at 65° F. to 75° F. for three days' time were in bad order at the end of that period, and it is also reported that poultry held for sixteen months were in rather poor condition for eating purposes, although not spoiled or inferior by any means. Practically perfect results were secured by storing poultry at a temperature of 10° F. above zero for four months, and the same experiment continued for eight months gave poultry which was in commercially perfect condition, although certain changes were noted which indicate that the fair life of the poultry as held in freezers at this temperature had been reached. At twelve months in the freezer the poultry was still in good condition, but proportionately poorer than at eight months.

It will be noted that these experiments show the possibilities of freezing and storing poultry at 10° F. above zero. This does not at all correspond with the claims of some of the big storage houses where it is said that temperatures below zero are necessary for best results. It is the author's opinion, and he has constantly held this opinion for some years, that 10° F. above zero is amply low for the freezing and storage of poultry, butter, game, etc., for periods which are represented naturally by the commercial limitations of the product. It is not ordinarily necessary to store any of these goods for more than six to eight months, but for this length of time 10° F. above zero for all practical purposes is just as good as 10° below zero.

Mr. R. H. Tait in a paper read before the "Missouri Car Lot Shippers' Association"* gives some interesting suggestions and facts which are always acceptable. He states that for poultry cooling two rooms are desirable, one double the size of the other; the larger room to be used for chilling and packing, and the other for storing after chilling. He also states that provision for fresh air circulation in these rooms is of importance, second only to refrigeration, and that with rooms properly designed this can be done without the use of fans, by placing the cooling coils in a bunker above the storage space. The temperature of the chill room, he says, should not go above 38° F., while putting in fresh poultry, and should be reduced to 32° F. After thorough chilling the birds should be moved to the storage room maintained at 32° to 30° F., and there held and packed.

Mr. Tait states that a packing station to handle from two to three cars of poultry per week should have a floor space of about 1,200 sq. ft. with a ceiling 10 ft. high, and that the two rooms used in connection with the poultry business should be arranged accessible to loading platform or railroad track, and recommends an enclosed flexible vestibule adjacent to car door to avoid exposure of cold goods to warm air during the

*_Ice and Refrigeration_, April, 1912, page 258.

summer. Such a plant with a capacity of two to three cars per week, he states, will cost from $6,000 to $7,000 including building, machinery, etc., and that it will require 15,000 gallons of water per day at temperature of 70° to 80° F. under severest conditions, and that it would not be practical to take this quantity of water from the city mains.

It might be remarked that where the quantity of poultry to be handled does not exceed two to three cars per week or an average of 8,000 to 10,000 lbs. per day, and where ice, either artificial or natural, is available at reasonable cost, the Cooper brine system, using ice and salt for cooling, is recommended in preference to a refrigerating machine. The cooling of poultry is a transient or intermittent service anyway, and there are many times during the heavy poultry shipping season when natural outside temperatures furnish all the refrigeration that is necessary. A plant equipped with the Cooper brine system can be built for very much less than the figures quoted above.

CHAPTER XXXI.

FREEZING AND STORING FISH.*

IMPORTANCE AND GROWTH OF THE INDUSTRY.

In the artificial freezing of fish and their subsequent retention in cold storage is found one of the most recent methods of food preservation, originating about forty years ago, and while it has acquired considerable importance in certain localities, its practical value is scarcely appreciated by the general public. It is applied in the various marketing centers of the United States, and to some extent in the countries of Europe and South America. Its greatest development and most extensive application exists along the great lakes, in freezing whitefish, trout, herring, pike, etc., about 7,000,000 pounds of which are frozen each year. On the Atlantic coast of the United States it is used in preserving bluefish, squeteague, mackerel, smelt, sturgeon, herring, etc., the trade in these "tailing on" or immediately following the season for fresh or green fish. On the Pacific coast large quantities of salmon and sturgeon are frozen and held in cold storage until shipped, the trade extending to all parts of America and northern Europe. At various points throughout the interior of the country there are cold storage houses where fishery products are held awaiting demand from the consumers. In Europe there is comparatively little freezing of fish, although the process is applied very extensively to preserving beef, mutton, etc., and the markets of Hamburg and other continental cities receive annually several million pounds of frozen salmon from our Pacific coast. In England large fish freezers were erected several years ago at Grimsby and Hull, and trawlers are in some cases supplied with refrigerat-

*By Charles H. Stevenson in *Ice and Refrigeration*, *February*, 1900.

ing plants where the fish are plunged alive into cold brine which freezes them solid.

During warm weather the temperature of the fish storage room can never be kept below 32° F. by the use of ice alone. While a temperature of 32° F. retards decomposition, the fish acquire a musty taste and loss of flavor, and eventually spoil. To entirely prevent decomposition the fish must be frozen immediately after capture, and then kept at a temperature of several degrees below freezing. The belief held by some persons that freezing destroys the flavor of fish is not well founded, the result depending more on its condition when the cold is applied and the manner of such application than upon the effect of the low temperature. Fish decreases less in value from freezing than meat does, but it is especially subject to two difficulties from which frozen meat is free; first, the eye dries up and loses its shining appearance after considerable exposure to cold, and second, the skin, being less elastic than the texture of the fish, becomes hard and somewhat loose on the flesh. Frozen fish is not less wholesome than fish not so preserved. The chemical constituents are identical, except that the latter may contain more water, but the water derived from injested fish has no greater food value than water taken as such. The principal objection to this form of preservation is the tendency to freeze fish in which decomposition has already set in, and the prosperity of the industry requires that any attempt to freeze fish already slightly tainted should be discountenanced. When properly frozen and held for a reasonable period, the natural flavor of fish is not seriously affected and the market value approximates that of fish freshly caught. The process is of very great value to the fishermen supplying the fresh fish trade, since it prevents a glut on the market, and it is also of benefit to the consumer in enabling him to obtain almost any variety of fish in an approximately fresh condition throughout the year.

DEVELOPMENT OF COLD STORAGE.

The first practical device for the freezing and cold storage of fish was invented by Enoch Piper, of Camden, Me., to whom a patent was issued in 1861. His process, based on the well known fact that a composition of ice and salt produces a much lower temperature than ice alone, consisted in placing the fish on a rack in a box or room having double sides filled with non-conducting material, and metallic pans containing ice and salt were set over the fish, and the whole inclosed. The temperature in the room would soon fall to several degrees below the freezing point of water, and in about twenty-four hours the fish would be thoroughly frozen. The fish were then covered with a coating of ice by immersing them a few times in ice cold water, forming a coating about ⅛-inch in thickness, after which the fish were wrapped in cloth, and a second coating of ice applied. In some instances they were covered with a material somewhat like gutta percha, concerning which much secrecy was exercised. The fish were then packed closely in another room well insulated against the entrance of warmth, and in which were a number of perpendicular metallic tubes, several inches in diameter, filled with a mixture of ice and salt to keep the temperature below the freezing point.

The process was also patented in the Dominion of Canada, and a plant was established at Bathurst, New Brunswick, in 1865, the output consisting almost entirely of salmon, a large proportion of which were imported into the United States. In order to hold the frozen fish in New York, while awaiting a market, Piper constructed a storage room in a shop on Beekman street, that being the first cold storage room for fish in the United States. The walls of the room were well insulated, and around the sides were two rows of zinc cylinders, ten inches in diameter at the top, and decreasing in size toward the bottom, connecting at the lower end with a drainage pipe. The cylinders were filled with a mixture of ice and salt, which was renewed whenever necessary. Whatever may have been the imperfections in his process of freezing, the system of storage was quite satisfactory, and differs little from that in use at the present time. Piper refused to sell rights to others for the use of his process, and after maintaining a monopoly of the business for three or four years his exclusive right to it was successfully contested by other fish dealers in New York, who applied it to storing other fish besides salmon.

The principal objection to Piper's process is that the fish are not in contact with the freezing mixture during the operation of freezing, and, consequently, too much time is required for them to become thoroughly frozen. Several devices have been used for overcoming this objection, among which are cov-

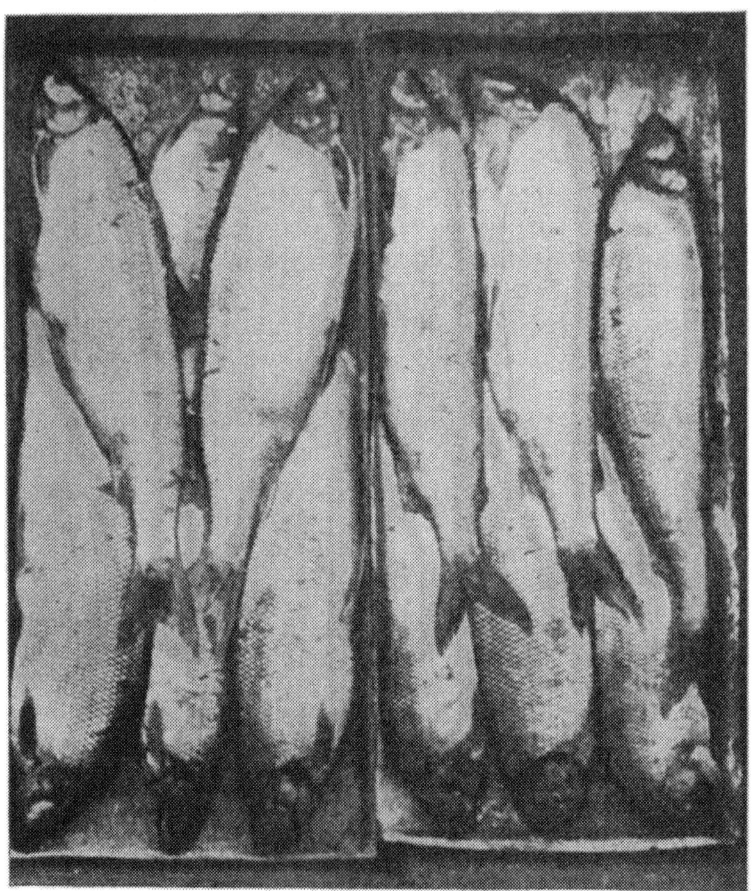

FIG. 1—PANS OF FROZEN WHITEFISH, SHOWING ARRANGEMENT OF FISH IN THE PANS.—DAVIS SYSTEM.

ering the fish with thin sheet rubber or other waterproof material, and packing them in the mixture of ice and salt.

The greatest improvement, and the one used almost exclusively when ice and salt form the freezing agency, originated in 1868 with Mr. William Davis, of Detroit, Mich., the description being as follows: Two thin sheet metal pans are made to slide one over the other, the object being to place the fish in one pan, slide the other pan vertically over it, and the box is then placed in direct contact with the freezing mixture. By having the box constructed in this manner, it is capable of being expanded or contracted to accommodate the size of whatever may be placed therein, and the top and bottom always be in contact with the articles to be frozen. After the fish are inclosed in the pans, the latter are placed in alternate layers with layers of the freezing mixture between and about them. When the fish are thoroughly frozen they are removed from the freezing pans and placed in cold storage at 10° or 12° F. below freezing.

As the trade developed the size of the storage rooms increased and improvements were adopted in the arrangement and form of the ice and salt receptacles, and in the method of handling the fish. But the freezing with pans immersed in ice and salt, as in the Davis process, and the subsequent storage in the manner used by Piper, continued without great modification until the introduction of mechanical refrigeration into the fishing trade in 1892. At that time ice and salt freezers and storage rooms existed at nearly all the fishing ports on the great lakes; eight or ten small ones were in New York City, and several were in use on the New England coast. Some of those on the great lakes were quite large, with storage capacity of 700 or 800 tons or more, and the aggregate capacity of all in the country approximated 8,000 tons. Cold storage houses fitted with ammonia machines had been established at various places along the coast and in the interior during the ten or fifteen years preceding, and in these some frozen fish had been stored. But the first establishment using a refrigerating machine for freezing fish exclusively was erected at Sandusky, Ohio, in 1892. The method of freezing in these establishments differs from the ice and salt process in that the pans of fish are placed on and between tiers of pipes carrying cold brine or ammonia instead of being immersed in ice and salt. In the storage rooms less difference exists, coils of brine carrying pipes taking the place of the ice and salt receptacles, the blocks of fish being removed from the pans and stored as in the older process.

DESCRIPTION OF ICE AND SALT FREEZERS.

The outfit of an ice and salt freezer consists principally of temporary stalls or bins where the fish are frozen, and insulating rooms where the frozen fish are stored at a low temperature. In addition to these there are ice houses, salt bins, freezing pans, and the various implements for the convenient prosecution of the business. The freezing bins are usually temporary structures within the fish house, and are generally without insulation. The walls of the fish house may form the back, while loose boards are fitted in to form the sides and front as the bin is filled, in the manner hereafter described. A better way is to construct the bin with permanent sides and back four or five inches thick, fitted with some non-conductor, with double or matched floor, and with movable front boards.

The storage rooms are commonly arranged in a series side by side and separated from each other by well insulated partitions, the capacity of the rooms ranging from 25 to 250 tons each. The outer walls of these rooms, as well as the floors and ceilings, are well insulated, made usually of heavy matched boards, with interior packing of some non-conductor of heat, such as planing mill shavings, sawdust, pulverized charcoal, chopped straw, rock wood, slag wood, etc. Most of the walls are sixteen or eighteen inches thick, filled with planing mill shavings or sawdust, and in some freezers the damaging effect of rats is obviated by placing linings of cement between the shavings and the board walls. Most of these loose materials have their economic drawbacks, chiefly because of their strong hygroscopic tendency, the material losing its insulating power and decaying, this decay also attacking the wood of the walls. Because of this, many of the storage rooms recently constructed are insulated by having the walls made up of a combination of rock or mineral wool, insulating paper, air spaces and inch boards.

The sides, and in some cases the ends of the room, are lined with the ice and salt receivers, consisting of galvanized sheet iron tanks, eight or ten inches wide at the top, narrowing to three or four inches at the bottom, and placed about four inches from the wall in order to expose their entire surface to the air in the room. These tanks open at the top, which extends above the ceiling, so that they may be filled without opening the storage rooms. At the bottom is usually a galvanized iron gutter, into which the water resulting from the melting ice flows, whence it is conducted through the floor of the room by a short pipe, protected from the entrance of air at its lower end by a small drip cup, into which the brine falls and runs over at the top. The ice and salt tanks must be cleaned from time to time in order to rid them of dirt and sawdust. Their capacity should be in proportion to the size of the room and the excellence of the insulation, and they should be large enough to render it unnecessary to fill them oftener than once a day, even in the warmest weather.

FREEZING BY MECHANICAL REFRIGERATION.

In the freezing houses using mechanical refrigeration there is, as customary with cold storage houses used for other products, a machinery room containing the boilers, compression pump or absorption tank, according to the system employed, brine pump, etc. Apart from these, and within well insulated walls, are the cold rooms, of which there are two kinds, one for the freezing of fish and the other for their storage after being frozen, the capacity of the latter being usually much greater than that of the former. In the freezing room the circulating pipes containing the cooling material are one-half inch to two inches in diameter, and arranged in shelves or nests with horizontal layers four or five inches, and sometimes ten inches apart, ranging from the floor to the ceiling, the entire room being occupied with these nests, except sufficient space for moving about. The temperature depends, of course, on the quantity of green fish and the progress of the freezing process; but with direct expansion, or using brine made of chloride of calcium as the circulating medium, a temperature of —10° F., or less, is obtainable. In this room the fish are frozen, and then they are removed to the storage rooms. These are constructed similarly to the storage rooms in ice and salt freezing houses, the only difference being that brine carrying pipes are substituted for the ice and salt receptacles. The pipes in the storage rooms are usually larger, but are not so numerous as in the freezing room. They are arranged at the ceiling, and sometimes about the upper side walls also.

In freezing fish, as in preserving most food products, close attention must be given to the economy of the process as well as to the excellence of the product, and the expense of the best process frequently prevents its use. To secure the best results, the stock to be frozen should be perfectly fresh and free from bruises and blood marks. It improves the appearance, and therefore increases the value, if the fish are graded according to size, but this is rarely done. All kinds of fish keep and look best when frozen just as they come from the water, with heads on and entrails in, and it is better that the fish be not eviscerated before freezing, except in case of very large fish, such as sturgeon. But since the freezers receive the surplus from the fresh fish trade, many have been already split and dressed. Generally, fish that are frozen with heads off and viscera removed are not strictly fresh, but this rule has several exceptions.

Whether round or eviscerated, the fish are first washed by dumping them into a wash box or trough containing fresh cold water, which is frequently renewed, and stirring them about with an oar-shaped paddle or cloth swab, to remove the slime, blood, etc. Some freezers consider it inadvisable to wash flat fish, because of their being too thin. From the wash box the fish are removed by hand and placed in the pans in such a manner as to make a neat and compact package entirely filling the pan, so that the cover will come in contact with the upper surface of the fish. It is desirable, when the size of the fish so admits, that the bellies be placed upward, since that portion has greater tendency to decompose, and, as the cold passes down, this arrangement results in freezing the upper portion of the block first, and also in less compression of the soft portion of the fish by removing the weight therefrom. It is also desirable to have the backs of the fish at the sides of the pan and the heads at the ends, so as to protect the blocks in handling, but this is by no means a uniform practice. In case the fish have been split and eviscerated it is desirable to place them slanting on the sides, but with backs up, so as to permit the moisture to run from the stomach cavity. Some freezers place herring and other small fish on their sides, two layers deep in the pans, while others place a bottom layer of three transverse rows, the end rows with the heads to the edge of the pan, and a top layer of two transverse rows laid in the two depressions formed between the bottom rows. In case of pike and some other dry fish a small quantity of water is sprinkled over them, since they do not ordinarily retain sufficient moisture to hold together when frozen, as is the case with most species. As soon as the pans have been filled and the covers fitted on they are placed in the sharp freezers, which have been described.

In those houses using ice and salt as the freezing medium the arrangement of the ice, salt and fish pans is as follows: The ice, after being passed through a grinder, where it is crushed into small particles, is mixed with salt in the proportion of from eight to sixteen pounds of salt to one hundred pounds of ice. The mixing is most conveniently done by scattering salt over each shovelful of ice as the ice is shoveled from the grinder to a wheelbarrow. Many varieties of salt are used, most houses preferring a coarse mined salt because of its cheapness. Others use finer salt because it comes into closer contact with the ice and results in a lower degree of cold and the more rapid freezing of the fish, although the mixture does not last as long.

The amount of ice and salt required in freezing a given quantity of fish depends principally on the fineness of the materials and the proportions in which they are used, and to a less extent on the outside temperature, the amount of moisture in the atmosphere, the size of the pans and the manner in which the fish are placed therein. The finer the ice and salt, the quicker the freezing and the consumption of the ice. A larger proportion of salt results also in quicker freezing. The most economical quantities appear to be about eighty-five pounds of salt and 1,000 pounds of ice to each 1,000 pounds of fish, although some freezers use much more salt and less ice. Much larger quantities of ice and salt are required during warm weather, and more is necessary also when the atmosphere is moist than when it is dry. Some of the ice and salt generally remains unmelted, and this may be used over again in connection with fresh materials, additional salt being mixed with it; and as it is weaker than new ice it should be used mainly at or near the bottom, the top of the pile taking care of the bottom, since the cold descends.

In making the freezing pile, an even layer of ice and salt, about three or four inches deep, is placed at the bottom, on which is laid a tier or layer of pans filled with fish, about three inches of ice space intervening between the pans and the sides of the bin. This is followed successively by a layer of ice and salt about two or three inches deep, and a layer of pans, the surface of each layer of ice being made even and smooth by means of a straight edge. Sideboards are placed as the height of the pile requires, and a wide board laid on the pile furnishes a walk for the workmen in placing the freezing mixture and the pans. Some freezers place the pans in double tiers between the layers of ice and salt, and in this case the thickness of the layers of freezing material must be increased. In some freezers a light sprinkling of salt is thrown on top of the pans as they are successively placed. The pile is built up as high as it is convenient for handling the pans of fish, which usually does not exceed six feet. A double quantity of the freezing material is put on top, and the whole should be covered with wood or canvas to exclude the air. The fish are usually frozen completely in about fifteen or eighteen hours, but they usually remain in the pile until the following morning, when they are ready to be placed in cold storage.

METHOD OF STORING THE FROZEN FISH.

Being moist, the fish are frozen solidly to each other and to the surfaces of the pans while in the sharp freezer. To remove them from the pan the latter is usually passed for a moment through cold water, which draws the frost sufficiently from the iron to allow the fish to be removed in a block without breaking apart. In one or two freezing houses the thawing of the fish from the sides of the pan is omitted, the cover being loosened and the block of fish removed by striking the pan at the ends and sides, after which the block of fish is dipped for a moment in cold water.

Considerable moisture adheres to the fish from its being dipped in water, and this being frozen by the surplus cold forms a coat of ice about one-fiftieth inch thick, entirely surrounding the irregular block. The process of freezing dries the fish to

FIG. 2—SHARP FREEZER FOR FISH.—PIPES HAVE BEEN CLEANED PREPARATORY TO RECEIVING A NEW BATCH OF FISH IN PANS.

some extent, the loss in weight amounting to about 2 per cent, but the ice coating adds about 4 per cent to the weight.

After the coating of ice has been applied, the fish are passed to the cold storage room, where they are arranged in neat piles, the blocks being placed vertically in some instances; but more frequently they are arranged horizontally in piles extending from the floor nearly to the ceiling. Strips two or three inches thick are laid on the floor to keep the fish slightly elevated, and allow the cold air to circulate underneath.

The quantity of ice and salt required in the establishments which use those materials in the storage rooms is dependent on the outside temperature and the excellence of the wall insula-

FIG. 3—SHARP FREEZER FOR FISH.—THE PANS IN THIS CASE CONTAIN STURGEON TO BE FROZEN.

tion, and is independent of the amount of frozen fish in the room, requiring no more freezing material to keep fifty tons of frozen fish at an even temperature than to keep two tons in a room of equal size. With 16-inch or 18-inch walls, well insulated, it requires the melting of about forty pounds of ice per day for each 100 square feet of wall surface when the outside temperature is 60° F., to maintain a temperature of 18° F. inside, this calculation leaving the opening of doors and the cooling of fresh material out of consideration. The temperature in the storage room should be constant, and about 16° or 18° F. is considered the most economical. Above 20° F. the fish are likely to turn yellow about the livers, a result generally attributed to the bursting of the "gall."

The storage rooms should be free from moisture, since the latter offers a favorable place for the settlement and development of micro-organisms of all kinds, which tend to mold the fish. To reduce excessive moisture, a pan of unslaked lime, chloride of calcium or other hygroscopic agency, may be placed in the room, the material being renewed as exhausted. If the storage rooms are very moist, they should be dried out before storing fish in them, this being readily accomplished by using a small gas, coke or charcoal stove. The storage rooms cooled by refrigerating machines may be dried by passing hot water through the pipes, which, of course, should, under no circumstances be done when there are fish in the rooms. In case of mold appearing on the fish, it might be well to try spraying them with a solution of formalin, consisting of ten parts of formalin and ninety parts of water, which should be used at the first sign of mold.

DETERIORATION OF FISH AFTER FREEZING.

All fish deteriorate to some extent in cold storage, depreciating both in flavor and firmness. The amount of this decrease is dependent primarily on the condition of the fish before freezing and the care exercised in the process of freezing, and, secondarily, on the length of time they remain in cold storage. The loss in quality during storage is due principally to evaporation, which begins as soon as the fish are placed in storage, and increases as the ice coating is sapped from the surface.

Evaporation proceeds at very low temperatures, though not so rapidly as at higher ones; even at a temperature of 0° F. the evaporation during two or three months is considerable. The heavier the ice coating the less the evaporation; but it is almost impracticable to entirely prevent it, and under ordinary conditions it amounts to about 5 per cent in weight in six months, but the loss in quality is greater than the loss in weight.

The most practicable method of restricting evaporation, other than coating with ice, is to wrap the fish in waxed or parchment paper and place them in shipping boxes, whose length and width are slightly greater than the blocks and deep enough to contain four or five blocks, or 120 to 150 pounds of fish.

Along the great lakes the most popular fish for cold storage are whitefish, lake trout, lake herring, blue pike, saugers, sturgeon, perch, wall eyed pike, grass pike, black bass, codfish and eels. In addition to these species, the great lakes freezers receive large quantities of blue fish and squeteague (sea trout) from the Atlantic. On the Atlantic coast bluefish, halibut, squeteague, sturgeon, mackerel, flat fish, cod, haddock, Spanish mackerel, striped bass, black bass, perch, eels, carp and pompano are frozen. Salmon, sturgeon and halibut are the principal species frozen on the Pacific coast.

Some varieties of fish are so very delicate that it is not deemed profitable to freeze them, especially shad, but even these are frozen in small quantities. Oysters and clams should never be frozen, the best temperature for cold storage being 35° to 40° F., and when stored in good condition they will keep about six weeks. As an experiment they have been kept twelve weeks, but storage for that length of time is not advisable. Caviar also should never be frozen, but held at about 40° F. Scallops and frogs' legs, however, are frozen hard in tin buckets and stored at a temperature of 16° to 18° F. Sturgeon and other fish too large for the pans are frequently hung up in the storage rooms by large meat hooks, and when frozen are dipped in cold water and stored in piles.

In some of the largest freezing houses on the Atlantic seaboard, which freeze and store fish as well as other food products, the fish to be frozen are simply hung up in the sharp freezer, the heads being forced on to the sharp ends of wire nails protruding from cross-laths arranged in series. After the fish are frozen they are removed and piled in storage rooms, where the temperature is from 15° to 18° F. When the handling of fish is of minor importance compared with other food products, they are generally placed on slat-work shelves in either a special freezing room or in a storage room where the temperature is kept below 20° F., or they are retained in bulk in baskets, boxes or barrels in the same room. But these methods are not productive of results even approximating those in the great lakes freezers.

The cost of cold storage and the deterioration in quality make it inadvisable to carry frozen fish more than nine or ten months, but sometimes the exigencies of trade result in carrying them two and even three years. In the latter case they are scarcely suitable for the fresh fish trade unless the very best of care has been exercised in the freezing and storage, and it is usually better to salt or smoke them.

The rate of charges in those houses which make a business of freezing and storage for the general trade is usually from a half cent to one cent for freezing and storage during the first month, and about half of that rate for storage during each subsequent month, depending on the quantity of fish. However, the cost of running a first-class plant at its full capacity is probably less than one-third, or even one-fourth, of the minimum above quoted, since it costs no more to run a storage room full of fish than one-fifth full.

CANADIAN BAIT FREEZING METHODS.

The Canadian Government, in promoting the organizations of associations for bait freezing and storage, pays a bonus of one-half the cost of such freezers under certain restrictions and regulations, and also $5.00 per ton for fish properly preserved each season, but the Government-aided product may not be sold commercially. Prof. E. E. Prince, Commissioner and General Inspector of Fisheries for Canada, describes in "The Fishing Gazette" the bait freezing methods of small plants, which cost from $500 to $2,000. There are two methods in use known as the pan system and the crate system. The pan method is essentially the same as already described in the foregoing, and is, doubtless, the best method. The crate freezing system is older and not as rapid nor as efficient. The pan system is described as follows:

"1. The fish are placed in galvanized iron pans 28x18x3 inches, made of No. 26 to 20 iron, and provided with a tight-fitting lid. Each pan holds 30 to 40 pounds of fish, and costs 50 to 60 cents.

2. The filled pans are transferred to an insulated freezing box or pen, with insulated sides and double-boarded floor. Insulating paper is placed between the boards. The front is closed by means of sliding boards, and the floor is pierced with drainage holes or outlets. A space of four inches must be left around each pan.

3. The pans are placed on a layer of sawdust covering the floor of the pen a few inches deep, upon which crushed ice and a little salt to a depth of five inches have been scattered.

4. The first tier of pans is then covered with four inches of crushed ice, mixed with one-sixth or less of salt. Successive tiers of pans and layers of ice and salt (four inches deep) are piled up to a height of five or six feet.

5. The top tier of pans having been duly covered with its layer of ice and salt, the empty salt bags are used as a cover.

In twelve to twenty-four hours the fish being moist, are frozen together in a solid cake in each pan. The pans are then dipped in water, the cakes of fish become detached and are dropped out, and are neatly piled in the storage room to be kept until required for use.

The process of crate freezing is as follows:

1. Forty pounds or fifty pounds weight of fish is placed in a lath crate or cage 24x18x3 inches.

2. The filled crates are passed into the freezing chamber for a period of twenty-four to thirty-six hours.

3. The fish in the crates, after being frozen, are transferred to the storage room and preserved until required.

The freezing chamber resembles in its essential features the storage room. It is not only insulated like the freezing pan in the "pan freezing" process, but the sides are formed of large freezing plates or tanks eight inches wide, passing up from the floor to the roof and through the ceiling, and fixed at right angles to the adjacent wall of the room. These tanks are filled with a freezing mixture of ice and salt, which can be placed in them without opening the freezing room. Between each tank projecting into the chamber above is an air-tight shutter, and an arrangement is made for draining away the overflow of brine. More salt is used in the freezer than in the battery of tanks in the storage room, and it is requisite that from one-third to three-quarters of a square foot of freezing surface should be provided for every cubic foot of space in the freezer."

CHAPTER XXXII.

KEEPING COLD STORES CLEAN.

CARE OF COLD STORAGE ROOMS WHEN EMPTY.

The care of cold storage rooms during periods of idleness, or when no goods are in storage, is of the greatest importance for the reason that good results in storage of goods depend largely on the condition of the storage room. With this fact in mind the author sent out a circular letter of inquiry to a number of cold storage warehousemen containing a list of questions which embrace the subject of whitewashing; whether it should be done by hand or machine; whether any other preparation is as good as whitewash; whether it would properly purify rooms for the storage of such goods as eggs after storing apples or other fruits; whether it is necessary to whitewash each year and also in regard to painting, at the time of whitewashing, the pipes or refrigerating surfaces which cool the room. Questions were also asked in regard to the methods of preparing whitewash and whether means of ventilating are provided at the time of whitewashing. Further information of a general character was solicited.

The number of replies received was rather disappointing, but some of the more careful and conscientious cold storage men gave detailed and very full information. It is evident from a majority of the answers received that comparatively little attention is given to the cold storage rooms when they do not contain goods. Cold storage rooms need as careful attention, although in a different way, when they do not contain goods, as when goods are stored therein. When the flow of refrigerating medium (usually ammonia or brine) is shut off at a time when there is frost on the pipes, this frost will evaporate in the form of air moisture, even though it does not actually melt, and cause the air of the room to become damp. Dampness with a comparatively high temperature will in time cause a growth of mold and a musty condition of the room. Systematic whitewashing with ventilation will kill this growth of mold, but it is much better to prevent a trouble of this kind than to overcome it after it has obtained a foothold.

As soon as the goods are removed from cold storage rooms the frost on cooling pipes should be removed and taken out of the room. If the fan system of air circulation is employed, with the coils all located in a coil room or bunker, this is a comparatively easy matter to attend to. Where the pipes are directly in the room, the resulting slop will necessarily cause the floor and walls to become damp to a greater or less extent. Moisture on floors of cold storage rooms should be taken up by throwing down dry sawdust or air slaked lime. It should be removed at once and not allowed to soak into the floor lining or insulation. A few barrels of dry sawdust should be on hand at all times for the purpose of soaking up melting frost or possible leakage from any cause. With the coil room and fan system the floor of coil room is usually water tight and properly connected with outlet to drain system so that damage to insulation cannot occur in this way.

After removing the frost from refrigerating pipes, measures should be taken to keep the rooms dry and pure. This may be done by exposing a quantity of quicklime in the room. It may be placed on the floor, but should not be placed on any wet spots unless it has already been air slaked and is in powdered form. It might under some circumstances cause the starting of a fire from the heat of slaking. Chloride of calcium placed on trays or pans or supported on a screen shelf above a water-tight pan, as illustrated in the chapter on "Uses of Chloride of Calcium," may be used to good advantage. Where the coil room and fan system are in use, chloride of calcium may be supported in the coil room as in the author's patented chloride of calcium process, or in any other suitable way, and by operating the fan a short time at intervals the room may be kept in a pure and dry state. During cool or cold weather it is a good plan to allow the air to blow through the rooms when it is dry outside and about the same or a little lower than the temperature of the room, What is still better is the cold weather ventilating system, which is described in the chapter on "Ventilation." With this system fresh air may be taken from outside the cold storage building and forced into the room in large quantities and the foul air from the room is allowed to escape through a suitable vent. The incoming air may be forced directly into the room without heating, or it may be heated to any required temperature by passing it over a steam coil or jacketed heater.

A few words in regard to the proper preparation of new cold storage buildings for the receiving of goods may not be out of place here. In the finishing up of a cold storage building it very often occurs that the work has to be rushed and enough time is not allowed for the proper whitewashing of the wood lining or interior surfaces of the room. This situation demands care and rapid work and advantage must be taken of all opportunities for whitewashing the rooms as fast as they are ready or as soon as a portion of their surfaces is ready. Keep men at work whitewashing following up the carpenters. By keeping the doors open and using the ventilating system intelligently, if one is installed, some of the rooms may be ready to receive goods as soon as the refrigerating equipment is ready to supply refrigeration. If no other means of properly drying are at hand, use chloride of calcium as illustrated in chapter referred to. In whitewashing cold storage rooms for the first time, it is advisable to apply first a thin coat of whitewash so that it may penetrate the wood as much as possible. It will also make a better ground for the second coat. The second coat may be somewhat thicker and should not be applied until the first coat is thoroughly dry.

WHITEWASHING AND MAKING WHITEWASH.

The proper drying out of whitewash in cold storage rooms is a difficult matter, owing to the inclosed nature of the rooms, which are usually provided with but one opening, also to the low outside and inside temperatures which usually prevail at the time of whitewashing. The cold weather ventilating system, already referred to, is of great assistance at such a time. By applying heat to the rooms and allowing the cold, moist air to escape as the dry, warm air is forced in, the whitewash may be dried very thoroughly. It is customary in some plants, especially in the larger cities where some of the rooms are in service during the greater part or all of the year, to dry out the rooms by placing a "salamander," or sheet iron heater for burning coke or charcoal, in the room. This is not a very scientific nor practical method, as the moisture driven out of the room in which the salamander is placed is conveyed to other rooms of the house or into the corridor to some extent; besides this, the salamander will dry out only a portion of the room at a time. The gas generated is also very objectionable and even dangerous to persons working in the room. In using a salamander it is best to light the fire and allow it to get well started before taking into the storage room. In this way a large part of the gas is avoided. For the most nearly perfect job of whitewashing from five to eight days are required to dry thoroughly. If the whitewash dries rapidly, as it may when a salamander is used, it will flake off and not be permanent. On the other hand, if it does not dry within a reasonable length of time, the water in same will soak into the wood and, in finally drying, the whitewash will have a dark or mottled appearance. Rapid drying, therefore, should be avoided as well as slow drying.

The importance of attending to the matter of whitewashing in new houses which are rushed to completion cannot be too strongly dwelt upon. The author has repeatedly come in contact with this situation and much time and effort have been expended by him in trying to get whitewashing properly done and at the right time. Those new to the business do not appreciate the importance of whitewashing and the necessity of looking after it carefully. Very bad results have in numerous cases followed the careless daubing on of whitewash, and allowing it to dry at its own pleasure. In some cases butter has been very strongly flavored in a way which could not be accounted for; again, eggs are damaged, and other goods to a greater or less extent, depending on their sensitiveness. If whitewash is plastered on the walls too thick and does not dry, the water contained therein penetrates the wood and may cause a fermentation, which leads to a peculiar bitter or strong smell in the room, which in turn will flavor the goods. If the case is an aggravated or serious one, mold will develop, and the serious nature of this trouble is too well understood to need description. Whitewashing should be done in the winter or during weather when the air is about as cold or colder outside than inside the storage rooms. It is then much easier to get the rooms dry. Bad effects have followed whitewashing during warm weather, because it is so difficult to get the rooms to dry properly.

It is a popular idea, and yet entirely wrong, that most anybody can prepare and apply whitewash. Of those who think they know how to whitewash, probably not one in ten knows how to slake the lime. This should be done in one of two ways, either of which is good. The author recommends the following: Take one-half bushel of lime and place it in a half-barrel (an oil barrel or vinegar barrel which has been cut down makes a good utensil for this purpose); pour on a small quantity of boiling water, barely sufficient to cover the lumps of lime; keep the lime well stirred clear to the bottom (a piece of one-inch gas pipe about five or six feet long is the best stirring stick). In case the lime is very quick, it should require two persons to slake the lime, one to pour on the water as needed and one to stir. The stirring should be kept up continuously from the time the lime begins to slake until it is reduced to a paste, and water should be added as fast as the lime slakes, so as to keep it at a rather thin, pasty consistency. It is very common to see lime placed in a barrel and water turned on and the lime allowed to slake itself. The result is that the whitewash is full of small pieces or lumps which are not slaked, but are burned as the result of water not coming in contact with the lime at the right time. It is not absolutely necessary that boiling water should be used, but unless the lime is quite quick, it facilitates the operation and results in more thorough slaking. Another method which may be employed is to place the lump lime on a cement floor and sprinkle water on slowly as the lime slakes. If this is handled carefully and attended to the result will be a finely-powdered slaked lime, which may be mixed with water to a proper consistency. The author does not recommend this method as compared to the one first described, as it is slower and there is much more danger of burning the lime and causing the whitewash to be lumpy.

A large number of those who replied to the circular letter of inquiry are using the Government formula for making whitewash, but one of the ingredients of this formula is rice boiled to a thin paste, which makes it seem difficult to the average person, and, further than this, the author does not believe in using any organic substance in preparing whitewash. For those who prefer the Government formula it is here given:

U. S. GOVERNMENT FORMULA FOR WHITEWASH.

Slake half a bushel of quick lime with boiling water, keep it covered during the process. Strain it and add a peck of salt dissolved in warm water, three pounds of ground rice put into boiling water and boiled to a thin paste, half a pound of powdered Spanish whiting, a pound of clean glue, dissolved in warm water; mix these well together and let the mixture stand for several days. Keep the wash thus prepared in a kettle or portable furnace and put it on as hot as possible with either painters' or whitewash brushes.

It is better to use the mineral substances, and the following has given good satisfaction under most circumstances: One-half bushel of lime, slaked with hot water, as previously described. When the lime is thoroughly slaked, add one-half peck of salt. It will be necessary to add more water as the salt is added, in order to keep the whitewash at the proper consistency; or the salt may be dissolved separately in as small an amount of hot water as will absorb it readily. The proper consistency for whitewash is a thin paste and it may be tempered as it is used. To each twelve-quart pail of whitewash, composed of lime and salt as above, add a good, fair handful of Portland cement and about a teaspoonful of ultramarine blue. The cement and blue should be added only as the wash is being used and should be thoroughly stirred into the whitewash; otherwise, when applied, it will be streaked. Cement is used for the purpose of giving the whitewash a better setting property so as to make it adhere better to the surface to which it is applied. The ultramarine blue is used simply to counteract the brownish color of the Portland cement. If white hydraulic cement is obtainable, it is better to use than Portland cement, and in this case the ultramarine blue may be dispensed with. It is, however, best to use a small amount, say half a teaspoonful to the pail, as a whiter surface results. The wash should be strained through a fine wire-cloth strainer before using, to remove the lumps if there are any present.

WHITEWASHING MACHINES.

The advisability of using whitewashing machines or spraying pumps in cold storage work has been an open question for some time. Of the replies received, about one-half recommend the use of the machine. Some say the machine will do the best work, but this is not the author's experience. There are some situations where the machine is a decided advantage; for instance, on overhead work, between open joists, or any surface which is difficult to get at with a brush. It is hardly possible to get as smooth and even a job with the machine as it is by hand, and, besides, a machine will necessarily put a good deal more whitewash on a given amount of surface than is put on with brushes. This is objectionable, for the reason that a heavy bed of whitewash on drying will flake off much more quickly. In some cases, those who use a machine go over it with a brush while still green in order to make it smooth and even. Another objection to a machine is that it will cause a mist in the air and the whitewash will spatter over any object in the room. A room must be entirely empty in order to use a machine. It should not, of course, be inferred that it would be practicable to whitewash a room while goods are stored in same, but it is necessary to clean a room out of *everything* that is liable to be injured by the whitewash in order to use a machine. The spray is also very uncomfortable for the operator. A moderately thin coat of whitewash on old work is as good for purifying purposes as a thick one; and for this reason hand-work is to be preferred to machine, as much less material may be applied. The more whitewash put on the more water to be gotten rid of in some way, and if the water is not removed promptly very bad effects may result, as already noted in discussing the drying out of cold storage rooms after whitewashing.

The author's impression is strongly in favor of hand-work, but it is not a desirable job for the man who has to do the work. It is probable for this reason that the machines are gaining headway. They have also been perfected to quite an extent during the past few years. There are a good many different makes of first-class machines on the market. The same machine that fruit growers use for spraying trees is available for whitewashing and the same machine is commonly sold for both purposes.

Good work in whitewashing should look well, be perfectly white or nearly so, should be hard and not liable to flake off or dust off onto the hands or clothing, and should have complete disinfecting and germ-killing properties. The slaking of the lime is the most important part of the operation and the success of same depends upon the care and attention given. Too much care can not be given to this detail, and cold storage men should see to it that whoever had this in charge looks after same conscientiously. Lime that is burned or drowned in slaking is not firm in texture when applied and is not as disinfecting nor fireproof as it should be.

PAINT FOR ROOMS AND PIPING.

There are many good cold-water paints on the market under various names which are advisable in some places for which whitewash is not well adapted, and many use them for all interior surfaces. For butcher's boxes or retail coolers especially they are preferred to whitewash, for the reason that they will not flake off readily. It is also good for doors and corridors of cold storage houses. Most of these cold-water paints are composed of secret ingredients, and some contain organic substances like glue, which makes their use inadvisable for cold storage purposes, except in special situations. Shellac is also largely in use for cold storage rooms, but it has no disinfecting or cleasing properties like whitewash. It makes a beautiful finish where the lumber in use has a good natural grain. Shellac has the advantage of being waterproof, and therefore walls may be easily washed at any time. It is, perhaps, unnecessary to state that any oil paint, or any other preparation with strong odor, has no place about the cold storage rooms or the corridors or other approaches thereto.

In connection with the whitewashing of rooms and their care during periods of idleness, it has seemed proper to take up the cleaning and painting of the pipes or refrigerating surfaces which cool the room. The answers to the questions covering this subject indicate that it is not customary among cold storage men to paint their pipes after they are once installed, and this is strictly in line with the author's ideas and experience on the subject.

There are two good reasons why the painting of pipes is not advisable after they are once put in place in the cold storage plant; first, it does not pay; second, it is dangerous. It does not pay, because after the pipes are once put in place a good job of painting cannot be done unless the coils are entirely removed from their supports so that they can be painted on both sides. The labor involved in removing the refrigerant, taking down the pipes, cleaning them and applying the paint is considerable, and the cost of the paint is no insignificant item. A good paint put on the pipes before they are set up in the cold storage house will protect them fairly for a period of from two to three years, more or less. Before the coils are set in place it is comparatively easy to paint them and it is recommended that coils should be painted when new. It is especially desirable to paint them at this time, as the pipes are clean and free from scale or rust. After the pipes become rusty from service, it is almost impossible to get them sufficiently clean so that the paint will adhere properly. Considering the low price of pipe and its comparatively long life when used with ammonia or chloride of calcium brine, it does not seem to warrant the expense. It is dangerous to paint pipes in a cold storage room for the reason that no paint known to the author is non-odorous or anywhere near it. The pipes should, therefore, be removed from the rooms for painting and allowed to dry and deodorize before they are returned to the cold storage room. This, however, is rather impracticable and it adds to the expense.

For painting pipes, various preparations have been used with more or less success. There are a number of patented and proprietary preparations on the market which are good and are sold at a reasonable price. Red lead and boiled oil is also an old stand-by for this purpose, but it is much more expensive than some of the preparations above mentioned. Boiled linseed oil without any pigment as a coating for refrigerating surfaces will give good protection from rust for a limited time, but the commercially prepared products will be found superior though somewhat more expensive.

CHAPTER XXXIII.

ICE BOXES AND REFRIGERATORS.

PRINCIPLES THAT SHOULD GOVERN CONSTRUCTION OF REFRIGERATORS.

The construction of refrigerators for domestic purposes, for butchers and other small users, is mostly in the hands of companies who manufacture them in large quantities. In the main they are only fairly well designed and very poorly insulated; and in a large number of cases poorly designed as regards the circulation of air. Besides the large manufacturers there are many small concerns, and the construction of small rooms for retailers, butchers, etc., is to a considerable extent in the hands of the local carpenter, contractor and builder. The points covering the design of such work will no doubt be of interest to those having occasion to build, and to users of refrigeration as well. An English writer* on this subject makes the following sensible suggestions:

The most usual location for the ice in a refrigerator is on the side of the box, and when this arrangement is in use (and it must be remembered that it is an absolutely necessary one in all cases in which the refrigerator is limited in height) it is best to keep the ice tanks or receptacles as high up as practicable, and to provide cold air ducts leading downwards to near the bottom of the refrigerator, thus insuring a sufficient air circulation.

When the ice tanks or receptacles are placed centrally in the box, in order to secure a uniform circulation of air throughout its length and width, it is necessary to provide warm air ducts which rise from the highest point in the cooling chamber to a level above that of the ice in the ice tanks or receptacles, and also cold air ducts from the bottom of the ice tanks or receptacles to a low level in the refrigerator chamber. This arrangement will be found fairly effective where the boxes are not of too large an area.

An arrangement which would probably be found to give considerably superior results to the above is the placing of the ice tanks or receptacles right at the extremities of the box. This location of the ice tanks or receptacles would give two means of producing cir-

*In the *Refrigerating Engineer*, London.

culating currents, and in this manner would of course tend to appreciably improve the refrigerating effect produced.

Whenever practicable, however, there can be no doubt that the most advantageous place for the ice is overhead. A refrigerator with a top ice tank or receptacle gives by far the best results in practice, that is to say of course provided that the warm and cold air ducts are properly placed. The first of these, or the warm air ducts, must be taken from the most elevated point in the cooling chamber up to a level somewhat higher than that of the ice in the ice tank or receptacle. The second, or the cold air ducts, should be led from the bottom of the ice tank or receptacle down to near the bottom of the refrigerator chamber. In this manner, a regular and continuous circulation of air will be maintained, and the warm, impure air will be forced to rise upward into the space above the ice in the ice tank or receptacle, the impurities becoming condensed upon the surface of the ice in the latter, and being carried away with the water resulting from the meltage of this ice.

There are few if any refrigerators in service which fulfill the above ideal conditions, even to an approximate degree. Seldom has the author seen anything in the shape of a cold air duct extending from the bottom of the ice chamber to near the bottom of the storage space, and the warm air duct from the top of storage space to near top of ice chamber is in many cases lacking. The principle of air circulation in refrigerated rooms is more fully set forth in the chapter on "Air Circulation in Cold Stores."

It is not to be wondered at that the small refrigerators are not properly insulated, as it can hardly be expected that the manufacturers are well informed on the subject when those who make a specialty of the cold storage business are not commonly familiar with its underlying principles. Further, competition between makers leads to a cheapening of construction, and as the appearance must be maintained, that part of the work not exposed to view must suffer. A refrigerator known to the author, when taken apart for examination, revealed no insulation at all, simply a two-inch air space. A dollar or two extra cost on the average refrigerator would be easily saved in one season where ice costs $5.00 per ton, put in the box. It is difficult to state exactly what material should be used and in what thickness and how applied. This can only be determined when character of work, cost of ice and service are known. The chapter on "Insulation" may profitably be studied by those interested in improved methods of insulating.

CHAPTER XXXIV.

REFRIGERATION FOR RETAILERS.

GENERAL SUGGESTIONS.

The conservation of dairy products consisting of milk and its manufactured products, butter and cheese, constitutes one of the most important uses to which refrigeration is applied. At present it is, in fact, difficult to imagine how a dealer in this class of goods, either as a wholesaler or a retailer, could do business without some form of cold storage, cooling room or ice box. In the United States even the smallest retailer has his refrigerator, and it is a rare exception to find such an establishment without one. Our European neighbors generally do business on a radically different basis, buying only a day's supply at a time, in the same way that vegetables and green groceries are handled. The losses resulting from this practice are considerable.

Without refrigerating facilities, goods must necessarily be purchased in a hand-to-mouth manner, obtain supplies daily so as to have them in a more marketable condition. A large aggregate loss results from this method. Goods purchased in a small way necessarily pay a higher percentage of profit to the wholesalers, and in some cases, also, much difficulty may be experienced in purchasing supplies daily, owing to failure of transportation from any cause, or a natural scarcity of goods in the open market. A retailer need not go into the cold storage business to the extent of putting away his season's supply (although many of the larger ones do this, and make a good yearly profit thereby), but every retailer, no matter how small, should have cooling space. If his business is properly systematized and advantage taken of his storage for perishable goods, a sure profit may be realized by buying in round lots at a time when the products are obtainable at the lower price.

The form of refrigerator suitable for the widely varying requirements of dealers doing a large or small business, may vary from the cheap ice chest which can be built by any carpenter for $10 to a completely equipped cold storage plant, which will keep dairy products in good condition for several months. There are several large retail establishments in the East operating what are known as "chain stores," the proprietors of which own and operate cold storage plants and in addition do some storing with the regular cold storage houses.

For an average business, the large refrigerator which only requires to be filled with ice once or twice a week, is in use. In general all retailers use ice only for cooling purposes, except the large users who do practically a cold storage business with their own goods. Ice will produce, when used in the ordinary way, a temperature of 38° to 50° F., and probably most of the larger refrigerators would show a temperature of about 40° F. to 45° F. in warm summer weather. This temperature answers nicely for temporary storage from day to day, or for two or three weeks, or even longer on some classes of goods, but for long storage of several months, only an apparatus that will give a control of temperature at all times should be used, such as the Cooper brine circulating system cooled by ice and salt described elsewhere, or a mechanical system of refrigeration.

The dairy products; butter, cheese, milk and eggs (eggs are not strictly speaking a dairy product, but are generally so called) are all more or less liable to deterioration. Butter, especially when exposed to the air and heat of summer, becomes rancid and unfit for use in a few days. In days gone by basement rooms or cellars were used very generally, and a cool cellar was a thing to be proud of so long as people did not know the value of ice or artificial refrigeration. A dairyman who would store his June butter in a cellar for winter use in these days would find it very unprofitable when compared with the results from a modern cold storage house. This would show beyond a doubt that modern methods must also be applied by the dealer if he is to meet competition and keep abreast of the times.

America has led, and is still leading the old world in refrigeration. Even the smallest retail dealers who handle butter and other dairy products, have large, well built refrigerators similarly constructed to those used for domestic purposes. Some of these are indeed a work of art, with beautiful glass fronts so arranged as to show off the goods to advantage and so constructed that they may be opened for cutting out the goods without exposing the interior of the room to warm currents of air. Some are arranged so that each package of butter or cheese rests on a pivoted table, one-half enclosed by glass. By turning this table half way around the goods are accessible and the glass front of the table is turned back, shutting off the interior of the refrigerator from outside air. These are in use very largely for butter and cheese, but may be used for other goods as well.

The facilities needed by retailers for the correct and economical handling and sale of dairy products depend largely on the quantity of goods to be handled, length of time to be cared for and climatic conditions of the locality in question. In the northern part of the United States and throughout Canada the use of natural ice is universal, as the ice crop is a certainty. In the southern states manufactured ice and mechanical refrigeration are a necessity.

Ice (either natural or manufactured), when used in a refrigerator or cooling rooms, will give a fairly dry air at a temperature of 40° to 45° F., providing proper arangements are provided for promoting a circulation of air over the ice. At this temperature well made fresh butter will remain firm in texture and will not lose quality to any great extent for several weeks. Cheese at this temperature will remain in good condition for a much longer period. Cheese is often retailed from the original package without the use of refrigeration, but during the warm weather of summer it will dry out and lose quality rapidly. It must also be sold very quickly when cut. Eggs may be kept for a few days or even longer at temperatures above mentioned.

Eggs and butter should be kept separated from fruits, meats or strong smelling cheese, as they give off odors very rapidly. Milk and cream are kept from day to day in a refrigerator to prevent souring, which takes place at high summer temperatures very quickly. Meat is one of the chief commodities sold from refrigerated rooms, and no retailer of meats can do business without refrigeration. The limit of time at which fresh meat may be stored in a temperature of 40° F. or thereabouts is two to four weeks. It is well known that meat kept at a temperature of the ordinary ice cooler for two weeks is in more palatable condition then when first killed if atmospheric conditions of the cooler are correct.

Small refrigerators may be purchased ready made and set up ready for business, and the larger ones which may be

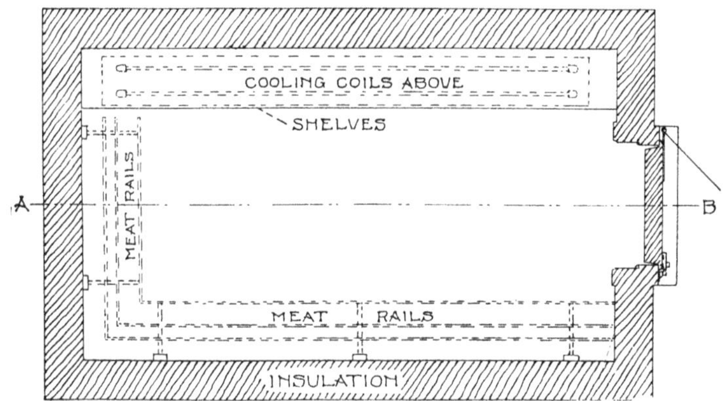

FIG. 1.—PLAN OF COOPER'S BRINE SYSTEM APPLIED TO BUTCHERS' BOX.

dignified by the name of rooms are to be had from the makers in sections to be put together by any carpenter. If a first class and economical job is wanted, a skillful cold storage architect or engineer should be employed to furnish plans, as it is well known that the average refrigerator is not half insulated, nor properly constructed. Regular cold storage rooms intended for storage of goods for long periods must be designed and arranged with care, and only a thoroughly competent architect should be employed for this work. The construction should be carefully attended to, and the rooms should be handled with the utmost intelligence, if good results are to be expected. Retailers should not go into this unless their volume of business is large enough to warrant it. It will pay much better to purchase supplies during the flush of the producing season, when the quality is best and price lowest, and store in a regular well-handled and thoroughly equipped cold

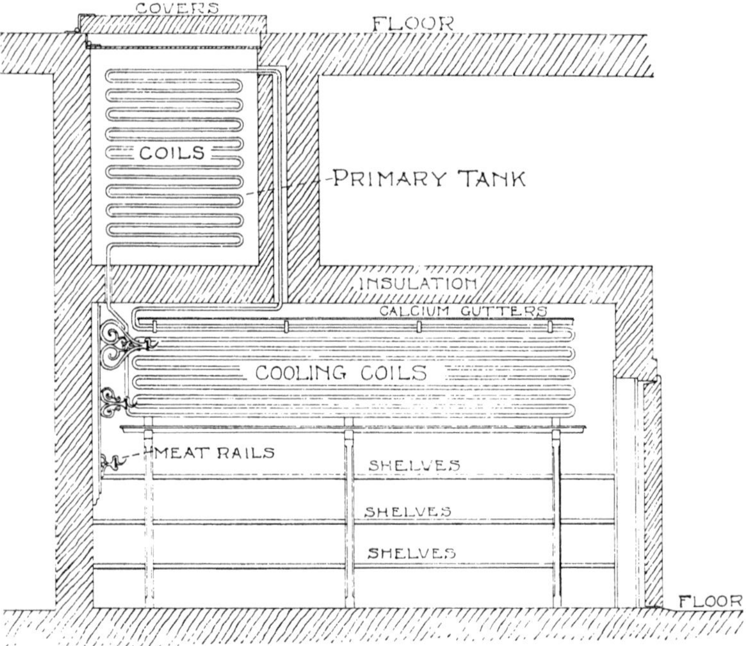

FIG. 2.—SECTION OF COOPER'S BRINE SYSTEM APPLIED TO BUTCHERS' BOX.

storage warehouse. It is better to do this, paying the moderate charges which are now made for such service than to put away goods for a long carry in a common refrigerator. A refrigerator cooled by ice is not intended for any such work. Its duty is the temporary safe keeping of goods and all dealers should have one for this purpose, and it should be used for this purpose only.

REFRIGERATION FOR BUTCHERS' BOXES.

One of the hardest services to which refrigeration can be applied is that of cooling small rooms, such as "Butchers' Boxes," as they are called. The continued running in and out of the room admits a large quantity of comparatively warm air, and at times when the atmosphere is damp on the outside, this leads to a condensation not only on the ceiling of the room, which naturally gets the first flow of warm air which rises as it enters the room, but also on the goods themselves. Much thought has been applied to the design of rooms for retail business and considerable improvement has been made in the utilization of ice (which is mostly used) for this purpose. The difficulty from condensation, however, cannot be entirely eliminated by any method of cooling without providing a vestibule or ante-room of some character, so that the condensation will occur in the vestibule and not in the room itself. The arrangement of the anteroom or vestibule is moreover a cumbersome affair, as it necessarily means that two doors instead of one must be opened and shut every time the room is entered. About the best that can be done is to locate the refrigerating surfaces so that the warm moist air from the outside as it enters the room will deposit its moisture on these surfaces instead of on the goods in storage. The trouble can also be obviated to some extent by providing a spring on the door so that it will close itself quickly when opened. Another very desirable feature is a tightly fitting door which will not stick or bind. There is at present no such satisfactory door for cold rooms, especially those of small size, as the patented doors now on the market, owing to rapidity and ease of operation.

The plan and section (Figs. 1 and 2) show the application of the author's brine system to the cooling of small rooms for butchers' boxes or for grocers or others requiring refrigeration in small units. This arrangement shows the system applied to a store room which is 13 feet high. This makes the cold room or refrigerator 9 feet in the clear, and allow access to the primary tank from the floor above, which is a decided advantage, as all of the rough work and muss of the icing is in this way entirely removed from the store itself. As will be seen by the plan, the pipes which cool the room extend across one side and to the ceiling. They also are near the door so that warm moist air entering from the inside will come in contact with the pipes and moisture be deposited thereon. The coils are provided with drain gutters underneath so that the drip from the pipes is caught without any spatter and led outside the room to the sewer. The cooling coils in the room, attached to the top pipe, as shown in the sections, are also provided with chloride of calcium gutters for the application of the Cooper chloride of calcium process for preventing frost on pipes and purifying and drying the air of the room. If necessary to keep the room sufficiently dry these gutters may be made extra large, providing in that way for an extra quantity of chloride of calcium to be supported thereon. The humidity of the room can be controlled at will in this way by using a greater or less quantity of chloride of calcium. This chloride of calcium process, as it is called, is described elsewhere in this book, as is also the Cooper brine system.

CHAPTER XXXV.

REFRIGERATION FROM ICE.

CHEAP, SAFE AND UNLIMITED REFRIGERATION.

The quantity of cold, (if the author may be pardoned for using so unscientific a term), which is present and active during the season of cold weather, is so enormous as to be staggering when presented in the form of actual figures. The practicability of storing this almost unlimited refrigeration during the winter for use during the heated period, has been only partially understood and very imperfectly developed.

It is well known that a great deal of ice is harvested during the winter and is put into structures which are known as ice houses, but just what any given quantity of ice means in energy and in its equivalent refrigerating capacity or heat units, is not generally known. It is, of course, admitted that ice is a good thing to have when the weather is hot, to cool drinking water and to maintain a house refrigerator at low temperature for the purpose of keeping a few table foods for a day or two, etc., but many seem to think that when it comes to a cold storage plant where foods are stored for longer periods that an expensive and complicated system is necessary to do the work. It is desired to show by the few figures which follow, the utter absurdity of rejecting Nature's refrigeration and substituting refrigeration by artificial means.

In the North Temperate Zone, take it for instance, from the latitude of the Ohio River and the southern boundary of Pennsylvania and the same general latitude westward to Nebraska, and north of this latitude and throughout Canada ice forms regularly each winter to a thickness which may be housed at nominal cost. Further south than that there are many localities where thin ice may be put up without much difficulty and without fail each winter. In the localities first mentioned the average thickness of natural ice forming on ponds, lakes and quiet and shallow bodies of water, would range in the southern part from a few inches up to two or three feet or even more in the northern part. Take therefore, an average locality like New York, Michigan and Wisconsin, and assume that the average thickness of ice formed each winter is 20 inches, and then as a matter of information and as a basis to work on and as a very common unit, take a square mile for purposes of calculation. This gives the following interesting and altogether surprising figures:

One square mile or 27,878,400 square feet of ice, 20 inches or 1 2/3 feet in thickness, assuming 57 pounds to the cubic foot or 95 pounds to each square foot (20 inches thick) would weigh 2,648,448,000 pounds or 1,324,224 tons.

Multiply 2,648,448,000, the number of pounds of ice in 1 square mile 20 inches thick, by 142 the latent heat or heat absorbing capacity of ice per pound, we get 376,079,616,000 British thermal units. Dividing this by 14,000 the number of B. T. U's. per pound, heating capacity of the best grades of steam coal, we find we have 26,862,830 pounds or 13,432 tons; the amount of coal that would be necessary to melt the ice on a surface one mile square and 20 inches thick.

It will be noted from the above that ice 20 inches thick which will form on a square mile amounts to the enormous total of 1,324,224 tons. If all the heat energy in coal could be applied to the melting of this quantity of ice on a basis of the above figures, it would require about 13,432 tons of coal to again reduce this quantity of ice to water.

Why stand idly by and allow the most of this valuable cold which nature provides during winter to evaporate and slip away? And then, as soon as warm weather comes, start coal fires for producing refrigeration by artificial means and begin to talk about fuel conservation at the same time.

The author's father once stated that he believed that "expensive steam driven machinery could not successfully compete with God Almighty and a Minnesota winter" in producing refrigeration. The above figures prove this to be true beyond a doubt, and the unlimited refrigeration which is going to waste each winter, not only in Minnesota, but in localities much further south, has not been fully appreciated. There can be no possible reason for using good heat producing coal, which represents a large amount of human labor, for the production of refrigeration, where it is possible to store up nature's refrigeration during winter in the highly concentrated form which is called ice. The storage of ice during winter is only in its infancy, and the author takes the risk of predicting that cold storage and refrigeration will in future, where ice can be obtained at all, be accomplished to a great extent by means of ice and not by means of the more or less expensive and complicated systems which cost a great deal of money to install and which require a great deal of care and attention in their operation, and which, even with the utmost care, are liable to get out of order when in greatest need.

Our foods are mostly grown or produced in the summer time during the heated period and stored and preserved in various ways for winter use. The cold or refrigeration which nature produces in winter can be stored and conserved for use during summer for the purpose of preserving perishable foods. The consuming of coal or other fuel to make refrigeration is like raising corn in a hot house. It is contrary to nature's laws and commercially impracticable, and therefore, the scope and possibilities are very limited.

It is now practicable to put up ice of very moderate thickness or to allow the ice to freeze right in the ice storage house, and improved appliances and methods for this purpose are decribed elsewhere in this book.

PRINCIPLES OF ICE REFRIGERATION.

A cold storage house may be successfully cooled by ice mixed with a small proportion of salt. Many persons who employ ice in an ordinary refrigerator or otherwise, are perhaps not fully aware that it may be employed with entire success for practical cold storage, even when placed in direct competition with the ammonia or other mechanical systems. Temperatures as low as from 5° to 10° F. are maintained in freezing-rooms, and eggs are held at 29° F. with a pure and dry atmosphere. These facts should establish beyond a question the possibilities of ice in the cold storage field. The system of natural ice cold storage which will produce these results is fully described further on in this chapter. Numerous plants are in operation which use manufactured or artificial ice with a small admixture of salt as a primary refrigerant. Artificial ice is as useful for this purpose as natural ice and for small plants is very desirable as compared with a small ice machine.

The immense natural ice crop is, for the most part, consumed in the temporary safe keeping of perishable products, which are stored in the common house refrigerator or the larger refrigerator of the retailer. Many cold storage houses utilizing natural ice are in operation, which give more or less satisfactory results: generally the latter. Some persons have an idea that a cold storage house is a room with sawdust-filled walls with ice in it, but there are many points about cold storage not understood by the average person. It is the purpose in this chapter to discuss the various methods of cold storage by means of ice so that the careful reader may discriminate between them and understand the underlying natural laws.

In discussing ice cold storage, it may be admitted at the outset that the use of ice in any form for the preservation of food products, like eggs, butter, cheese and fruits, for what is known as long-period storage, has fallen into disrepute, owing to defects in the older systems. There are reasons for this, although the idea that the ammonia system is so much superior has been carried to an extreme not warranted by the existing facts. The real reason why the ammonia system has a better reputation is that natural ice has usually been misapplied to the work of cold storage, that is, it has been improperly used. The problem of cooling storage rooms by utilizing the stored refrigeration of the winter months in the form of natural ice has had the attention of many persons, among them the author and his father before him. Several systems had previously been developed with varying success, but it is believed that up to the time the "Cooper System Gravity Brine Circulation" was first put in service, no system was in existence which could successfully compete with the ammonia or other mechanical systems.

The use of ice as a refrigerant was long antedated by the use of natural refrigeration, which may be obtained in cellars or caves. It is well known that at a depth of a few feet below the surface, the earth maintains a comparatively uniform temperature of about 50° F. to 60° F. during all seasons of the year. This temperature varies somewhat, but above would cover a great majority of cases in any northern latitude where snow falls, and as compared with a summer heat ranging from 70° F. to 90° F. it will be readily observed that this natural low temperature of the earth is of considerable service in retarding decay and the natural deterioration of perishable products. By digging beneath the surface of the earth a cellar was formed which would produce results in refrigeration which were quite satisfactory during the early history of the perishable goods business, but would hardly withstand the critical test to which goods from modern cold storage houses are subjected. With the advent of the natural ice trade, ice came into use for household and other refrigerating purposes. *Ice is at present and will probably always remain the most practical means of placing concentrated refrigeration at the disposal of the comparatively small consumer.* It seems that prior to the nineteenth century the great cooling effect to be obtained from a small quantity of ice was not known nor appreciated by the world at large. The preservation of natural ice was likewise not thought practicable for a time sufficiently long to allow of its use as a cooling agent during the heat of summer. With a knowledge of the cooling power possessed by the earth during warm weather the first ice houses were constructed below ground, without provision for drainage. The result of such an arrangement is easy to understand. Now ice men are careful to build above ground and provide good drainage as being necessary to the successful keeping of the ice. The first ice house did not provide protection for the ice, other than a roof overhead; all ice houses now employ sawdust or some other non-conductor of heat to protect the ice from contact with the air, and prevent the penetration of heat. Ice stored in the underground ice houses was mostly melted by July, while ice stored in a modern ice house may be kept until fall with a meltage of only ten or fifteen per cent. The evolution of the modern ice house from the underground pit has been gradual, and was not made all in one jump. It seems remarkable that the loss from meltage in the house is now so little, and this is accounted for only by considering the tremendous amount of refrigeration which is stored up in a small quantity of ice, and a knowledge of proper means for protecting same. (For further information on ice harvesting and storing and the construction of ice houses, see separate chapters on these subjects.)

The refrigerating value of ice as compared with an equal weight of cold water at 32° F. is as 142 is to 1. That is, ice has 142 times as much cooling power in passing from ice at 32° F. to water at 32° F., as an equal weight of water in passing from 32° F. to 33° F. It has perhaps been noticed that ice forms quite slowly even in extremely cold weather. This is because the water must give up a large amount of heat before it will become ice. The natural bodies of water are quickly reduced in temperature to about the freezing point by a cold spell of weather in the fall, but the freezing of the water into ice at the freezing point (32° F.) is quite a different matter. This natural phenomenon is accounted for by what is known as latent heat. It is this latent heat in water which makes it so slow to freeze, and when once frozen, makes the ice so slow in melting, as the same latent heat which is given off in freezing must be absorbed from surrounding objects before the ice will melt into water. To fully understand this it is necessary to become familiar with the unit of measurement used in determining quantity or amount of refrigeration produced by melting ice, and the relation between heat and cold.

Heat is a positive quantity, that is, possesses character, so to speak, while cold is simply the absence of heat. It follows, therefore, that any unit of measurement applicable to heat will also measure refrigeration. If heat is extracted from any object it becomes cold, and it becomes cold in exactly the same amount or proportion as the heat is absorbed. The quantity of heat absorbed is measured by the British Thermal Unit, generally abbreviated to B. T. U. One B. T. U. is equal to the raising in temperature of one pound of water one degree, as shown by an ordinary thermometer. The standard American thermometer is named after its originator, Fahrenheit, and measurements by this thermometer are usually abbreviated to a simple F., to distinguish from some other thermometers in use. In writing temperatures the F. is placed after the degree mark. We would say then that one pound of ice in changing from ice to water at 32° F. absorbs 142 B. T. units. When a pound of water is raised in temperature from 32° F. to 33° F., only one B. T. U. is absorbed. In other words ice in melting has 142 times the refrigerating value that the same weight of water has when raised in temperature 1° F. This latent heat of liquefaction, as it is called, explains why ice melts so slowly, and why a comparatively small quantity will perform such a large refrigerating duty.

When used for cold storage purposes, the temperatures ice alone will produce are limited. As the melting point of ice is 32° F., the temperature which can be obtained in a room cooled by ice only must necessarily be somewhat higher. The lowest practicable temperatures are about 36° F. to 38° F. during warm weather. By mixing finely crushed ice with a small proportion of salt the melting of the ice is hastened, and a much lower temperature results. This is caused by the great affinity which salt has for water. When salt comes in contact with ice this property causes it to extract the water from ice rapidly, reducing it from a solid to a liquid, causing a rapid production of refrigeration or rather the absorption of heat. A pound of ice will do a given amount of work in refrigeration regardless of whether it is melted naturally at 32° F. or at some lower temperature in combination with salt. The lowest temperature obtainable with a mixture of ice and common salt is slightly below zero, Fahrenheit. This is directly in the mixture. A room cannot be cooled as low as this with ice and salt. By using chloride of calcium salt mixed with crushed ice a temperature many degrees below zero may be obtained. This salt costs about double what common salt does, and is not at present in use for freezing purposes.

From tests conducted by the author it is evident that chloride of calcium cannot be successfully used for practical freezing purposes in a freezing mixture of ice and salt. While the tests referred to are not accurate or conclusive it seems evident that chloride of calcium when first applied to ice has the property of generating heat which consumes the ice rapidly. The ultimate temperature is extremely low but the consumption of ice in proportion to the actual refrigeration produced is too great to make the use of calcium practicable. It is possible that some special preparation of calcium, perhaps the calcined calcium in granular form may be better adapted and will produce different results than the commercial calcium which was used in the experiment referred to.

Moisture in cold storage rooms has been the source of much discussion and solicitude among cold storage operators, and a knowledge of the action of this condition in rooms artificially cooled, and its relation to temperature, will assist in our present study. When a storage room is cooled by ice only, the higher the temperature at which the room is held the dryer will be the atmosphere, and the better will be the circulation. This statement is general, and may be modified by exceptional conditions. A moderately dry air and a good circulation are necessary to successful cold storage, but with these two conditions must go, as an imperative adjunct, a low temperature if good results are to be obtained. It has been stated already that the lowest dependable temperature with ice only was 36° F. to 38° F. Comparatively few products are now stored in a temperature above 32° F. to 34° F., and a large bulk of the business is handled at a temperature ranging from 30° F. to 32° F. It is therefore evident that ice alone will not produce temperatures sufficiently low for the handling of a successful cold storage business. Temperatures sufficiently low can be obtained only by ice mixed with salt or by the use of refrigerating machinery. As before stated, in a room cooled directly from ice, the nearer the temperature of the storage room approaches the temperature of melting ice, the poorer will be the circulation, and the higher per cent of moisture the air will contain. Circulation of air within a storage room is caused by a difference in weight of air in different parts of the room. The air in immediate contact with the ice is cooler and heavier, and therefore falls to the bottom of the storage room. The warmer and lighter air at the top of the storage room at the same time rises to the ice chamber. As long as the difference in weight and temperature exists, circulation will take place. The principle underlying air moisture is quite complicated, but may be understood by a little study. It is well known that when warm, moist air is circulated in contact with a cold surface the moisture will be condensed upon the cold surface. This is illustrated by the so-called "sweating" of a pitcher of ice-water in warm, humid weather. This same action takes place in every cold storage room. When the room is cooled directly by ice, moisture contained in the comparatively warm air of the storage room is continually being condensed on the cold surface of the ice. As the air becomes nearer and nearer the temperature of the melting ice, less and less moisture will be condensed, and the air becomes in consequence more and more saturated with moisture. If it were possible to cool a storage room to 32° F. with ice melting at 32° F., the air of the room would be fully charged with moisture, and totally unfit for the storage of any food product. If a room is cooled to 35° F. with ice melting at 32° F. the per cent of moisture in the air would be 91 per cent of what it would be if the room were cooled to 32° F. in the manner above indicated. If the room is cooled to 38° F. the air would contain 79 per cent, and if the room be cooled to 40° F., it would contain 70 per cent. In the actual practice these air moistures would be somewhat higher, owing to the presence of moisture which is continually given off by the goods in storage. Even the temperatures with their corresponding percentages of air moisture as here stated are known to be too high for the successful preservation of food products for long periods of three months and upwards, and even for shorter periods results will not be as perfect as with a dryer atmosphere and lower temperature. Further than this the circulation, temperature and humidity in a room cooled by ice only are largely dependent on outside weather conditions. The temperature will, of course, be higher during the hot weather of summer. The humidity is, as we have seen, controlled by the temperature of the air in the room, as is also the circulation. When the temperature outdoors during fall and winter is at or near the melting point of the ice in the storage room (32° F.) no circulation will take place. The air will become very damp and impure from the moisture and impurities given off by the goods in storage, the goods will mold and decay rapidly. This is a condition to be met with in every house which is cooled by placing natural ice in direct contact with the air of the storage room. (Further information on the relation between humidity and air circulation see separate chapters under these headings.)

EARLY SYSTEMS OF ICE COLD STORAGE.

Reasons have been given why natural ice, as generally used, will not produce satisfactory conditions for the storage of food products for long periods. This information will enable the reader to fully understand the weak as well as the strong points of the various systems here described which utilize ice as a refrigerant. Ice alone may produce useful and even satisfactory results if the goods need only to be carried for a period of one, two or even three months, but where it is desired to erect a building with the idea of handling a variety of products for long storage and with intention of building up a permanent business, the old primitive methods of overhead, or side, or end ice, will result in disappointment and loss. This has been the history of at least nine-tenths of the public cold storage warehouses cooled in this way. If those who contemplate embarking in the business cannot build a house which will carry the various products successfully, it is better to keep out of the business altogether. The author has had occasion to remodel and even tear down cold storage houses in which ice was the only refrigerant, and in not a single instance known, has a house, operated in this way, been able to build up a substantial and profitable business for its owner. Quite a number of such houses are now in use, and a few are being put up at the present time, but they are mostly operated for private use for one or two products only, and for comparatively short time storage. They do not give successful results when used for sensitive goods like butter and eggs.

The first application of natural ice to the preservation of food products was that of placing goods directly in contact with the ice, in a similar manner to the method now employed in shipping fish or poultry or in cooling melons for temporary holding. This method can be employed for but few products, because the goods become wet and water soaked. The air in such a chamber has not the benefit of the purifying and dry-

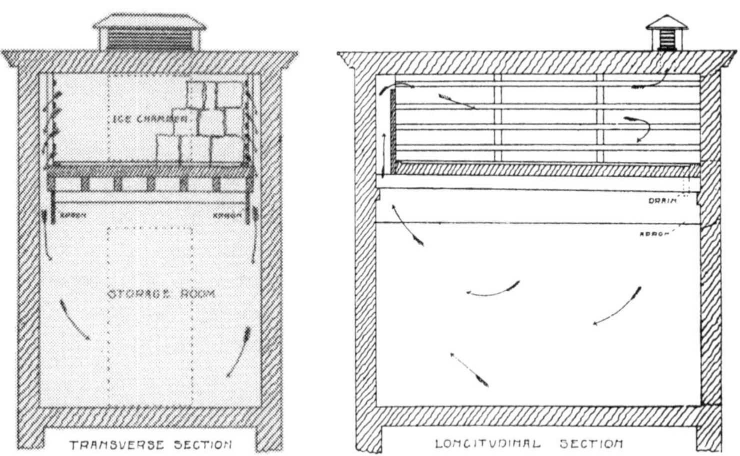

FIG. 1.—THE FISHER SYSTEM—SECTIONAL VIEWS.

ing influence of circulation, and goods in condition favorable for such action mold and decay rapidly. As an improvement on this method, it was natural to separate the goods from the ice, by placing the ice at one end or side of the chamber and the goods at the other, and not in contact with each other. The wetting of the goods is thus avoided, but when the goods are not placed in contact with the ice, they are of course carried at a somewhat higher temperature. No circulation of consequence is present, and the air becomes moist and impure very rapidly. Improving on the side or end icing plan, a two-compartment refrigerator was constructed, with the ice above and the goods to be preserved stored below. By providing openings for the flow of cold air from the ice down into the storage compartment, and for the flow of comparatively warm air up into the ice compartment, a circulation of air was produced, which was the first really important principle discovered in cold storage work. Air is purified and dried by circulation under proper conditions. The reason for this is discussed in the chapter on "Air Circulation." The first successful ice cold storage houses were built with ice above the storage chamber, and a large majority of those still in use are of this general plan, with, of course, many modifications. As before stated, they are useful mostly for short-time storage. When placed in competition with a house equipped with a system which gives positive control of circulation, moisture, temperature and purity of the atmosphere, they soon lose business and fall into disuse. Many patents have been issued on the various systems of ice cold storage. A few only of those system which have come to the author's attention will be briefly described, with the idea of showing the development of ice cold storage, and also that the reader may form some impression as to the relative merits and weak features of the different systems which have been more or less prominent in the past.

The Fisher System.—One of the oldest systems of ice cold storage and one on which many houses have been erected, is the "Fisher System," (See Fig. 1.) The points of this system which are covered by patent are not known to the author, but the main essentials of the houses as constructed by Fisher, were an ice chamber located above a storage room with an insulated waterproof floor separating the two. Openings were provided for the circulation of air from the ice chamber to the storage room, and flues from the storage rooms to the top of the ice chamber. One who is familiar with the operation of this system says that Fisher's houses, when new, would do fair work, but when they became old the results were very bad. None of these houses known to the author are now in operation. The principle was very simple, and as good results might be obtained by this system as with a majority of the later ones using ice only.

The Wickes System.—The "Wickes System" has been largely introduced among certain lines of trade, more particularly in the refrigerator car service. It is claimed that several

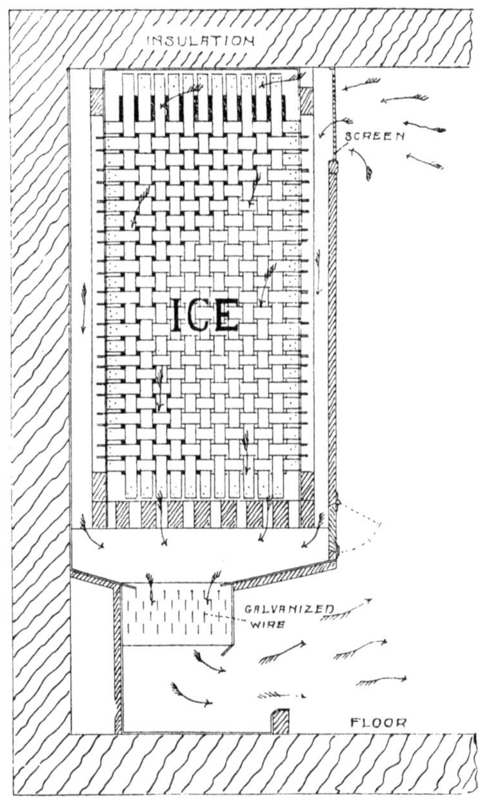

FIG. 2.—THE WICKES SYSTEM.

thousand of the Wickes cars were in constant service. The Wickes company some years ago installed a number of cold storage plants, but it is believed that they do not now recommend their system for such use. The devices which make up the Wickes system (see Fig. 2) consist of a basket-work ice bunker, composed of galvanized iron strips. Attached to the strips where the air flows into the ice bunker are projecting tongues, which, it is claimed, give largely increased cooling and moisture-absorbing surface, which dry and purify the air more thoroughly. Where the air flows out at the bottom of the ice bunker, it passes down over a network of galvanized wire, which is kept cold and moistened by the water dripping from the melting ice above. These devices which have been added to the ordinary construction of the ice box no doubt add somewhat to the efficiency of the system, but are scarcely worth their cost. Any system like the Wickes, employing side or end icing, must be greatly inferior to the overhead ice sys-

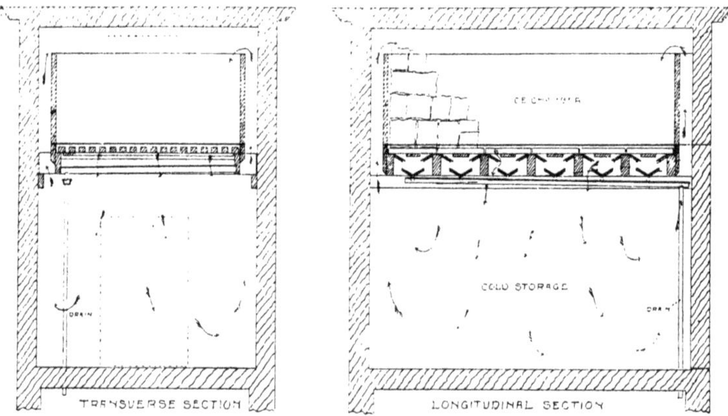

FIG. 3—THE STEVENS SYSTEM—SECTIONAL VIEWS.

tem, because the circulation of the air becomes stagnant when the ice is reduced in the ice bunker. The temperature also rises under these conditions, and unless a very large ice bunker is provided and the supply of ice fully maintained it is not possible to produce as low temperatures as with an overhead ice system.

The Stevens System.—A good many houses have been erected on what is known as the "Stevens System." (See Fig. 3.) This differs somewhat from other systems of overhead icing in having an arrangement of fenders and drip troughs forming an open pan over the entire floor of the ice room, except at the ends and sides, which are left open for the flow of warm air upward from the storage room. The cold air from the ice works down through the open pan. The pan is formed by a series of gutters suspended between the joists and

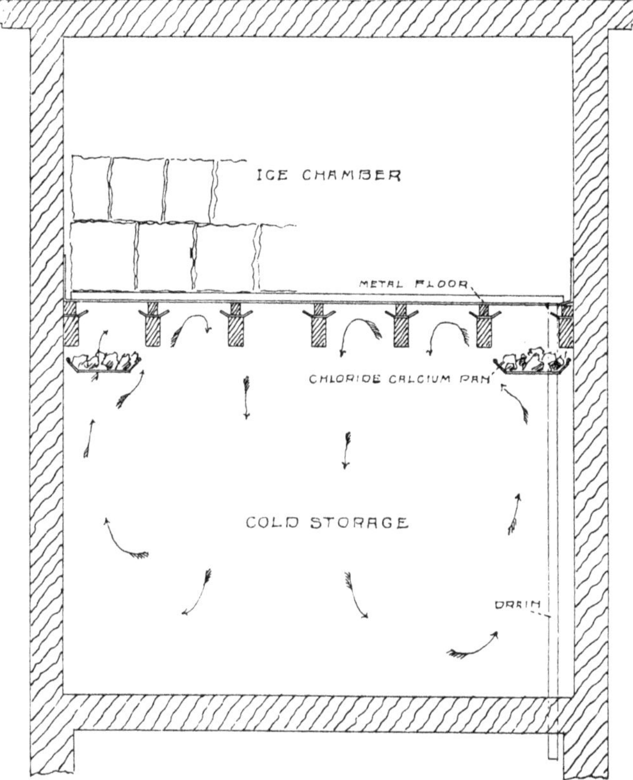

FIG. 4.—THE NYCE SYSTEM—SECTIONAL VIEW.

capping pieces over the joists to cause the water to drip into the gutters, at the same time allowing a circulation of air between gutters and capping pieces. Those who have used the system state that trouble resulted from spattering of water from the troughs. This system has the advantage of maintaining fairly uniform temperatures, regardless of the amount of ice in the ice chamber. Quite a number of these old houses are still in use. The results obtainable are not essentially different from those to be had by other overhead ice systems.

The Nyce System.—The system invented by Professor Nyce is one of the old-timers still to be found in use. In this system (see Fig. 4) the cooling effect of melting ice, and the drying and purifying effect of chloride of calcium, are depended upon to produce the desired result. It is an overhead

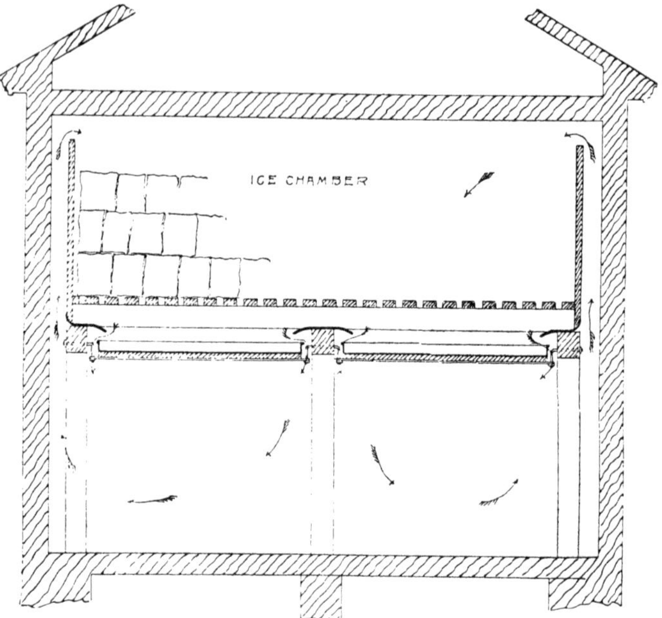

FIG. 5.—THE JACKSON SYSTEM—SECTIONAL VIEW.

ice system, but the air is not circulated from the ice chamber into the storage room. The storage room is cooled by contact with the metallic ceiling of the storage room, which also forms the floor of the ice chamber. Professor Nyce no doubt studied out this system from having observed the bad effects which result in the ordinary overhead ice cold storage during cool or cold weather. To absorb the moisture which is given off by the goods and from the opening of doors, the well-known drying qualities of chloride of calcium were used. The results obtained by cooling and drying a room in this way were quite satisfactory, and compared favorably with any of the other ice systems in general use. The patents on this system have long ago run out, but the system was not sufficiently successful to encourage its general use, and so far as known, no new houses of this kind are being built at present.

The Jackson System.—The "Jackson System" of overhead ice cold storage is one of the most general in use, and it is claimed that over three hundred houses have been constructed. The system (see Fig. 5) is extremely simple, and the chief patent is on a removable pan suspended under an open ice floor. It is, of course, an overhead ice system, with air circulating from the ice chamber down into the storage room. The spaces between the joists supporting the ice are left open, and aprons of galvanized iron protect the girders which support the joist, and conduct the drip to the removable pans before referred to. In some cases cylindrical tubes or tanks of galvanized iron are provided. These are filled with ice and salt for the purpose of reducing the temperature still lower than is possible with the ice alone. The use of tanks in a room provided with a circulation of air from the ice cannot result in any great benefit to the rooms, as the circulation is retarded or stopped, and a pollution of the air results to a considerable extent. Tanks of different shapes and sizes are used in a number of systems, and will be considered by themselves in another paragraph. The "Jackson System," so-called, is principally a pan hung below the ice joist so as to promote a circulation of air from the ice chamber into the storage room. Other devices as simple will accomplish the same result. Nothing new of consequence has been added to this system for a number of years, but a few houses are being installed on this plan, largely because it has been advertised and pushed in former years.

The Dexter System.—The Dexter patents cover a much more complicated apparatus than any system or prior invention which utilizes ice as a refrigerant. The "Dexter System" of indirect circulation is a very ingenious device. (See Fig.

6.) It consists of a series of air flues between the exterior and interior walls of the cold storage room. The cold air from the ice chamber flows through another set located outside of the first set. This effectually prevents the penetration of outside heat, and makes the regulation of temperature comparatively

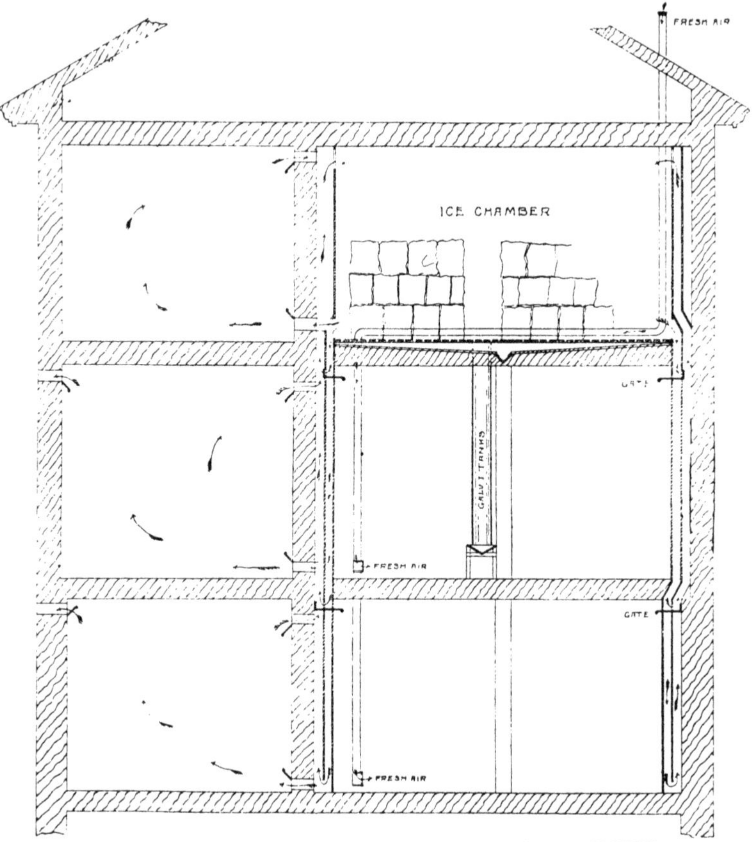

FIG. 6.—THE DEXTER SYSTEM—SECTIONAL VIEW.

easy, even in warm weather. This is practically like putting one cold storage room inside of another. Dexter uses also the galvanized tubes or tanks filled with ice and salt for bringing down the temperature to the desired point. The circulation of air within a room cooled in this way is sluggish, and the air too moist for most products which are generally placed in cold storage for safe keeping. Dexter also has patents on a method of circulating air from the ice chamber down through or around tanks filled with ice and salt, into the storage room, but the writer is not aware that these devices have proven to be possessed of any particular merit or that they have been brought into general use. Other patents have been taken out on a scheme for constructing an ice floor or pan. This has been found leaky in a number of cases, and has been removed and built over. Still other patents are on a system of ventilation, and a method of insulating the ends of joist where they enter the walls of a building.

Any system of cooling storage rooms in which the air is circulated directly from the ice has the constant trouble with dampness of the ice room or bunker. Moisture always condenses on the ceilings or side walls of the ice receptacle, and mold results very soon. The air circulating over the molded surface carries mold spores into the storage room. The goods stored therein suffer in almost every case. A house which has been in service for some time may be very bad in this respect, especially during cool weather of fall or early winter, as the temperature is lower and the air of storage room more moist. Dampness of ice room also causes decay of woodwork and insulation.

The Direct Tankage System.—There are or have been a number of cold storage houses, cooling rooms and freezers refrigerated by what the author calls the "Direct Tankage System." This system consists simply of placing metal receptacles filled with ice and salt in the room to be cooled. There are several forms of tanks in service, the more common of which are the square cornered or rectangular tanks, the thin tanks, or what are sometimes called "freezing walls," and the cylindrical or round tanks. Usually these tanks are made of galvanized iron. They may be made of a thickness of iron ranging from gauge 18 to gauge 24 metal. Gauge 20 iron is usually the best to use. These tanks are almost invariably filled from the top through the ceiling, or what would naturally be the floor of the room above. They have, however, in some extreme cases been filled from the side, either from without or from within the room.

The rectangular or square tanks, as at first employed, have gradually gone out of use, because they are difficult to make and keep in shape and, as built in a number of cases, were so large that the meltage of ice would be largely near the tank sides, and very little towards the center. Tanks of this class have been used which were as large as three feet in their smallest dimension, and as the meltage was almost entirely within eight or ten inches of the outer surface, the waste of space and lack of economy are at once apparent.

The thin or flat tanks, which are sometimes called "freezing walls," as usually constructed, are only about four to ten inches in thickness, and are sometimes narrower at the bottom than at the top. These of course are iced from the top, and many fish freezers, built years ago, did good service when equipped in this manner. One serious objection was that only one surface of the tank was available to any considerable extent for cooling service, as the back or that portion of the tank near the wall received comparatively little air circulation, in fact, in many cases the back of the tank was placed directly against the wall of the room with no space left between. The construction of these tanks, also, is difficult.

Furthermore, any flat surface when used for a purpose of this kind has a tendency to bulge outward, owing to the pressure of the ice and salt within. The result is that the tanks become leaky and will rust out rapidly.

The cylindrical tanks are very much the best of the three kinds mentioned. They are easy to make, and owing to the cylindrical shape will not readily get out of order and are much more practical than either the freezing walls or the rectangular tanks. It has been found in actual practice that in producing refrigeration through a metallic surface from the meltage of ice, where one side of the metal is exposed to the air of the cool room, and the other has ice and salt in direct contact, comparatively little refrigerative effect is obtained from the ice lying more than six inches away from the exposed metallic surface. Cylindrical tanks, therefore, have been built of a diameter of eleven inches, that being a size readily constructed mechanically and one which will give best results when used for freezing or cooling. In the illustration of the Dexter System (page 449) may be seen the cylindrical tanks suspended from the floor of the ice chamber.

The direct tankage system, while it has been in use quite extensively, is not at present being installed to any great extent, as its disadvantages are many. The nastiness and muss occasioned by the icing of tanks through the ceiling of the storage room is in itself sufficient to condemn the system. The continuous slop resulting from handling the ice and salt upon the floor above the storage room will result in the rotting and decay of timbers and insulation in a comparatively short time. It will also readily be seen that the great amount of space wasted by thus icing the tanks is a serious drawback to this system. Practically nearly as much space is required for the mere charging of the tanks as is available for refrigerating purposes. Another disadvantage of the direct tankage system is that it is wasteful of space in the storage rooms, as the tanks do not present as much surface to the air of the room proportionately as does iron piping in the form of coils. The tanks in the room are also sloppy and wet and the pan underneath is liable to become choked up and overflow on the floor of the storage room. Further than this it is extremely difficult to regulate the temperature of a room with this system, owing to the fact that there is no control or balance on the refrigerating effect. Directly after charging the tanks, the temperature will run down and then slowly rise until the next time of charging. The author has worked with this system for a number of years and has abandoned its use entirely in favor of the Cooper brine system cooled with ice and salt, which is described further on in this chapter.

John A. Ruddick, Dairy and Cold Storage Commissioner, Department of Agriculture, Ottawa, Canada, has this to say regarding the disadvantages of the direct tankage system: "I am doubtful, after some years' experience, if it is the best system to recommend. The cylinders are not always kept full, causing insufficient and irregular refrigeration, and excessive dampness is likely to result because of insufficient air circulation or because of the moisture from the cylinders whenever the ice is allowed to melt off the outside of the cylinders."

WHY THE AMMONIA SYSTEM IS SUPERIOR TO ICE.

These various systems of ice refrigeration which have come into general use during the past thirty-five years, have been briefly outlined and commented on by the author, so that the reader may comprehend, roughly, the history of and the reasons why ice refrigeration has not given satisfaction when placed in competition with the mechanical systems which are now generally understood to be the best for all purposes. It is now necessary for us to make an investigation of the "ammonia" or "mechanical" systems, (see chapter on "Systems of Refrigeration") when allied to cold storage, in order to ascertain in what vital particular this system surpasses the old-time ice systems. In visiting such a cold storage warehouse, we find a building with insulated walls not differing from those of an ice cold storage. The interior we find divided by insulated partitions into separate rooms for various products; goods having a strong or disagreeable odor being carefully isolated from delicate goods like butter and eggs. In this respect the ammonia cold storage has the advantage over the old style ice cold storage, as the latter, even if divided into different rooms, generally has an ice chamber common to all, making contamination of one product from another probable. Each room of an ammonia cold storage is equipped with a coil or coils of piping placed on the walls or any convenient location. Through this piping flows a liquid or a gas at a low temperature. This cools the piping, which in turn cools the air of the storage room. The surface of the pipe being at a low temperature, frost accumulates on the pipe. This frost is moisture which is taken from the air of the room. The low temperature of the pipe thus causes a constant drying of the air of the room. The main difference between a room cooled in this way and one cooled by ice, is that it is much dryer, because cooled by frozen surfaces at a temperature which will collect moisture from the air of the room in the form of frost. In three houses out of four, no circulation of air is provided for, nor means for supplying fresh air. If we pursue our investigation further and enter the machine room, we find a complete steam plant, with which we are all fairly familiar, and much other machinery and apparatus besides, which takes a bright engineer some time to successfully master in all its details. This, then, is the average "ammonia" cold storage, as seen by an outsider. The real and only reason why such plant produces better results than the average ice system, aside from a control of temperature, may be summed up in the two words "DRY AIR." It is now purposed to describe a system which has all the advantages of the ammonia system in the respect of producing a dry atmosphere in the storage room, and yet has the advantage of the ice systems in being simple to operate, economical and sure against breakdown.

THE COOPER SYSTEM OF BRINE CIRCULATION.

It has already been pointed out that it is impossible to produce a dryness or humidity of air in a cold storage room cooled by ice, beyond a percentage which is fixed by the temperature of the room. That is to say, practically no control of humidity is possible in such a room. Further than this, the air in an ice-cooled room is almost invariably moister than in a room of the same temperature cooled by pipe surfaces. In a room cooled by frosted pipe surfaces, the moisture which is given off by the goods, and that which finds its way into the room when doors are opened or otherwise, is frozen on the pipes in the form of ice or frost. This is because the pipes have a temperature below the freezing point of the moisture in the air, causing the moisture to freeze on the surface of the pipes, and leading to a greater drying of the air than where ice is the cooling agent. Not only will pipe surfaces at a temperature below the freezing point of the air moisture produce a dryer room, but they will also produce a lower temperature, and make the control of temperature possible. It has already been stated that the reason why the ammonia cold storage houses produced better results aside from a control of temperature, was their ability to give a dryer air. A system which will utilize ice as a primary refrigerant and yet give a dry air and low temperature, would then necessarily be able to compete on an even basis with the ammonia or mechanical systems of refrigeration.

With a due appreciation of the facts as stated above, there was begun a series of experiments to demonstrate the possibilities of ice refrigeration, and the refrigerating apparatus now known as the Cooper systems is the result. At the time of beginning these experiments, the house experimented with was a nearly new one, equipped with what was at that time supposed to be the very best and latest system of ice refrigeration (Dexter System), and at that time the writer was familiar with nearly all the prominent systems of ice cold storage, as already described. It was thought that if brine cooled by the ammonia system and circulated through pipes for cooling storage rooms would give better results than ice, the same results might be produced by cooling brine with a mixture of ice and salt, and circulating the brine through pipes in the same way. To demonstrate the practicability of the idea, a small room was fitted up for a test. An insulated tank was constructed, in which was placed a pipe coil surrounded by ice and salt. Another coil was placed on the wall of the room, and the two connected together. A pump driven by an electric motor, caused the brine to flow from the coil in the tank through the coil on the wall, and then again through the tank coil continuously. A temperature of brine ranging from 12° F. to 18° F. was readily obtained, and the experiment was such a marked success, even with this crude apparatus, that it was extended to two other rooms, larger than the first. This time the pipes were so arranged that a partial circulation of brine would take place without the operating of the pump, but still another trial was necessary to fully demonstrate that the system could be operated entirely without a pump; that is, by the natural or gravity circulation of brine. This was obtained in a manner similar to the circulation of water in the hot water heating systems used in heating buildings. In the Cooper gravity brine system, the tank which contains the ice and salt, and the tank coils or primary coils, as they are called, are located at a higher level than the secondary coils which do the air cooling in the rooms. Fig. 7 shows the arrangement of coils in use. When the tank is filled with ice and salt, the brine standing in the primary or tank coil is cooled by contact with the ice and salt which surrounds the pipes, to a lower temperature than the brine contained in the secondary coils, and consequently flows down into the secondary coils. At the same time

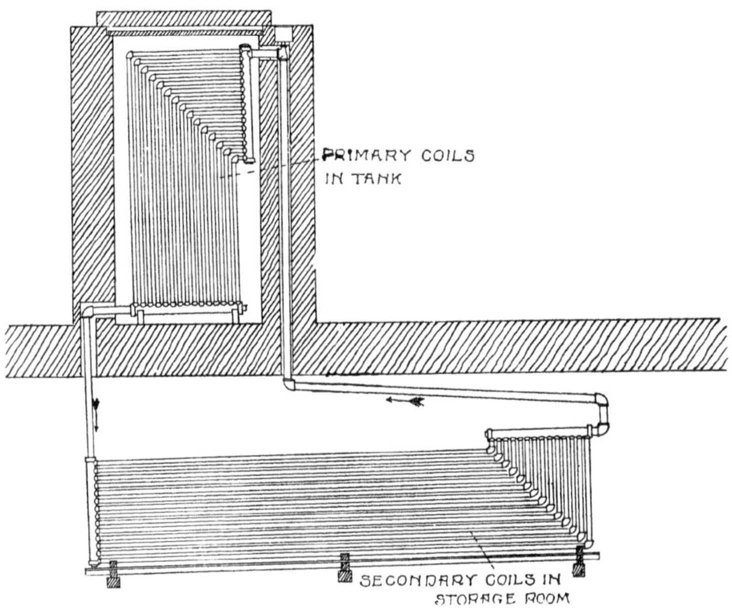

FIG. 7.—DIAGRAM SHOWING ARRANGEMENT OF BRINE COILS IN COOPER'S BRINE SYSTEM.

the brine from the secondary coils rises into the primary coils, where, as it is cooled, it repeats the circuit in the direction shown by the arrows. The term "gravity," as applied to this system of brine circulation, refers to the cause of circulation which is owing to the difference in specific gravity (weight) between the cold brine in the primary coils and the comparatively warm brine in the secondary coils. The temperature of the circulating brine will range from 0° F. to 20° F. or

25° F. It is comparatively easy to cool a room to 10° F. or 12° F. with the Cooper brine system. Regular temperatures as low as 5° above zero, Fahrenheit, have been maintained without difficulty. The brine which circulates in the primary and secondary coils is usually a solution of chloride of calcium, which is used in preference to common salt brine for the reason that it rusts the pipes less and will not freeze as readily. The circulating brine is entirely independent of the brine which runs out of the tank as a result of the mixture of ice and salt. The refrigeration in the waste brine is utilized for cooling purposes by running it through a coil of pipe of suitable size at any convenient place in the building, and it is afterwards led to the sewer. The chloride of calcium brine, on the other hand, remains always in the pipes, the only loss being from leakage, which is, of course, very small. It will be appreciated, by experienced persons, that this system of cooling is simple in principle, very unlikely to get out of order, and when once in operation, will continue as long as the supply of ice and salt is maintained in contact with the primary coils in the tank. In operation it is usually necessary to fill the tank but once each day with ice and salt, and the circulation will remain continuous and automatic through the twenty-four hours. The ice in the tank will melt down one to four feet per day, depending on how hard the apparatus is being worked, and it is only necessary to refill with enough ice and salt to keep the tank full. (See directions for operating further on in this chapter.)

Rooms cooled by the Cooper brine system are subject to precisely the same drying and purifying influences as are rooms cooled by any of the mechanical systems of refrigeration. The moisture and impurities in the air are, to a great extent, frozen on the surface of the pipes, and temperatures are easily controlled. In applying ice to cold storage work by this system, the ice has no more connection with the air of the storage room than if it were miles away. The ice is, in fact, generally placed in an ice room of cheap construction, built independently of the cold storage rooms. Where the ice room is already built, it is only necessary to build the cold storage rooms alongside of the ice house, equip them with cooling apparatus and means for getting the ice to the tank containing the primary coils. Even the location of ice house is not important. The cold storage house may be located in the center of the business district, and the ice on the ice field. The necessary quantity may be hauled each day. This is entirely practicable, and its feasibility has been demonstrated in several different localities.

Except in plants of very small size, the ice is usually crushed and elevated to the tank by machinery. This saves much labor, and results in better work. The machine for crushing the ice is generally located at or near the floor of the ice room. The ice is fed into the crusher through a chute, which is made in sections. As the ice is worked down in the house, a section is removed to bring the top of the chute about on a level with the top of the ice. The ice is first chopped into irregular pieces of twenty or thirty pounds or less, then shoveled into the chute, which drops the ice into the crusher. From the crusher, in pieces not larger than a hen's egg, it drops into a bucket elevator which raises it to a point near the tank, and somewhat above it, where the elevator dumps the ice into an inclined tube terminating in a flexible spout. The flexible spout is pivoted, and will deliver ice to any part of the tank without shoveling. The only hand labor necessary on the ice is the chopping of the ice and shoveling it into the chute. Two men will easily handle four tons of ice an hour in this way. Four tons of ice a day will cool a storage house with forty carloads capacity, during average summer weather. The section, Fig. 8, on following page, shows the ice-handling apparatus in conjunction with the other parts of a fully-equipped storage house. It may be noticed that the storage rooms have no communication with the ice house, and that the cooling is effected by circulating the air of the rooms in contact with the secondary coils of the Cooper brine system.

The storage rooms are cooled and the temperature regulated directly by the above named system. This may be done by placing the secondary coils of the brine system directly in the room, but a better method is by using the forced air circulation system, especially if the rooms are fairly large. For this purpose, the secondary coils are placed in a room by themselves, known as the coil room. A fan draws the air from the coil room and distributes it to all parts of the storage room. The air is returned to the coil room by being drawn off at the

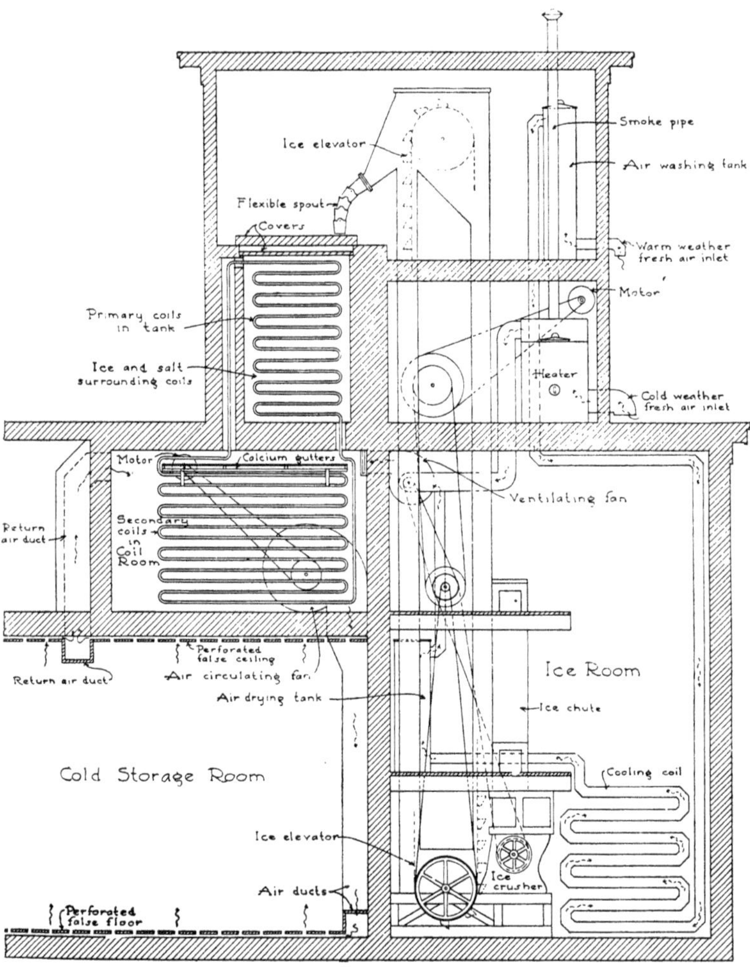

FIG. 8.—ILLUSTRATING THE COOPER SYSTEM OF COLD STORAGE. This diagram is for the purpose of showing the different systems clearly, but is not in good proportion, as the storage room is shown much smaller than it is. The apparatus is really quite small as compared with the rooms.

top of the storage room through a perforated false ceiling. Depending on various conditions, either perforated side air ducts or perforated false floors are used for distributing the air uniformly throughout the room. (In the chapter on "Air Circulation" are discussed the various phases of air circulation and the relation of refrigerating surfaces thereto, describing the various designs by the author for improving air circulation. See also chapter on "Ventilation" and "Uses of Chloride of Calcium" for other simple devices which are installed in combination with the Cooper brine system as illustrated in Fig. 8.)

DISCUSSION AND DIRECTIONS AS APPLIED TO THE COOPER BRINE SYSTEM AND OTHER ICE AND SALT SYSTEMS.

It has been thoroughly demonstrated by the author that tanks of a greater length than 10 feet were usually unnecessary and the additional length when used was practically wasted. In practical operation ice in the tank does not settle or melt down generally, more than from one to three feet per day of twenty-four hours. All that is required, therefore, in the length of the tank is that it should be sufficiently long to allow the salt brine which is dripping down through the ice to become thoroughly diluted and expend its ice-melting power before reaching the bottom. While the largest amount of refrigerating duty is accomplished where the ice and salt are in direct contact with one another, near the top of the tank, yet there is considerable refrigerating value or ice-melting power in the brine which results from a union of the ice and salt. This brine trickles down through the finely crushed ice in the tank and has the same action on the ice as the salt, only to a lesser extent. As the brine becomes more and more dilute, it has less and less value in this respect and finally possesses practically none, if the tank is of sufficient length.

As before stated, the practical limit of length for the tank is ten feet, although for some purposes a length of twelve feet may give more satisfactory results. The author has had in service for freezing purposes tanks which were sixteen feet in length. With tanks of this length the meltage in the lower three or four feet is very small. In fact, the bottom six feet of these tanks will not do any considerable amount of work. The further down in the tank, the less meltage of ice there will be, depending on the temperature at which the room is being carried.

The work of charging the tanks with ice and salt is comparatively simple, but at the same time there is a chance for the exercise of considerable skill and sound common sense. It is an old idea in connection with freezing ice cream that the ice and salt should be filled into the freezer in alternate layers instead of being thoroughly mixed together. There is no question at all but that this is a mistake. The more thoroughly the ice and salt can be mixed together, the better the freezing action to be obtained, and the greater the economy of ice and salt.

If the ice and salt are filled into the tank in alternate layers there are two bad results which interfere with the practical and efficient operation of the plant. One is that the salt will cake and form lumps, which will probably go clear to the bottom of the tank without entirely dissolving; another is that quite a large portion of the refrigeration which is developed where the ice and salt do come in direct contact is used in the freezing together of the ice which has no salt mixed with it. This is especially true at the top of the tank, where there would be very little or no brine dripping down through.

A greater amount of work may be done with a given size of apparatus if the ice and salt are thoroughly mixed together. Greater efficiency may also be obtained; that is, more work out of a given quantity of ice and salt.

The proper way to charge the tanks of the Cooper brine system, therefore, is to thoroughly mix the ice and salt together. In a few cases this is done before the ice is filled into the tanks, but a better way is to salt the ice as the tanks are being filled. If the ice is crushed by a machine and the tanks are filled through a spout, this is a comparatively easy matter. Where the ice has to be crushed by hand it should be broken as finely as possible and no pieces of ice larger than a man's fist allowed to go into the tank. This is sometimes difficult to do where the ice is broken with a sledge hammer or an axe, but with care big pieces may be avoided. It might be stated here that the finer the ice is broken the better is the action obtained from the salt; that is, the less salt it will take to do a given amount of refrigerating.

In refilling tanks it is well to first put on a certain amount of salt, whatever is required, and then settle the honeycombed ice already in the tank by stirring with a stirring stick. After the tank is filled a small amount of salt may be placed on top and thoroughly stirred into the ice by the use of the stirring stick. This stirring stick may be constructed of a piece of 1¼ by 3 inch hard wood with a tapering point at one end, smoothed off to a handle at the other end, and made about four to six feet in length.

The tanks of the Cooper brine system should be cleaned out at least once a year, as a certain amount of mud and dirt will accumulate at the bottom even with the most careful handling of salt and cleanest possible ice employed. In filling the primary tanks of this system the instructions above may be followed closely and may be much more readily carried out, as plenty of space is available for the stirring of the ice and salt. The best way to proceed in filling the primary tanks of the Cooper brine system is to detail two men in the tank house, one for salting the tank, and the other for stirring the salt into the ice. If only one man is available the ice should be handled slowly so he may get salt fully stirred into the ice. In this way a very thorough mixture can be obtained and economy will result. The flexible spout which feeds the ice into the primary tank can be arranged so that it may be held in any position by the use of a rope. It is not necessary that this should be held in the hand.

The direction and remarks regarding the Cooper brine system as above, apply equally to any tank system employing ice and salt. No exact directions can be given as to quantity of ice and salt required for the reason that the conditions are constantly changing. It may be remembered as a rule, however, that the quantity of ice melted, goods cooled and temperatures produced are in almost direct proportion to the amount of salt used in the tank. The more salt, the more ice melted, the more refrigeration produced, the more goods cooled, or the lower temperature obtained. Pay no attention whatever to the amount of ice being used. It is the salt on which you depend for gauging the operation of the house. As a guide it may be said that from five to fifteen pails (25 lbs. each) will be used on a primary tank of the Cooper brine system which has a dimension of about 4x5 feet at the top and 10 feet deep, under average conditions.

KIND OF SALT TO BE USED.

The kind of salt to be used is a question which always comes up in the operating of an ice and salt refrigerating system. That most suitable for use will depend on locality of plant to some extent. The old salt wells at Syracuse, N. Y., have for years manufactured a solar or sun evaporated salt which has a coarse cubical grain and which is useful for freezing purposes. It is not quite as good, however, as the better grades of rock or mined salt for the reason that it dissolves more quickly. The rock salt is more dense in structure and dissolves slower. The rock salt is mined from the earth much the same as coal, and is crushed and screened to various sizes. That known as "CC" and No. 1 and No. 2 is best adapted for freezing purposes. The "CC" is finer grain, about like wheat, and is good where the apparatus needs to be crowded or pushed for capacity. The No. 1 is good for even temperature and easy work and No. 2 only in deep tanks where the capacity does not need to be crowded or pushed in any way. Crushed rock salt is now obtainable from mines in Western New York. Louisiana, Kansas and more recently along the St. Clair River in Michigan. There are doubtless many other localities where salt will be mined as the demand increases. The price is always low and it is doubtful if it will ever go higher as improved methods of mining and handling make it one of the cheapest natural products. Locality, of course, governs price but very few localities have prices higher than $7.00 per ton in bulk, and $5.00 per ton to $6.00 per ton is quite common.

In some cases where operators of cold storage plants have run short of regular freezer salt they have substituted common barrel salt. The results from same are very unsatisfactory. Steam evaporated salt, especially of fine grain dissolves so quickly that the refrigerating effect is very small. It might be noted in this connection that the greatest refrigerating effect is at the point of direct contact between the ice and salt, and the brine resulting from a union of ice and salt has comparatively little refrigerating or ice melting value.

CHAPTER XXXVI.

ICE STORAGE UNDER REFRIGERATION.

HISTORICAL.

The storage of ice in refrigerated rooms or houses is comparatively new, and there are no methods or types of construction or arrangements of apparatus which may be called standard.

Ice storage originated with the storage of natural ice, and as everyone knows, natural ice has been stored for many years in most any sort of a structure with or without insulated walls, but relying largely or wholly for protection on packing in sawdust or other material of a porous nature to prevent the penetration of heat. Artificial or machine-made ice was at first stored in the same way and for comparatively short periods, or it was stored in an insulated room without refrigeration and for a few days at a time. Later, storage rooms for artificial ice were refrigerated in connection with ice making plants, and brine or ammonia piping run from the freezing tank for the purpose of cooling the room, and these rooms were mostly used simply as in-and-out rooms to protect a few days' supply, or possibly to give a reserve stock of ice in case of breakdown to the machine, or to take care of extraordinary demand in extremely hot weather.

As the demand for ice became larger and the capacity of the manufactured ice plant was taxed, it was appreciated that ice might be stored in a comparatively large ice storage room during cool weather when the demand for ice was comparatively slack, and thus the real ice storage houses for storing ice under refrigeration were developed. At first they were very crude affairs, being simply rooms with piping, but during recent years there has been some tendency to systematize construction and there have been some very large ice storage houses built for artificial ice.

First experience with the storage of artificial ice was, in general, unsatisfactory for several reasons, and there is yet a prevailing opinion in many places that artificial ice cannot be successfully stored. One of the chief reasons why results were bad at first was that sufficient insulation was not used and the ice not properly stored in the room and the piping arrangements were neglected. At the present time there is no greater difficulty in the storage of artificial ice for practically unlimited periods, than there is in the storage of other goods which are held under refrigeration to the limit of their natural life.

At first ice was piled into the ice storage room in a solid mass directly on the floor and tightly against the side walls and the result of this, in many cases, where poor insulation was employed, was that even though the temperature near the ceiling where the piping was located was maintained below the freezing point, yet the ice melted on the floor and on the sides where the cold air could not get to it. Another effect of poor insulation was that the cooling pipes, arranged generally on the ceiling, absorbed too much moisture from the air, causing a drying out of the ice, which lead to a honey-combing, which, when the ice was removed from storage and exposed to warm air, would result in its splintering and falling to pieces; hence the impression that artificial ice could not be successfully stored gained much headway.

There are several points in connection with the storage of ice under refrigeration (and this applies to natural ice as well as to artificial) which can be set down as a basis to work on, and we may list them as follows:

1. Suitable insulation.
2. Ample piping properly located.
3. Proper packing of ice in the room.
4. A temperature below the freezing point in all parts of the room.

The first requisite, insulation, is subject to much discussion, as what one man would call prime insulation, another would not; but it may be stated here that the average insulation as applied to cold storage houses and also to ice storage houses is usually not more than half enough. Unless good insulation is used a much larger piping equipment must be provided, and this in turn means a greater drying of the air of the room, and this leads to a tendency to an evaporation of the ice, causing the honeycombing and splintering above referred to. Good insulation, therefore, is necessary as a matter of economy, as well as to successful storage. Those who advocate the use of expensive high grade insulating materials without reference to the character of goods to be stored and the type of building, usually for the ice storage plant, advocate a thickness of insulation of from three to five inches. This is really absurd from an engineering standpoint, as there is no material known which has sufficient insulating value to give what might be called prime insulation in this thickness. Cheaper materials and more of them would be better economy.

The arrangement of piping in an ice storage room is usually given small attention and generally is arranged where most convenient. The ceiling is sometimes covered with piping, and then whatever is left over to make a full complement is distributed around the side walls, and there are many ice rooms where the cooling coils are located on the side walls only. The correct arrangement of piping for an ice room is on the ceiling, and ordinarily sufficient piping may be located on the ceiling to cool the room if the insulation is adequate. In very high rooms of from 50 to 60 ft. or more it might be advisable to locate a portion of the piping on the side walls as well as on the ceiling. Locating the piping so as to produce a circulation of air is what is desired in a cold storage room for the storage of perishable food products, but contrary to this piping in an ice room should be located with regard to preventing a circulation of air to any marked extent, as circulation dries the air, and this tends to evaporation or drying out of the ice. Theoretically the correct way of cooling an ice storage room would be with an indirect air circulating system either by using a fan or possibly by a gravity circulation, with a thin cold air circulating space within the insulated wall. There is no heat to be extracted from the goods stored, and it is only a question of intercepting the heat which leaks through the insulation, and this could better be done by an indirect air circulating system than any other way.

In storing ice in the room the former practice was, as above stated, to pile it in a mass promiscuously without regard to the maintaining of temperatures throughout the room. This, as above stated, leads to a meltage of ice on the sides and floor, and this in turn brought out the idea to store ice on strips placed on the floor, usually 2x4's, and in many cases the ice was also stored with strips between the tiers. Considering the fact that the ice itself has no heat to be taken up, and assuming that the ice goes into the room below the freezing point and in a perfectly dry condition, it is only necessary to store it in the room so that the heat which leaks through the insulation will be absorbed by the cooling pipes before reaching the ice. As the amount of heat coming through the floor insulation is small as compared with the side wall and ceiling insulation, it is usually not necessary to store ice on strips on the floor, as the irregularity of the ice cakes will allow sufficient circulation of air through the ice to prevent meltage. Depending on how the ice is piled and the height and size of the room it may be necessary to leave a space of 2 to 4 inches around the sides of the room so that the heat coming through the wall can find its way to the cooling coils. Artificial ice made in cans is larger at one end than at the other, and the storing of every alternate cake in the opposite direction is common, but it is better, if cooling pipes are located on the ceiling, to store the cakes in the same direction, so as to leave some small amount of space for circulation of air. The cold air will then drop down through the middle of the mass of ice and circulate out to the side walls, and thus take up the heat coming in and rise to the cooling pipes. In piling ice in the house it should be arranged so that it does not pack too tightly, for the reason stated above. Sometimes it may be necessary to store the ice on 4-inch strips on the floor as well as leave from 2 to 4 inches all around the sides of the room, and in very large rooms an air circulating space through the center may be necessary. No exact rule which would apply to every ice storage house can be laid down, but one or two seasons' experience will be necessary to determine the very best course to pursue.

Temperature in the storage of ice is important only in that it must be below the freezing point of water in all parts of the room. If suitable strips are arranged for the right circulation of air, as above stated, a temperature of 28° F. in the room is all that is necessary, but if the circulation of air is not penetrating, the temperature of the room should be proportionately lower in order to maintain all parts of the ice pile below the freezing point. In this connection it may be remarked in passing that the conductivity of ice is very low indeed, and if ice is piled tightly unless there is a low temperature at the ceiling, the heat of the earth coming through the floor under a fairly solid mass of ice will cause meltage as it is impossible for the refrigeration to do its work. In some houses it may be necessary to carry a lower temperature in warm weather than when comparatively cool.

It is desirable to have the ice storage room as near a cube as practicable, but it is not, of course, well to make a room too high on account of danger of collapsing in a high wind. Sixty feet is a suitable height for ice storage providing the walls are sufficiently well designed and braced. Care must be also taken that a suitable foundation is provided as the weight of 60 feet of solid ice produces severe pressure, and it will not do at all to have this load on newly filled earth. In case of unequal settlement the ice might shift to an extent which would bulge the walls or possibly wreck the house. The author was the first to recommend an ice storage room as high as 50 feet, and this was used for natural ice in connection with the storage of ice for use in a Cooper brine system plant. Later the author recommended the construction of a storage room for manufactured "plate" ice to be 60 feet in height, and many of the new plants are being built of this height. There seems no practical reason why a room cannot be made still higher providing a good foundation can be secured and correct design to withstand wind pressure is adopted. The old style ice stor-cost of construction as well as increased loss of refrigeration or age rooms and houses were generally built of a height ranging from 24 to 40 feet, and but very few of them more than 36 feet in height. This means a large roof exposure and increased ice meltage. What is said here applies to the storage of ice in an insulated room and with no covering material on the ice, and in a room which is held under refrigeration so that there is no meltage.

A suitable vestibule or forecooler should be provided in connection with every refrigerated ice storage room into which

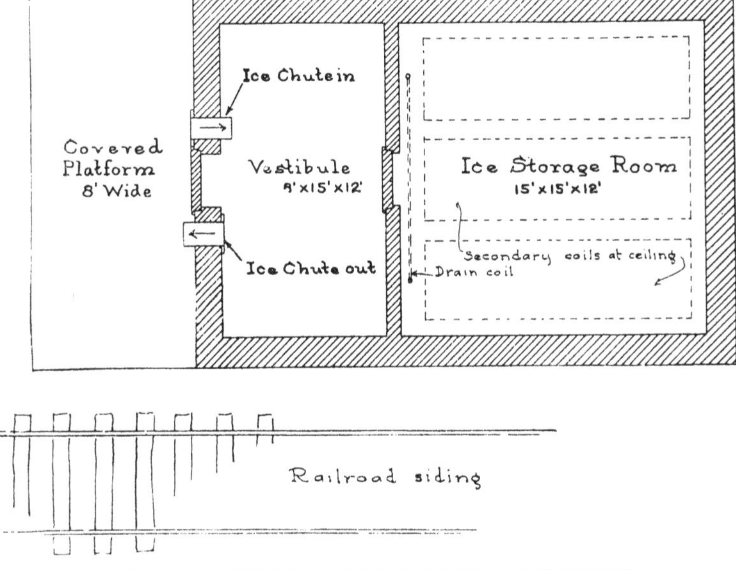

FIG. 1.—FLOOR PLAN FOR ICE STORAGE HOUSE.

the ice is run as fast as taken from the tanks, and where the temperature is maintained below the freezing point, so that the ice will dry off nicely before it is stored in the main room. This is of the utmost importance to best results. It is also necessary to prevent penetration of outside air to the storage room, and this can only be done by using the automatic ice doors or chutes for putting the ice into the room and for removing it. These patented automatic doors are quick in action and close tightly and no other device should be used.

The above statements are based on the storage of what in the ice trade is known as "can" ice, which is frozen in galvanized cans suspended in tanks of brine, and is considered the most difficult to successfully store. "Plate" ice, frozen on metal plates from one side only and more slowly frozen, is denser and has much the same character as natural ice . "Natural" ice frozen by nature, we are all familiar with, but natural ice is just now beginning to be stored under refrigeration. The next ten years will see some large developments along this line. Ice storage under refrigeration—any kind of ice—is only beginning, comparatively speaking.

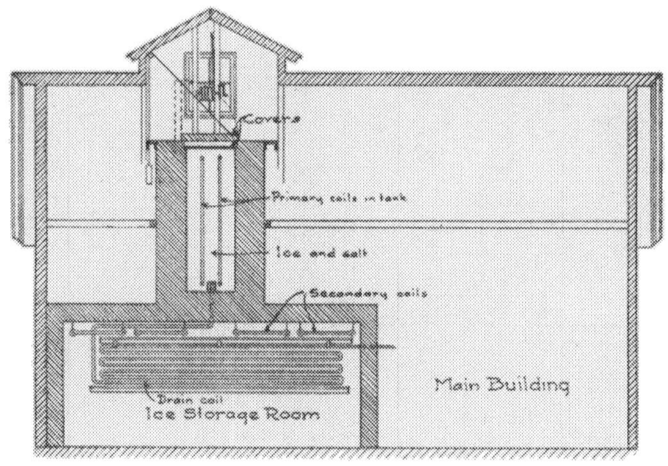

FIG. 2.—LONGITUDINAL SECTION.

A BRANCH PLANT ICE STORAGE.

The plan shown in Fig. 1, was designed for the storage of a carload or two of artificial ice at branch plants where the ice is shipped in cars from a central factory. The plant as arranged is equipped with the Cooper brine system, using ice and salt for cooling for maintaining a temperature of 28° F. to positively prevent the meltage of ice while being stored in the room.

The main ice storage room, 15x15x12 feet is protected from direct access of warm air by a vestibule 8x15 feet and

12 feet high. This vestibule is not refrigerated. The cakes of ice are passed in and out by means of the ice chutes shown and the vestibule can be used for the temporary storage of small quantities of ice. Sufficient ice may be hauled out into this room for loading a wagon, and the vestibule may be used for the temporary storage of perishable goods, like fruits, etc.

A convenient platform eight feet wide is intended to give access to teams on one end and one side to railroad track on the other end.

Good construction in a building of this kind will pay for itself in a very short time, and the plan shown in Fig. 1 may

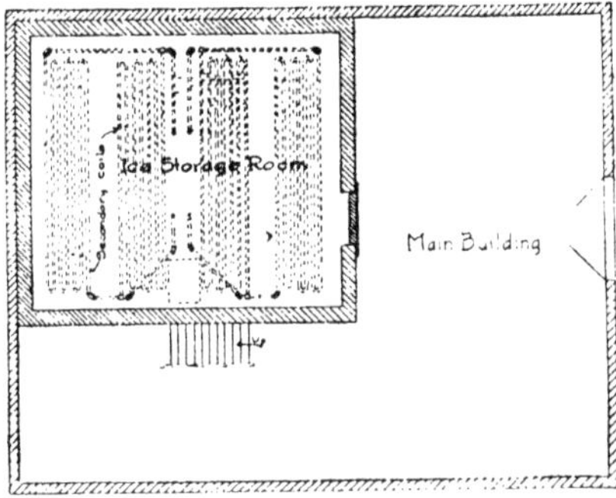

FIG. 3.—MAIN FLOOR PLAN.

serve as a model to those who handle ice at outlying points and at some distance from the place where it is made.

ANOTHER BRANCH PLANT ICE STORAGE ROOM.

The illustrations Figs. 2, 3 and 4, show another application of the Cooper brine system to the cooling of a storage room for artificial ice, and this arrangement is applicable to many situations where ice is shipped in by the carload or hauled from a central plant to outlying storage houses for distribution. In the present case the ice storage room is shown within a main building which is used for office and other purposes, and the ice storage room, therefore, occupies but a small portion of the total available space in the building. The ice storage room is quite well protected from outside exposure as there are but two walls of the room adjoining exterior walls of the building.

The general plan as shown is applicable to almost any size or capacity of room, but the one here shown is 7½ feet and 16x18 feet, inside dimensions. This would store approximately 50 tons of ice, but deducting for space occupied by cooling coils and assuming that ice would not be piled more than 6 feet in height, the comfortable capacity of the room would be from 35 to 40 tons or what would now be a maximum or very large car of ice. It may be here suggested that it is far more economical of space, cost of plant and cost of operation to make a room much higher than the one shown. For instance: A room 16x18 feet could just as well be 15 to 20

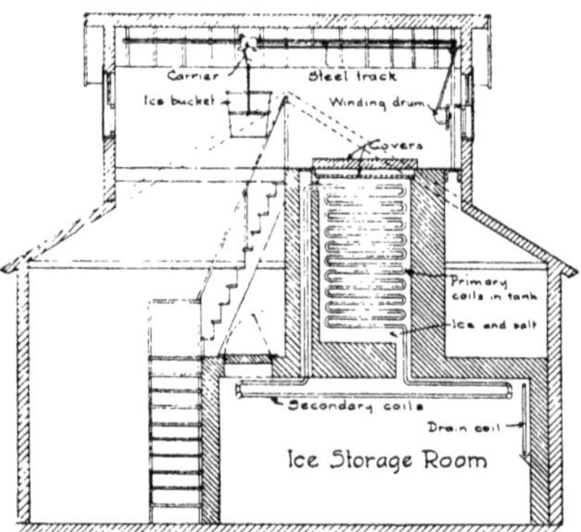

FIG. 4.—TRANSVERSE SECTION.

feet in height as the cost of refrigerating would be very little more than a room 7½ feet in height.

The arrangement of coils shown in the plan is very economical of space. With practically the entire ceiling covered with coils it prevents to a great extent the circulation of air in the room. The primary tank of the Cooper brine system stands directly above the ice storage room, and ice for supplying same is secured through a trap door in the ceiling of the ice room, the ice being crushed in the room and hoisted to the top of the tank in a bucket. A hand winding drum is provided and a carrier and track for delivering the ice directly over the primary tank. With good insulation the amount of ice required for refrigerating is comparatively small.

ADVANTAGES OF LARGE ICE STORAGE CAPACITY.

The advantages of large ice storage capacity with a comparatively small ice making capacity have been discussed at length and both methods have their advocates. It would seem, however, that the advantages in favor of large ice storage capacity are so great that this will be the coming method. R. P. Kehoe in *Power* gives some figures which may be interesting in this connection as follows:

First, a 100-ton ice making plant, costing $84,000; daily operating expenses, $98.50; total yearly expenses, $30,075; estimated profit, $14,925; percentage of profit to investment, 17.8%. Second, a 60-ton can ice making plant and 5,000-ton ice storage, refrigerated, costing $75,000; daily operating expense, $67.00; yearly expense, $31,060; estimated profit, $13,940; percentage of profit to investment, 18.6%. Third, a 60-ton plate ice making plant and 5,000-ton ice storage, not refrigerated; investment $100,000; daily operating expenses, $41.50; total yearly expense, $23,840; estimated profit, $18,660; percentage of profit to investment, 18.66%.

Mr. Kehoe states that his figures are based on wooden structures with cheap insulation and a 50 per cent yearly load factor. Other figures may be made on this subject which will show still better results from large ice storage capacity, and we believe that when actual figures are taken from such a plant they will show the economy. Ice storage is being more and more generally used in connection with ice making plants.

CHAPTER XXXVII.

HARVESTING, HANDLING AND STORING NATURAL ICE.

THE GENERAL ICE CROP.

It is not expected that this chapter will be of much assistance to the experienced ice harvester, but those new in the business and persons having a comparatively small amount of ice to house may be able to obtain some information in regard to the methods used, and select such tools and devices from those described as will best suit their particular needs. Natural ice has been talked down, legislated against and generally speaking has come to be regarded as a back number for cold storage purposes, but a large percentage of the perishable goods stored in the two northern tiers of states and in Canada are stored in structures cooled with natural ice and are likely to be for many years to come, and the harvesting, handling and storing of the natural ice crop is therefore of sufficient importance to warrant a fair description. In the states which are in about the same latitude as New York and Minnesota and throughout Canada, a failure of the ice crop is unknown, and ice forms quite regularly to a thickness of from ten to twenty inches. In Pennsylvania and Iowa and the states in the same latitude and isothermal conditions, ice is usually harvested of a thickness ranging from six to twelve inches, sometimes thicker.

Before the introduction of the ice machine, natural ice was harvested as far south as Tennessee and Missouri and in the mountain regions of Virginia and North Carolina. In some cases this is still done, but the crop is uncertain, and as the ice is thin it is expensive to harvest. Probably the thickest ice on record is harvested at Winnipeg, Manitoba, Canada, where it reaches a thickness of forty inches, at times even more, and almost invariably of excellent quality. The lake ice harvested in Minnesota and Wisconsin is almost marvelous in its purity and brilliancy. Ice has been cut in these states during three successive winters, eighteen or more inches in thickness, free from snow or white ice, and clear and transparent as spring water. Lake Superior ice, owing to the beautiful, clear water from which it is frozen, is of excellent quality. It is on record that ordinary newspaper print has been read through a cake of Lake Superior ice twenty-nine inches in thickness. The immense harvests of the Kennebec and Penobscot rivers of Maine and the Hudson in New York, are of national reputation. Ice from these rivers is used largely among the populous coast cities of the East, and before the advent of the ice machine, was used extensively in the Southern states. The shipment of ice south has now practically ceased, and even some of the chief cities of the North Atlantic seaboard now use the manufactured article to a large extent.

Ice of a thickness of from ten to sixteen inches handles well and cuts up economically if used for retailing by wagon—a thickness of twelve or fourteen inches being probably the most desirable. It is not of course always possible to get the thickness desired owing to the exigencies of the weather during harvest. The maximum thickness which is formed in the locality where harvested, also necessarily limits in this direction. In southern locations it is difficult to get ice thick enough, while further north the ice often becomes too thick to handle to best advantage. To get a good quality of ice into the house at a low per ton cost is the serious problem of the ice harvester during the winter. To the end that advantage may be taken of favorable conditions of the weather and other related matters, these should be closely studied.

COST OF HARVESTING AND HOUSING ICE.

No dependable figures that may be relied upon to apply to any specific case can be given as to the cost of ice delivered in the ice house, owing to local conditions, which are of necessity different in every instance. Ice was housed in a Lake Michigan town in Wisconsin some years ago for the seemingly impossible cost of six cents per ton. The conditions were ideal for the cutting of ice, and were as follows: House on lake shore; steam hoist, with low fuel cost, for hoisting ice directly from water into house; no snow to contend with; perfect ice harvesting weather with temperature ranging from zero to twenty degrees above; ice of a uniform thickness of eighteen to twenty inches; labor cost 75 cents per day for experienced men. It may be noted that these exceptional conditions are exceedingly rare, so that the cost as here given probably could not be duplicated at this time, but by taking the above as a basis for calculation, it is possible to estimate approximately the cost of harvesting under conditions varying from the above.

Ice cut and handled during fairly favorable weather and hauled not more than a mile, may be housed in northern latitudes for twenty-five cents per ton, in comparatively large quantities, perhaps somewhat less. Further south, with ice much thinner and contending perhaps with more or less snow, rain, or thawy weather, the cost will be from two to four times as much. Should the house be situated at the ice field, the cost may be reduced ten to fifteen cents per ton, or more, according to the length of haul avoided. It is assumed in these estimates that no haul will exceed four miles.

The cost of hauling ice depends also greatly on whether the ice is hauled on runners or wheels, as a much larger load may be hauled on runners. A fall of snow sufficient for sleighing is therefore a boon to the harvester whose house is located at some distance from the field. The snow must of course be removed from the field, but this is more than offset by the improved facility afforded for transportation. An excessively heavy snowfall, however, may add much to the cost of harvesting, as the ice has to be uncovered. Should a heavy rain follow the plowing and making ready of the field, the rain being perhaps followed by sleet and snow, the ice harvester's lot is not a happy one. Not only must the work be done over again, but perhaps the recent fall of snow must be removed, or the snow ice resulting planed off. Other minor items, like loss or breakage of tools, and contingencies which come up from time to time, influence the ultimate per ton cost of ice delivered in the house.

CARE AND PREPARATION OF THE ICE FIELD.

Before undertaking to harvest a supply of ice the harvester should inform himself regarding the legal and sanitary regulations of his locality. He should be fully satisfied that the field is lawfully his property, and that all Board of Health and other rules are fully complied with. Most of the larger cities and many of the smaller ones have quite stringent ordinances regu-

FIG. 1.—STARTING CHISEL.

lating the harvesting and sale of ice. When used for refrigeration or cooling purposes only and not for family use, ice can usually be cut from any source. Some cities, however, will not allow ice to be harvested for any purpose whatever from waters suspected of pollution by sewage or otherwise.

The selection in the first place and the care of the field prior to harvesting, are both essential for securing a good quality of ice, and an economical cut. The prompt removal of snow from the surface of the field as fast as it falls constitutes the chief labor of preparing the field for harvest. It is seldom that a field of ice freezes sufficiently thick to cut without one or more snowfalls upon it. Flooding, or "wetting down," the ice, is resorted to by some with the first fall of snow, especially in the southern tier of natural ice states. When the ice is in-

FIG. 2.—RING CHISEL.

tended for family trade this process should not be resorted to, as all dirt and impurities lying on the surface of the ice are frozen on and become imbedded in the ice.

The "wetting down" process consists simply in flooding the surface of the ice, which saturates the snow with water so that it may be frozen into ice, protects the under strata of clear ice from thawing weather, and serves to increase the thickness of the ice rapidly. A snow ice coating is also thought to make the cake tougher and less liable to break in cutting and handling. The "wetting down" is accomplished by a gang of men armed with narrow bladed ice chisels. A starting chisel (Fig. 1), or ring chisel (Fig. 2) may be used. The men should proceed in a row across the field, punching holes at intervals of say six feet, and working at a distance apart from six to twenty-five feet, depending on the thickness of the ice and amount of snowfall. A small hole only is necessary. "Wetting down" should be done on a cold, still day, when it is reasonable to suppose that the wet snow will be frozen solid. In comparatively warm climates, where the natural ice crop is

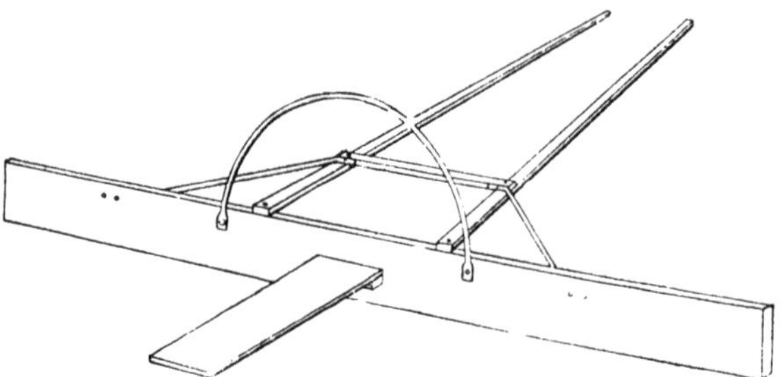

FIG. 3.—HOME-MADE SCRAPER.

precarious, a fall of snow must be dealt with promptly by "wetting down" or removing from the field. As small a quantity as an inch of dry snow greatly retards the freezing, and the surface of the ice should therefore be kept free from the protecting snow blanket.

In northern latitudes flooding is not often resorted to, and the snow is removed largely to prevent the formation of snow ice in case of a thaw or rain. Should a rain come on with snow on the ice, the snow becomes saturated with water, which when frozen makes snow ice. Snow ice also results from a thaw when snow lies on the surface of the field. It will thus be seen that in some localities snow ice is desired and in others it is avoided. Snow ice is porous and white, because it contains air in fine cells. Its presence lessens the selling value of the ice, but does not interfere with its refrigerating value. Perfectly clear ice is desired and readily obtained in the North, but natural ice free from snow is seldom seen in the southern tier of ice states. Any heavy fall of snow must necessarily be removed before the marking out and plowing can be commenced, and a field of ice perfectly free from snow is desirable at all times. An experienced ice harvester will know how to proceed under these different weather conditions and varying stages of the harvest, and these must be taken into consideration at all times if the novice would proceed intelligently.

For the removal of snow from the field, various devices are in use, depending on the magnitude of the work in hand. Good progress can be made on a small field by the use of a hand

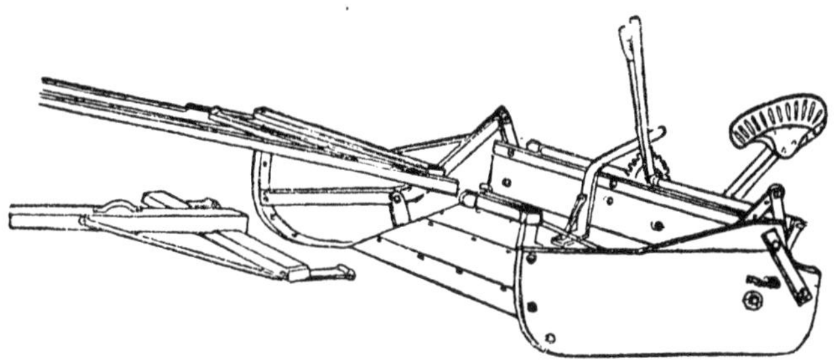

FIG. 4.—IMPROVED SNOW SCOOP SCRAPER.

scraper or large snow shovel, especially where the snowfall is dry and light. This method is also useful where the snow is to be removed from ice which will not bear the weight of a horse. For general use in harvesting small crops in northern latitudes, the home-made scraper illustrated in Fig. 3 will be found of service. It is easily and cheaply made, and can be made of any desired size to suit the work in hand. An oak plank, two or three inches thick and ten to sixteen inches wide may be used, of any length up to twelve or fourteen feet. A piece of ⅛x1½ inch iron fastened to the lower edge will improve the efficiency and wearing qualities greatly. A small scraper of this kind may be fitted with shafts for one horse and the larger ones with a pole for two horses. A small one may be constructed to be operated by two men. A handle of round iron flattened and screwed to plank, as shown, is useful in swinging the scraper into position or in lifting over banks at the dump and as a means of holding on. If preferred, a rope may be attached in a similar manner for this purpose.

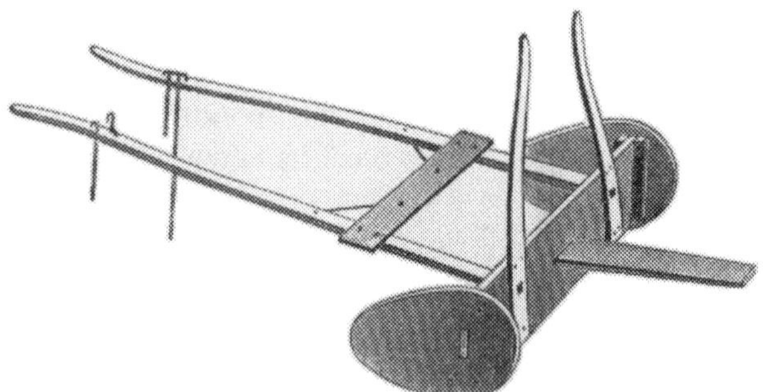

FIG. 5.—SIMPLE SNOW SCRAPER.

The larger and more durable cleaning-off scrapers which are used on larger fields may be purchased from the manufacturers. Fig. 4 illustrates a very good machine for this purpose. Fig. 5 is a common form to be purchased at a low cost. Its

FIG. 6.—SCOOP SCRAPER.

operation is similar to the home-made scraper shown in Fig. 3. Where the snow is heavy or deep the scoop scraper illustrated in Fig. 6 is used. These range from six to eight feet in width, depending on the character of the work. After the heaviest snow is removed the cleaning-off scraper may be put on for removing the loosened snow. Should a thick crust form on the snow some expedient must be resorted to for loosening

FIG. 7.—ICE AUGER.

it, so that it may be scraped; a disc harrow or a modern ice field cultivator will sometimes be found useful for this purpose.

As the snow is scraped from the ice it is generally best to remove it to some distance from the place of cutting, either to

FIG. 8.—MEASURING IRON.

the shore, or far enough from the field to prevent the ice from "flooding," either before or after cutting commences. Where the field is located on a large body of water, the snow is sometimes scraped into piles or windrows known as "dumps." The "dumps" may be hauled away on sleds or with the scoop

FIG. 9.—FIELD ICE PLANER.

scrapers or self dumping scrapers, or they may be allowed to remain on the ice. If allowed to remain on the ice a deep groove is sometimes plowed around the "dump," the weight of the snow causing the ice in this place to break loose and sink beneath the level of the cutting field. This method is not resorted to except on large fields and in case of an exceptionally heavy fall of snow.

A careful harvester will observe the thickness of his field from the time it will safely bear his weight, and will know from day to day the exact thickness he can depend upon, so that when the time comes he may act promptly. The thickness is accurately determined by the use of an ice auger (Fig. 7), and measuring iron (Fig. 8). The measuring iron has inch

FIG. 10.—ELEVATOR PLANER.

marks on it, and is bent up on the end, so that it can be inserted through the hole made by the ice auger and drawn up against the under side of the ice. The thickness of snow ice, if any, may be noted at the top. For more accurate work three holes may be bored, forming a triangle, and slanting toward each other at the bottom; a small saw is used for cutting the triangular plug by sawing from hole to hole.

If sufficient snow ice or dirty ice is present to be a detriment to the quality of the crop, in the northern latitudes, it is generally removed before the ice is housed. This may be done by the use of the snow ice planer on the field, or by the elevator ice planer as the cakes pass up the incline. Where the endless chain elevator is not in use, the snow ice must of course be removed on the field. The field planer (Fig. 9) is used in connection with the marker with swing guide. The planer is usually set to remove two inches of ice at a time, as a

FIG. 11.—ICE CHIP CONVEYOR.

smoother job results than where a deeper cut is made. If it is necessary to remove more than this, a second or third grooving and planing takes place. The best job of planing may be done by using a 21-inch guide on the marker and using a check gauge, by which the groove is cut to the exact depth of the snow ice to be removed. Then the plane being twenty-two inches wide, and the knife set at the bottom of the guide plates, will lap over one inch on the planed portion, removing the marked grooves completely, and leaving the surface as smooth as new ice. An improved ice field cultivator requir-

FIG. 12.—HAND ICE PLOW.

ing no marking has now largely superseded the above method.

A marker with guide which can be adjusted from twenty-

FIG. 13.—LINE MARKER.

two to twenty-one inches is very convenient for use in planing. The chips of ice resulting are removed in the same way that a

FIG. 14.—MARKER PLOW WITH SWING GUIDE.

heavy fall of snow would be. The chips being very heavy make the planing of snow ice on the field a very expensive operation. Where the harvest is of sufficient magnitude to warrant, the use of the elevator planer (Fig. 10) is greatly to be desired. This will remove any thickness of snow ice, reduce the cakes to the same thickness and leave the upper surface of the ice corrugated, which will prevent breakage when removing ice from the house. A chip conveyor (Fig. 11)

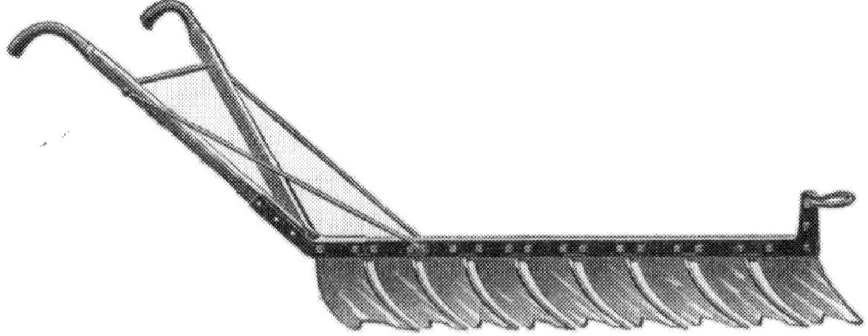

FIG. 15.—FIELD ICE PLOW.

removes the ice chips and slush a distance from the elevator, and is almost a necessity in conjunction with the elevator planer.

HARVESTING THE ICE.

With the field clean, free from snow and of the desired thickness, the marker is put to work. It is best to start the

FIG. 16.—SWING GUIDE ICE PLOW.

marking plow by stretching a strong line between two stakes driven into auger holes in the ice about 200 feet apart, to serve as a guide. As all following marks are made from the first, it is important that this should be straight. A long plank as a straight edge is used to guide the hand plow (Fig. 12) or a line marker (Fig. 13) may be used as a substitute. Either is followed by the regular marker (Fig. 14) with guide which goes over the field, cutting grooves parallel to the first. The marker is used only for the first grooving, the

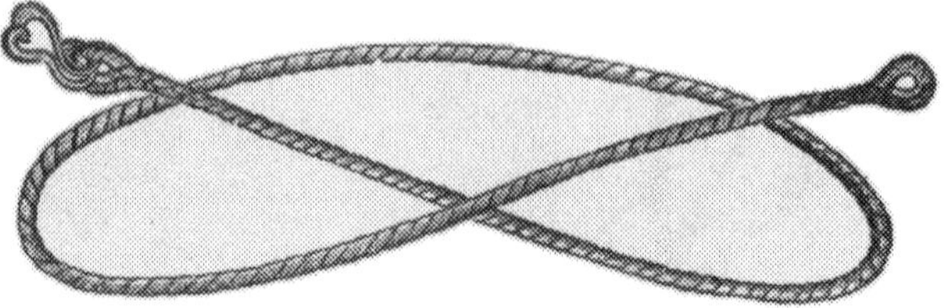

FIG. 17.—ICE PLOW ROPE.

greater part of the cutting being done by the deeper field plow (Fig. 15). In marking out the first groove the operator should take care to hold the marker upright to prevent cutting irregular shaped cakes. After the first groove is made the guide on the marker runs in this groove, gauging the distance of the second. This is repeated over the entire field.

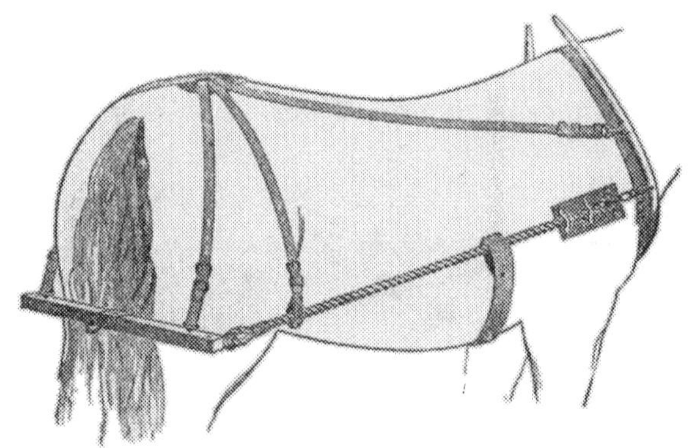

FIG. 18.—ICE PLOW HARNESS.

It is important that the cross marking should be at right angles to the first, or parallel marking, for which purpose a large wooden square ten or twelve feet long is used. By this method it is comparatively easy to have the cakes square. Cakes 22x22 inches or 22x32 inches are the common sizes.

Marking and plowing may be done with one machine, where the ice field is small. The swing guide plow (Fig. 16) is the one used for this purpose. After the ice is marked out the guide is removed and the field plowed over as with the regular field plows. Swing guide plows are generally made with seven teeth and either six, seven or eight inches in depth.

A well equipped harvester has several different plows for the different purposes. Following the marker a six-inch nine-

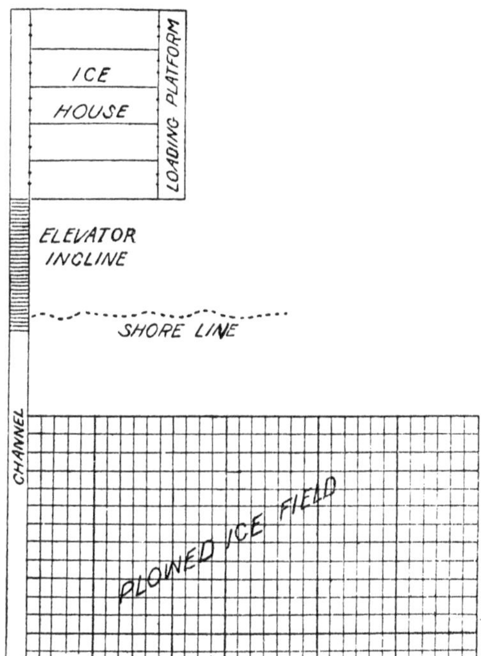

FIG. 19.—DIAGRAM FOR ICE HARVEST.

tooth plow is run in the marker grooves, making these about five inches deep. Following immediately behind is another plow, eight inches deep, with eight teeth, making the grooves seven inches deep. This is deep enough for ten or twelve-inch ice, but if the ice is fourteen or sixteen inches thick, still another plow follows, ten inches deep with six teeth, making the grooves about eight or eight and a half inches deep.

Should the plows be somewhat dull, perhaps this depth is not reached, and a second plowing with the ten-inch plow becomes necessary, probably making the grooves nine or ten inches deep, which is sufficient for ice sixteen inches thick, or even more.

The headlines in which the large floats are to be barred off are run deeper, some of the large companies having a

FIG. 20.—ICE SAW.

twelve-inch, five-tooth plow for this purpose; still others deem it economical to use a fourteen-inch plow on very thick ice. Where the harvest is comparatively small, a number of the plows mentioned may be dispensed with, even to doing the total cutting with the swing guide plow (Fig. 16). A set of plows commonly used by the smaller harvester consists of a marker, an eight-inch, eight-tooth plow and a ten-inch plow. If the ice to be cut does not exceed twelve inches the ten-inch plow may be dispensed with; a marker and a nine-inch

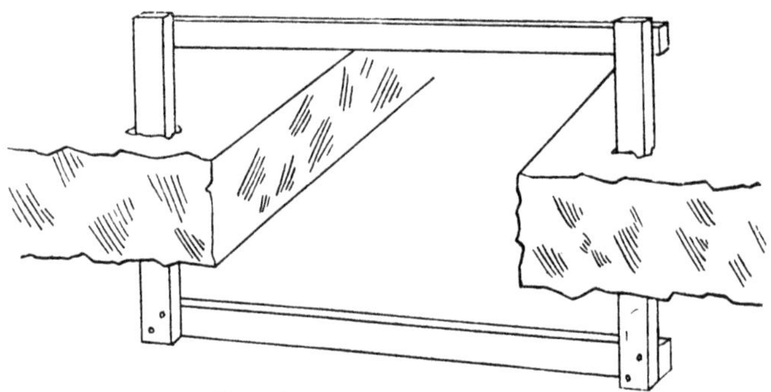

FIG. 21.—CHANNEL BRACE.

seven-tooth plow is used as a set also, and is a favorite with the small harvester.

The plow rope (Fig. 17) by which markers and plows are drawn, should be nine or ten feet long. This prevents the front end of the plow from rising and causing a "chatter," or irregular cutting. Many, however, use shorter ropes or none at all. The regular plow or grooving harness with whiffle-tree well

FIG. 22.—CAULKING BAR.

elevated as shown in Fig. 18, is more convenient and easier on horses than the ordinary harness. Generally speaking, about

FIG. 23.—BREAKING BAR.

half or two-thirds of the thickness of the ice should be cut through by the plow; but not less than four inches of ice should

FIG. 24.—KNOB HANDLE FORK BAR.

be left between the bottom of the groove and the water below. Four inches of solid ice is necessary to safely bear the weight of

FIG. 25.—RING HANDLE FORK BAR.

a team of horses. Too much ice should not be plowed in advance of the housing capacity; enough for two or three days is

FIG. 26.—RING HANDLE SPLITTING FORK.

ample; then in the event of a thaw or rain labor is saved, as the grooves freeze up very quickly.

FIG. 27.—WOOD HANDLE CANAL CHISEL.

No matter how small the harvest of ice, floats of some size are used, as they facilitate the floating of the ice to the channel where they are separated. A float consists of a number of cakes of ice, usually from fifty to one hundred, and if floated some distance they are made much larger. The size on large fields is determined by the deep grooves already referred to as forming headlines for floats. The channel to the elevator extends across the end of the field. The deep grooves for sawing

FIG. 28.—STEEL HANDLE CANAL CHISEL.

are located about twelve or fifteen cakes apart and run lengthwise of the field, while barring-off grooves run in the opposite direction, or parallel to the channel and are from four to eight cakes apart. By barring off the longest side of the float much sawing is saved.

FIG. 29.—KNOB HANDLE 3-TINED FORK BAR.

Fig. 19 shows a diagram of the layout of an ice field, house, channel, etc. The location of the channel for floating the cakes to the incline should, of course, be selected before marking out and plowing the field. The channel should be plowed with a deep groove on each side, and the ice removed by sawing out with the hand ice saws (Fig. 20). Or, if plows are not plentiful, the channel may be sawed out while the field is being plowed. The ice from the channel may be sunk under the ice along the sides of the channel, as it is usually more or less irregular and broken.

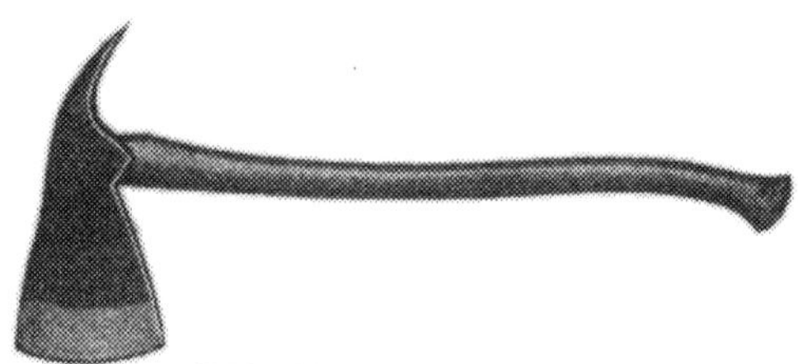

FIG. 30.—HOUSE AXE.

In sawing out for a channel the cakes should be sawed

slightly narrower at the top, so that they may be readily sunk under the channel sides. Any broken or odd shaped pieces which come into the channel should also be sunk in the same manner. This disposes of them easily, and as this broken ice

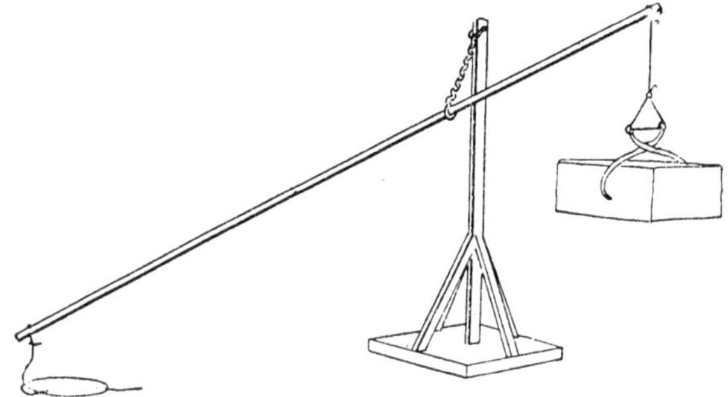

FIG. 31.—SIMPLE ICE LIFT.

freezes to the under side of the ice field it aids greatly in supporting the channel sides, which have a strong tendency to flood from the continued weight and travel. Where the field

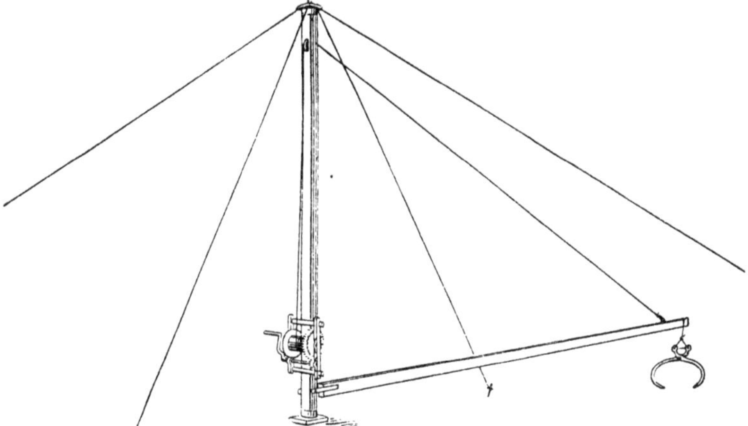

FIG. 32.—WINDLASS OR CRAB HOIST.

is on a river, or where the channel is long, it may be necessary to put braces across to prevent the channel from closing. A simple device of this kind is shown in Fig. 21. It should extend the same distance above and below the ice, and be out of the way of passing cakes. Water sprinkled around the uprights where they pass through the ice will soon freeze solid and make a strong anchorage.

In breaking out the floats from the planed field, it is best to select only a sufficient area for the days' pack. The grooves in the field adjoining this area are calked tightly with chips to

FIG. 33.—INCLINED SLIDE AND TABLE.

prevent the water running into and freezing in the grooves. The caulking bar (Fig. 22) is used for this purpose. With the ice saws the grooves at the end of the selected area are sawn through and a float is broken off by striking into the groove at the back, in several places, with the barring-off tool, or breaking

FIG. 34.—TABLE WITH SLIDE AND DRAW ROPE.

bar (Fig. 23). The fork bars (Figs. 24 and 25) are likewise used for this purpose. The splitting fork (Fig. 26) is also much used for barring off thick ice, and is a general favorite for the purpose, even on moderately thin ice.

The floats at the channel are broken up into strips, or small floats of single or double rows of cakes, and when these are in the channel they are separated into single cakes. For this purpose the channel chisels (Figs. 27 and 28) are used. When the grooves are much frozen the three-tined fork bar (Fig. 29) is used to good advantage. When the weather is frosty and the grooves in good condition the ice will cleave very accurately from top to bottom of the grooves; but if the weather be soft

FIG. 35.—JACK GRAPPLE.

and the grooves badly frozen, it is often necessary, on thick ice, to use the house-axes (Fig. 30) to trim up the cakes. It is only possible to do this on a comparatively small harvest where the ice is hauled out on a table before loading. This house-axe trimming is impossible where the endless chain elevator is used.

FIG. 36.—ILLUSTRATING DIRECT PULL ACROSS TABLE

When trimming with the house-axe it is best to hew the cakes a trifle narrower at the bottom, as the ice will then loosen much easier from the house and with less breakage.

The methods of removing the cakes of ice from the water are so numerous that the ice harvester may easily select the one

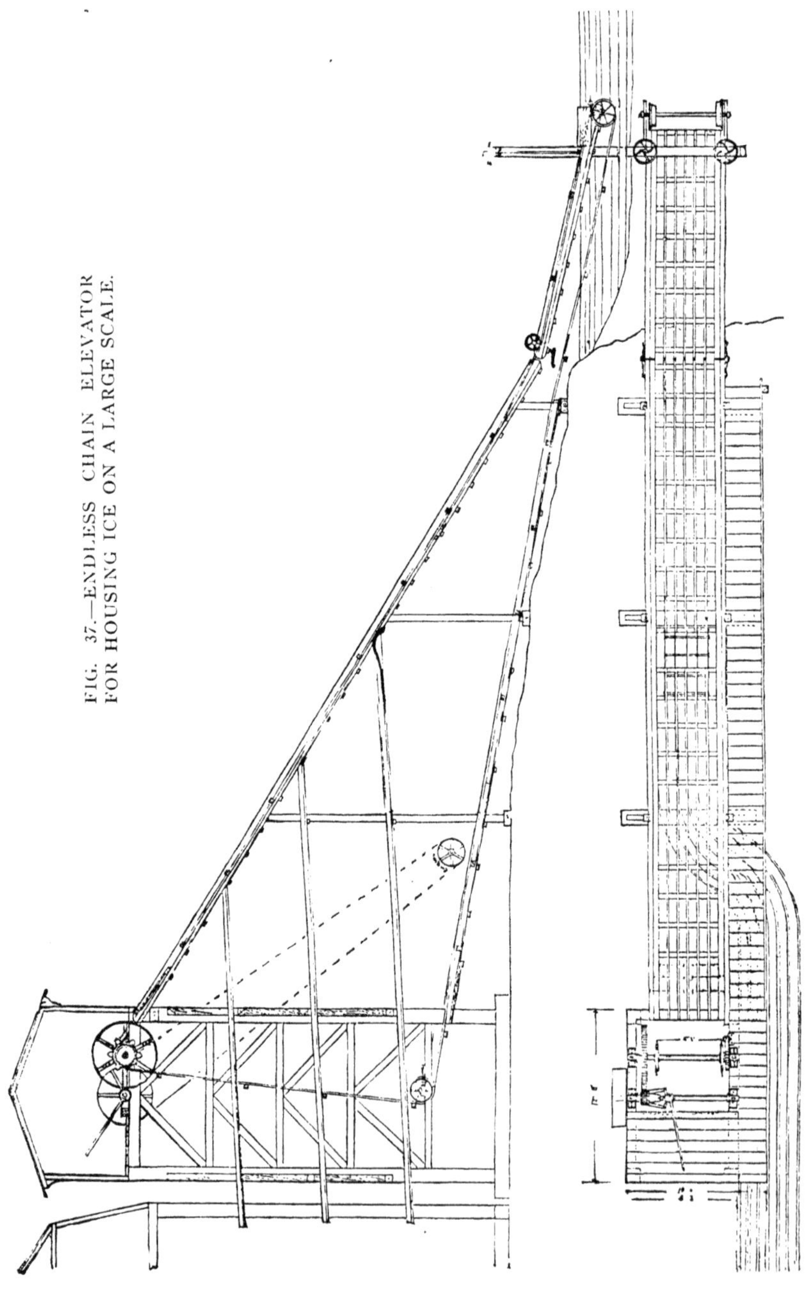

FIG. 37.—ENDLESS CHAIN ELEVATOR FOR HOUSING ICE ON A LARGE SCALE.

best adapted to his needs. For the handling of a small harvest of less than one hundred tons an inexpensive rig must of course be selected, but when housing several thousand tons or more the most improved endless chain elevators make a great saving in the cost per ton. Two men with tongs will pull a small cake of ice from the water, but some simple device is generally to be preferred even for the filling of a farm ice house of ten to twenty-five tons capacity.

A simple and easily portable rig for raising ice from the water and placing it directly on the conveyance is shown in Fig.

FIG. 38.—HOISTING CRAB.

31. It consists of a simple lever or pole, supported on a post set in a base or platform. The lever is supported from the top of the post by a rope or chain giving play enough so that the cakes may be lifted and swung over the sleigh or wagon. The necessary leverage for lifting any size of cake may be obtained by adjusting the chain at the required point on the pole. A rope attached to the long end of the pole enables the operator to secure a lift which would otherwise be impossible. Fig. 32 shows a rig frequently used, especially in some parts of the West. It will raise the ice with little effort and deposit it directly on the conveyance, but has the disadvantage of not being easily transported, and is very slow in action.

The inclined slide and table (Fig. 33) is the most common device in use for removing the cakes from the water and placing them in a position to be easily loaded. Two active men with ice hooks will pull out on the table a great many cakes per day, but quite often a horse is employed, in which case a draw-rope is used, that passes through a pulley fixed to a cross-bar above the table (see Fig. 34). The jack (Fig. 35) is also used for this work. Sometimes the horse or horses pull directly across the table without using the pulley; two horses, working both ways and using a grapple on both sides of the incline, will haul out a

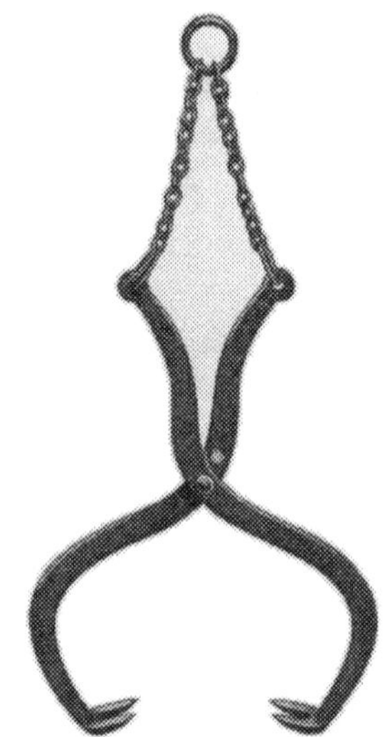

FIG. 39.—HOISTING TONGS.

surprising number of cakes, enough to keep busy a large num- of teams. Fig. 36 shows a good arrangement of table on shore and a direct pull across the table. Where a table is used, it should, to facilitate handling, be slightly higher than the conveyance.

HOUSING AND PACKING THE ICE.

The endless chain elevator already referred to, may be purchased from the manufacturers with almost any variation to fit individual needs, and is a necessity for the economical housing of ice on a large scale. Fig. 37 shows an apparatus of this kind.

Some of the large companies harvest and place ice in the houses at an almost incredible speed with these improved facilities. It is on record that 720 tons of ice per hour have been transported from the water to the houses by a single apparatus.

Where ice is hauled from the field to the house, the simplest method in use for elevating into house where a very small amount is stored, is the inclined slide, up which the ice may be pushed by two men with ice hooks. The hoisting crab (Fig.

FIG. 40.—SINGLE GIG ELEVATOR.

38) with hoisting tongs, (Fig. 39), together with the slide, may also be used, or the single gig elevator as shown in Fig. 40. In this cut it is shown raising ice directly from the water. It is also well adapted to handling ice delivered by conveyance. A double gig elevator, operated by means of a hoisting engine, makes a first-class rig for moderately large houses, and where the amount of ice is sufficiently large, the regular endless chain elevator with bars, same as used for removing ice from the

water, is largely in use. Hoisting tongs (Fig. 39) are in some localities largely in use for housing ice, and are used for lifting cakes directly from the water to the chute conducting it to the house; usually two pairs of tongs are arranged so that one pair goes down as the loaded pair goes up. This is a comparatively slow process, but it is a good outfit where small quantities are handled.

FIG. 41.—STARTING CHISEL.

The method of storing ice in the house should be governed by the purpose for which it is to be used. If the ice is to be used for cooling purposes in the old overhead ice cold storage house, and none of it to be removed, it should be packed as closely as possible, and the joints between the cakes calked or packed with chips, using the calking bar already illustrated in Fig. 22. This method is satisfactorily employed where the ice is not to be removed from the house, but in other cases it is not

FIG. 42.—EDGING UP TONGS.

to be recommended, as the ice freezes together quickly as soon as the top tier begins to melt. When the ice is to be removed from the house it is best not to pack it too closely.

There are several ways of packing, any of which will make it possible to remove the ice from the house with very little labor or trouble. Where the ice is quite thick the cakes may be hewn narrower at the bottom, as already suggested, and the cakes stored as closely as they will pack. With thinner ice it is best to leave a space of one to three inches on the sides of the cakes all around. Care must be taken to have the seams in a straight line in each direction. The starting chisel (Fig. 41) is useful for this purpose. Should the cakes be of different thicknesses, as when harvested from a running stream, they should be adzed off to an even thickness, if this work has not already been done by the snow ice plane or the elevator planer.

No matter what method of storing is used, the successive tiers of ice should be so placed as to break joints, the object being to bind the ice into one solid body and prevent it from caving or spreading. If this simple rule is followed, pressure on the sides of the house is avoided. Disastrous results have followed the careless packing of ice. Ice 22x32 inches is very good for breaking joints, as one tier may be placed in one direction, and the next in the opposite. Where the 22x22 inch cakes are stored, it is best to harvest some double-sized cakes for binding purposes. Many harvesters do not break joints oftener than every six or eight feet but "every tier broken" is better and safer. Where some kind of covering is used, usually the two top tiers of ice in the house are packed closely together to prevent the covering from working down into the seams. The more modern method of ice storage is to have the rooms insulated in the floor, side walls and ceiling and then no covering of any kind is necessary on the ice.

Some harvesters pack ice largely on edge, placing only enough on the flat side to form a binder to prevent the ice from moving. The small edging-up tongs (Fig. 42) are much used for this method of storing. The main advantage claimed for edge storing is that for a given space used, ice will loosen much more easily from the house and with less waste. One tier on edge and one flat makes a good combination for easy loosening.

For covering ice in the old style house, shavings, sawdust, straw or hay is used. Salt or marsh hay is thought best for the purpose. Ice dealers, as before stated, sometimes use covering material, but for cold storage uses it is not customary and really undesirable on account of the extra labor required.

It should be borne in mind in every case that where ice is to be removed from the house for sale or use, chips made in the house during the filling of same should be thrown out and not chinked into the ice. Where ice is chinked the chips melt first, running down into the seams of the lower tiers, freezing there and forming a solid body of ice, difficult to remove without much labor and breakage.

The prevailing idea that thick ice will keep better and longer than that which is comparatively thin, is erroneous. Regardless of the thickness of the ice, the cakes in the interior of the pile do not melt until exposed to the action of the air, the meltage being almost wholly on the top, sides and bottom of the mass. When ice is put into the house in quite cold weather, it will take the temperature of the outside air when exposed during transit to the house. If the house is filled with ice at the temperature of the air, say at 20° F., the first ice to melt is at the top of the house, and the water from the meltage runs down into the joints between the cakes of ice lower down in the pile. These being at a temperature somewhat below the freezing point of water, the meltage from above is frozen into ice, in some cases cementing the cakes into a solid mass, as above described. Ice removed from the interior of the house in the fall generally shows no signs of meltage whatever.

TOOLS FOR HARVESTING AND HANDLING ICE.

The following lists are given as a guide to those who are unaccustomed to cutting ice. The five lists here given, with the size of the harvest for which each is suited, are offered as a basis on which the new beginner may form an estimate for his own particular conditions.

Set No. 1.—Suitable for use in harvesting up to 100 tons.

1 ice plow with swing guide.
1 splitting chisel.
2 ice hooks.
2 pairs ice tongs and 1 4-foot saw.

Set No. 2.—Suitable for harvesting 100 to 1,000 tons.

1 ice plow with swing guide.
1 breaking bar—pad end used as calking chisel.
1 splitting chisel.
1 4-foot saw.
1 grapple—to raise up incline—or 1 market tongs if sweep arrangement is used.
1 plow rope.
1 line marker.
2 to 6 ice hooks.
3 tongs.

Set No. 3.—Suitable for harvesting 1,000 to 2,000 tons of ice, using six to ten men and two horses; hoisting with one grapple.

1 8-in. swing guide plow.	1 plow rope.
1 breaking bar.	1 line marker.
1 calking bar.	2 to 3 doz. 4½-ft. ice hooks.
1 bar chisel.	1 to 6 doz. 6-ft ice hooks.
1 No. 2 splitting chisel.	1 to 12 doz. 14-ft. ice hooks.
2 5-foot saws.	1 12-in. top gin.
1 grapple and handle.	1 12-in. wharf gin.

Set No. 4.—Adapted for harvesting 2,000 to 5,000 tons of ice, using ten to fifteen men and three or four horses; hoisting with two grapples.

1 3½-in. marker, 22-in. Sw. Gd.	2 grapples and handles.
1 9-in. plow (or 8-in.).	2 plow ropes.
1 No. 1 splitting fork.	1 line marker.
1 breaking bar.	1 doz. 4½-ft. ice hooks.
1 calking bar.	1 to 6 doz. 6-ft. ice hooks.
2 bar chisels.	1 to 6 doz. 14-ft. ice hooks.
1 No. 1 splitting chisel.	2 12-in. top gins.
3 5-ft. saws.	2 12-in. wharf gins.

Set No. 5.—Outfit for harvesting 10,000 to 15,000 tons of ice, or more, engaging, say, fifty men and four horses; hoisting with incline elevator, and filling three chambers at once.

1 3½-in. marker, 22-in. sw. gd. (extra 32-in. guide for 22x32-in. ice.) (Extra 44-in. guide for 22x44-in. or 44-in. sq. ice.)	6 bar chisels.
	1 No. 1 canal chisel.
	2 No. 2 splitting chisels.
	6 5-ft. saws.
1 6-in. 7-tooth plow.	4 plow ropes.
1 8-in. 7-tooth plow.	1 scoop net.
1 10-in. 6-tooth plow.	1 auger.
1 6-in. hand plow.	1 measure.
2 No. 1 splitting forks.	4 doz. 4½-ft. ice hooks.
1 No. 1 fork bar.	1 to 4 doz. 8-ft. ice hooks.
2 calking bars.	1 to 2 doz. 12-ft. ice hooks.

The quality or number of tools required is largely governed by the speed with which it is desired to harvest the crop. The sets listed above are for average work; if fewer men are employed the sets may be decreased, and for rapid work increased. It is of course desirable to get the ice housed as quickly as possible to avoid changes in the weather, snows, etc. Many, however, prefer to harvest slowly, with a small crew of men, so as to keep their hands at work during the winter, in which case, of course, they run the risk of having their ice break up because of mild weather before they have their houses filled.

CHAPTER XXXVIII.

ICE STORAGE HOUSES.

STORING ICE AND SNOW IN PITS.

By freezing, water expands so that eleven volumes of water become about twelve volumes of ice. Consequently the specific gravity of ice is less than that of water, and ice will float on water. When water is transformed into ice its temperature is not changed, but remains at the "freezing point" so long as it remains in contact with water. So also when ice is melting, the temperature remains at 32° F. until all the ice is transformed into water. By freezing, the latent as well as the sensible heat of the liquid is liberated, and when the ice melts a certain amount of heat is absorbed, being taken from the surroundings.

Snow is equal to ice in refrigerating value, and a pound of dry snow has the same cooling effect as a pound of dry ice, but if the ice or snow contain water, their cooling effect is correspondingly reduced. If, for instance, one-tenth part of the ice is water, there only remains nine-tenths to be melted, and the cooling effect is reduced correspondingly. Usually, however, ice harvested in a thaw does not contain to exceed 3% of water, and its cooling effect is nearly equal to that of dry ice. On the other hand, "frozen ice" (ice below the freezing point of 32° F.) requires but one-half the heat required by water to raise it to the freezing point.* Even during a hard frost the ice on the surface of the water is only at 32° F., and while harvested it is more or less submerged in water at 32° F., so that its temperature will rarely be much below the freezing point, except when carted for long distances in very cold weather. Supposing it is put into the ice house at ten degrees below the freezing point, it only takes five heat units to bring it to the freezing point,

*So stated by the late Prof. N. J. Fjord, of Copenhagen, Denmark.

and its cooling value is therefore only equal to one-half that of water through the same range of temperature. It follows that it is of comparatively small moment whether ice is harvested in a thawing or freezing condition. The difference in its value varies only about 5%.

It is more important, however, that the ice be packed closely in the house. A solid block of ice, a foot cube, weighs

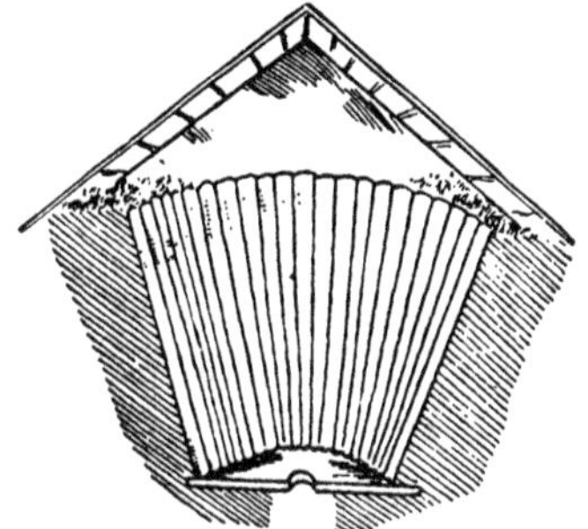

FIG. 1.—INTERIOR OLD STYLE ICE CELLAR.

about 57 pounds, but a cubic foot of the ice house will hold only of:

Ice thrown in at random, about........................30 to 35 lbs.
Ice thrown in and knocked to pieces..................35 to 40 lbs.
Ice piled loosely.......................................40 to 45 lbs.
Ice piled closely and chinked with fine ice...........45 to 50 lbs.

The limits in ordinary practice are usually between 40 and 50 pounds, a difference of 20%. The same amount will melt in the ice house whether the ice is packed loosely or carefully. Suppose 15 pounds per cubic foot would melt in the summer,

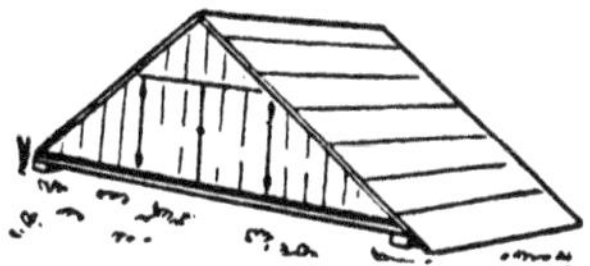

FIG. 2.—ROOF OF OLD STYLE ICE CELLAR.

there would be left only 25 pounds, where there was originally 40 pounds, but 35 pounds when 50 pounds were stored. The difference in the ice left would therefore be 40%. So it is evident that it pays to pack the ice well and fill the house to its utmost capacity, consistent with ease in removing, cost of the ice, and the purpose for which the ice is to be used.

EVOLUTION OF THE MODERN ICE HOUSE.

The common use of ice is comparatively recent, and the modern ice house is therefore of recent development. History

FIG. 3.—MODERN ICE PIT.

records that the Romans made use of a form of underground cellar or pit to preserve snow, which was used for cooling beverages during the heated term. A similar receptacle has been used in many places in this country, especially in the South, and may still be met with in remote and thinly settled neigh-

FIG. 4.—CONSTRUCTION OF MODERN ICE PIT.

borhoods. Figs. 1 and 2 show the outline of the construction adopted in the old style, and the construction of the modern ice pit is seen in Figs. 3 and 4.

PRIMITIVE CONSTRUCTIONS.

The first commercial ice houses were built below the surface of the ground, but at present all are constructed above ground, for the reason that drainage is more easily secured, and the ice is more easily removed from the house. The protection afforded by the earth is of comparatively small value when the disadvantages of storing below ground are taken into consideration. Nevertheless, in places where ice forms only one, two or three inches in thickness, or where snow is housed to be used for cooling purposes, the ice pit has its sphere of usefulness. Mr. J. W. Porter, of Virginia, gives the following interesting information, which, among other things, shows that one of our most esteemed presidents was a progressive and up-to-date man:

> Pits are dug in the ground, of such size and depth as is desired to hold from thirty to fifty loads of ice. The shape is an inverted truncated cone. The walls are lined with slabs of wood, split or sawed, or they may be walled with brick or stone. My own is 14 feet deep, 18 feet across the top and 10 feet across bottom, walled up with stone and then lined with boards standing on end. A one-story tool room projecting beyond walls two feet is erected; a very common way is to have a half pitch shingle roof start from sills laid outside of walls, with door to pitch in and take out. After filling, it is leveled fine and filled with clean straw or forest leaves. The ice is rapidly gathered in pieces and shoved in from the wagon, with much less labor than cutting, laying and packing which would be impracticable. Sometimes when ice is not produced, great snow balls are rolled and pitched in and trodden. Upon "Issentiallo," where Jefferson lived and died, within a rifle shot from where I write, is such an ice house, built by Jefferson, which is 54 feet deep and is still used for ice or snow.*

In packing snow in the ice house, it is advisable to have it thoroughly wet when it is put in. More cooling material can be packed into the same space when the snow is wet all through than when dry and frozen, because it may be tramped together and packed more nearly solid. It is then possible to get 50 pounds of wet snow into each cubic foot of space, 44 pounds of which is dry and as durable and good in every respect as 44 pounds of solid ice. Many people think that the snow should be frozen, but that is a mistake. If it is dry, wet the snow as it is stored or wait until it rains. When it is thoroughly wet it is time to harvest and pack it.

The water is expelled by trampling, and drains off, leaving comparatively dry cooling material, which is as effective and keeps practically as well as an equal amount of dry ice. One active man can pack and trample together 500 to 1,000 cubic

*From Green's "Fruit Grower."

feet of snow a day, and with this insignificant amount of labor, snow may be used to the same advantage as ice. On the other hand, of newly fallen, light snow thrown into the ice house and carefully trampled, only 25 to 30 pounds can be packed within a cubic foot, and it will keep no longer than wet snow.

Ice may be put up and protected from the heat of summer at very small expense. The simplest method is "stacking," which consists simply of piling up the ice and enclosing it with a fence-like structure, leaving space between the ice and the fence for a couple of feet of sawdust or other filling material. No roof is put on, but the ice is covered with a goodly quantity of the same material that is used for the sides. This method is not practicable for a small harvest as the wastage is too great, but where a thousand tons or more are put up in this way, the meltage is sometimes surprisingly small, perhaps no greater than 20% to 30%. Some ice dealers fill their houses and put up a certain amount in stacks as well, using the stacks first. This method is, of course, only possible where ice can be cut of sufficient thickness to tier up regularly and could not be used where it was desired to put up thin ice or snow, as with the ice pit. It is at best only a crude makeshift and can not be recommended except in case of insufficient capacity or temporary disabling of ice house. Sometimes it is desirable to put up ice in this way while awaiting the completion of the cold storage house in which it is to be used.

By the smaller users a variety of means are employed. We have known farmers who selected sloping ground that would have good drainage and then put down some old rails and covered them well with straw. On this foundation the blocks of ice were placed, and when the weather was freezing they would pour water over the ice and thus freeze the entire mass into one huge block. This was then well covered with straw and boards and a temporary roof put over it. Ice thus packed on the north side of the barn by one farmer furnished the family ice for making butter and ice cream during an entire summer. Ice may be kept piled in a heap on a 2-foot thick layer of sawdust or peat, and covered with the same material.

CONSTRUCTION AND INSULATION OF ICE HOUSES.

It is a common idea that the insulated walls of an ice house should have air spaces, which if "dead"—that is, all connection with the outside air prevented—are supposed to be fully as good or even better insulators than the same space filled with sawdust or other filling material. This is a mistake. In a vacant space between a cold and warm wall, a circulation of the air will always take place, conducting the heat from the warm wall to the cold one. If such space is closely packed with dry chaff, sawdust, mill shavings, or a like material, the circulation, while not entirely prevented, is greatly retarded. Of course tight walls effectively stop circulation and prevent, to an extent, conduction, and several partitions of paper or boards in the wall are therefore useful, but the "dead air space" itself is of comparatively small account. It is important that the insulating material with which the space is filled, should be dry, and however well it is packed, there will always be a slight circulation, the air passing down along the cold side of the wall, and up on the outer or warm side, and unless the outer surfaces of the wall are air tight, moisture will find its way in, will be deposited on the cold side of the wall, and will gradually saturate the insulating material. In such cases it may be advisable, if convenient, to take the insulating material out occasionally to be dried before it is replaced, or it may be entirely renewed. The moisture which collects in the material nearest the inside wall, is generally supposed to pass through the woodwork from the ice. It is, however, really due to a circulation of air, as stated, and which can not be entirely prevented. The reader is referred to the chapter on "Insulation" for further information and details.

FILLING WITH ICE.

In filling an ice house care should be taken to have the ice piled in such a manner that in melting or shrinking it will not press upon the walls. This is easily accomplished by having the floor slightly pitched towards the center of the house, then there is less danger of ice sliding towards the outer walls. Disastrous results have sometimes occurred from this cause. If covering material is used on top of the ice it should be inspected frequently and any holes found must be filled at once. Bad meltage toward the center of the pile may cause a portion of the ice to break away and damage the house.

In refilling an ice house or the ice room of a cold storage plant, it is best to cut away any portion of ice remaining in the room which has melted in an irregular way, and remove it from the house. This applies to the top layer of ice and the sides. Fill around the old ice with the new, adzing off so that both are level at the top and form a level bed on which to begin refilling. Do not attempt to fill up the spaces left from meltage by throwing in irregular shaped pieces or fine chips, as they have no sustaining power and when the weight comes on them will settle and may result in a wrecked or badly sprung building. A case is in mind where the chips and loose ice were used for filling and after the house was filled it was found necessary to remove a considerable amount of ice at great expense and stay up the front of the building with heavy timbers. This job cost nearly as much as the total cost of filling the house with ice.

WASTE OF ICE IN HOUSE.

Waste of ice in an ice chamber is largely caused by meltage from the top, the sides and bottom. Under proper ice house conditions no serious waste ever takes place inside a pile of ice. The melting from the sides, bottom and top is caused by incomplete insulation. During the summer in some houses in Denmark (the experiments on which the following figures are based were made in Denmark, and in applying them to this country proper allowance must be made for difference in climatic conditions; they are too high for average conditions in natural ice territories of the United States) the waste from the bottom may vary from one foot to five feet according to more or less careful insulation. If the ice house is provided with an absolutely tight floor, laid on a thick layer of dry sawdust, the bottom waste rarely exceeds eight to twelve inches during the year. On the other hand, if the ice is piled in the house on the bare ground the waste may reach five feet. Placed on a layer of two feet (after being pressed together by the weight of ice) of sawdust or peat, the ice heap will not be wasted from the bottom to the extent of more than one to one and one-half feet. The causes of waste from the top and sides are, first, circulation of air; second, penetration of heat through walls and loft.

Circulation of air is produced by cracks or openings near the floor through which cold air escapes, being replaced by warm air entering at the top of the house and striking the ice on its downward passage. Such circulation is prevented by having the walls as tight as possible, especially near the bottom. It is of less consequence whether the house is more or less tight at the top, if only the cold air can not escape at the bottom. This fact also shows the importance of having the door or doors to the ice house as high up on the walls as convenient. In a well-built ice house but little waste is caused from a circulation of air coming into the house from the outside.

The main source of waste is the penetration of heat through the insulated walls. Experiments have shown that in ice boxes of the same construction and all exposed alike, the ice melted in the following proportions according to the insulating material used, chaff (cut-up straw) being considered the standard, and the ice melted in the ice box insulated by that material being expressed by the figure 100:

Material	
Cotton dried in a warm room	79
" " on a loft	88
Husk of barley, dried on loft	90
Husk of wheat, dried on a loft	92
Husk of oats, dried on a loft	94
Leaves " " " "	96
Chaff " " " "	100
Husk of rice " " " "	101
Wheat straw " " " "	110
Saw-dust " " " "	114
Peat, dry " " " "	116
Saw-dust, green	170
Peat, moist	260
Saw-dust, thoroughly wet	260
Peat " "	320
Loam " "	560
Sand " "	630

From this it is evident that the more moisture there is in the material the better it conducts the heat, and the poorer it is as insulating material. The difference in the value, as non-conductors, of the materials usually at hand is comparatively small, so that material should be used which is most easily procured, be it husk of any grain, or chaff, or sawdust. Only see to it that it is dry. For the bottom under the ice, however, chaff or leaves, or husks should not be used, as these easily ferment, develop heat, and rot. Sawdust or mill shavings is usually the best available material for the bottom layer. Branches of spruce or the like may also be used to advantage.

The waste from top and sides of the pile of ice depends upon the temperature outside and upon the proportion of surface inside of the house as compared with the ice capacity. As the result of many careful experiments with large and small ice houses, Professor Fjord, of Denmark, established a law according to which the daily waste in a well built ice house, for every 100 square feet of inside surface is 1.7 pounds for each degree Centigrade of average heat. Thus in a house of 1,000 square feet inside surface, in thirty days of an average temperature of 15° C. (59 F.) the waste would be 1.7 pounds x 15 x 30 x 10. equals 7,650 pounds.

If there is 45 pounds of ice to the cubic foot, the waste would be 170 cubic feet, and if there is only 35 pounds to the cubic foot, the waste would be 220 cubic feet. In Denmark, the yearly waste would be about 45 pounds for every square foot of the inside surface, or if there is 45 pounds to the cubic foot, 1⅛ cubic feet; and with only 35 pounds to the cubic foot, 1 2/7 cubic feet.

If the house is filled with ice to its fullest capacity, the balance of ice left, i. e., the house full, less the yearly waste, which represents the ice that can be taken from the house during the year (it makes very little difference whether much or little is taken first or last, provided some is to be kept the year around, for whether there is much or little left in the house, the amount melted in a day is practically the same) varies according to the size of the house, and it may be calculated from the accompanying table, the headings of the last three columns representing the amount of ice packed in each cubic foot of the house.

In this table the waste from the bottom is calculated at one foot. If the bottom is poorly insulated, more waste should be calculated, as mentioned before. Supposing the bottom waste to be two feet instead of one foot, an additional waste of 8 1/3% of the ice harvested must be expected in a pile twelve feet high.

	Inside of Ice House.			Balance Left in a Well-Built Ice House, Cu. Ft.		
No.	Dimensions, Feet.	Surface, Sq. Ft.	Volume, Cu. Ft.	45 Lbs.	40 Lbs.	35 Lbs.
1	10x10x10	600	1000	400	325	229
2	12x12x10	768	1440	672	576	453
3	13x13x12	962	2028	1066	946	791
4	15x15x12	1170	2700	1530	1384	1196
5	18x18x12	1512	3888	2376	2187	1944
6	20x20x12	1760	4800	3040	2820	2547
7	25x20x12	2080	6000	3920	3660	3326
8	30x25x12	2820	9000	6180	5828	5374
9	40x25x12	3560	12000	8440	7995	7423
10	50x25x12	4300	15000	10700	10163	9471
11	60x25x12	5040	18000	12960	12330	11520
12	80x25x12	6520	24000	17480	16665	15617

By means of these figures it is possible to calculate the size of an ice house needed for any purpose in which the amount of ice required is known. In the United States a waste of more than 20% is considered excessive, and in the larger houses from 10% to 15% is commonly figured. Professor Fjord's figures here given represent too great a wastage, but as they are the only known data obtainable they are used as a basis and are given for what they are worth.

SIMPLE FARM ICE HOUSE.

For a good simple plan for a farm ice house, that given below has been designed by the author. It will be found cheap to construct and thoroughly practical. The advantages of a supply of ice on the farm which will last through the summer are well understood by those who are provided with an ice house. Those who have never put up ice should arrange to do so during the next harvesting season.

Once tried, and the advantages of a supply of ice in hot weather experienced, it will become a permanent rule to house ice every winter. A systematic course can then be followed, and the use of labor-saving tools and methods which expedite the work employed. While securing the ice is the chief consideration, no one should be content with anything short of the best methods obtainable; this is a necessity during mild winters, when the crop must be secured speedily or not at all.

The ice house as here illustrated in Fig. 5, by plan and section, is twelve feet square outside and eleven feet high. After

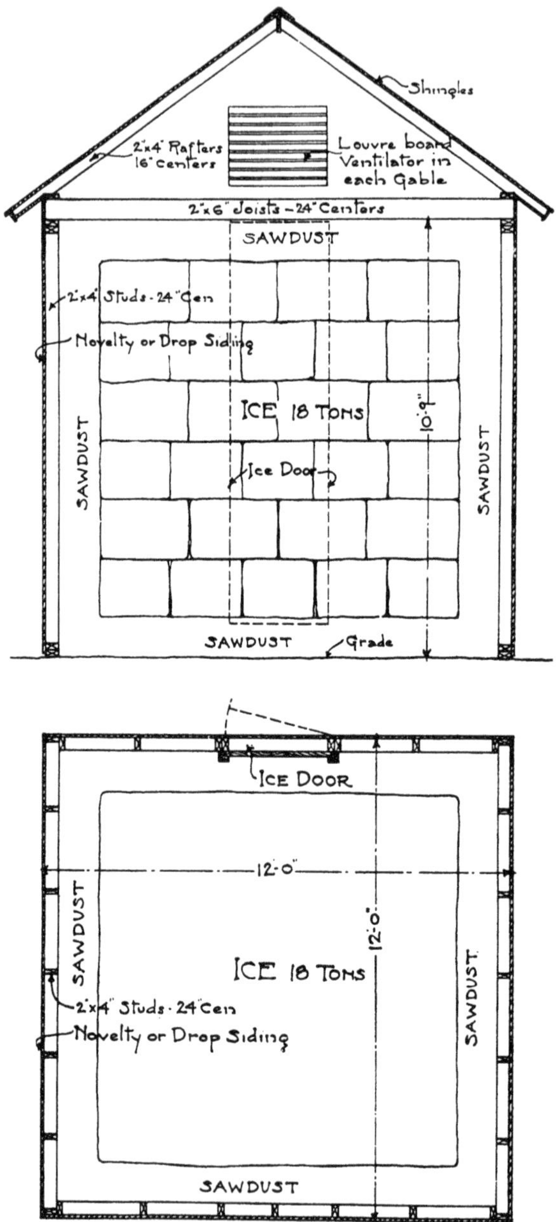

FIG. 5—SECTION AND PLAN FOR SIMPLE FARM ICE HOUSE.

allowing for a foot of sawdust or other filling material at top, bottom and sides, about eighteen tons of ice can be stored in it. If the house is to be built on a sand or gravel soil where drainage is good, no precaution need be taken in regard to the drainage. If, on the other hand, the house is built on a clay soil, it would be advisable to excavate a few inches and fill with coarse gravel or pounded stone, and if necessary, a porous drain tile may be laid through the center of the house and carried to a low place outside for conducting away the meltage from the ice.

The sills consist of double 2 x 4's on which are erected 2 x 4 studding, 24-inch centers. These are topped with a double plate of two 2 x 4's on which rest 2 x 6 joists, 24-inch centers. The studs are boarded up outside with novelty or drop siding. There is no inside boarding, the sawdust being allowed to fill the space between the studs. The roof is constructed of 2 x 4 rafters, 16-inch centers, boarded and covered with shingles. In each gable is a louvre or slat ventilator for the purpose of allowing free circulation of air. One of these may be made removable or hung on hinges to allow access for covering the ice with packing material. The ice door should be built in two or more sections hinged to open outwardly.

On the inside, pieces of 2-inch plank are placed to keep the sawdust or other filling material away from the outer doors. As the ice is removed from the house the pieces of plank are removed as necessary. The actual material for constructing this ice house will cost, under average conditions, from $30 to $40, and after figuring labor, the entire cost of the house should not exceed $50.

This plan is subject to modification as to size and construction. By using 2 x 8's or 2 x 6's for studs a larger house may be constructed with the plan otherwise unchanged. If the joists above ice are objectionable, they may be omitted by using heavier studs and rafters, in which case the ice door is extended up into gable, making top sections of door in the form of a slat ventilator.

FILLING THE HOUSE.

If ice-cutting tools are not available, it is no reason why

you should be discouraged. With two or three cross-cut saws, an axe or a pointed bar, two or three ice hooks and a pair of tongs, the house can be filled. It is desired to secure a more extended kit of tools for next season, it would be well for several farmers to combine and exchange work in filling their ice houses. (See chapter on "Harvesting, Handling and Storing Ice" for further particulars.)

The standard size of an ice cake is 22 by 22 inches or 22 by 32 inches. From 40 to 50 cubic feet of ice-house measure will represent a ton. If the ice is packed solid, 40 feet is correct, whereas if it is packed an inch or two apart, as some prefer, 50 feet is about right.

When filling ice into the house about a foot of sawdust, chopped straw or mill shavings should be placed under the first tier of cakes.

Leave at least a foot of space inside the studding all around, which should be filled with packing material as the ice is put in, and it is also advisable to fill on top of the ice with a foot or more of sawdust or up to the top of the joists. It is advisable to cut the cakes of ice as regular in shape as possible, oblong rather than square. In this way each alternate tier can be reversed so that the joints will be broken, as shown in section. This will bind the ice together and prevent it from sliding or breaking apart.

As the ice is removed from the house, see that the remaining ice is kept covered with sawdust, and if any holes appear, fill them at once. If dry sawdust is not available, straw, marsh grass or mill shavings or tanbark may be used. Whatever is used should be tramped down solidly.

Ice houses are sometimes built with double walls, with a space of one to two feet wide between, firmly filled with dry sawdust or other similar material. It is cheaper and serves the purpose well to pile the ice without a floor directly on the ground, on a thick layer of sawdust or brush wood. Good drainage must be provided, though sometimes in a porous soil no direct outlet for water is needed. The outside wall should rest on a stone foundation built up slightly above the ground, on the top of which the wall may be built of studding and matched boards, the inside wall should also be made of matched boards, the space between being filled with closely packed insulating material. On the beams a loft floor should be laid of loosely packed boards, which may be removed while the house is being filled with ice. On the loft floor a layer of insulating material should also be laid. The entrance should be placed near the top of the wall and be provided with double doors which may be furnished with windows to allow light to enter.

It often occurs that an ice house may be placed inside of another building, for instance in the corner of a barn. Instances are known where ice has been successfully kept in a hay mow or under a straw stack, by providing an opening with double doors. Where a room is built within a building, it is best to construct the sides, floor and ceiling double, with some non-conducting filling material between. The walls should be at least one and one-half feet thick, and the ice inside of the room protected by a foot or two of marsh hay or clean straw. This is to be kept in place as the ice is removed. In building inside of a structure used for other purposes provision should be made for draining away the drip from the melting ice, so that this may not serve to rot the timbers or injure the foundation of the building.

The following is a description of a house which has given good service to its owner: The ground was tiled thoroughly for drainage and a shallow surface gutter made all around the outside. The foundation was hollow, square building tiles, 10 x 10 x 36 inches, being used. The sills, 2 x 8, were doubled and lapped and well bound at the corners. The 2 x 6 studs were toe-nailed to the sills, so that the sills projected two inches over studs on the outside; girths, 2 x 4, were spiked two and one-half feet apart horizontally and flatwise on the studs, so as to be flush with plates and sills. The weather boards were put up and down and battened. The lining of 1 x 12-inch boards was nailed horizontally on the studs so an 8-inch air space was left, and one inch of said space left open at the top for the escape of warm air. (The author believes that if this 8-inch air space was filled with a good insulating material like dry sawdust or mill shavings better protection would be afforded.) For free circulation and to accelerate the escape of hot air a ventilator was placed in the middle of the ridge of the roof and an opening left in each gable end close under the roof. The door extended from three feet above the ground level to the level of the eaves, and was placed on the up-hill side of the ice house. There was a small door in the gable to receive the last two layers.

The following description is of an ice house intended for a somewhat larger harvest than the preceding and the building is more thoroughly constructed. It also presents a better appearance architecturally. The foundations and floors are of cobble stones to provide drainage. The course of cobble stones should be a foot or eighteen inches thick, and project slightly above the surface of the ground. The sills are double 2-inch stuff ten or twelve inches in width; the studding of 2 x 10 or 2 x 12 set 24-inch centers; the rafters or roof joists to be of sufficient strength, depending on the size of the building and kind of material employed. Floor joists of 2 x 8, or 3 x 8 may be used, or a floor may be laid down loosely on the sawdust which is filled in over the cobble stones to a depth of a foot or more. If floor joists are used the floor should be of 2-inch stuff, laid open at joints to allow meltage to drain readily. The studs are boarded with matched lumber outside and inside, and the space between filled with insulating material, preferably of sawdust (perfectly dry) or mill shavings, well rammed down. The rafters should likewise be boarded underneath and filled, or ceiling joists may be run across at the top of the studs forming a floor and an attic. This space should be suitably ventilated by slat ventilators in the gable ends over attic floor. If an attic floor is put in, the rafters need not be filled, the attic floor being filled instead. The general description of a model creamery ice house a little further on may be consulted in connection with the above. It is not necessary to place hay or straw on the ice where all the surfaces in the building are insulated as above described, and the room may be filled full to the ceiling or within a few inches of same. A little experience will show whether or not a covering of any kind is necessary or advisable.

The above methods of constructing ice houses, with one exception, are not minutely described with drawings because of their simplicity and because individual ideas and judgment can best modify them to suit local conditions. The information given will enable any experienced carpenter, or even a person

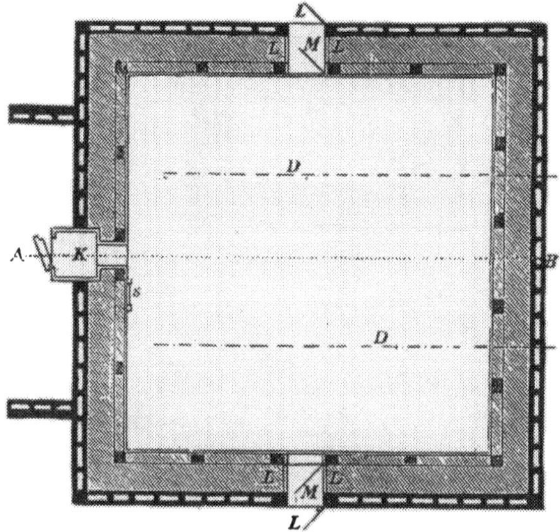

FIG. 6—PLAN OF DANISH ICE HOUSE.

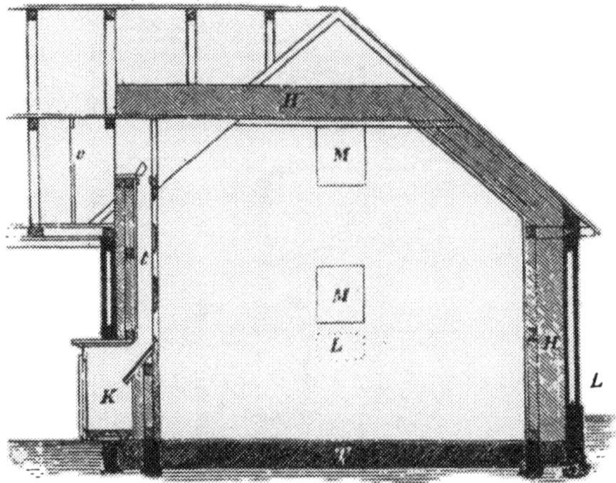

FIG. 7—SECTION OF DANISH ICE HOUSE.

ordinarily familiar with tools, to take the material at hand, and erect a structure of suitable size and character to meet the conditions. The houses already described are not well adapted for ice

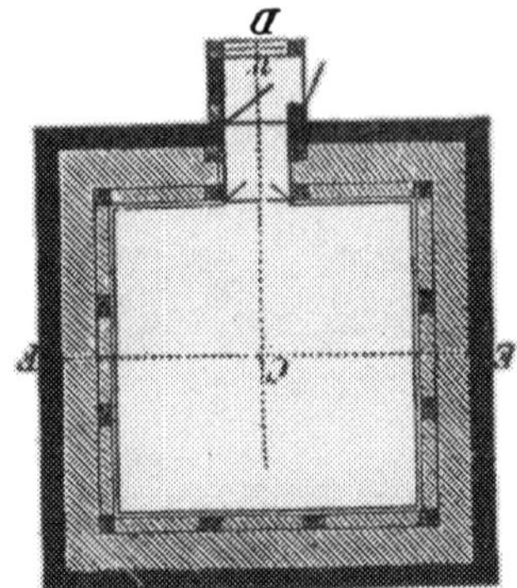

FIG. 8—PLAN SMALL DANISH ICE HOUSE.

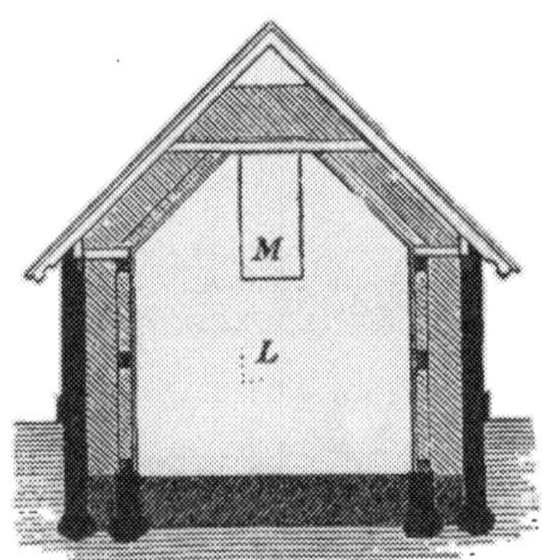

FIG. 9—SECTION SMALL DANISH ICE HOUSE.

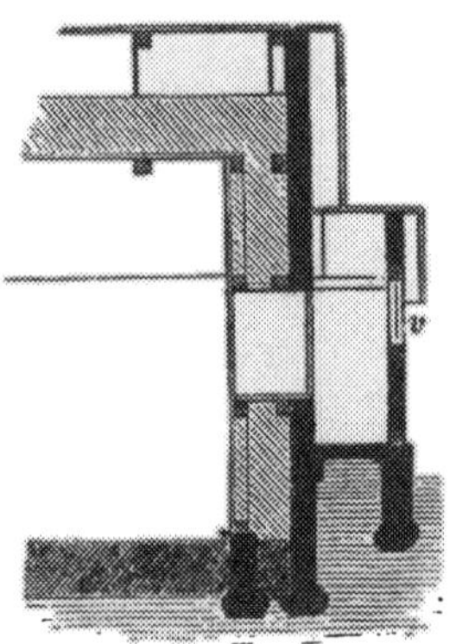

FIG. 10—SECTION ON C-D SMALL DANISH ICE HOUSE.

houses intended to hold more than 100 to 150 tons of ice. For larger houses one of the designs described further on is more suitable. In any case the amount of money which can be profitably expended on an ice house depends largely on the cost of delivering ice to the house. If it costs but twenty cents per ton to store the ice in the house, it is not advisable to spend much money in building a house to preserve same; it would be better business policy to build a cheap house, making it larger to allow for greater meltage. On the other hand, if the ice in the house costs from seventy-five cents to a dollar per ton, it would pay well to build in a first-class manner after the best plans obtainable. Between these two extremes are all the variations which may be met by considering cost of ice and cost of construction. These remarks apply equally well to ice houses of any capacity. The greater the cost of the ice the more money can be profitably expended in protecting same from meltage when housed.

Although a similar construction may not be advisable or practicable in this country, on account of the expense, yet a description of the Danish methods herewith given, may be of interest. The designs shown have been particularly recommended for creamery use.*

The illustrations taken from Bernhard Boggild's Maelkeribruget represent an ice house as usually built for creameries or dairies in Denmark. The larger one (Figs. 6 and 7) will hold 15,000 cubic feet of ice, and is built in connection with the creamery. Fig. 7 is a section through A-B of Fig. 6. The inside lining is of matched and varnished boards placed vertically, on the outside walls, but horizontally on the partition and ceiling. D represents the drainage; M is doors for putting in ice; L, doors for renewing the sawdust; V, window; T represents peat or sawdust on the floor; H, chaff, husk, sawdust, or other insulating material in the hollow walls. All doors are made as tight-fitting as possible by tacking cloth on the edges. A, small hall, K connects the ice house with the creamery, the ice being thrown out through the flue, t, through the upper door, later through the lower ones.

The small ice house, Figs. 8, 9 and 10, will hold 1,650 cubic feet of ice. Fig. 9 is a section through E-F, and Fig. 10 is a section through C-D of Fig. 8. M is a door through which the ice is put into the house; L, a door for renewing the sawdust; v, window.

For the benefit of our readers we subjoin an estimate of the materials needed for, and the cost of building these two houses. This estimate was made by an American builder.

COST OF BUILDING.

Large Ice House of Brick	$1,343.30
Large Ice House of Wood	1,044.70
Small Ice House of Brick	389.50
Small Ice House of Wood	393.50

*Abstracted from J. H. Monrad's *Dairy Messenger*.

The outside walls are represented in the illustration as built of brick. In the estimates, the cost is figured for brick as well as for wood. It will be noticed that the small house costs about the same whether built of brick or wood, while for a large house wood is the cheaper. Of course, wages and the price of material differ in various sections of the country, and these figures can only be a guide, which we trust may be useful to dairymen contemplating the erection of ice houses.

MODEL CREAMERY ICE HOUSE.

The plans and details shown in Figs. 11, 12 and 13, represent a building designed by the author for a model creamery ice house. The cost of this building is very much less in proportion to its capacity than that for any of the buildings constructed according to the customary Danish practice as described in the foregoing. The walls also are much thinner and it is quite probable that the ice will not keep as well under the same condition of outside temperature; at the same time the difference in the melting of ice in the house insulated as detailed and the Danish houses would probably not exceed 20% if the ice were carried through to the end of the season without removing any from the house. In Northern latitudes, where natural ice may be stored cheaply at a cost ranging from 15 to 50 cents per ton, it is not good practice to invest too much money in an ice house for its protection. It is better to build the ice house a little larger to allow for additional meltage. This model ice house is intended to be insulated with mill shavings in the floor, ceiling and sides, and the ice is not to be packed in any way. This is a very decided advantage over the older methods of storing ice, as when the ice is clean, without any hay, sawdust or chaff thereon, the labor of removing from the house and applying to the purposes for which it is to be used is probably not more than one-half what it is where some covering material is employed. No doubt those who have had experience in digging ice out of the old style ice house will appreciate this method of construction. The cost of construction, too, is not greatly in excess of what the old method would be.

Referring to the plans, it will be seen that the house is 27 feet 8 inches long by 17 feet 8 inches wide, inside measurement, giving an outside measurement of 20x30 feet. The

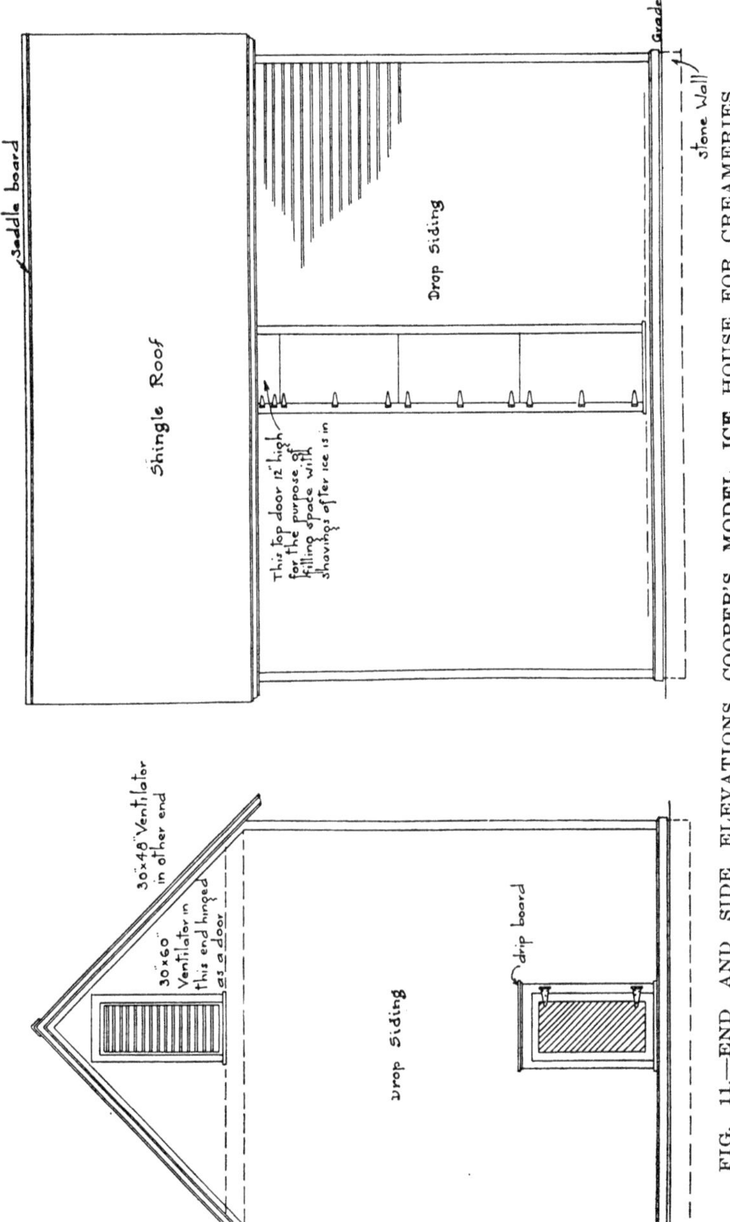

FIG. 11.—END AND SIDE ELEVATIONS COOPER'S MODEL ICE HOUSE FOR CREAMERIES.

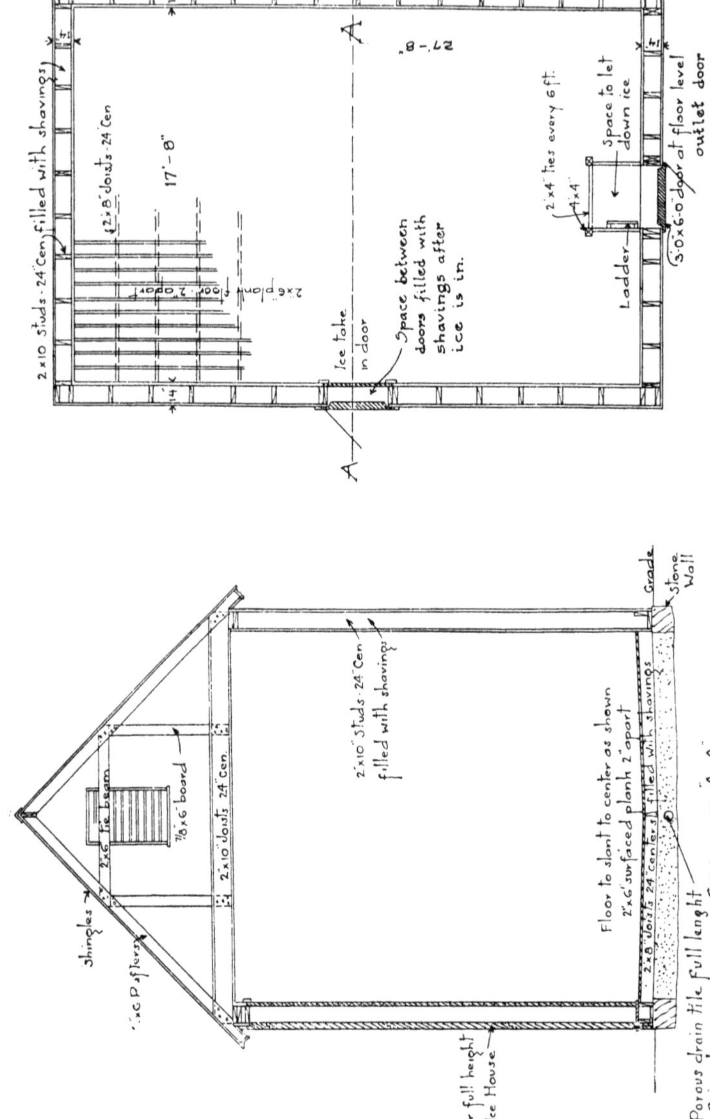

FIG. 12.—SECTION AND PLAN COOPER'S MODEL ICE HOUSE FOR CREAMERIES.

height is 20 feet, making the house very nearly a cube and with comparatively small outside exposure. The space inside is a rectangle, as it is intended to fill the house just as close to

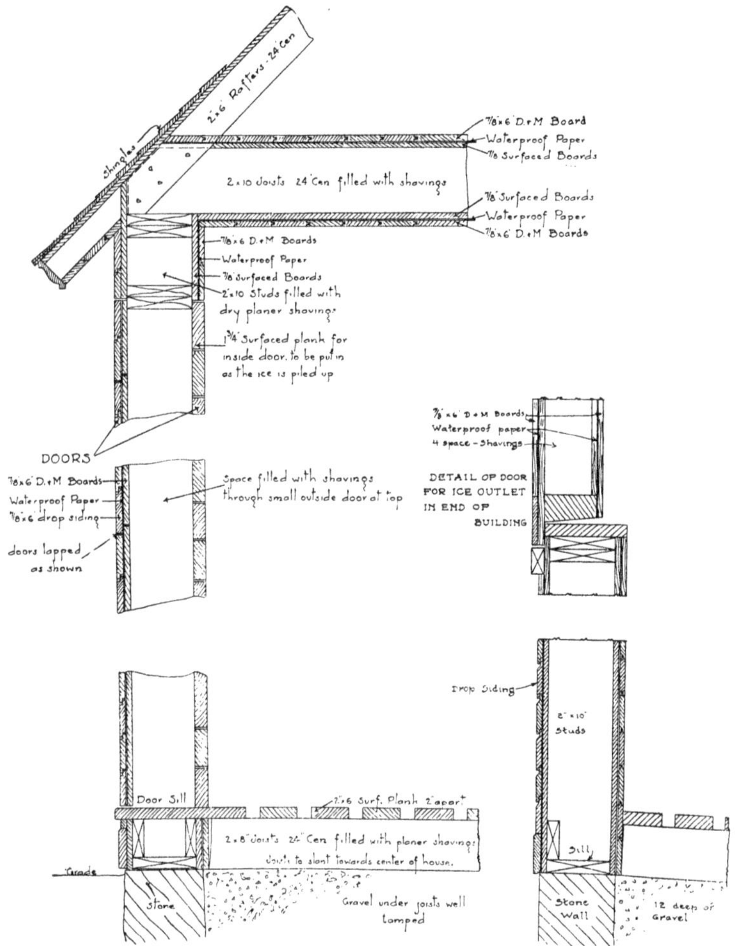

FIG. 13.—DETAILS OF CONSTRUCTION COOPER'S MODEL ICE HOUSE FOR CREAMERIES.

the ceiling as possible, and this may readily be done, as the outside doors open up to the top of the ice storage space. The roof is what is known as a "half pitch roof" and is provided with ventilators at both ends, one of which is in the form of a door for entering the attic space. This allows for circulation of air through the space above the ice and prevents to a large extent the penetration of heat from the roof.

The house as planned shows doors for icing at the side and removing from the end, but both the filling and removing doors may be placed in any location desired. The door for removing ice is fitted with a frame work within the room which leaves a space in the body of ice which may be utilized for lowering ice from the top of pile or it may be removed through the filling door. A ladder is located in the space inside the door for ascending to the top of the ice. There is no opening whatever through the ceiling of the ice room into the attic. There is an old popular idea that ventilation over the ice is necessary, but this is only true where the ice is covered with a packing or non-conducting material like sawdust, hay, etc. It should not be employed where ice is not covered, as it is unnecessary and results in melting the ice badly.

As shown in the drawings, the walls of the building are erected on a stone foundation which is carried down to a sufficient depth. A deep foundation is not necessary, however, as the entire weight of the ice rests on the ground and not on the foundation of the building. Good drainage may be provided by filling in gravel between the walls of the building to form a floor to the depth of twelve inches. If the soil should be of clay, it would be well to lay a porous tile drain through the center of the house connected with some drainage point. If the soil is sandy or gravelly, the layer of gravel and tile drain may be dispensed with entirely.

The floor is formed by placing 2 x 8 joists 24-inch centers and slanting them slightly towards the center to the drain tile. On the top of the 2 x 8 joists is laid a floor consisting of 2 x 6 plank laid two inches apart. The space between the joists is tc be filled with mill shavings, tightly rammed. Mill shavings are specified all the way through this building, although other materials like dry sawdust, cut straw or tan bark might be used. Mill shavings are, however, to be preferred. The floor of the ice house is entirely independent of the walls of the building and is intended to be laid last, and the floor joists not set over the foundation wall. This allows the ice to settle independently of the building. The side walls are constructed by laying a sill of 2 x 10 stuff on the stone wall. On this sill are erected the 2 x 10 studding, placed 24-inch centers. As planned, these studdings are double-boarded inside and out, with water-proof insulating paper between and filled with mill shavings. If it is desired to cheapen the cost of the house, the studding may be single-boarded on each side with paper underneath, care being taken not to tear the paper when filling the space with shavings.

The roof is designed to be of shingles, but it may be of any other material. The doors where the ice is to be filled into the house are made in sections and hinge on the outside. Removable plank pieces are placed inside and the space between filled with shavings.

Ice houses of the character similar to the one here described have been built of a capacity of 1,000 tons; and, considering the cost, are very successful in every way. The estimated cost of the house which is described is, under ordinary conditions where lumber can be obtained from nearby mills, $300.00 to $400.00. If lumber is expensive and labor high, and ice may be housed cheaply, the house may be cheapened by leaving out the gravel and stone foundation wall and single boarding the studs inside and outside instead of double boarding. The ceiling joists likewise can be single-boarded, and in case of necessity the floor can be left out entirely. The details for a house of this kind depend upon the location and must be selected by the builder or architect in order to provide proper protection for the ice and at the same time not have the cost too high.

MODEL COMMERCIAL ICE HOUSES.*

The accompanying plans show in detail the construction of two ice houses on the most modern ideas in the art of ice house building. The endeavor has been to devise a house which will, according to the opinions of experienced men, not only offer every facility for putting up the ice quickly and economically

---

*From *Ice Trade Journal*. Plans by Gifford Bros., Hudson, N. Y.

and removing it in quantities as desired at lowest possible cost, but also to preserve the ice for the longest possible periods with the least possible waste, and at a cost for construction which is not prohibitive.

As two different uses are made of ice houses, differing considerably in their effect upon the ice itself, it is necessary to accommodate the house to the purpose. Some houses are used for retail delivery only, and a room is opened several times a day perhaps and ice removed in small lots of a few tons during a number of weeks. Such a house should be planned differently from those from which the ice is taken by the car or boat load as fast as it can be broken out and stowed away. The first plans, Figs. 14 to 19, inclusive, are for a house for the former purpose, and Figs. 20 to 29, inclusive, are for a house from which large shipments are made and the house soon emptied.

The first plans are for a twelve-room, single-posted house. The capacity is 6,400 tons. The house is 126 feet 11 inches long by 83 feet 6 inches wide and 30 feet 2 inches high from sill to plate. The outside walls are 14 inches thick, the partitions are 11 inches wide and the middle walls, dividing the interior of the house into two sections, each of which is again divided into six rooms, is 14 inches thick, just as are the outside walls. Each room is, then, 40x20 feet in the clear, and has a capacity of 535 tons.

The posts are 3 x 12 inches and 4 x 12 inches, all 30 feet long. They are covered on the outside by 1-inch novelty siding, laid horizontally, and on the inside by 1-inch matched boards, laid diagonally. It is recommended that a high grade of building paper be used on the inside and outside of the posts under the novelty siding and the matched boards. The space between posts should be filled with either mill shavings, fine dry sawdust or dry tan bark well packed down. These three materials are recommended in the order named.

The posts beside the doorways are each 4 x 12 inches for additional strength at these points. The partition posts are 3 x 10 and 30 feet long, that width being sufficient for the inside of the rooms. The plans call for 1-inch matched boarding put on diagonally on one side of the partition only. If thought desirable for better insulation, the boards may be put on both sides and the space filled with sawdust or shavings, as in side walls. In some houses boards are put across between partition

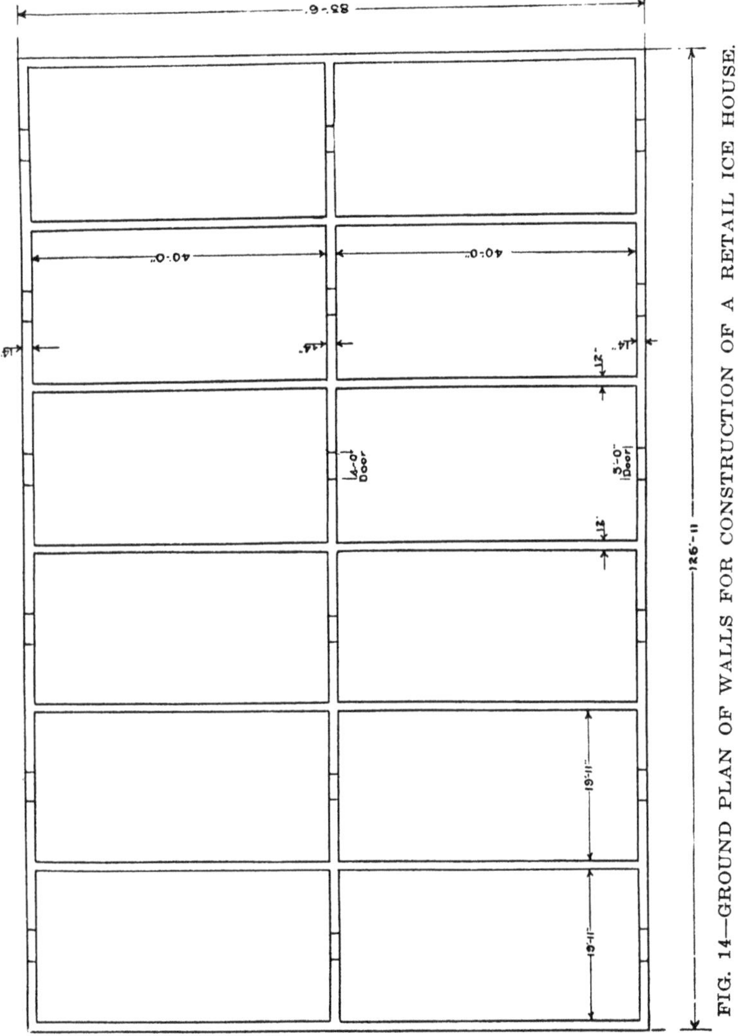

FIG. 14—GROUND PLAN OF WALLS FOR CONSTRUCTION OF A RETAIL ICE HOUSE.

posts half way up and only the space from the boards to the top of the partition filled in.

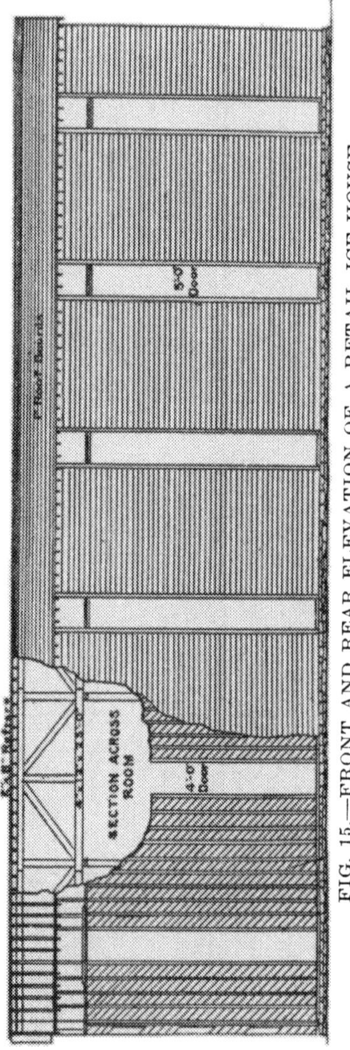

FIG. 15.—FRONT AND REAR ELEVATION OF A RETAIL ICE HOUSE.

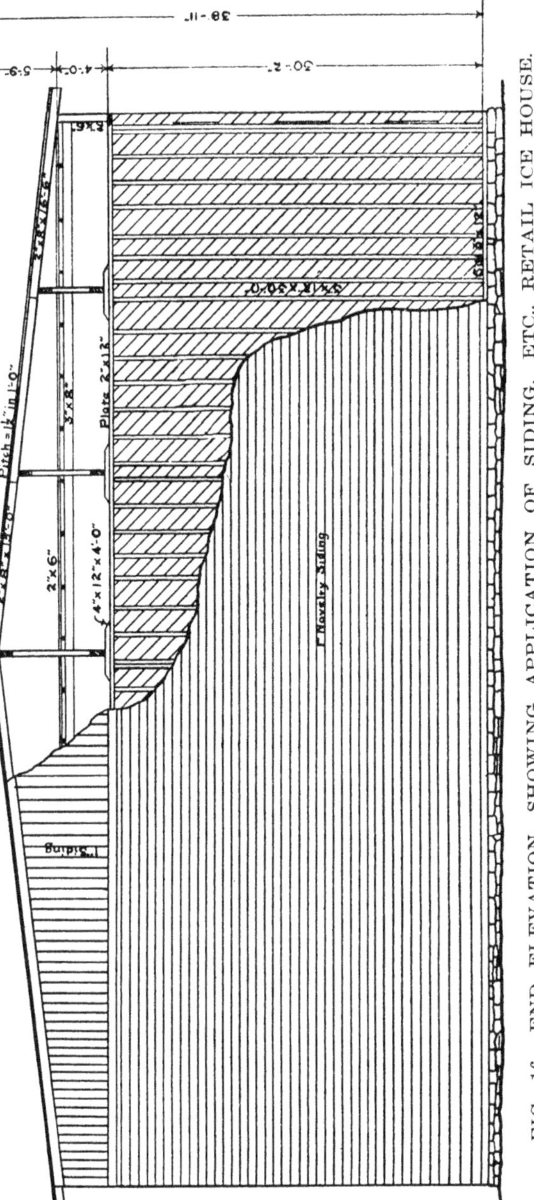

FIG. 16.—END ELEVATION, SHOWING APPLICATION OF SIDING, ETC., RETAIL ICE HOUSE.

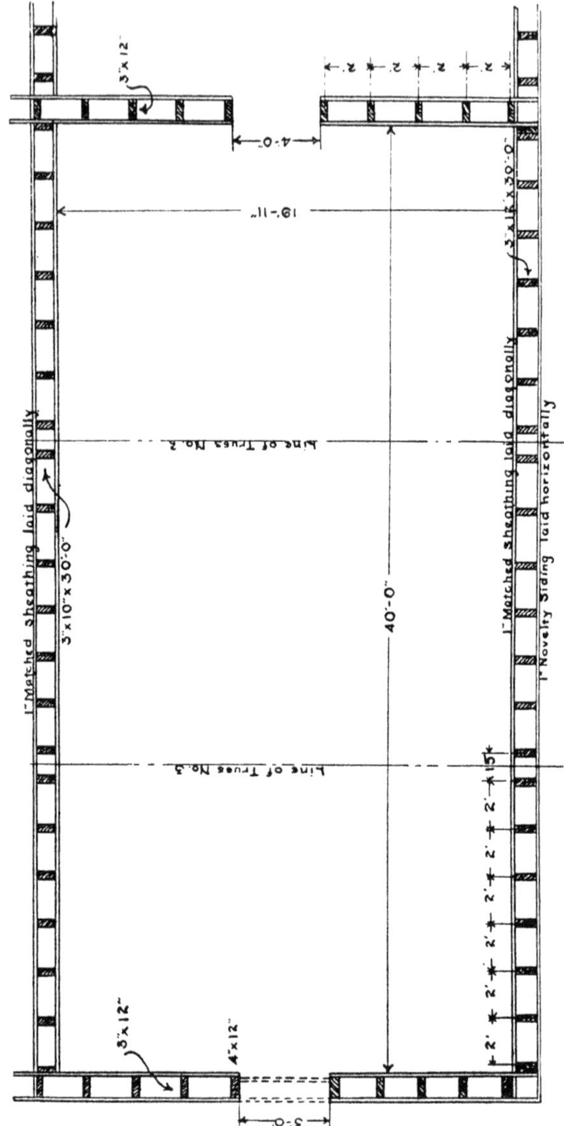

FIG. 17.—DETAIL OF POSTS FOR CORNER ROOM OF RETAIL ICE HOUSE.

The diagonal boarding on partition walls and inside of front and side walls is called for by the plans, as that method of covering adds some to the strength of the structure, but it is more wasteful of lumber than horizontal boarding, and the latter may be substituted for the diagonal at the discretion of the one doing the building.

The posts are placed 24 inches apart, center to center, for greater strength, but 30 or 36 inches is considered by many as not too great a distance. Such changes in detail may be made in several places, as will be readily seen from the plans, in the interests of economy, but they do not affect the general condition materially. Such departures from the plans are not, however, recommended, as the better the house is built the better it will keep the ice and the longer it will last without repairs. Parsimony does not pay in ice house building in the long run.

The detailed instructions for boarding and filling in the doorways are sufficiently elaborate without further description. The drawings give the sizes and position of the trusses, so that no difficulty should be experienced in understanding the roof construction and supports.

If of any advantage in filling, as it probably would be, except in very unusual natural surroundings, doorways four feet wide may be cut in the middle partition at the back of each room, so that the back rooms may be filled through the front. To get rid of chips and give more light, doors are also frequently cut in the center of end walls and the partition walls at the sides of the rooms. As no loft floor is provided for in the plans, this being considered unnecessary, the openings under the eaves are left free for ventilation.

With these brief explanations, the drawings should be easily comprehended not only by any builder, but by any ice man, however unfamiliar with building operations; and by their aid, no matter how inexperienced, he should be able to erect a house at once moderate in cost and correct in structure and one most likely to give excellent results.

Nothing has been said in connection with the first house as to the foundation and the much mooted drainage question, but as advice on these important points will apply as much to

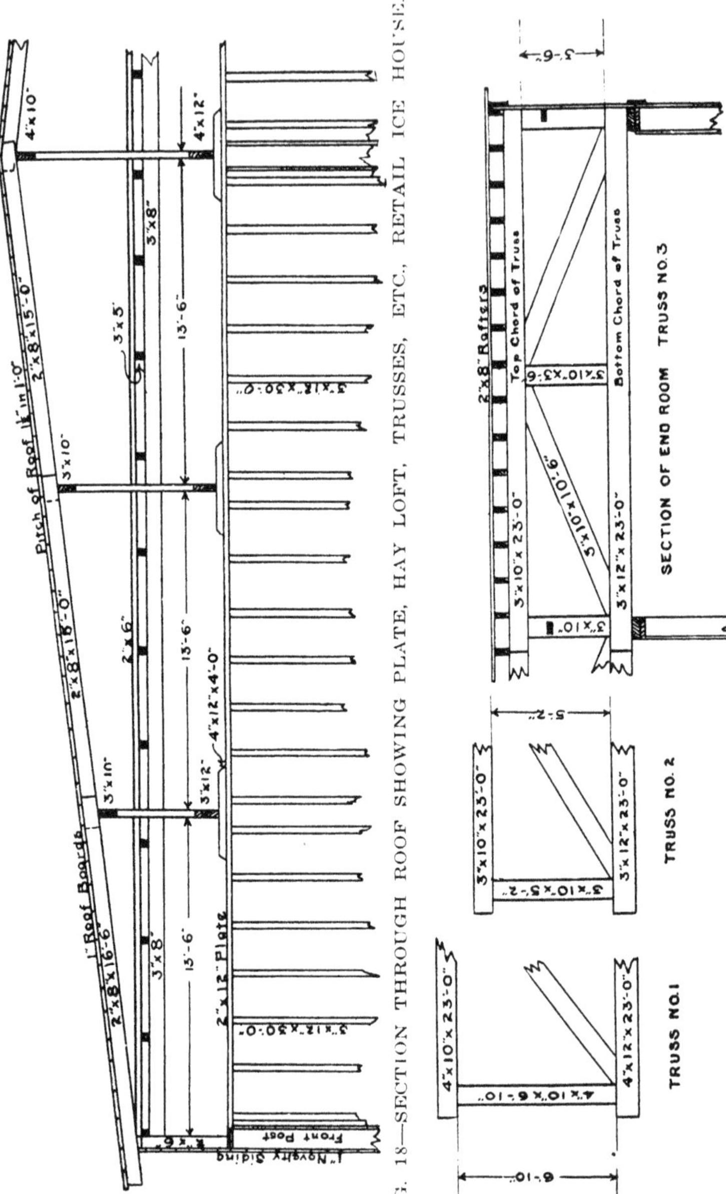

FIG. 18—SECTION THROUGH ROOF SHOWING PLATE, HAY LOFT, TRUSSES, ETC., RETAIL ICE HOUSE.

FIG. 19.—DETAIL OF TRUSSES FOR CONSTRUCTION OF RETAIL ICE HOUSE.

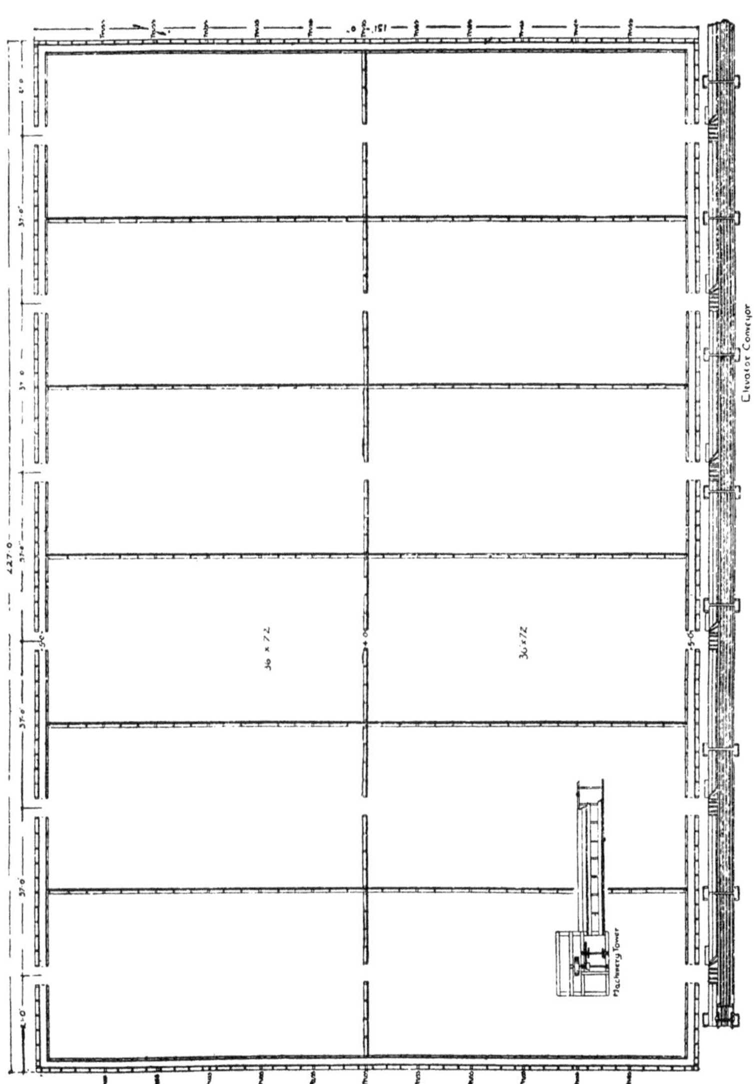

FIG. 20.—DIAGRAM SHOWING FOUNDATION PLAN FOR A LARGE WHOLESALE ICE HOUSE.

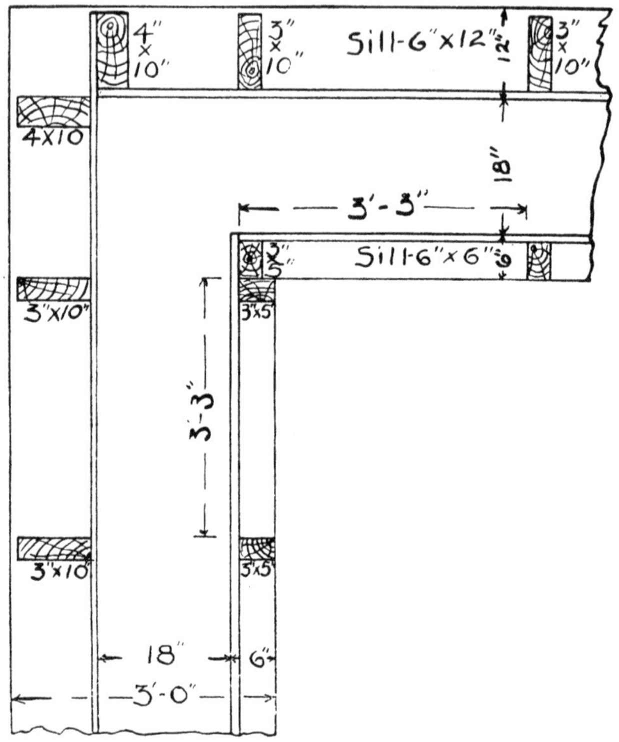

FIG. 21.—DETAIL PLAN OF CORNER, SIDE AND REAR WALLS.

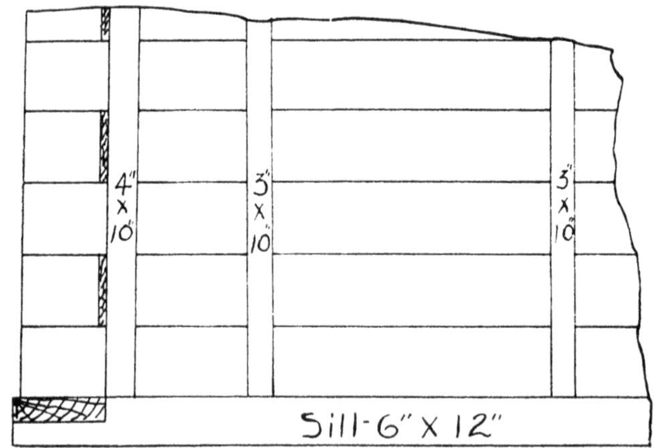

FIG. 22.—ELEVATION OF CORNER, FRONT AND SIDE WALL.

the second house, now to be considered, as to the first, they will be touched upon now.

The foundations should be of stone laid in cement, and the same should be liberally applied, both inside and outside, so as to prevent air and dampness from finding its way in and undermining the stored piles of ice. Most of the damage done by the shifting and sliding of ice, by which side and end walls are pushed out of position, is due to imperfectly protected foundations. These walls are not intended to support or hold up the

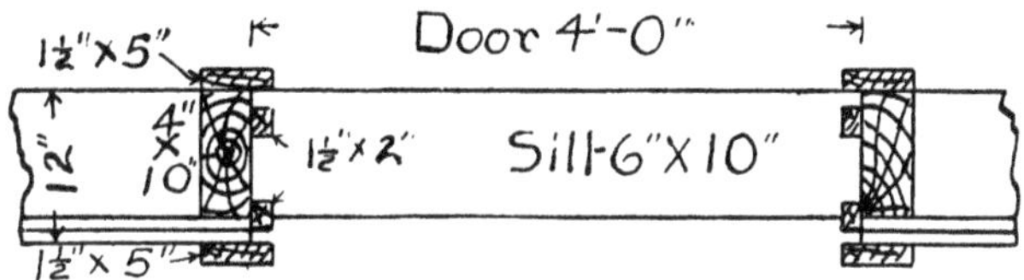

FIG. 23.—DETAIL OF PARTITION.

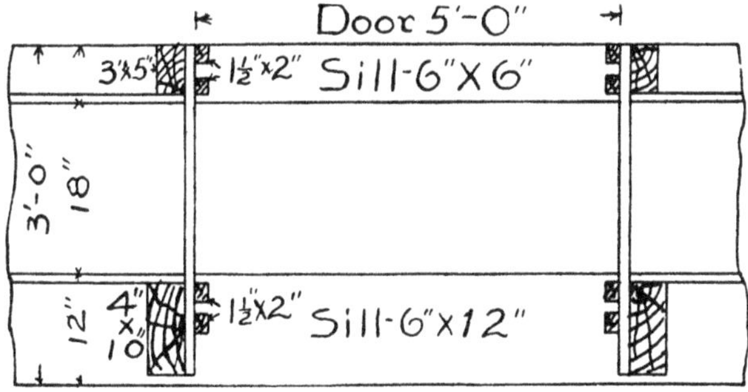

FIG. 24.—DETAIL PLAN FRONT AND REAR DOORS.

ice. That is supposed to stand square and true, and will if the various tiers are reversed so as to tie the whole in an erect position, as it should be. Some ice men reverse every other tier and some every second or third one. The first plan is considered the safer and better practice, as affording better security to the ice house. In a house built on these plans, sawdust between the walls and the ice is entirely unnecessary, if good quality of paper and well dried and tamped filling is used between the posts. The sills should be laid in cement of course.

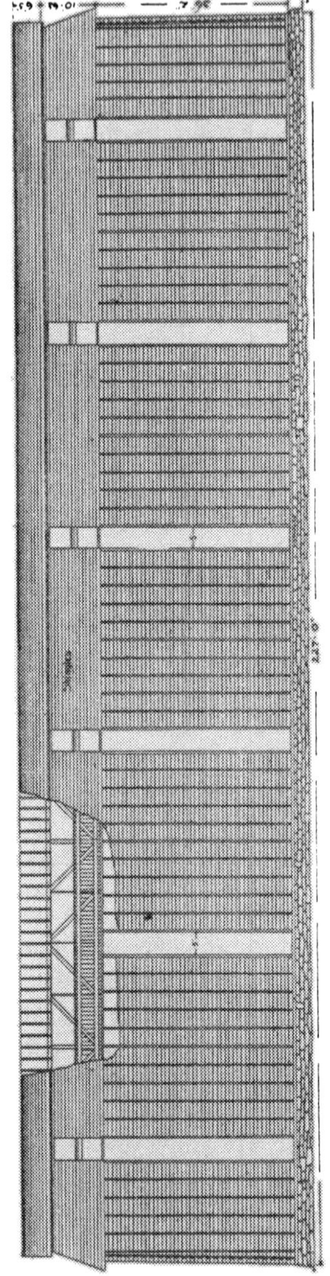

FIG. 25.—FRONT AND REAR ELEVATION OF LARGE WHOLESALE ICE HOUSE, BASED UPON FORTY-FIVE CUBIC FEET PER TON.

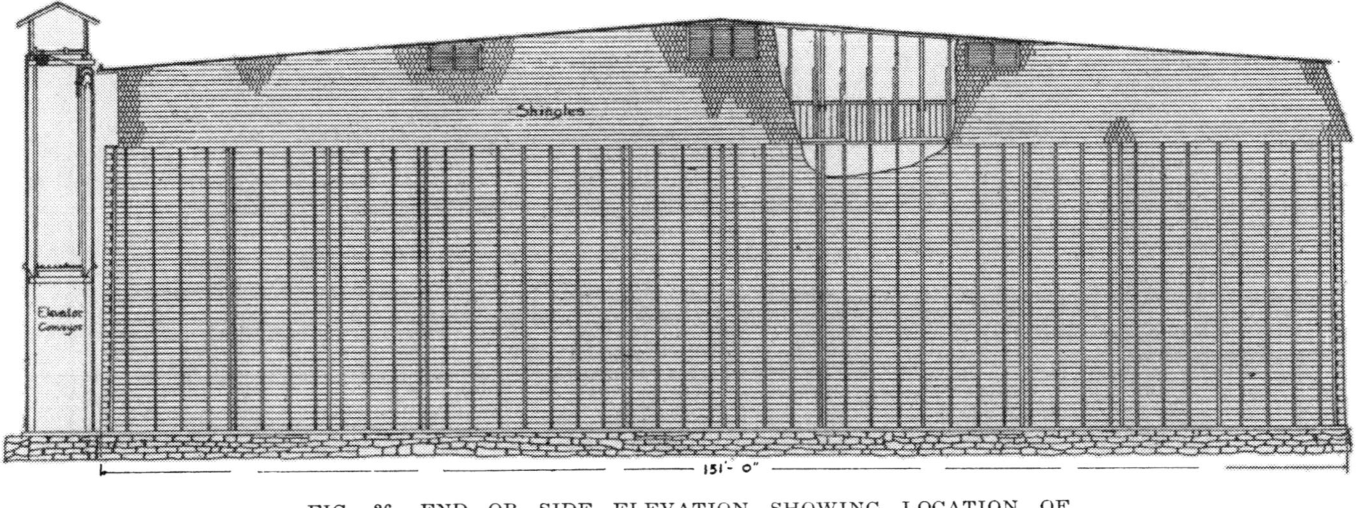

FIG. 26.—END OR SIDE ELEVATION SHOWING LOCATION OF SWINGING GALLERIES, LOUVRES FOR VENTILATION, ETC., FOR LARGE WHOLESALE ICE HOUSE.

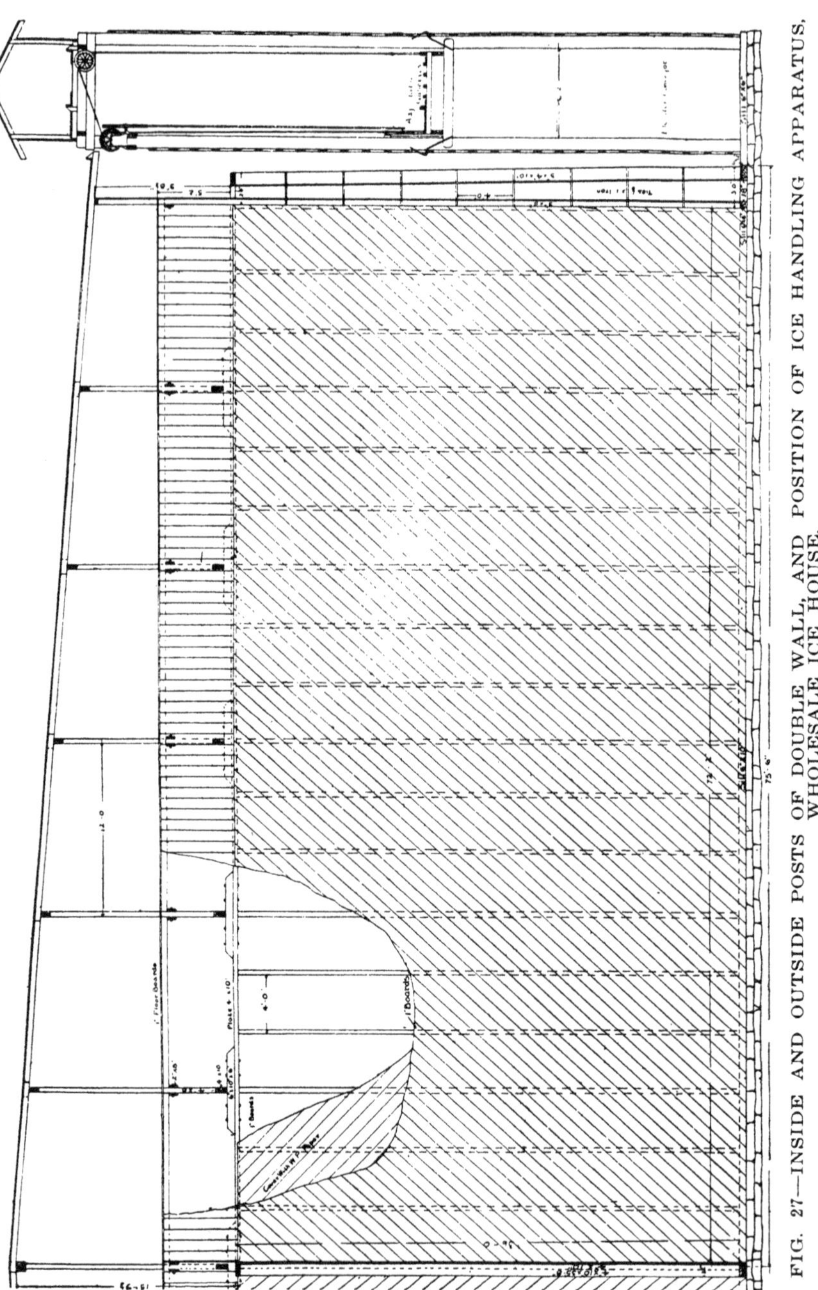

FIG. 27—INSIDE AND OUTSIDE POSTS OF DOUBLE WALL, AND POSITION OF ICE HANDLING APPARATUS, WHOLESALE ICE HOUSE.

The matter of putting in drains is such an open question—experienced men differing entirely as to their value and their construction if used—that it should be determined entirely by the geological structure of the land where the house is to be erected in each case. A gravelly or sandy soil with a substratum of blue clay is desirable, but not always to be found or made. If a drain of any kind is used, its construction requires the very greatest care, because where water can flow out air not only may, but almost invariably does, find an easy access. Hemlock boards laid loosely along the floor will serve both to distribute the weight of the ice and to allow the meltage to settle into the ground.

Figs. 20 to 29 are plans for a house of 24,888 tons; 227 feet long by 151 feet wide, and 36-foot posts, divided into 12 rooms, each 36 x 72 feet clear, and each of 2,074 tons capacity. This house has double walls. The construction is of 1-inch novelty siding, one layer of paper, 3 x 10 posts, one layer of paper, 1-inch matched boarding, 18-inch air space, 1-inch matched boarding, one layer of paper, 3 x 4 posts, one layer of paper, 1-inch matched boarding. The spaces between posts in both walls to be filled with either mill shavings, sawdust or tanbark, as in first house. The partition walls consist of 1-inch matched boarding, 3 x 10 posts and 1-inch matched boarding. They may be lined with paper and filled for entire or part height if desired. Whenever any partition or wall is filled with any insulating material, each side of the posts should be covered with a good quality of building paper, not only for additional insulation, but for protection of the filling from dampness, which destroys its usefulness. The outer and inner walls are supported by 1-inch iron ties, which hold them rigidly. The posts at the doorways are 4 x 10. The sills are 6 x 12 and 6 x 6 for the outer and inner walls respectively. The posts are distanced four feet center to center in the walls.

This house requires from its size and use a loft floor, which the plans indicate is to be placed five feet two inches above the plate, so as to allow room for stowing ice to the plate and covering with hay. Hinged doorways should be provided in the floor over each room to allow easy access, light, etc.

Ventilation is afforded by the doorways, which are carried up above the plate to the roof and provided above the loft floor with doors on hinges, which may be opened or closed according to season, wind direction, etc. The roof is carried out beyond the eaves, and openings left between the rafters for additional ventilation. There should be no openings of any kind into the

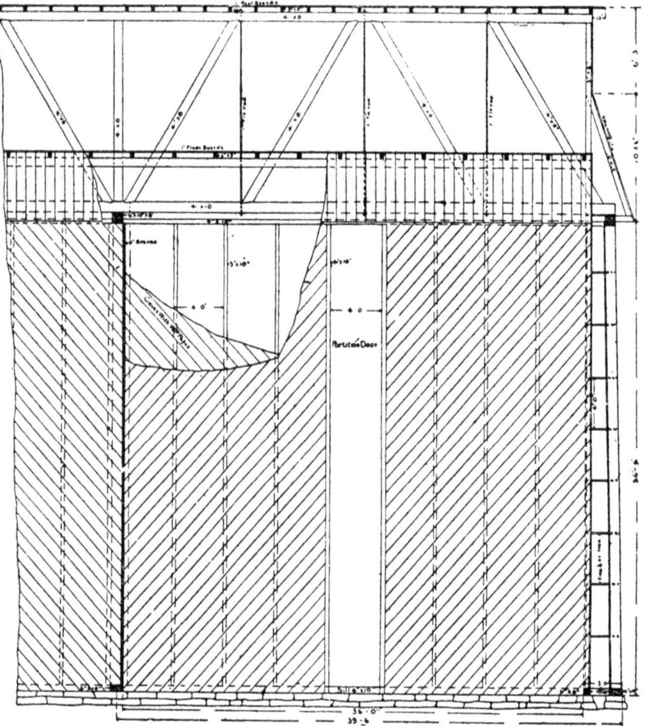

FIG. 28.—INTERIOR OF WHOLESALE ICE HOUSE SHOWING TRUSS SUPPORTING ROOF.

ice rooms. Plenty of air is needed through the loft to break up radiation through the roof, but none should be permitted to enter the ice chambers. The whole purpose of the construction of this house is to keep air from penetrating, in even the slightest amount, to the ice. Above the plate, plenty of air; below it, none whatever. Openings may be cut in the roof and scuttles put on which may be readily opened and closed if the loft is found too warm with the doors open and the eave spaces free.

COLD STORAGE IN CONNECTION WITH ICE HOUSE.

A cold storage plant may be operated in combination with the storage and sale of natural ice at a comparatively small cost.

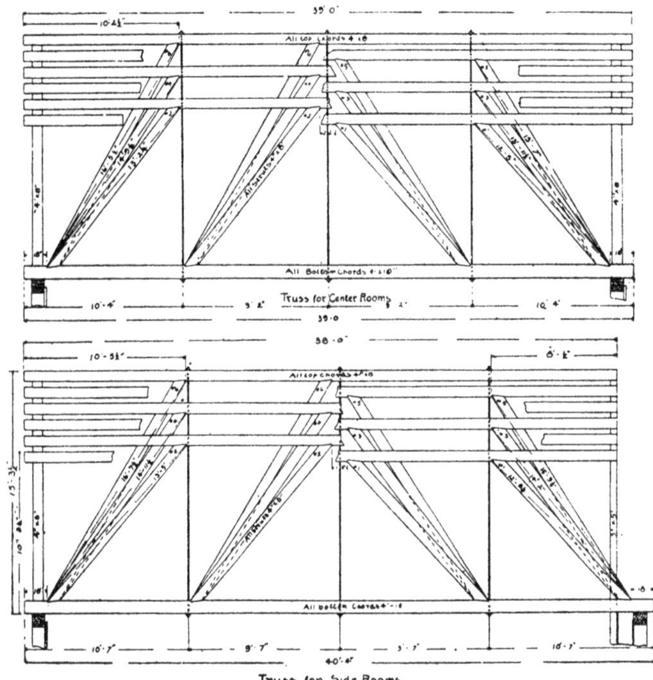

FIG. 29.—DETAIL OF TRUSS CONSTRUCTION OF WHOLESALE ICE HOUSE.

There are many localities, especially in large towns and the smaller cities, where the natural ice dealer could work up a good business in the cold storage line, especially as he sells ice to many produce dealers who would become his customers for cold storage space as well. In many cases where an artificial ice plant has been installed, a cold storage house has been erected as an adjunct and made a success. There is no reason why the same thing cannot be done with the natural ice business. Heretofore owing to the fact that there has not been a successful system of refrigeration which could be operated by the use of natural ice, nothing of any consequence has been attempted along this line, but with the introduction of the Cooper brine system, using ice and salt as the refrigerant, equally good re-

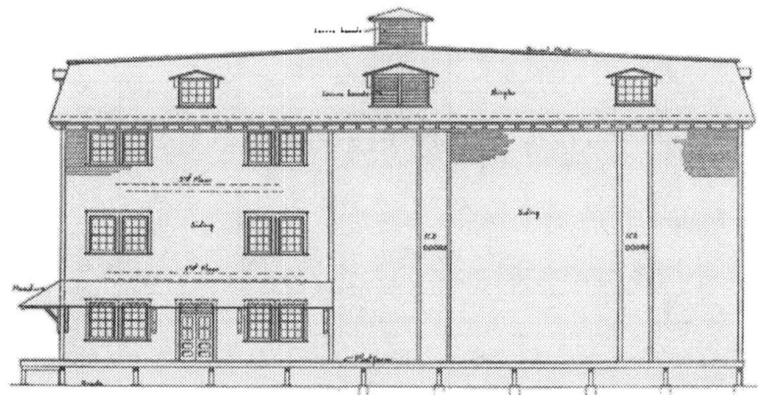

FIG. 30.—FRONT ELEVATION COMBINATION ICE AND COLD STORAGE HOUSE.

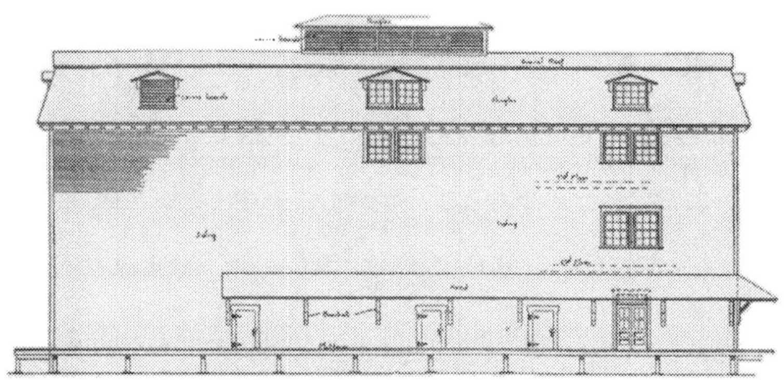

FIG. 31.—SIDE ELEVATION COMBINATION ICE AND COLD STORAGE HOUSE.

sults may be obtained down to a temperature of 10° F. as by the mechanical systems of refrigeration. Many of these plants are now in operation and uniformly successful. This system is fully described elsewhere in this book.

Figs. 30 to 32 show the outlines of the plan, sections and elevations of the combination ice storage and cold storage house which has been designed by the author. These plans may of course be adapted to any location. The cold storage part may be built in any shape and of any size and for any purpose. It may be built in one, two or three floors, or more if desired. The plant, as designed and illustrated in the sketches, shows the cold storage part, with its ice house, as separate from the regular ice storage rooms. This is for the reason that the ice produced in many localities is not a very sure crop and in other cases it is contaminated by sewage, etc. The ice used for the cold storage part of the building need not be from a pure source, as it is used for cooling purposes only or for icing cars. The ice stored in the regular ice storage rooms is supposed to be from a pure source, and may, if necessary, be shipped in cars.

In connection with the cold storage part of this plant an ice crusher and ice elevator are employed for handling the ice. The spout leading from the elevator is arranged so as to deliver ice at several points on the railroad track for the purpose of icing refrigerator cars with crushed ice. Cars may be iced with this device with great rapidity, and it is consequently economical, as the ice need not be handled except to feed it into the ice crusher in large pieces. The ice storage rooms of this plant are insulated with mill shavings, are separated into two divisions and protected inside and outside by the best grade of insulating paper. The floor and ceiling are insulated in about the same manner, and no packing or covering material is used on the ice. Ice men will appreciate the advantage of this method, as it saves a large amount of labor, and that, too, at a time when labor means a good deal to the ice man. It is calculated that the meltage of ice in this house will not exceed 15%, and under ordinary conditions should not exceed 10% or 12½%, which is about as small as any ice man can figure on.

As shown by the plan, the receiving room for the cold storage department is located on the corner and a platform runs around three sides of the building. This makes the loading and unloading of goods from cold store and cars a comparatively small matter, and, as the receiving room fronts on the

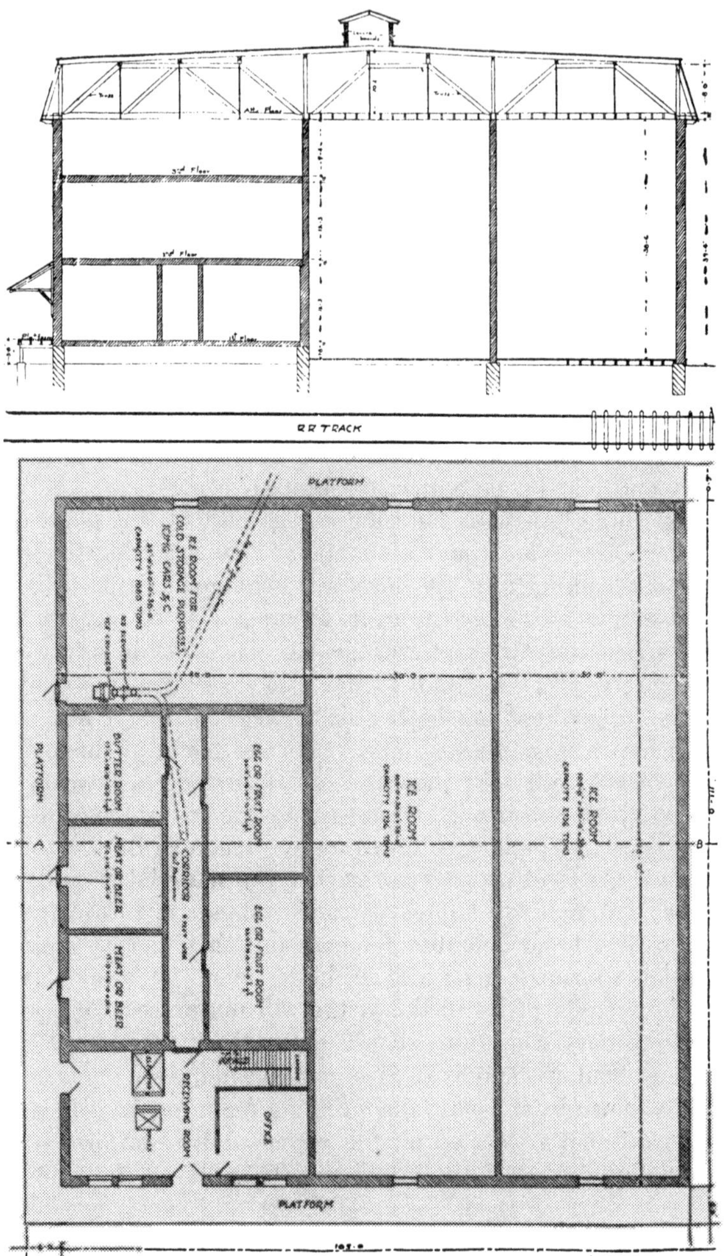

FIG. 32.—FIRST FLOOR PLAN AND TRANSVERSE SECTION ON A-B COMBINATION ICE AND COLD STORAGE HOUSE.

street, makes deliveries to and from wagons equally simple. The office, which is located in the receiving room, is intended to be used not only for the cold storage department, but for the ice business as well, and is so located that the ice business may be readily looked after. In this plant no basement is constructed, and the cold storage plant is all on the first floor. It will be noted, however, that the second and third floors are reserved and may be insulated and equipped with cooling apparatus as the business is extended. An elevator serves the second, third and attic floors, and the attic over the cold storage and also over the ice storage rooms may be utilized for ordinary warehouse purposes.

The cold storage rooms, as divided, are intended for eggs or fruit, one room for butter and two rooms for meat or beer. These meat or beer rooms open directly onto the platform, making them accessible from the railroad siding and from the street. These rooms may be rented to the meat or beer agencies and the owner need not have anything to do with the same except to regulate the temperature. The insulation of the cold storage department consists of mill shavings and hair felt properly divided and protected by the best grades of insulating paper. The butter room, meat or beer rooms are cooled by piping placed directly in the room. The egg or fruit rooms are cooled by the forced circulation of air, the coils for which are located over the corridor. A ventilating system furnishes fresh air at any season of the year to all of the rooms. The "Cooper systems": brine system, forced air circulation, ventilating system and chloride of calcium process, with which the plant is equipped, are fully described elsewhere in this book.

CONCRETE ICE HOUSES.

The subject of concrete ice houses is one prolific of discussion, and there has actually been several such houses built, but the author failed to locate a concrete ice house of modern construction, wherein the side walls, floor and ceiling are insulated and wherein the ice is not covered with some protective material which has been even a moderate success. It is, of course, practicable to build an ice storage house of concrete of the old style, where the ice is placed in the room a foot or more from the walls, and the space between the ice and the walls filled with sawdust or other similar material; but this method of ice storage is rapidly going out of use for anything except the smaller purposes and, therefore, it is hardly worth while to discuss concrete ice house construction except for such small uses. Further than this, there is no question but what small ice storage houses will in future be built according to the modern idea of complete insulation on side walls, floor and ceiling, rather than the protective covering and filling between the ice and the walls as in the days of old.

Concrete ice houses, if they ever become practical, structurally and mechanically, are very many years in the future, and it will be necessary to introduce some very greatly improved design to make them commercially feasible. Concrete construction is all very well in its place, and concrete is a most useful material when properly applied, but its application does not belong to ice house construction at the present time except for foundations only.

TIGHT LOFT FLOOR CONSTRUCTION.

The model ice house for creameries, plans and details of which are shown elsewhere in this chapter, has the "tight loft floor," which has within a few years been advocated by a few of the more progressive ice men. A "tight loft floor" means nothing more than an insulated ceiling, and the expression "tight loft floor" is used by ice men for the reason that they have been accustomed to a loft in their ice storage houses. A loft is really unnecessary and if an ice house with a flat roof were built, insulation could be applied right in the roof and no loft would be necessary. An ice storage house insulated not only in sidewalks and floor but insulated in the ceiling as well has been advocated by the author for many years, and he was the first one doubtless to use this construction. It is good common sense and saves money in construction as well as resulting in the saving of ice meltage.

SUGGESTION FOR IMPROVED CONSTRUCTION.

The author having originated the so-called "tight loft floor" construction referred to in the previous paragraph now advocates the omitting entirely of the attic which is commonly a part of an ice house. A flat roof with a pitch of not more than a half inch to the foot can be made to support sufficient insulation so that it is not necessary to have a space at the top of the building such as is ordinarily contained between the attic floor and the roof. It is entirely practicable to use sufficient insulation in the roof to make up for the protection which the attic ordinarily gives, and this is the construction commonly employed and advocated by the author.

As a still further advance in the methods of storing ice, both artificial and natural, the maintaining of temperatures in the ice room below the freezing point so that no meltage occurs is now suggested as wholly practicable and desirable. This is especially true for natural ice where the cost is pretty high on account of shipping by freight or for some other reason, and where natural ice is cut from a pure source there is no reason why it is not worthy of as good treatment and storage as artificial ice. The reader is referred to the chapter on "Ice Storage Under Refrigeration."

CHAPTER XXXIX.

THE COOLING OF INHABITED BUILDINGS.

COOLING DWELLINGS IN THE TROPICS.

Much has been said on this subject, but comparatively little has been actually done. This is not because the problem is difficult of solution mechanically or practically, but rather because the cooling of inhabited buildings has not been considered necessary up to the present time. In the tropics it would be very desirable to cool buildings during a considerable portion of the year, but in the temperate zone the overheated period consists, ordinarily, of only a few days or two or three weeks at most, and up to the present stage of civilization people have preferred to accustom themselves to the high temperature as best they can and endure it, rather than make the necessary outlay for artificial means of cooling.

At the Second International Congress of Refrigeration held at Vienna in 1910 there were presented three different papers on the subject of cooling inhabited buildings, and these three papers were written from three different viewpoints. The first by J. F. H. Koopman of Holland considered the problem of the cooling of living rooms in the tropics, and his conclusions are summed up under three headings as follows:

First: The wind pressure is greatest in the warmest hours of the day.

Second: The differences between maximum day temperature and minimum night temperature are greater in the warm months than in the cold months.

Third: The relative humidity of the air is least during the warmest hours of the day.

Mr. Koopman concluded that for a sufficient and economical cooling of buildings in the tropics conditions were necessary which he outlined under several headings as follows:

(a) The building must be faultlessly insulated in all respects and provided with well-closed double windows and doors (the latter with so-called air vents).

(b) With the exception of the air inlets and outlets, the house must be quite closed; thereby human health and durability of furnishings will be safeguarded.

(c) Walls, floors, ceilings, as also ducts, should be of stone of the greatest capacity for absorbing heat. This material must not pollute the air.

(d) The storage ducts are placed on the ground floor, or if there is not enough room there, or other reasons make it necessary, in the attic.

(e) The airing during the day is aided by the wind, the inflow openings being provided with pressure heads and the outflow with suction heads.

(f) The ventilating plant should be so designed and operated that there is a slight super-pressure in the rooms during the day, so that no warm outer air can enter. Care must also be taken that the air introduced into the rooms does not deposit moisture.

(g) During the night the air is for the most part circulated through the storer, and a small percentage only goes through the building, so that not only cooling of rooms but also walls is achieved, so that this second storing assists the day cooling. The circulation of the night air is done by a fan, but in special cases it can also be done by deflectors if there is sufficient wind. The quantity of air passing through a bedroom in one hour should not be more than six times the contents of such room, to avoid unpleasant draughts.

(h) Where dry night air and clean water are available evaporation of water should be employed in the ducts to increase the storage. Here care must be taken that all water sprayed in the ducts evaporates during the night, so that the moistening of the day air may be avoided. For the same reason the use of water during the day is objectionable.

(i) If one has cool, clean water, but not in sufficient quantities for completely effecting the air cooling, such cooling can be used additionally. A moistening of the air will only take place if the air be exceedingly dry, otherwise the cooling of the air will generally be combined with a decrease of its moisture.

Mr. Koopman may be congratulated on his summary of the requirements, and these may be taken as covering the points necessary to consider in the artificial cooling of inhabited spaces. His suggestion that a building must be faultlessly insulated and provided with well-closed double windows and doors would, the author believes, hardly be required in the north temperate zone. Besides faultlessly insulated does not mean anything particular, and as a matter of fact, a comparatively simple and cheap form of insulation would answer very nicely. Some of the larger buildings, in fact, need not have insulation at all as ordinarily considered from a cold storage standpoint. It is, however, positively necessary to see that windows and doors are tight and kept closed, and that the cooling or ventilating of the space be through suitable inlets aud outlets. Mr. Koopman's section "c" is understandable only when it is known that he proposes to store up some refrigerating effect in the stone walls, etc.

COOLING OF LIVING ROOMS IN AFRICA.

Mr. Bourgoin, a Naval Engineer, under the title of "Cooling of Living Rooms in Africa" deals particularly with the problem as it applies to the Soudan. His cooling scheme is by the dripping of water over a material which gives it a large surface, and the absorption of heat by evaporation is utilized for the cooling of air, a fan being used to increase the cooling effect. He also suggests the use of compressed air for cooling as well as the use of ice and mechanical refrigeration. His description is rather long and technical, but as he gives calculations showing energy expended and results obtained his paper is useful for reference purposes.

COOLING OF STOCK EXCHANGE, NEW YORK.

Henry Torrance, Jr., of New York City, in a paper entitled "Refrigeration and Ventilation of Inhabited Places" gives a description of the absorption refrigerating plant installed by the Carbondale Machine Company for cooling the large Board of Trade room of the New York Stock Exchange. About 40,000 cu. ft. of air per minute was cooled by passing it over brine pipes cooled by refrigerating machines. The air is first cooled to about 60° F. and then warmed slightly to about 15° below that of the outside atmosphere. The plant is operated only as required by high temperature weather conditions. The calculations show that the flow of heat into the room amounts to only 54 tons of refrigeration per day, while 276 tons of refrigeration were required to obtain the necessary result.

The magnitude of this installation is shown by the following data:

Size of room	1,240,000 cu. ft.
Temperature of room	75° to 78° F.
Outside conditions	85°—85% Humidity
Cubic feet fresh air blown in per minute	40,000 to 50,000
Maximum tonnage required to handle this work	276
Usual tonnage	200 to 220
Cubic feet in building per maximum ton	4,500
Square feet exposed area in building per maximum ton	670
Humidity of room runs	60% to 70%
Condensed water running off pan below bunker coils	2,000 lbs. per hr.

OTHER INSTALLATIONS FOR COOLING LIVING SPACES.

Several other refrigerating installations have been made for the cooling of telephone exchanges, schools, hospitals, theaters, banks, office buildings, etc. These installations have been described and their operation quite generally commented on by the public press, but as before stated, people generally are not as yet educated up to the desirability of artificial refrigeration of living rooms during the heated term. While we all "swelter" and suffer with the heat more or less each season it seems that this discomfort with its accompanying loss of efficiency, is quickly forgotten as soon as the heated term is past. Even those people who could well afford to make any reasonable investment for a few days comfort each year do not seem to be sufficiently interested to install the necessary apparatus.

The cooling of school buildings and educational institutions during the summer season has been suggested. Ordinarily students are given a vacation of three months or so during the summer, as much on account of the difficulty of concentrating on the work in hand as because the vacation is really necessary. Some of our prominent educators after considerable experience with special classes during "summer schools" are advocating what is known as the "all the year round" plan of education. Normal schools and colleges keep many of their buildings in use during the summer quarter and much of the best and most highly useful work is often done during this period. If refrigeration could be applied to school rooms during the extremely heated term students could work up to their maximum efficiency and not exhaust their vitality, as they do under unfavorable conditions of temperature and humidity at the present time. There are enormous investments in high school, college, normal schools and university buildings throughout the country, and it would really seem that by increasing this investment by a small amount the capacity of these institutions for usefulness could be greatly increased and for the reasons outlined.

There is really nothing difficult about the problem of cooling rooms to a comfortable and livable temperature during weather when the temperature ranges above 80° F. It is a simple engineering problem, and it depends purely on having the building or rooms suitably arranged to start with and understanding the natural laws governing refrigeration. Most any well-grounded refrigerating engineer can cool any given space to any temperature required and do it practically and economically. It is only a question of giving him means to do it with. Generally speaking it is impractical to take a building as ordinarily constructed and refrigerate it with success. The building should be designed with this in view. A prime requisite is that all doors and windows be kept closed during the period of refrigeration. Ventilation must be supplied through the means of cooling, and the Fan System is preferable in all cases. We look for increased interest in this matter and it is certainly only a question of time when all the better class buildings will have means of cooling in hot weather as well as means of heating in cold weather.

The author may be pardoned for enthusing over the proposition for the reason that this comment is being dictated in an office temperature of 100° F.

CHAPTER XL.

ACCOUNTING.

ESSENTIAL FEATURES.

The essential feature is a record book in which entries are made when goods are received, and when goods are delivered. The auxiliary books consist of an "In" receipt book on which the original entry of delivery to the warehouse is made and an "Out" receipt book on which the original delivery entry of goods going out is made. In addition there is a negotiable warehouse receipt book. This simple set of books is subject to endless variation depending on the complication and character of business handled, and these books it should be borne in mind in no way form a part of a double entry system of bookkeeping, but are purely books of memorandum and record.

When goods are received at the warehouse for storage, an "in" receipt is given showing date, lot, number, and, if practicable, the room and section of house where stored. An entry of this kind should give number of packages with their weight if storage is charged by weight and any special marks if necessary. Goods as received are commonly entered on the left hand side of a double page in the storage record and as delivered out are entered on the right hand side, and thus a subtraction of the entries on the right hand page from the entries on the left hand page will give the quantity of goods of any particular mark, or lot number, or kind, in storage for that particular account at any time.

Auxiliary books to the storage record consist of what may be called an "In" receipt book on which is entered a tally of the goods as above outlined, and from which the entries into storage record are made. The "In" receipt book is commonly in duplicate or possibly in triplicate and one copy is either given to the drayman when the goods are delivered or mailed to the owner. They are numbered consecutively and the number of the "In" receipt may conveniently correspond with the lot number of the goods as they are received. If a customer requires a negotiable warehouse receipt he may return the "In" receipt as evidence that the goods were delivered by him to the storage house and receive a negotiable receipt which carries title to the goods in storage, a form of which is shown in Fig. 1.

The "Out" receipt book may be similar in form and arrangement to the "In" receipt book, and is used essentially as in the "In" receipt book only this book checks deliveries from the house to the owners, and from the out receipt book entries are made on the storage record book. When goods are delivered out, the person receiving them signs the "out" receipt stub and takes one copy by which to check his load.

Charges for storage may be made on the storage record and posted to the journal monthly, or charges may be made and posted as goods go out. This is, however, a mere question of arrangement between the cold storage house and the owner of the goods. It is becoming more and more common to render bills for storage each month, and it is really no more than fair. The storage house, for instance, which stores eggs chiefly, otherwise would have no income until the stock began to go out in the fall of the year.

The laws of the various states compel those who are doing a storage business for others or what is known as a public storage business to keep a storage record, and to have a printed form of receipt, and as the perishable goods business is now handled, a negotiable warehouse receipt is positively necessary as goods in storage are used largely as collateral for bank loans.

An important feature of handling goods for storage is to mark them distinctly if they have not already a distinctive mark, and as a matter of safety it is better to number each lot as received consecutively for the reason that the lot number will then designate the time when stored, and in case packages should be used only once there would be no liability of confusion. Should customers object to having the lot number stamped directly on the package, tags may be attached and lot number stamped on the tags. Positive identification of every package of goods and its location in the house is a very important feature of warehouse accounting.

The above gives in outline the simple requirements of an accounting system for cold storage houses. There may be a considerable variation from the details without omitting the essentials. Some of the large houses use a card system exclusively, while others have some such system as is employed by the Milwaukee Cold Storage Co., Milwaukee, Wis., blanks of which are shown herewith and which President John A. Hill has outlined as follows:

On receipt of a consignment, receiving clerk fills in triplicating blanks with date, name of owner, quantity and description of goods, weight if necessary to show it, lot number, location in warehouse, owner's marks, railroad car number, and his signature. The charge bill No. 2, printed on yellow paper, and No. 3, printed on green paper, are kept in receiving clerk's office, and original accompanies the goods to storage room, where quantity is rechecked by "Jones," who returns No. 1 to receiving clerk, who files all three copies in general office. Here the charges are filled in, and the storage rate; then No. 3 is given to teamster, if goods came in by wagon, or sent to owner, as his receipt. The original, after entering the lot on the storage record, is filed, and the yellow charge bill is entered on a weekly statement, the total charges on weekly statement being taken to ledger account. All charges on goods in and out for a week are shown on the weekly statement, and vouchers for each item mailed with the weekly statement.

On goods going out the general office fills in the date on white triplicating sheets, also the quantity and description of goods, lot number, location, owner's name and name of firm getting the goods, if other than owner. The plain, unprinted and punched copy is retained in general office in case of loss of others; the original and No. 2 go to delivery clerk, who fills the order, and signs his name to original under "checked." Teamster signs original "John Green," and takes No. 2 for his load check by owner. Original is returned to general office, where time, rate and charges are filled in, and amount entered on weekly statement. Delivery is charged off stock record, completing the transaction.

This system works satisfactorily for our needs, but the No. 3 green sheet, which is the non-negotiable receipt, should be worded to conform to the warehouse laws of the State where used.

ACCOUNTING FOR A COMBINED ICE MAKING AND COLD STORAGE PLANT.

Combined ice making and cold storage plants are increasing in number rapidly, and surely the ice plant is able to op-

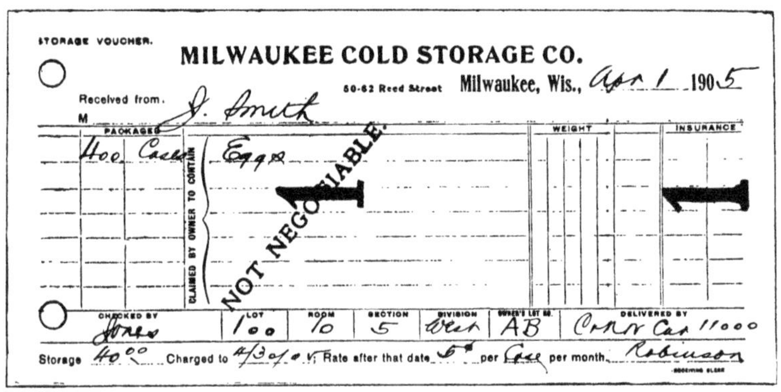
STORAGE VOUCHER.

MILWAUKEE COLD STORAGE CO.

50-62 Reed Street Milwaukee, Wis., Apr 1 1905

Received from M J. Smith

PACKAGES	CLAIMED BY OWNER TO CONTAIN	WEIGHT	INSURANCE
400 Cases	Eggs		

NOT NEGOTIABLE. 1

CHECKED BY	LOT	ROOM	SECTION	DIVISION	OWNER'S LOT NO.	DELIVERED BY
Jones	100	10	5	West	AB	C.N.W. Car 11000

Storage 40.00 Charged to 4/30/05 Rate after that date 5¢ per Case per month. Robinson
RECEIVING CLERK

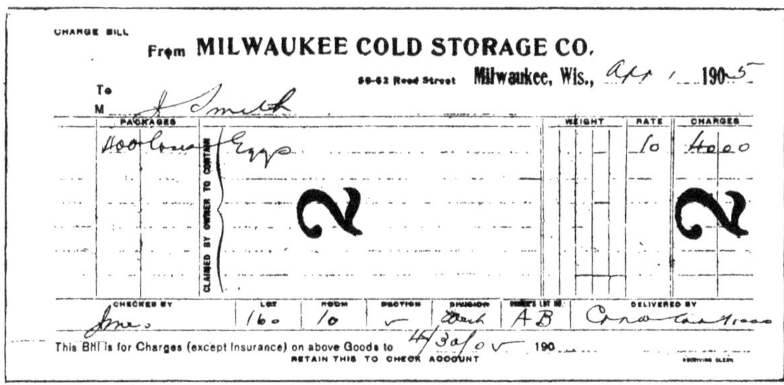
CHARGE BILL

From MILWAUKEE COLD STORAGE CO.

50-62 Reed Street Milwaukee, Wis., Apr 1 1905

To M J. Smith

PACKAGES	CLAIMED BY OWNER TO CONTAIN	WEIGHT	RATE	CHARGES
400 Cases	Eggs		10	40.00

2

CHECKED BY	LOT	ROOM	SECTION	DIVISION	OWNER'S LOT NO.	DELIVERED BY
Jones	100	10	5	West	AB	C.N.W. Car 11000

This Bill is for Charges (except Insurance) on above Goods to 4/30/05 190...
RETAIN THIS TO CHECK ACCOUNT
RECEIVING CLERK

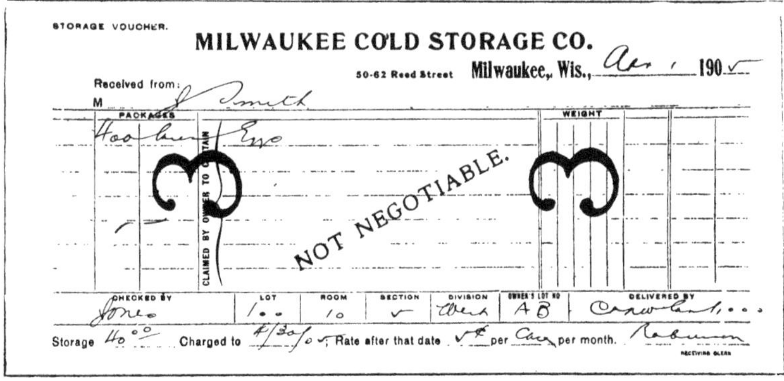
STORAGE VOUCHER.

MILWAUKEE COLD STORAGE CO.

50-62 Reed Street Milwaukee, Wis., Apr 1 1905

Received from M J. Smith

PACKAGES	CLAIMED BY OWNER TO CONTAIN	WEIGHT
400 Cases	Eggs	

NOT NEGOTIABLE. 3

CHECKED BY	LOT	ROOM	SECTION	DIVISION	OWNER'S LOT NO.	DELIVERED BY
Jones	100	10	5	West	AB	C.N.W. Car 11000

Storage 40.00 Charged to 4/30/05 Rate after that date 5¢ per Case per month. Robinson
RECEIVING CLERK

FIG. 1.—ORIGINAL, DUPLICATE AND TRIPLICATE FORM OF "IN" RECEIPT BOOK.

WITHDRAWAL VOUCHER

Milwaukee, 9/15 1905 MILWAUKEE COLD STORAGE CO.

	REC'D			LOT	ROOM	SEC	WEIGHT	STORAGE FROM	TIME	RATE	CHARGES
Ent'd	4/1	50	cs Eggs	100	10	5		4/30	5	25	12 50

DELIVERED ON ORDER OF J. Smith TO Brown & Co.

APPROVED [illegible] CHECKER Jones RECEIVED THE ABOVE DESCRIBED PROPERTY John Green

WITHDRAWAL VOUCHER

Milwaukee, 9/15 1905 MILWAUKEE COLD STORAGE CO.

	REC'D			LOT	ROOM	SEC	WEIGHT	STG. FROM
		50	cs Eggs	100	10	5		

This Copy is to be taken by Teamster, and retained by owner to check current account. **2**

DELIVERED ON ORDER OF J. Smith TO Brown & Co

APPROVED CHECKED RECEIVED THE ABOVE DESCRIBED PROPERTY

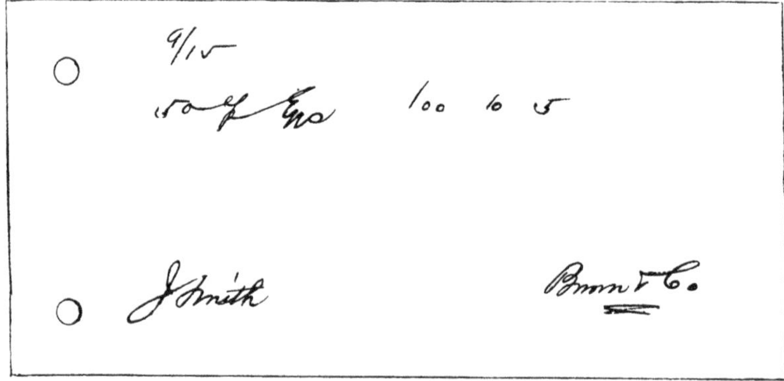
9/15

50 cs Eggs 100 10 5

J Smith Brown & Co.

FIG. 2.—ORIGINAL, DUPLICATE AND TRIPLICATE FORM OF "OUT" RECEIPT BOOK.

erate a cold storage plant in connection at much lower cost than an equipment of machinery operated for cold storage alone. If the ice manufacturer could only know how much he is actually making out of his cold storage department, doubtless more attention would be given to developing this end of the business. The following from one of the author's most esteemed friends will no doubt prove valuable to those who also "want to know." While the method in finding cost of ice making is doubtless subject to some correction for time of year, and changing efficiency of machine, etc., yet no better way is apparent, and a close approximation is so much better than the methods of estimate and guess work commonly employed.

As you know, we operate ice-making plants, and in connection therewith cold storage houses. When the cold storage end of the business amounted to but little, it was our custom to credit in our monthly cost sheets all the earnings of the cold storage houses to ice making. This, at certain times of the year, made a considerable reduction in the apparent cost of making ice, but it was perhaps a good enough method of accounting until the cold storage business grew to greater magnitude. But, as this was gradually taking place, we became somewhat dissatisfied with the fluctuations shown in our ice cost, due to the full amount of the cold storage earnings being credited to the cost of ice making. We saw that this process was giving us a somewhat fictitious cost for our ice, which, in months when we were doing a heavy cold storage business was shown to be much less than in other months when we were making just as much ice, and under just as favorable conditions, excepting for these cold storage earnings. While these earnings added to the profit of our company, they had, beyond the actual cost of furnishing the refrigeration, nothing to do with the true cost of making ice.

We then began to grope about for some method by which we could make a proper charge for refrigeration, to our cold storage department, this same item, the actual cost of the refrigeration only, to be credited to ice making, and not any profit which there might be in the cold storage, just as though it were owned by a separate concern. We obtained from various sources what information we could as to the generally accepted cost of refrigerating a given number of cubic feet to certain temperatures. But, we imagined that our insulation was perhaps better than the average, and that other conditions with us might be so different from those prevailing in other plants that we wished to have some rule based on our own experience which would give us a proper charge for refrigeration to be made against our cold storage houses, and a corresponding credit to be made to our ice-making department. It so happened that we were able to operate at ice making alone, during months when there was practically nothing in our cold storage houses. In this way, we obtained the cost of fuel when we were running at different speeds. Let us say, for example, that we found we spent so much for fuel when we were making half our capacity of ice, and so much, when we were making our full capacity of ice. This also gave us the fuel cost per ton of ice made under these different conditions. Then, in months when we were doing work in the cold storage department, whether we were making much ice or not, we took the amount of fuel consumed, and compared this with the amount used in one of such months as described above. In this way, we take the quantity of fuel consumed and say that this quantity would have made, upon the ratios determined as above, so many tons of ice. Then, we find that we did make so many tons during the month. And hence, the equivalent of the balance, in tons of ice, at the cost price shown for the month, is the charge which we set up against the cold storage department for refrigeration. This amount is credited to ice making, by our cost sheet showing that we made the number of tons actually produced, and that the equivalent of the number of tons arrived at as above, was furnished the cold storage department in refrigeration. This number of tons is added to the amount actually produced, and then this total is divided into the sum expended in operating the ice department. This department thus gets credit for the refrigeration furnished the cold storage, and that department only pays actual cost for this refrigeration.

We also have separated our investment in cold storage buildings, and land occupied thereby, from the investment items shown for our factory, and we keep a separate record for all items of insurance, interest, taxes, depreciation, repairs, and labor for these two departments, so that each has its own system of accounting, just as though operated by different owners. This seems to us to give an accurate and fair charge based upon our own actual experience month after month, and not upon a theoretical rule as to the cost of refrigeration, even though that may have been made after long experience, in many plants operating under varying conditions.

In these days of modern accounting, other managers of ice making and cold storage plants may have been confronted by this same problem, and I am giving you the above for what it may be worth to these men.

CHAPTER XLI

THERMOMETERS.

HISTORICAL.

Correct temperatures being the most important matter to be considered in the successful refrigeration of perishable goods, it follows that some accurate instrument must be used for measuring same. The common instrument in use is called a thermometer. The word thermometer is of Greek derivation and means a *measure for heat.* This instrument is constructed on the well-known principle that heat expands all bodies. After Fahrenheit, the name which our common thermometer bears, and who was a native of Dantzig, failed in business, he turned his attention to mechanics and chemistry.

He began a series of experiments for the production of the thermometer. And it is owing to his determination to succeed, and to his loyalty to the conviction that he must give to the world the instrument which has proved so serviceable to mankind, that we are enabled to have a definite way of speaking of hot, or very hot; cold, or very cold. For his first instrument, Fahrenheit used alcohol. But before long he became convinced that a more suitable article to use in the glass tube was the semi-solid mercury. By this time, about the year 1720, Fahrenheit had moved from Dantzig to Amsterdam. And here, in the capital city of Holland, he made the mercury thermometer.

The basis of Fahrenheit's plan was this: To mark on the tube the two points respectively at which water is congealed and boiled, and graduate the space between. He commenced with an arbitrary marking, beginning with 32 degrees, because he found that the mercury descended 32 degrees before coming to what he thought the extreme cold resulting from a mixture of ice, water and sal ammoniac. In 1724 he published a distinct treatise on the conclusions that had resulted therefrom.

Not long afterwards, Celsus, a Swedish scientist, produced the centigrade thermometer, which suggested the graduation of 100 degrees between the freezing and boiling point. Reaumur, a French scientist, also proposed another graduation, and one which has been accepted by the French.

The mercury thermometer graduated to the Fahrenheit scale, with the freezing point of water at 32° F., is the most common, and, in fact, practically the only instrument in use in America. The accuracy of some of the cheaper thermometers cannot be depended upon. The ordinary cheap thermometer where the figures are stamped on the scale and with graduations separate from the stem are inaccurate sometimes to the extent of five degrees to ten degrees. It does not pay to use cheap thermometers for cold storage purposes. Often those given as advertising matter are in use, and some years ago an eastern firm sent out a large number of such thermometers as prizes. One of these was used for reading outside weather temperatures. It was finally discovered that the readings were more than twenty degrees out of the way. This is exceptionally bad, but it is frequently the case that the thermometers are two, three or four degrees incorrect, and, in fact, the greater part of the common thermometers are two or three degrees or more off. Of course, where a number of thermometers are in use, it is not probable that any one of them could be very much off without it being found out, and it is easy to hang them all in one room and make a comparison. On receiving a new lot of thermometers it is a good scheme to hang them up side by side in the cold storage room with a thermometer which has been in use and known to be fairly accurate. Then if any one of the new thermometers is incorrect it may be returned to the manufacturer. Thermometers are not expensive, even the high-class ones. At a cost of from $15 to $24 per dozen, special cold storage thermometers graduated, say, from zero to 70° F., may be obtained. The best ones have the graduations (scaled to read to one-half of a degree) etched on the stem and the figures on a metal or enamel scale. In this way, if they are correct when tested they cannot become materially incorrect without being broken. These thermometers cost more than those which are not etched on the stem, but are more reliable. It is important that a thermometer for cold storage rooms should be graduated with one-half degree marks. Then a slight variation in temperature may be quickly noticed. Some

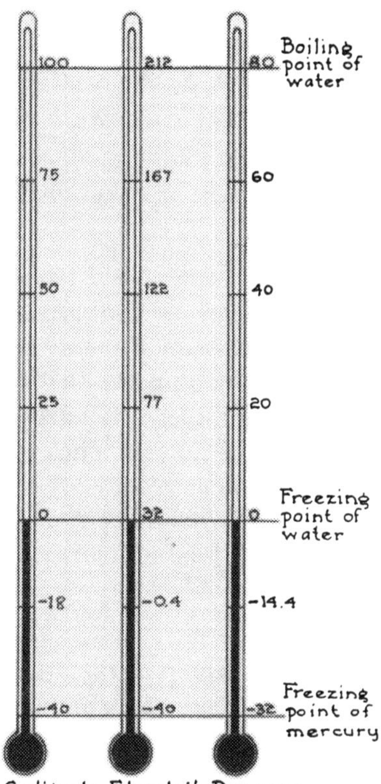

FIG. 1.—COMPARISON OF CENTIGRADE, FAHRENHEIT AND REAUMUR.

thermometers are only graduated on one-degree or two-degree divisions, in which case it is difficult to notice a variation of less than one-half a degree to one degree. It is a good plan to gather up thermometers not in use and have one certain place for hanging them in a cold storage room. Then they may be compared, and if any of them are injured or inaccurate from any cause it will be discovered. It is a good plan to make such comparison once each year, perhaps in the winter or when few goods are in storage.

COMPARISON OF CENTIGRADE, FAHRENHEIT AND REAUMUR

°C. to °R., multiply by 4 and divide by 5.

°C. to °F., multiply by 9, divide by 5, then add 32.

°R. to °C., multiply by 5 and divide by 4.

°R. to °F., multiply by 9, divide by 4, then add 32.

°F. to °R., first subtract 32, then multiply by 4 and divide by 9.

°F. to °C., first subtract 32, then multiply by 5 and divide by 9.

On the Fahrenheit thermometer, the freezing point is 32 degrees above zero, while on the Centigrade thermometer the freezing point is at zero. The size of the degree on the Fahrenheit instrument is smaller than on the Centigrade, the boiling point of water being represented by 212 degrees on the former as compared with 100 on the latter. Thus it will be seen that on the Centigrade thermometer the difference between the freezing and boiling points is 100 degrees, while on the Fahrenheit thermometer it is 180 degrees, or the difference between 32 and 212.

A Fahrenheit degree is only five-ninths of a Centigrade degree, and accordingly a degree of the latter is nine-fifths or one and four-fifths times the size of a Fahrenheit degree.

In working out the temperature in terms of the Centigrade thermometer, assuming that the Fahrenheit thermometer registered 32 degrees below zero, it would first be necessary to add the 32 degrees below zero to the 32 degrees above zero, because the freezing point starts at 32, and the sum would be 64 degrees. As a Fahrenheit degree is only five-ninths the size of a Centigrade degree, 64 should be divided by nine, giving a result of 7.11, which, multiplied by five, would give 35.55, the number of degrees below zero on the Centigrade thermometer.

THERMOMETER SCALES COMPARED.

On page 769 a diagram is shown with the thermometer scales in common use compared. The Fahrenheit scale is used most exclusively in America, while the Reaumur and Centigrade scales are in common use in foreign countries. The freezing and boiling point of each are made to show the figure used at these points, and thus a rough comparison may be made.

CONVERSION OF THERMOMETER DEGREES.

C.	F.	R.
10	50.0	8.0
9	48.2	7.2
8	46.4	6.4
7	44.6	5.8
6	42.8	4.8
5	41.0	4.0
4	39.2	3.2
3	37.4	2.4
2	35.6	1.6
1	33.8	0.8
Zero	32.0	Zero
—1	30.2	—0.8
2	28.4	1.6
3	26.6	2.4
4	24.8	3.2
5	23.0	4.0
6	21.2	4.8
7	19.4	5.6
8	17.6	6.4
9	15.8	7.2
10	14.0	8.0
11	12.2	8.8
12	10.4	9.6
13	8.6	10.4
14	6.8	11.2
15	5.0	12.0
16	3.2	12.8
17	1.4	13.6
18	0.4	14.4
19	2.2	15.2

RECORDING THERMOMETER.

The recording thermometer has come into use rapidly during the past few years, and has a useful purpose in many places. For the private cold storage plant which is equipped with a perfect system of refrigeration in which the temperature is closely under control, it is a matter of great satisfaction to the

FIG. 2.—SELF-CONTAINED TYPE RECORDING THERMOMETER.

owner to make a record of the perfect temperatures produced and maintained. For the concern or man who depends on hired assistants or engineers to run the plant and control the temperatures, a record of the work done is almost a necessity. For the plant doing a public cold storage business, and where a mere clerical or watchman record is liable to be questioned by dissatisfied customers, a continuous record by a recording thermometer sets all doubt at rest, and may prove useful in a court of law.

There are various styles and makes of recording thermometers on the market, and the illustration shows that known as the Self-Contained Type Columbia Recording Thermometer.

A chart graduated from 10 degrees below zero to 50 degrees F. above zero might be used for freezers, and one graduated from 20 degrees F. above to 70 degrees F. above zero for high-temperature cold storage rooms. An instrument making one revolution in seven days would be preferable, although one making a revolution in twenty-four hours might be used if it could be given close attention.

A type of recording thermometer known as the "Distance Reading Thermometer" can be applied where the dial may be located at any reasonable distance from the space in which the temperature is desired to be recorded. Flexible steel tubing is used for making the connection.

CHAPTER XLII.

MISCELLANEOUS.

INTRODUCTORY.

The following miscellaneous information and facts bearing on cold storage matters is added here in the form of notes or paragraphs for the reason that a larger part of it cannot be properly classified under the regular chapter headings of the balance of the book. The various products which are placed in cold storage and treated here in a brief manner are not generally as important as those which are treated under the various chapter headings. No doubt, as the business is developed, many of the products here mentioned will become more generally important and the information obtainable will be sufficiently in detail so that they may be treated as a separate chapter. There are many products not mentioned in detail which are given in the alphabetical list of correct temperatures for cold storage.

RURAL SITE FOR COLD STORE.

The advantages of a location somewhat remote from the large cities as the site of a cold storage house for perishable products, especially the more sensitive, like butter and eggs, is well known, and it is the author's opinion that the tendency is more and more toward the establishment of storages for the greater bulk of products and perishable goods in the country at or near the locality where produced. The pure country atmosphere is far better than the gaseous and comparatively polluted air of our large cities. This is especially true during cool or cold weather when the purifying effect of the refrigerating surfaces is very small or entirely absent. It has been remarked to the author by a prominent produce dealer that eggs and butter would keep better in cold storage in the air where they were produced than they would if shipped to a distance and stored in a different atmosphere. There is considerable question about this but there seems to be no question whatever about the advantage of a rural location for the perfect keeping of perishable goods.

Further than this operating expenses will be less, and goods will be much fresher and in more perfect condition for storage when delivered at the storage house. The owner also has the advantage of being able to ship to the market where the best price is to be had, and the goods are at all times under his oversight and control.

SHAVINGS AND SAWDUST AS INSULATION.

The comparative value of sawdust and mill shavings for insulation purposes has been a much discussed problem among cold storage men. Thoroughly dry sawdust is beyond doubt a better insulator than mill shavings, but thoroughly dry sawdust is practically out of the question in anything like sufficient quantity to make it a commercial practicability. The greater part of the sawdust available is wet or green just as it comes from the mills, sawed generally from wet logs from the water. Mill shavings on the contrary are generally fairly dry, being taken from the surface of the lumber which dries out very quickly after being sawed. Mill shavings though partially green or damp are far better to use than sawdust which is even slightly so. If shavings are thoroughly rammed into the wall they will not settle down in drying out, as they are elastic and will hold their place. Sawdust on the contrary cannot be packed sufficiently tight so that it will not settle down when it dries out, leaving open spaces in the insulated wall. Dry sawdust, in case it becomes wet or damp from any cause, at any time after it is placed in the wall, is liable to heat or ferment and disintegrate or dry rot, and so lose its insulating value. Baled mill shavings which are received in a damp condition may be much improved by shaking out the bales and allowing them to lie exposed to a free circulation of air. Stirring them occasionally will also facilitate the drying process. They also may be kiln dried without opening the bales if facilities are available. (Chapter on "Insulation" gives some further information.)

PAINTING METAL SURFACES.

The question as to whether it is advisable to paint galvanized iron piping, galvanized iron tanks and other iron and steel surfaces, whether galvanized or not, is one which comes up very often in cold storage practice. In many cases it is not advisable to paint, as it is cheaper to renew the piping, etc., than it is to go to the expense of painting periodically. In other cases it is a positive detriment to paint piping in a cold storage room. An odor may be created which cannot be easily removed.

Whether or not it is advisable to paint metal surfaces depends a great deal on whether they are readily accessible and easily cleaned or not. Good galvanized surfaces should stand well without painting and this is one of the reasons why galvanized surfaces are used instead of black. For painting metal surfaces use nothing but the very best obtainable preparations. A number of these are on the market and sold at a reasonable price. A good homemade paint may be prepared from red lead and boiled linseed oil. It will require about twenty-five pounds of dry lead to a gallon of oil. A pound of lamp black to each twenty-five pounds of lead will give a rich dark brown color which is more agreeable than the natural color of the red lead. (See also chapter on "Keeping Cold Stores Clean.")

DEODORIZING COLD STORAGE ROOM.

Questions are frequently asked the author in regard to properly deodorizing cold storage rooms which have been used for the storage of vegetables, fruit, etc., so as to make them suitable for the storage of sensitive products like butter, eggs, etc. In some cases it is possible to do so, in others not. If rooms have been used for some years for some strong-smelling product and have not been properly aired out, whitewashed, etc., it is hardly possible to disinfect them sufficiently to make them available for the storage of eggs or butter. Thorough ventilating and whitewashing will do a great deal, however, to improve rooms in a bad condition. (See chapter on "Keeping Cold Stores Clean.")

WHITEWASHING.

Whether or not it is necessary to whitewash cold storage rooms each year is a problem which comes up very often. The most thoroughgoing cold storage manager insists that rooms which are used for the storage of eggs must be whitewashed every year while the rooms are empty, and the author recommends that it is a very good practice to whitewash all other rooms yearly as well as egg rooms. It is perhaps not absolutely necessary but if a rule is established of yearly whitewashing it will be attended to, whereas if it is understood that it is only necessary to whitewash occasionally, the rooms may not be whitewashed for several years, or possibly not at all. The expense is comparatively small and the work is usually done at a time of year when there is very little for the help to do in the regular line of business. Practically speaking the cost of whitewash is nothing, and it is a sort of insurance against must and mold. It will not absolutely prevent must and mold, but if carefully done it will at least demonstrate the fact that these troubles are not caused by the bad condition of the interior wood work of the room itself.

HUMIDITY.

The influence of a large or small quantity of goods stored in a given space on the humidity of such space is not generally considered or understood. The pipes which cool a storage room act as moisture absorbing surfaces. These moisture absorbing surfaces are constantly in action, as it requires the same amount of refrigeration to maintain the temperature of an empty room as it does a room which is filled with goods. If a room is only partly filled with eggs, for instance, the humidity of such a room will be much lower (i. e. much less moist) than would be the same room filled to its capacity. This point is of no consequence where there is no attempt to regulate humidity, but in a modern and progressive plant these things should be watched closely. A room partly filled with eggs and carried through the season will certainly turn out shrunken or evaporated eggs as compared with a similar room which is carried through the season filled to its capacity. Owing to the fact that a room filled with goods will be much more moist than one only partly filled, in some cases where circulation is imperfect this will lead to must or mold. A complete and thorough system of air circulation makes the control of conditions of cold storage rooms very simple. Such a room may be filled with goods just as full as possible without any bad effect resulting.

SPACE REQUIRED FOR STORING VARIOUS PRODUCTS.

In estimating the storage capacity of a cold storage room or house it is frequently convenient to know the space required by different products. The following is given as the author's personal experience: A sixty-pound tub of butter will require about 2½ cubic feet of space; a case of eggs containing thirty dozen requires about 3 cubic feet of space; a sixty pound box of cheese requires about 2 cubic feet and a three-bushel barrel of apples from 8 to 10 cubic feet. These figures allow a reasonable margin for what is known as "piling alleys;" that is, space leading from the door back through the piles of goods so that different lots may be accessible. The actual cubic space required therefore is somewhat less than the figures given. These figures have, however, been found in actual practice to work out quite closely. The space required by any particular product may be ascertained by finding the cubic space required for each package and adding thereto from 15 per cent to 25 per cent. In making such calculation it is necessary to figure the actual space occupied by each package as it will be placed in stowing in the house. It is perhaps unnecessary to state that the capacity of a given room depends greatly on how carefully the goods are piled. The careless man may waste a large amount of space which might be saved by careful work.

TRUCKS FOR HANDLING GOODS FOR STORAGE.

The correct method of handling goods, when delivered by car or by wagon at the platform, into the storage rooms, and again from the storage rooms to the car or wagon and platform, appears to be a simple matter; nevertheless, there is a vast amount of valuable time wasted by not adopting the most practicable and labor-saving method. In smaller plants a hand truck with two wheels on which may be handled three or four tubs of butter or cases of eggs is best adapted to the work providing the distance which is to be covered is not too great. These hand trucks are very convenient for the handling of small loads even in comparatively large plants, and therefore should be provided even where the larger trucks are generally used. In comparatively small plants a four-wheel truck 30 inches wide and $5\frac{1}{2}$ feet long is the most convenient means of handling goods in and out of the storage room. A truck of this size is all that one man can conveniently handle when it is fully loaded with goods. In larger plants a wider and somewhat longer truck may be used, in which case it is necessary that the doors and corridors be proportioned wider to allow proper turning of same. The four-wheel trucks are generally of two types. One with two large wheels about in the center and a smaller wheel or caster at either end, and another kind with two large wheels a short distance from one end and two castors at the other end. If there are inclines which must be traveled this latter type is by far the best and they are now generally used. Trucks should be provided with a hand rail at one end only. For hand trucks the regular cheese truck of a somewhat modified type is the most convenient, as boxes, barrels, tubs and cases may be readily handled on same. The foot of a cheese truck is generally half round. By special order these can be made square and should be placed at such an angle with the frame that the truck will stand up without leaning against anything when not in use.

STOWING GOODS IN COLD STORAGE.

In piling goods in the refrigerated room there are two objects which should be borne in mind. The goods must be so stacked or piled that they may be subjected to the best possible conditions, and they must be so stacked or stored that they will occupy the least room consistent with the keeping of them in good condition. It is also desirable to handle goods in such a way that they may be easily stored and quickly removed as they are wanted. It is, however, hardly necessary to speak of this, especially as most warehouse foremen will see to it without being told, that he does not do any more work than is necessary in getting the goods in and out. He is much more likely to sacrifice his space unnecessarily, or carelessly pile goods, so that they may be injured while in storage.

Goods may be materially injured by piling to a great height without properly providing for sustaining the weight which tends to crush the lower tiers, or by piling goods so closely in a compact mass that the air cannot circulate between and around the packages. A familiar instance of damage from crushing will be seen in almost every warehouse when apples are being removed in the spring of the year. If apples are piled more than five or six tiers in height they should be piled one barrel directly above another with a 2 x 4 inch strip on each end between each tier so that the weight of the upper tier of barrels is supported on the heads of the barrels and not on the center or bilge. Damage from too close piling often occurs when eggs are stored without placing strips between the cases and leaving spaces on the sides and ends. Directions for each separate product cannot very well be given here, but by bearing in mind the principle which makes it necessary to pile goods so that the air may circulate freely, the cold storage manager can determine for himself what is necessary in connection with the particular product he is handling. Goods like eggs, cheese, apples, oranges, or any products which give off moisture, must not be piled closely. If such goods are piled so that the air cannot circulate freely through them they are liable to become moldy and musty from the collection of moisture in the centre of the pile. On the other hand goods like butter, canned fruit, dried fruit, etc., cannot be piled too closely, as these goods do not give off moisture or at least it is not necessary that they give off moisture in order that they be preserved properly. The same is true of the greater proportion of goods which are actually frozen, like poultry, fish, etc. Where carefully kept, as much as possible, from contact with air the better the results, generally speaking.

What is said above will make it clear to the reader why a forced circulation of air is better than a gravity circulation, both as regards economy of space in storing goods and the best results obtainable. The chapter on "Air Circulation in Cold Stores" treats this subject thoroughly.

REMOVING GOODS FROM STORAGE.

The reasons for the sweating of goods when removed from cold storage to the comparatively warmer outside atmosphere are not generally understood. This phenomenon is caused by a condensation of the moisture on the cold surface of the goods and may be prevented by protecting them from direct contact with the warm outside air. A method which has been used with success is to pile the goods closely on the receiving room floor and cover them tightly sides and top with a tarpaulin or heavy canvas like a wagon cover. It will take somewhat longer for the goods to acquire the temperature of the outer air but they may be warmed in this way without sweating. In extremely warm weather it may take thirty-six hours or possibly forty-eight hours, but in comparatively cool weather, if goods are removed at night and covered in this way, they may be ready for delivery the next morning. This method of handling is only possible where sufficient notice may be had in advance and is particularly useful for the removing of eggs from cold storage during a warm spell in summer or early fall. This method of treatment prevents the depositing of moisture or "sweating" and the goods are warmed more slowly, which aids greatly in their preservation. Those who operate their own cold storage plant in connection with the produce business will find this method very useful and beneficial as they can generally anticipate their needs. For the removing of eggs before candling it is especially desirable, as it always musses and soils the eggs to handle them while damp, to say nothing of the actual damage to the quality likely to occur.

SLOW COOLING OF GOODS FOR COLD STORAGE.

Economy of cooling goods which are to be stored at low temperatures, by stages, is not perhaps well understood. Take for instance: Butter which is now sometimes stored at zero and below. It is not only far more economical but it is better for the goods to cool gradually than to place them immediately in the room which is carried at the extremely low temperature of zero. If a temporary cooling room at a temperature of say 25° to 35° F. be provided, where the butter could be run in temporarily before placing in the sharp freezer, it would be easier to maintain a low temperature in the freezer and at the same time result in a better carry in the stored goods. It also makes it unnecessary to provide a large amount of piping in the permanent sharp freezing room. What is said here cannot be applied to poultry or other goods which deteriorate rapidly if not frozen. Poultry, for instance, as generally frozen in the large cities, is sometimes from one to two weeks killed before it is finally placed in storage. Under these conditions it is necessary to freeze as quickly as possible. Any product, however, which does not deteriorate rapidly is kept better, when cooled slowly when placed in storage and warmed slowly when removed from storage, than the reverse. Some storage people have an idea that if goods are frozen quickly the original flavor and aroma will be preserved, which, if the goods are thawed slowly, will be better retained than if cooled or frozen by stages. This idea is erroneous so far as it applies to any goods known to the author.

STORING VARIOUS PRODUCTS IN THE SAME ROOM.

There is a strong temptation in the comparatively small cold storage plant, say, for instance, one which is operated in connection with an ice factory, and where one or two or a few rooms at the most are available, to store several different products in the same room. This is in fact done as a matter of regular practice and is one of the reasons why the small auxiliary cold storage plant is not considered a success. Satisfaction cannot be given the owners of a miscellaneous line of goods which are all stored together and at the same temperature. There are but few classes of goods which may be successfully stored in the same room and at the same temperature for comparatively long periods with good results. For instance: Butter requires a lower temperature than cheese, and fruit and vegetables a higher temperature than cheese, so they cannot be successfully stored in the same compartment for any length of time. Generally speaking, each product should be stored by itself, but for short periods of a few days, it is customary to use a room for the storage of several different products like cheese, butter, fruits, eggs, etc. If a small quantity of eggs or butter is stored in a room with a large quantity of fruit and vegetables they are very likely to absorb an odor within a very few days. Nothing definite, of course, can be stated in this respect, as conditions vary widely, especially as to ventilation. For anything like regular cold storage purposes it is not only advisable but almost absolutely necessary to provide different rooms for different products. This, however, may be qualified to some extent by storing fruit and vegetables together under favorable conditions. Dried fruits, nuts, flour, etc., known as grocers' sundries, are also stored in the same room at a temperature varying usually from 35° to 45° F. Any temperature under 45° F. will keep them in fair condition. See proper heading for temperature at which to store various goods.

MOLD IN COOLING ROOM.

Troubles from mold forming on the walls or ceiling of meat rooms or other small temporary storage rooms which are used by retailers and others is quite frequent. This is caused in a large number of cases by a lack of circulation of air in the room; by improperly locating the ice bunker or cooling pipes; or by improperly locating a door, or the excessive opening of same. Mold always results from a surplus of moisture and comparatively high temperature. If the circulation is inferior the air near the ceiling of the room becomes charged with moisture and this may be deposited on the ceiling or side walls. This will quickly cause a growth of mold. If a door is left standing open in warm humid weather the warm air rises to the ceiling of the room and is condensed thereon. This also causes mold. If the cooling surfaces and the door into the room are properly located with reference to each other, the warm air which comes in when the door is opened will come in contact first with the cooling surfaces. Mold which has formed may be removed by wiping with a damp cloth. Small retail rooms may be whitewashed, or if it is desirable to wash them out frequently, shellac finish is best.

STORING EGGS AND LEMONS.

It is absolutely unsafe to store eggs and lemons in the same building. This has been done in a few cases without damage to the eggs, but it cannot be recommended for the reason that some heavy losses have been sustained from the eggs becoming flavored with the odor of lemon. Eggs which have become flavored even slightly with this odor are almost unsalable and will not be taken by the best class of trade. Oranges as well as lemons will cause this trouble. The citrous fruits, as they are called, after being in storage for some length of time give off a gas which is very penetrating. It has even been claimed that this gas would penetrate a solid brick wall. This is hardly probable, but the fact remains that it is very dangerous to store citrous fruit in the same building with sensitive goods like butter and eggs. The best large houses have separate buildings detached from their main building for the storage of citrous fruits and other odorous products, and the first-class smaller houses have rooms independent of their main building. In the designing of houses for the handling of general products including fruits, etc., the author recommends, and in his regular practice plans to have, a room of this character which is entered from a separate outside entrance. The room may possibly be in the same building but it is better to erect it as an addition to the main building.

STORING VEGETABLES IN CELLARS.

Vegetables are not usually placed in cold storage, but better results may be obtained from cold storing than by storing in the old fashioned way in the cellar. A few notes on cellar storage, however, are here given as a matter of general information. A suitable cellar for the storage of vegetables during winter must be clean and should either have a cement floor or a clean sand floor. It should have a few openings to the outside atmosphere which are provided with curtains to exclude the light. The correct temperature for most vegetables is a few degrees above the freezing point, say from 35° to 40° F. It is very difficult or impossible to regulate the humidity of a cellar. It should not be damp so as to promote mold, neither should it be so dry as to cause a drying out and shrinkage of the stored products. If the cellar is damp it is well to use a pail or two of lime, which must be renewed from time to time or as soon as it absorbs moisture enough to make it fine and powdery.

The lime will not only absorb dampness, but it will absorb the unpleasant odors and purify the air in the cellar. A cellar should be ventilated from time to time during the winter, as the outside temperature will permit. Too much ventilation will destroy the flavor of vegetables and cause them to dry out and shrivel. This, however, applies more to ventilation during cold weather and is not true to as great an extent during the fall or spring. A ventilator extending from the floor of the cellar to a few feet above the ground outside is sometimes used, but such an arrangement is more or less inoperative and it is better to depend on the opening of windows as opportunity presents itself.

Do not store vegetables in too large bulk. It is also best to keep each variety by itself. They should be well matured and gathered before they are chilled or frozen. If gathered on a warm day, allow them to cool before placing in the cellar. Onions and potatoes are best stored on shelves or in bins. Pumpkins and squashes require more air and must be kept dryer than the softer vegetables like carrots, turnips, beets, etc. Onions keep best without removing the tops. Pumpkins and squashes should be placed on a shelf near the top of the room and should not touch each other. Inspect them frequently, and when one becomes soft it can be used or removed.

Cabbage may be wrapped in newspaper, packed in barrels and stored in the coolest part of the cellar. If it is desired to keep potatoes, beets, carrots, turnips, etc., for late spring use pack them closely in boxes or barrels, fill in between and cover with sand or garden soil.

CABBAGE IN COLD STORAGE.

For the best results in the cold storage of cabbage, they should be grown especially for this purpose. Late planted cabbage, which barely close the heads before frost, keeps much better than early cabbage. The "Holland Seed" variety is known to be very satisfactory as a good keeper. It is essential that only good firm heads be accepted for storage. It is desirable that the cabbage be "trimmed" before storing about in the same manner as is required for shipping to market. The cabbage should be either packed in crates not more than 2½ feet in height or they should be piled on racks spaced about this distance apart. A fair circulation of air is necessary to the best results. Racks are easily constructed by erecting upright sides, placing cross pieces six inches wide horizontally on cleats. The cross pieces should only be close enough together to prevent the cabbage from dropping through. The racks may be of any width and height when constructed in this way, and it is customary to allow two or three tiers of cabbage to rest on each set of cross pieces. A thickness of 3 to 4 feet of cabbage would not be objectionable if a thorough system of air circulation is installed.

Owing to the fact that cabbages are composed largely of water and are of a porous nature, it is necessary to provide a good circulation of air throughout the storage room and ample moisture absorbing surfaces. This may be done by the use of chloride of calcium in cold weather. Ventilating the room by introducing outside air will prevent an undue accumulation of moisture and gases. Care must be taken in ventilating in this way to prevent freezing the goods by admitting air which is too cold or by causing dampness or too high a temperature by letting in warm air. A fan system of air circulation and a thorough ventilating system is very desirable, and the best results cannot be obtained without them. (See chapter on "Air Circulation and Ventilation.")

It has been found that a temperature of 31° F. will produce the best results in cold storing cabbages, 32° to 36° F., however, have been used with success, but if the higher temperatures are employed it is especially necessary to hold them reasonably steady and uniform in all parts of the room. Under favorable storing conditions, and when stored in a well equipped house, cabbages may be carried from fall until spring with a shrinkage of not more than 5 to 10 per cent, which is practically nothing as compared with the shrinkage experienced when storing in the old style frost-proof house without refrigeration.

COLD STORAGE OF ONIONS.

This vegetable may be successfully cold stored for long periods, and if carefully harvested and properly cared for prior to being placed in the cold storage room, they may be taken out of storage in good condition as late as May or June following the season of production. It is best not to store in large bulk, but rather in trays, or crates, or racks, so that the air may circulate freely. A temperature of 32° F. is thought best for the storage of onions and the room should be only moderately dry. Onions do not freeze easily and require a temperature several degrees below the freezing point of water (32° F.) to materially injure them, providing they are thawed slowly. In some cases onions have been stored where they have been purposely frozen so as to keep them in better condition, but this practice is not recommended, as better results are obtainable by keeping a uniform temperature in cold storage.

COLD STORAGE OF CELERY.

The storage of celery varies a great deal in different places. In some of the large cities celery is kept in storage the year round. Generally, however, celery is stored early in the fall or on maturity of the plant, and the goods are not placed in storage until the ground freezes in the fall. A temperature of as near the freezing point as possible is recommended. Celery storage rooms are usually held at from 32° to 34° F. The keeping qualities vary with the different varieties and condition of the goods. Hardy varieties and "green top" keep well for three months or possibly more. Some of the varieties will not keep longer than from one to two months. Celery after being trimmed, or what is known as "dressed goods," will not keep for any length of time. A few days only being about the limit.

COLD STORAGE OF FRUITS.

Cantaloupes.—This fruit may be stored with good success at a temperature of 33° F., but for a few days' storage only a temperature of 35° to 40° F. is sufficiently low. A great deal depends on the condition of the goods when received as to how long they may be carried, but from four to six weeks is the extreme limit under favorable conditions, and they must be in prime and sound condition to be held more than a few days or a couple of weeks. They are usually stored only for a short period to tide over a temporary glut in the market.

Bananas.—These are not considered a cold storage product as they are generally received in an unripe condition and are only cold stored in exceptional cases for a short period to prevent ripening. The rapidity of ripening depends on the temperature in which they are stored. Some recommend a temperature as low as 40° to 50° F., but usually a temperature of 55° to 60° F. is considered suitable for the temporary or short period for which they are stored. Care must be taken that they do not get too low in temperature or they will become chilled and turn black and therefore unsalable.

Oranges.—These will keep best for from one to three months at a temperature of 34° for the average fruit. The fan system of air circulation is best, and a ventilating system which will supply pure, cold and dry air to force out the gas which accumulates from the citrous fruits is also beneficial. This will also lessen the danger of contaminating other goods, such as butter and eggs. Do not store these products in the same building with citrous fruits. See what is said elsewhere in this chapter on storing "Eggs and Lemons."

Lemons.—This fruit should have the same general treatment as oranges, but a temperature of 38° F. is considered as low as is safe for average fruit. A lower temperature is detrimental and will cause them to decay.

Melons.—These are put in storage but very little, and for short holding only, the value of the product and limited possibility of keeping not as yet warranting a large business in this line. By removing from the vines very carefully, cutting the stem an inch or so from the melon, then shellacing the stem and wrapping the melon in wax paper, they may be stored from two to three months. Care must be taken not to bruise or mar the rind of the melon. Provide racks and store as loosely as possible to prevent bruising and crushing. A temperature of from 34° to 36° F. is considered best for long period holding, and a temperature of 40° F. is sufficiently low for short periods.

Plums.—This fruit is extremely perishable and not generally considered as cold storage goods except for a few days at a time to tide over an overstocked market. By rigid attention to quality of stock and providing the best facilities for cold storage good results may be obtainable on a comparatively long time carry. Green gage plums have been kept in good condition for a period of ten weeks at a temperature of 32° F.

Cherries.—Quite perishable and can only be stored for comparatively short periods at best. A temperature of 32° to 34° F. is recommended.

Strawberries.—It is practically out of the question to cold store this fruit. They are only placed in refrigerated rooms at a temperature of 40° to 50° F. to prevent rapid ripening and deterioration. Some experimenting has been done in freezing strawberries and holding in this condition for a long period. The actual results cannot be given and it is doubtful if this is commercially practicable. Nevertheless, some experimenting along this line might prove interesting.

Currants.—These may be kept from four to six weeks at a temperature of from 32° to 34° F. The red varieties keep better than the black or white currants. They should be protected from the air by paper coverings.

DRIED FRUIT IN STORAGE.

Dried fruit has been stored in large quantities of late years for the purpose of preventing loss of weight by evaporation during warm weather and to prevent mold and fermentation. One of the objects of storing is also to prevent the development of insect life. Certain forms of grubs or worms germinate at comparatively high temperatures and damage the stock. A temperature of from 40° to 45° F. is sufficient to prevent this, and is in common use for these products, but a temperature of from 36° to 38° F. is said to absolutely prevent any insect life from maturing or germinating. If a room is available at a somewhat lower temperature there is no damage to most goods of this class even as low as 25° F.

FLORISTS' GOODS IN COLD STORAGE.

Florists are storing quite a variety of goods at the present time and a constantly increasing variety is being stored each year. The following are a few of the common goods which are placed in storage for safe keeping. Lily-of-the-valley pips at a temperature of about 25° F., storage period November until spring (see chapter heading); Chinese, Japanese and Bermuda lily bulbs stored at a temperature of 34° to 36° F., storage period from fall to spring; wild smilax, temperature 32° to 34° F., through the winter season; galaxia leaves at 25° F.; lucothia, at a temperature of 30° F., through the winter season.

REGULATION OF PLANT GROWTH BY REFRIGERATION.

Plant growth may be regulated by maintaining proper temperatures. This principle is applied in retarding the development of bulbs and flowering plants so as to produce blossoms at any time of the year as the demand may be. Lily-of-the-valley, for instance, is made to blossom at Easter time, and roses which naturally blossom in summer, are made to blossom at Christmas. Potted plants are held at a temperature of 30° to 35° F., and then transplanted to the comparatively high temperature of the green-house at such a time as will bring them to flowering or fruitage at the date desired entirely independent of outside weather conditions. The possibilities along this line are being developed as rapidly as the demand warrants.

REFRIGERATION APPLIED TO THE SILK INDUSTRY.

One of the serious obstacles to be overcome in applying refrigeration in the silk industry is the danger that the silk worm will hatch from the eggs at the time when the mulberry leaves have not reached sufficient maturity to constitute a proper food. Refrigeration has been applied to prevent premature hatching of the eggs, if on account of a backward season or for any other reason the matured mulberry leaves are not to be had. A temperature of about 32° F. will retard the hatching of the eggs at will and without affecting the silk worm in any way whatever.

EXPERIMENTS IN GRAIN GROWING.

It is reported from Stockholm, Sweden, that experiments are in progress whereby it is expected to produce a hardier grain for severe weather conditions. These experiments have been undertaken owing to failure to secure a grain which would be hardy under the severe climate of Norway and Sweden.

Canadian and other grains have been sown, but have not shown proper seed producing qualities after having been thoroughly tested.

Paul Hellstrom, Chief of the Government Biological Institute at Luela, has undertaken the experiment whereby he expects to harden oats, barley and other grains so as to make them sufficiently hardy to withstand a considerable degree of frost. Green-houses have been erected, which are in reality cold storage houses, in which the plants may be subjected to the lowest temperatures they can stand without being frozen. In this way the seed that matures from the most hardy plants will be used for propagation. By repeating this operation, and gradually lowering the temperature, it is expected that after five or six years' freezing, the nature of the grains will have been so materially changed that they will stand several degrees of frost without damage. The experiment looks reasonable and should meet with success, but it will require very delicate handling in order to regulate the temperatures sufficiently close for the purpose required.

The possibilities along this line, not only for propagating grain, but in propagating other hardy plants and fruits are unlimited, and the experiments referred to will be watched with a great deal of interest. This matter is mentioned here as a suggstion to those who are interested in the production of hardy fruits and plants.

MANUFACTURE OF PARAFFINE.

The application of mechanical refrigeration in the manufacture of paraffine was the first use of mechanical refrigeration in connection with a manufactured product. Paraffine wax is extracted from paraffine oil by cooling same to the correct temperature to cause the wax to crystallize. In this way paraffine wax of varying densities or melting points may be extracted. The mechanical details of the apparatus may be any arrangement which will present a surface at the correct temperature which may be immersed in a tank of paraffine oil. A revolving surface from which the paraffine may be scraped off as it revolves is desirable.

COLD STORAGE AND FREEZING TEMPERATURES.

The temperatures given opposite the various goods named below may fairly be stated to give the average of the best present practice. In some cases comparatively little is known of the correct temperatures at which the various products should be stored, as no tests have been made. For those goods with which the author is familiar, embracing the more important of the perishable products which are cold stored, temperatures are given which he believes to be correct as applied to average practice. For many of the special or uncommon goods the best obtainable authority has been consulted. Owing to the small amount of accurate information available until recently, the present list shows some marked changes from the temperatures regarded as correct a few years ago. No doubt future changes will be made, but hardly to the same extent.

Arbitrary temperatures are given for each commodity; but the condition of the goods, length of time to be stored, conditions of air circulation, humidity, etc., are factors in determining the best suitable temperature. The following list of products and temperatures should be considered as a guide only, subject to change to meet varying conditions under which goods are stored:

COLD STORAGE AND FREEZING TEMPERATURES FOR VARIOUS PRODUCTS.

Products	Deg. F.	Products	Deg. F.
Apple butter	42	Cheese (long carry)	35
Apples	30	Chestnuts	34
Asparagus	33	Chocolate dipping room	65
Bananas	58	Cider	32
Beans (dried)	45	Cigars	42
Beer (bottled)	45	Corn (dried)	45
Berries, fresh (few days only)	40	Corn meal	42
Buckwheat flour	42	Cranberries	33
Bulbs	34	Cream (short carry)	33
Butter	14	Cucumbers	38
Butterine	20	Currants (few days only)	32
Cabbage	31	Cut roses	36
Canned fruits	40	Dates	55
Canned meats	40	Dried beef	40
Cantaloupes (one to two months)	33	Dried fish	40
Cantaloupes (short carry)	40	Dried fruits	40
Carrots	33	Eggs	30
Caviar	36	Ferns	28
Celery	32	Field grown roses	32
		Figs	55

COLD STORAGE AND FREEZING TEMPERATURES FOR VARIOUS PRODUCTS—CONTINUED.

Products	Deg. F.
Fish, fresh water (after frozen)	18
Fish, not frozen (short carry)	28
Fish, salt water (after frozen)	15
Fish (to freeze)	5
Frogs legs (after frozen)	18
Fruit trees	30
Fur and fabric room	28
Furs (undressed)	35
Game (after frozen)	10
Game (short carry)	28
Game (to freeze)	0
Ginger ale	36
Grapes	36
Hams (not brined)	20
Hogs	30
Hops	32
Huckleberries (frozen, long carry)	20
Ice cream (few days only)	15
Ice storage room (refrigerated)	28
Japanese fern balls	31
Lard	40
Lemons (long carry)	38
Lemons (short carry)	50
Lily of the valley pips	25
Livers	20
Maple sugar	45
Maple syrup	45
Meat, fresh (ten to thirty days)	30
Meats, fresh (few days only)	35
Meats, salt (after curing)	43
Mild cured pickled salmon	33
Milk (short carry)	35
Nursery stock	30
Nuts in shell	40
Oatmeal	42
Oils	45
Oleomargarine	20
Onions	32
Oranges (long carry)	34
Oranges (short carry)	50
Oxtails	30
Oysters, iced (in tubs)	35
Oysters (in shell)	43
Palm seeds	38
Parsnips	32
Peach butter	42
Peaches (short carry)	50
Pears	33
Peas (dried)	45
Plums (one to two months)	32
Potatoes	34
Poultry (after frozen)	10
Poultry, dressed (iced)	30
Poultry (short carry)	28
Poultry (to freeze)	0
Raisins	55
Ribs (not brined)	20
Salt meat curing room	33
Sardines (canned)	40
Sauerkraut	38
Sausage casings	20
Scallops (after frozen)	16
Shoulders (not brined)	20
Strained honey	45
Sugar	45
Syrup	45
Tenderloin, etc	33
Tobacco	42
Tomatoes (ripe)	42
Veal	30
Watermelons (short carry)	40
Wheat flour	42
Wines	50

REINFORCED CONCRETE IN COLD STORAGE CONSTRUCTION.

The use of reinforced cement concrete as a material for the construction of cold storage warehouses has come into prominence during the past ten years. Owing to the reduced price of cement, concrete makes an economical material for building construction of many kinds, but for cold storage houses it is not so well adapted as to some other purposes, and the difficulty of securing satisfactory insulation with concrete construction is not the least of the troubles encountered by refrigerating engineers.

A type of construction which is more flexible and perhaps better in every other way is to employ brick for exterior walls and reinforced concrete for posts and floors only. Even with this type of construction the conduction from one floor to another through the floors and posts makes insulation a serious problem. This is especially true where a great difference in temperature is necessary on different floors of the same building. In large plants this can be overcome by insulating separately the low temperature rooms and placing them in a separate section of the building.

Another important disadvantage of concrete construction is the practical impossibility of making any important changes in the arrangement of rooms, equipment, etc., after the building is once built and insulated. This applies particularly to a building which is used for workroom and other purposes not requiring refrigeration, as well as for cold storage. While concrete will, doubtless, find application in many places and especially in connection with large cold stores, it is doubtful if this material will in the near future come into general use for comparatively small plants. The saving in insurance, which is the chief advantage, is, in small plants, more than offset by the disadvantages mentioned above and by the increased cost of construction.

TOBACCO AND HOPS IN COLD STORAGE.

The American Warehousemen's Association in response to inquiries received sent out letters to its members relative to the storage of tobacco and hops, and the suggestions made in connection therewith are interesting and may be summarized as follows:

Tobacco.—"We have stored tobacco in bales at temperature of 38° to 40° F. with satisfactory results.

"We have handled Havana and Sumatra tobacco in bales, always carrying it at a temperature of 32° F. At this temperature there seems to be just enough moisture in the air to keep the tobacco in good shape, yet not enough to allow it to mold. It must be stored in a separate room from other commodities on account of the penetrating odor which will certainly spoil dried fruits, eggs, butter, and even apples will absorb it to a considerable extent.

"We have never stored tobacco except when made up in cigars, and this we held at 26° F. with good results."

Hops.—"The proper way to store them is to see that there is at least three or four inches space between each bale in piling. They are required to be kept in an absolutely dry room. The temperature should be held closely about 32° F.; any considerable change is apt to cause mold and render them unfit for use.

"Hops are generally stored at a temperature of about 32° F. If held closely at this temperature results should be satisfactory.

"We have had considerable experience with hops. They are a delicate thing to store, and must be kept in very dry atmosphere, and should not be stored with other goods, as they readily take the smell and taste of other goods, and the odor of the hops is likely to prove injurious to the other goods; are very susceptible to mold if any dampness is in the room. Usual rate of storage about ⅛c per pound per month.

"We store hops at a temperature of 32° F. They come in bales of about 350 pounds, and storage charge is ⅛c per pound per month. We have had no difficulty in turning out hops in satisfactory condition.

"Have stored both domestic and imported hops for a number of years with good success. We store at a temperature of 38° to 40° F.; the imported bales are stood on end, and the domestic bales piled up on sides three high."

PRE-COOLING OF CELERY.

The Florida Vegetable Growers' Association has recently been conducting some experiments in pre-cooling or temporary cold storing celery before shipment, and as reported the scheme seems to be a success as compared with the old method. It seems that there is considerable natural heat of fermentation arising from celery, and the pre-cooling tends to check this, as it also tends to check "blight" and what is known as "black-heart"; both troubles caused doubtless by fungus diseases, which are held in check by low temperature. We may look for pre-cooling to be applied to celery as well as other products which deteriorate rapidly at ordinary temperatures.

THE BREATHING OF FRUITS IN COLD STORAGE.

The deterioration or destructive processes which take place in fruit are greatly retarded by low temperature, but these destructive processes are operative and just as certain at low temperatures as at high temperatures, and this action has been likened to the breathing of animals for the reason that fruits absorb oxygen and give out carbon dioxide or what is commonly known as carbonic acid gas.

Fruit before picking and while still attached to the twig has its food supplied to it, but just as soon as the fruit is picked, with nothing to make good the losses, the destructive processes commence. There is, therefore, a constant reduction in weight and vitality. Cold storage makes the absorption of oxygen and the transpiration of carbon dioxide much slower as they are the result of complicated chemical action and all chemical actions progress much slower at low temperature. The loss of weight from fruit in cold storage is not due entirely to the mere drying out of water, but to the natural progressive starvation or destruction referred to.

This action in the life of fruit was the subject of a bulletin of the New Hampshire Experiment Station, published in February, 1908, entitled, "The Respiration of Fruit." Reference is also made to Bulletin No. 142, Bureau of Chemistry, United States Department of Agriculture, entitled, "Studies on Fruit Respiration." As scientifically considered these publications are quite interesting. As long as fruit remains on the parent branch it is alive and growing, but immediately when picked it becomes inert, begins to die, and the action of cold storage is merely to retard the dying processes and postpone the ultimate decay which is the natural end of all fruit.

TOPICAL INDEX.

www.ingramcontent.com/pod-product-compliance
Lightning Source LLC
LaVergne TN
LVHW011338110826
845153LV00012B/455/J
9781427612571